Canterbury College
D0414622

Understanding Animal Breeding

Richard M. Bourdon
Colorado State University

Prentice Hall
Upper Saddle River, NJ 07458

Library of Congress Cataloging-in-Publication Data
Bourdon, Richard M.
Understanding animal breeding/Richard M. Bourdon. —1st ed.
p. cm.
Includes index.
ISBN 0-02-312851-8
1. Animal breeding. I. Title.
SF105.B67 1997
636.08'2—dc20 96-27484
CIP

Acquisitions Editor: *Charles Stewart*
Editorial Assistant: *Kathleen Linsner*
Managing Editor: *Mary Carnis*
Production Management: *Cindy Hass, Carlisle Publishers Services*
Director of Manufacturing & Production: *Bruce Johnson*
Manufacturing Buyer: *Ilene Sanford*
Marketing Manager: *Debbie Yarnell*
Cover Design: *Miguel Ortiz*
Formatting/page make-up: *Carlisle Communications Ltd.*

Simon & Schuster/A Viacom Company
Upper Saddle River, New Jersey 07458

Printed in the United States of America

10 9 8 7 6 5 4 3 2 1

ISBN 0 02-312851-8

Prentice-Hall International (UK) Limited, *London*
Prentice-Hall of Australia Pty. Limited, *Sydney*
Prentice-Hall Canada Inc., *Toronto*
Prentice-Hall Hispanoamericana, S.A., *Mexico*
Prentice-Hall of India Private Limited, *New Delhi*
Prentice-Hall of Japan, Inc., *Tokyo*
Simon & Schuster Asia Pte. Ltd., *Singapore*
Editora Prentice-Hall do Brasil, Ltda., *Rio de Janeiro*

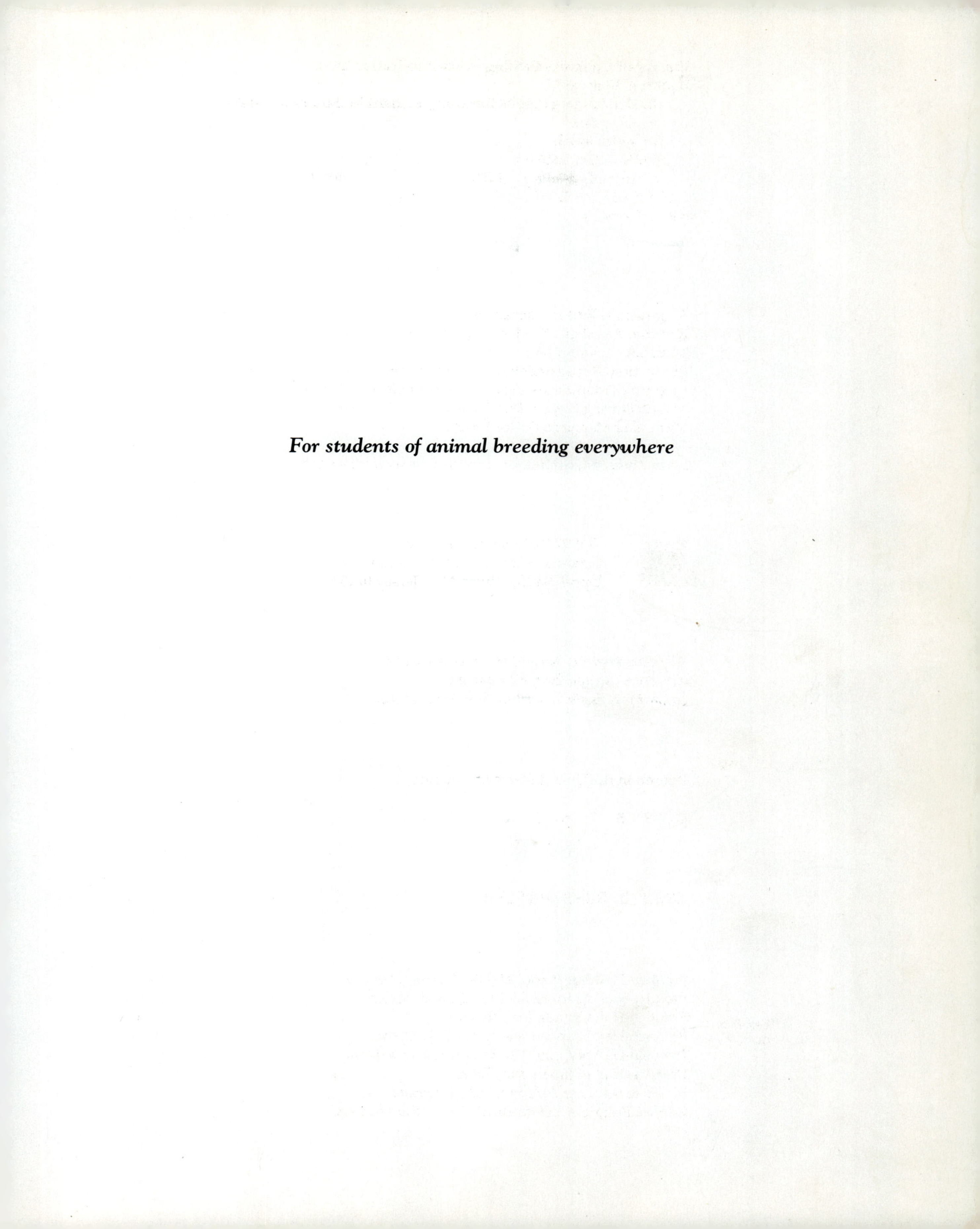

For students of animal breeding everywhere

Contents

PART I

ANIMAL BREEDING FROM THE TOP DOWN

PART II

ANIMAL BREEDING FROM THE BOTTOM UP

CHAPTER 11 GENETIC PREDICTION 212

CHAPTER 12 LARGE-SCALE GENETIC EVALUATION 242

Preface

Animal breeding is a fascinating discipline. It is entirely logical. It can also be very abstract. Anyone who truly understands the difference between a breeding value and a gene combination value, or who can explain why relatives resemble each other in highly heritable traits but not in lowly heritable ones can attest to the abstract side of animal breeding. For those of us blessed with abstract reasoning skills, this aspect of animal breeding makes it all the more challenging and exciting. But for the rest of us, abstraction creates difficulty and breeds discouragement. As a rule, students of animal science tend to be less comfortable with abstract concepts than, say, math majors. They would rather deal with something concrete, something tangible, like a reproductive tract in a physiology laboratory. All the parts are there: the ovaries, the uterus and the cervix. You can see them and feel them with your hands, and they become real in your mind. But you can never touch a breeding value.

Knowing this, I tried to structure my undergraduate course in a way that would make the conceptual side of animal breeding easier for students to understand. In order that students have appropriate written information on the subject, I handed out copies of my lecture notes at the beginning of each semester. They were well received. They were, however, just notes—I had written them for me, not for the students.

Hence this book. My goal in writing it is to create a text that is unintimidating. It *is* a textbook, but I hope it does not *feel* like a textbook. It is meant to be a learning text, not simply a reference book. It is designed not just to present information, but to teach how that information is best perceived (and, in some cases, how it should *not* be perceived).

Understanding Animal Breeding is primarily intended for use in an introductory course in the principles of animal breeding. Students using the book will benefit from previous exposure to classical genetics and statistics, but such experience is not critical—this book assumes little. Breeders in the field (as opposed to academics) may find the book useful as well. I have tried hard to keep the book from getting too advanced. It presents the concepts I think a sophisticated breeder should understand—and not much more.

Understanding Animal Breeding is really two books in one. The main body of the text contains a minimum of mathematics and is designed for all readers, but especially for those who are put off by equations and Greek symbols. (It is not, however, *completely* equation or symbol free.) For those (like me) who feel more comfortable with a concept if they can see it demonstrated or proven mathematically, I have provided a more mathematical treatment in boxed sections which appear in most chapters. To make sure that the book could be read without the mathematical sections and still make sense, I wrote the entire non-mathematical text before adding in the math sections.

I can think of three ways to read this book. If you are interested only in the concepts and would rather avoid math, skip the boxed sections. If math is second nature to you, read this text as you would any other—do not skip any part of it. Maybe the most productive way to read the book combines these approaches. At any one sitting, read the non-mathematical text first. When the concepts seem reasonably clear, go back and study the math behind them.

In many subjects, especially abstract ones, understanding the material often reduces to understanding definitions. This is certainly true in animal breeding. For this reason, I have highlighted key words and phrases in bold print in the text and defined them in the margins. The definitions reappear in the glossary. If you are using this book in a course in animal breeding, one good way to study is to concentrate on the definitions. If you truly understand all the definitions in the glossary and can use them in context, you understand animal breeding.

A word of caution about definitions: In writing and revising this book, I learned that animal breeding terms, like words in general, take on different meanings in different times and different contexts. Sometimes this results from sloppy usage—I am as guilty as anyone—and sometimes it reflects a justifiable shift in meaning. In cases where a change in meaning seemed more to contaminate a good concept than improve it, I stuck closely to the original definition. In other cases, I adopted the change in meaning, sometimes providing more than one definition. I have paid little attention to or omitted altogether a few animal breeding terms that I felt, for one reason or another, ought to be put out of their misery, and have gone so far as to invent some terms that I thought needed inventing.

The exercises at the end of each chapter are designed to reinforce understanding of the concepts presented in the text. The exercises are of two kinds: study questions and problems. The study questions provide a way for you to test your comprehension of the general concepts presented in each chapter. The problems tend to be mathematical in nature and often require that you read the boxed sections of the book in order to solve them. I make no claim for the practicality of the problems. They do not represent a sample of the tasks that a real-life animal breeder would encounter every day. Some are downright silly. Yet they can be very revealing. Practice may not make perfect, but it helps.

In order to understand the jargon of animal breeding, it is impossible to avoid statistics altogether. And in all fairness, some statistical concepts have real utility for non-statisticians. I have therefore included an entire chapter on statistics and their application to animal breeding. But even here there are boxed and non-boxed sections. The more conceptual treatment of statistical topics appears in the main body of text, and the more mathematical treatment in the boxed sections.

Most students are visual learners. If an idea can be represented by a visual image, understanding comes more easily. For this purpose I have included lots of diagrams in this book, especially in the sections explaining statistical concepts.

Traditionally, textbooks begin with the most basic theoretical concepts, then build on them in a step by step fashion until at some point application becomes apparent. I have chosen to reverse this approach. I prefer to stress application first, then explain the theory needed to answer applied questions. In this way the theory becomes more meaningful because its usefulness is already evident. I have not been entirely consistent in presenting application before theory in this book—that is not always possible—but you may notice that many concepts do not appear until they are truly needed.

I have also tried to present the big picture before the details. Part 1 is entitled *Animal Breeding from the Top Down.* It examines the fundamental questions faced by animal breeders and explains in a very general way the tools used to answer those questions. Only after animal breeding has been viewed from this very broad perspective does Part 2: *Animal Breeding from the Bottom Up* review the essentials of classical genetics. The next two sections describe the basic tools of animal breeding, selection and mating, in more detail. Part 5: *New Techniques, Old Strategies* contains two rather different chapters: one on the potential effects of biotechnology on animal breeding and one containing practical advice for breeders. This last chapter is completely applied in nature, although it assumes knowledge of the theoretical concepts presented earlier. It serves to tie things together.

Despite all the emphasis on application, ***Understanding Animal Breeding*** is *not* a text on applied animal breeding. It is a book about underlying principles. You can learn from it the precise meaning of an expected progeny difference, but do not expect it to explain what to look for in a herd sire.

Simply-inherited and polygenic traits, though subject to the same Mendelian mechanisms, are treated quite differently in animal breeding. Students, typically weaned on classical qualitative genetics, often fail to appreciate this. To help drive the point home, I have included a whole chapter on simply-inherited and polygenic traits and the distinction between them. Furthermore, the sections of the book on selection and mating each begin with a chapter involving simple-inheritance, and the remaining chapters in each section deal almost exclusively with polygenic material.

As a book on animal breeding in general, ***Understanding Animal Breeding*** is supposed to be applicable to any domestic species, and I have attempted to use many different species in the examples. But I must admit to beef cattle experience and bias. I tend to emphasize the animal breeding technology (EPDs, etc.) found in cattle breeding, which just happens to be the most advanced technology around. This may be frustrating for some, especially for those whose interests are in recreational or companion animals. To a dog breeder, for example, an EPD—if it is not mistaken for a social disease—is a thing of fiction. What is important to remember, however, is that the underlying principles of animal breeding are the same for all species—only breeding technologies differ.

Acknowledgments

Many people played a part in the making of this book. Foremost are my students over the past ten years or so. They unknowingly inspired me to write a text, and as guinea pigs in my pedagogical experiments, they made clear to me the concepts that are the most frustrating and need the most attention. From them I learned some approaches that usually work and some others that clearly do not.

I am indebted to all who received the manuscript or portions of it. Their suggestions were always helpful. I am especially grateful to Bill Hohenboken and my wife, Lucie. Bill's written comments rivaled the manuscript in sheer volume. He not only made *Understanding Animal Breeding* a better book, but, in some instances, saved me a lot of future embarrassment. Lucie read the book from beginning to end and took me to task when my writing became cryptic. She has little interest in the subject of animal breeding—a fact that makes her sacrifice all the more impressive, and, in a way, also made her the model reviewer.

I want to thank my colleagues at Colorado State and all those at Prentice Hall for being patient with me. This project took much more time than I ever anticipated, and I had to abandon or put on hold many other important tasks.

Finally, I thank Lucie, Carrie, Hansell, Lee and Allie. Maybe now that the book is done, they can get my full attention.

R.M.B.
Fort Collins, Colorado
August, 1996

PART I

Animal Breeding from the Top Down

There are two fundamental questions faced by animal breeders. The first asks, What is the "best" animal? Is the best Labrador retriever the one with show-winning conformation or the one with exceptional retrieving instinct? Is the best dairy cow the one that gives the most milk; the one with the best feet, legs, and udder support; or the one that combines performance in these traits in some optimal way? These are matters of intense debate among breeders, and, in truth, no one has all the answers. The question is an important one, however, because the answers that breeders decide upon determine the direction of genetic change for breeding operations, breeds, and even species.

The second question asks, How do you breed animals so that their descendants will be, if not "best," at least better than today's animals? In other words, how are animal populations improved genetically? This question involves genetic principles and animal breeding technology, and is the subject of most of this book.

The next two chapters examine these questions from a broad perspective. After reading them, you should have a good feel for what animal breeding is all about.

CHAPTER 1

What Is the "Best" Animal?

"Best" is a relative term. There is no best animal for all situations. The kind of animal that works best in one environment may be quite different from the optimal animal under another set of circumstances. There are no hard-and-fast rules for determining the most appropriate animal for a given situation, but there is a general method that can provide—for breeders motivated by sustainable profit, anyway—an educated guess. The method has been dubbed the "systems approach." It requires a detailed knowledge of traits of importance and how performance in these traits interacts with such factors as the physical environment, management policies, costs, and prices. These things vary with species and breed and also depend upon the structure of a breeding industry and a breeder's place within that structure.

TRAITS, PHENOTYPES, AND GENOTYPES

trait
Any observable or measurable characteristic of an individual.

When we describe animals, we usually characterize them either in terms of appearance or performance or some combination of both. In any case, we talk about **traits.** A trait is any observable or measurable characteristic of an individual.

Some examples of *observable* traits—traits we would normally mention in describing the appearance of an animal—are coat color, size, muscling, leg set, head shape, and so on. Some examples of *measurable* traits—traits we would likely refer to in describing how an animal has performed—are weaning weight, lactation yield, time to run a mile, etc. There are hundreds of traits of interest in domesticated animals. Many of them are specific to a species or breed. Staple length, for example, is a measure of the length of wool fiber. It is a useful trait for wool sheep, but not for animals that do not produce wool.

phenotype
An observed category or measured level of performance for a trait in an individual.

Note that in none of the examples of traits mentioned above is the appearance or performance of a particular animal described. An animal may be red and weigh 576 pounds at weaning, but *red* coat color and *576-pound* weaning weight are not traits—the traits are simply coat color and weaning weight. *Red* and *576-pounds* are observed categories or measured levels of performance for the traits of coat color and weaning weight. They are **phenotypes** for these traits.

Students and breeders often confuse traits and phenotypes. It is not uncommon to hear statements like: "Foul temperament is a common trait of this line." The trait, of course, is temperament. "Foul" describes a phenotype for this trait. More examples of traits and phenotypes are given in Table 1.1.

TABLE 1.1 Examples of Traits and Phenotypes

Trait	Possible Phenotypes
Presence of horns	Horned, polled, dehorned
Height at withers	16 hands, 14–2
Yearling weight	850 lb, 1,225 lb
Placing	First, third, last
Shell color	White, brown
Quarter-mile time	19.3 sec, 20.8 sec
Calving ease	Assisted, unassisted
Litter size	5,11,14

Breeders tend to use the word *phenotype* when referring to an animal's appearance, often giving the impression that phenotype *means* appearance. In fact, an animal has as many phenotypes as there are traits to be observed or measured on that animal. As you can see from Table 1.1, phenotypes can describe much more than appearance. Consider the phenotype of 19.3 seconds for quarter-mile time. Clearly 19.3 seconds does not describe appearance in any way. It is a legitimate phenotype, however. We often use the word *performance* instead of phenotype for traits that are measured rather than observed with the eye. In this book, I use the terms *phenotype* and *performance* interchangeably.

As animal breeders, we are mainly concerned with changing animal populations genetically. From a breeding standpoint, therefore, we want to know not only the most desirable phenotypes, but the most desirable **genotypes** as well. That is because an animal's genotype provides the genetic background for its phenotypes. Mathematically,

genotype
The genetic makeup of an individual.

$$P = G + E$$

where P represents an individual's phenotype, G represents its genotype, and E represents **environmental effects**—the effects that external (nongenetic) factors have on animal performance.[1] In other words, an animal's phenotype is determined by its genotype and the environment it experiences.

environmental effect
The effect that external (nongenetic) factors have on animal performance.

The word *genotype* is used in several different ways. We can speak of an animal's genotype in general, referring to all the genes and gene combinations that affect the array of traits of interest to us. An example used in the next section of this chapter involves a "tropically adapted" genotype. In this case, the genotype includes all the genes and gene combinations affecting heat resistance, parasite resistance, and any other traits that make up tropical adaptation. This sense of the word *genotype* is generally implied in this chapter. Animals with similar genotypes (as genotype is defined here) are said to be of the same **biological type.** This does not mean that they are genetically identical—they are just more alike than animals of a different biological type. For example, animals that are tropically adapted, though they may vary considerably, can be considered a single biological type.

biological type
A classification for animals with similar genotypes for traits of interest. Examples include heavy draft types (horses), prolific wool types (sheep), large dual-purpose types (cattle), and tropically adapted types (many species).

[1]Technically, the equation $P = G + E$ is oversimplified. For the purposes of this discussion, however, it will do fine. For more precise versions of the equation, see Chapter 7.

We can also speak of an animal's genotype for a particular trait, referring to just those genes and gene combinations that affect that trait (e.g., heat resistance). Or, as you will see in Chapter 3, we can limit the definition of genotype even further. In any case, the genotypes of our animals' descendants are what we can change with breeding methods. Favorable changes in genotypes result in improved phenotypes.

ANALYZING THE SYSTEM

What is the best animal? To answer this question is to know what traits are of primary importance and what phenotypes and genotypes are most desirable for those traits. Most breeders, if they have any experience at all, have some opinion about the key traits and better genotypes. A Thoroughbred breeder, for example, might describe her perfect animal as ". . . fast, but with enough endurance and heart for the longer distances, and easily rated." A beef cattle breeder's version might be, ". . . easy calving, with as much growth as possible for the birth weight, and moderate milk." There are probably as many opinions of this sort as there are breeders, and for the most part they are quite subjective.

How do you develop a sense of the important traits and best genotypes in a more objective way? The answer lies in understanding that the genotype of an animal is just one part of a much larger **system.** A system is a group of interdependent component parts. Examples of systems abound. An internal combustion engine is a system. So is a corporation or a family. Animal-related systems vary from large, extremely complex systems such as an entire animal industry, to smaller (but still complex) systems like an individual ranch or farm, to the small and comparatively simple system comprised of an owner and his pet.

system
A group of interdependent component parts.

To see how a system works, consider the system that is a single farm. The components of this system could be categorized in a number of ways. One choice would be to list them under the following headings:

- Animals (genotype)
- Physical environment
- Fixed resources and management
- Economics

The *animal* category contains the characteristic genotype or genotypes—there may be more than one—of the animals on the farm. On a dairy farm, for example, a typical genotype could be described as having small size, low feed intake, moderate yield, and high butterfat content. A contrasting genotype might have large size, high intake, high yield, and low butterfat.

Physical environment refers to those elements of the environment over which humans exert little control. Examples of physical environmental factors include weather, altitude, soils, and quality and quantity of native forages. For some production systems, physical environment is extremely important. Range cattle and sheep often exist under conditions little different from those of their wild ancestors. They must deal with the vagaries of the physical environment every day.

Other species are little affected by physical environment. Dairy cattle and hogs that are confined indoors and fed harvested feeds are literally and figuratively insulated from nature. So are most pets.

Fixed resources include things like the size of the farm, the ability of the farm to grow supplementary feeds, and available labor. *Management* involves all the policies implemented by the farmer. Some examples are level of supplementary feeding, health care, and the length of time animals remain on the farm.

Economics refers to the costs of farm inputs like feed, labor, and supplies, and the prices for farm outputs—in this context, the animals themselves. Related economic factors include farm equity and long- and short-term interest rates.

interaction
A dependent relationship among components of a system in which the effect of any one component depends on other components present in the system.

If you think about it, it should be clear that the many components within these categories of the system **interact** with each other. That is, the effect of any one component depends on other components present in the system. For example, the best preventive health program (management) depends on the kinds of pathogens in the area (physical environment) and the costs of vaccines, dewormers, etc. (economics). To determine which health program is the most cost-effective, you must have knowledge of alternative programs, local pathogens, and treatment costs and understand how treatment programs interact with these other factors to affect profitability. Similarly, the best genotype depends on the local environment, the management practices in use, and the costs of inputs and prices of animal products. To determine the best genotype, you must have knowledge of environmental, management, and economic components and understand how they interact with genotype to affect profitability. The key, then, to determining the traits of importance and optimal genotypes for those traits is a thorough analysis of the entire system and an understanding of the many interactions among components of the system.

Genotype by Environment Interactions

genotype by environment ($G \times E$) interaction
A dependent relationship between genotypes and environments in which the difference in performance between two (or more) genotypes changes from environment to environment.

From a breeding standpoint, the most revealing interactions are those that involve the genotype of the animals. For many species, **genotype by environment interactions** play a critical role in determining the most appropriate biological type for a given environment. Genotype by environment ($G \times E$) interactions occur when the difference in performance between two or more genotypes changes from environment to environment.[2]

A classic example of the interaction between genotype and physical environment involves animals that are genetically adapted to temperate locations versus animals that are genetically adapted to tropical areas. "Genetically adapted" to a location means that animals have evolved in that location over many generations and, as a result, carry the genes that allow them to survive and thrive there. The interaction is depicted graphically in Figure 1.1.

In the temperate environment, the temperately adapted genotype outperforms the tropically adapted genotype, but both types perform quite well. The tropical environment is considerably more stressful due to extreme heat, humidity,

[2]We usually understand environment (E) to mean physical environment, but it can be interpreted more broadly. Fixed resources and management or even economics can be considered "environment."

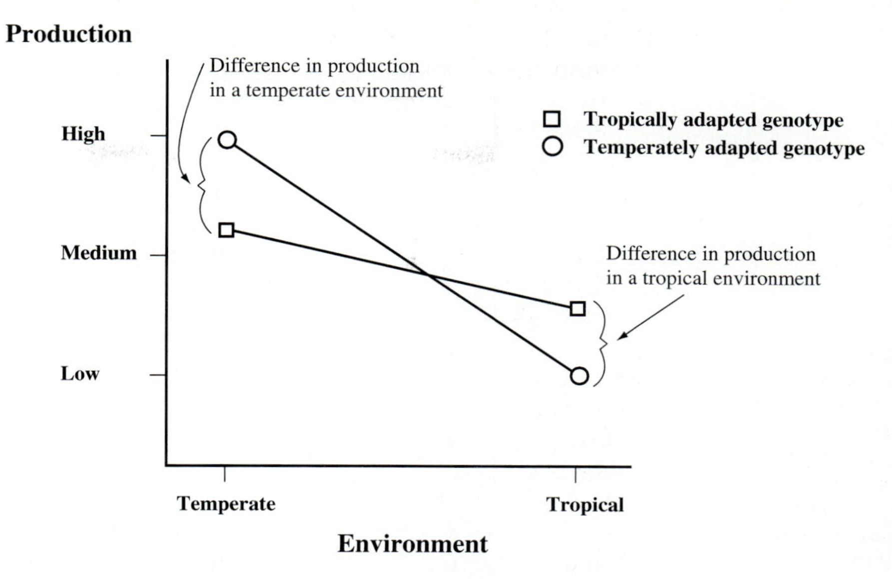

FIGURE 1.1 Example of a genotype by environment interaction.

and insects and other parasites, and both genotypes produce at a lower level. But the loss in productivity is much less for the tropically adapted genotype than for the temperately adapted genotype. This is probably because the tropically adapted type is genetically resistant to heat and parasites. The example fits the definition of a $G \times E$ interaction because the *difference* in the performance of the two genotypes is not the same in both environments.

What lesson can be learned from this example? What traits are important and what genotypes are optimal for these traits? Clearly, high levels of heat and parasite tolerance are important in tropical environments. In temperate environments, these traits assume much less importance. Understanding the nature of this $G \times E$ interaction does not completely answer the question, What is the best animal? but it is a beginning.

Common Misconceptions about Interactions

Interactions are sometimes hard to conceptualize. One device you can use that will aid your understanding of interactions is to phrase the description of each interaction using the words *relative* and *depends*. In the case of a $G \times E$ interaction, you could say that the *relative* performance of different genotypes *depends* upon the environment. For a management by economics interaction, you could say that the *relative* profitability of different management policies *depends* on the costs involved.

Interactions are always graphable, and if you can graph them on paper or in your mind, understanding them becomes easier. In the graph of a $G \times E$ interaction, the vertical axis represents some measure of outcome: performance, production, profit—whatever is appropriate for making comparisons. The horizontal axis

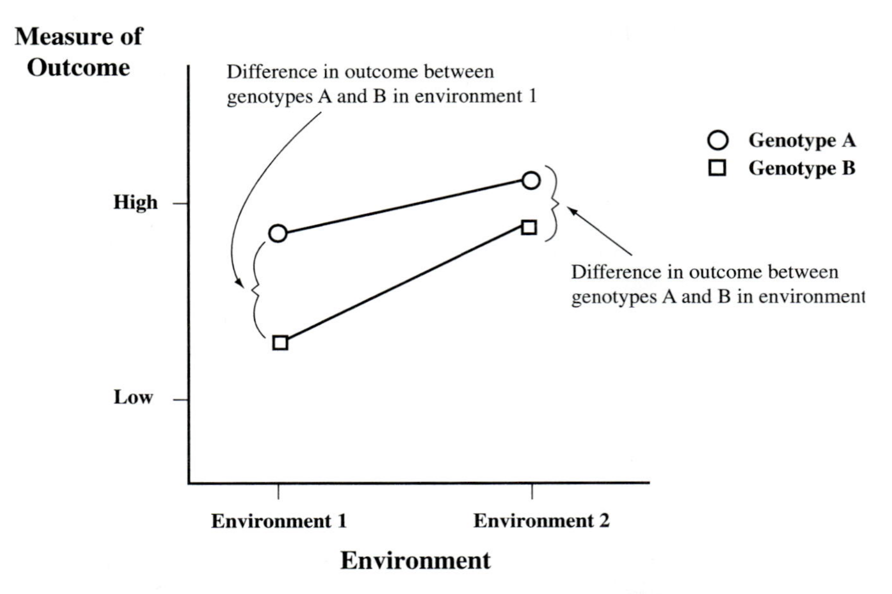

FIGURE 1.2 Generic example of a genotype by environment interaction.

contains the different environments. The connected symbols represent different genotypes. A generic example is shown in Figure 1.2. The usefulness of graphing is not limited to just $G \times E$ interactions—any interaction can be graphed.

For a $G \times E$ interaction to be graphable, there must be at least two genotypes and at least two environments. A common mistake is to consider only one genotype in two environments. For example, animals grow faster in an environment where feed levels are high than in an environment where feed levels are low (see Figure 1.3), but this alone does not constitute a $G \times E$ interaction. Rather, it is an example of an environmental effect—simply the effect that nongenetic factors (in this case, levels of feed) have on animal performance.

To have a $G \times E$ interaction, we need a second genotype and must show that the *difference* in performance between genotypes is not the same across environments. Compare Figures 1.2 and 1.3. Note how different the graph of an interaction appears from that of an environmental effect.

Another common mistake in conceptualizing $G \times E$ interactions is to consider two genotypes to be distinctly different when, in fact, they are the same. In the temperately versus tropically adapted example, the two genotypes were indeed different because they were *genetically* adapted to different environments. This implies generations of selection from which evolved genetically different types, each uniquely adapted to its environment. Different **breeds** —races of animals within a species—are often genetically adapted to different conditions. In beef cattle, the Brahman breed is a tropically adapted type, and the Hereford breed is a temperately adapted type. For a counterexample, consider two sets of Thoroughbreds, one raised and trained (but not *evolved*) at high altitude and the other raised and trained at sea level. We would logically expect the horses adapted to the thin air of high altitudes to run faster at high altitude than the horses reared at sea level. Performance of both types at low altitude would probably be similar. This is an example of an interaction, but not a *genotype* by environment interaction

breed
A race of animals within a species. Animals of the same breed usually have a common origin and similar identifying characteristics.

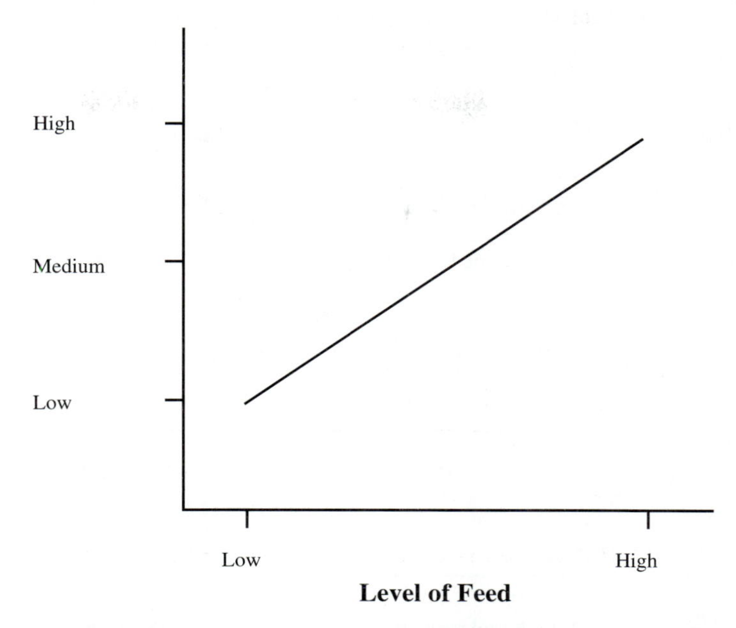

FIGURE 1.3 Example of an environmental effect, in this case the effect of level of feed on growth rate. Note how different this graph appears from the graph of an interaction in Figure 1.2.

because there is just one genotype. The two sets of Thoroughbreds are not genetically different—they have simply experienced different training conditions. Instead of being *genetically* adapted to two different environments, they are *environmentally* adapted. The interaction here is really a training environment by racing environment interaction.

The $G \times E$ interaction depicted in Figure 1.1 is an extreme example. Not only does the difference in performance between the two genotypes change from one environment to the next, it is mathematically positive in one environment and negative in the other. The genotypes actually rank differently in the two environments. This extreme form of interaction appears graphically as a crossing of the lines. Most interactions are less obvious. For a $G \times E$ interaction to occur, the difference in performance between genotypes may change only a little from one environment to the next. Graphically, the lines need not cross—they just cannot be parallel. Examples of different types of interactions are shown in Figure 1.4.

Other Interactions Involving Genotype

Understanding the major interactions between genotype and physical environment can be a big help in determining the best animal for a given situation, but other interactions involving genotype can be equally revealing. Genotype × management and genotype × economics interactions are often important. Sometimes a particular genotype is superior under a particular kind of management—say, intensive, high input management, but not under less intensive management.

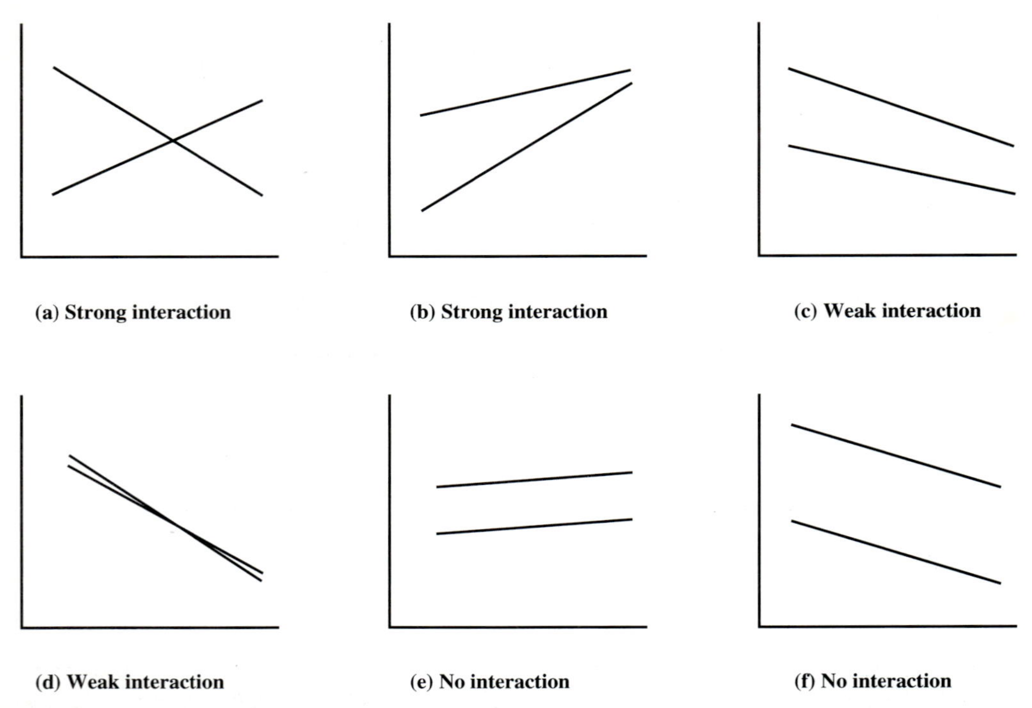

FIGURE 1.4 Examples of different types of interactions. If these were $G \times E$ interactions, the vertical axes would represent some measure of animal performance, the horizontal axes would indicate levels of environment, and the lines themselves would correspond to different genotypes: (*a*), strong interaction—large change in performance differences and reranking also; (*b*), strong interaction—large change in performance differences, but no reranking; (*c*), weak interaction—small change in performance differences, no reranking; (*d*), weak interaction—small change in performance differences, reranking; and (*e*) and (*f*), no interaction—no change in performance differences.

Similarly, one genotype may be the best when labor costs are high, but not when labor is cheap. Interactions involving genotype can be very complex, and to really understand them is to know a great deal about an animal industry.

BREEDING OBJECTIVES AND INDUSTRY STRUCTURE

In the process of determining the best animal, you might ask, Best for whom? The answer to this question depends on the structure of a breeding industry and a breeder's place within that structure. Most breeding industries can be thought of as having a pyramidal structure: a relatively few elite breeders at the top selling breeding stock to a larger number of multipliers who in turn sell animals to a great many end users.

germ plasm Genetic material in the form of live animals, semen, or embryos.

The pyramid suggests a flow of **germ plasm**—genetic material in the form of live animals, semen, or embryos—from the top down, the elite breeders producing the most advanced animals, breeders at the multiplier level replicating those animals, and end users benefiting from the genetic improvement occurring at the higher levels. Ideally, breeders at each level try to produce animals that will

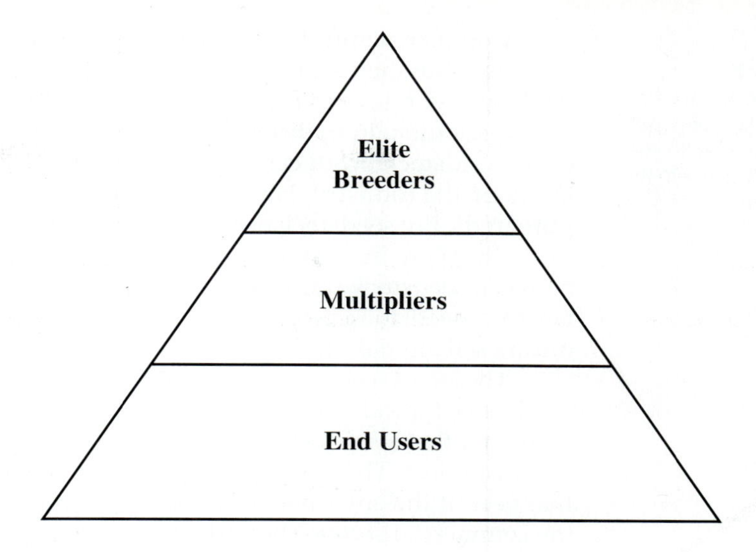

breeding objective
A general goal for a breeding program—a notion of what constitutes the best animal.[3]

be in the greatest demand by their customers at the next level down, with the ultimate result that the best animal is the animal that is the most useful or profitable for the end user. Thus, breeders at all levels tailor their **breeding objectives** to meet the needs of end users.

Traditional Livestock Species

commercial producer
An animal breeder whose primary product is a commodity for public consumption.

Who are the end users? In traditional food and fiber producing species (sheep, cattle, swine, and poultry), the end users are **commercial producers.**[4] These are the animal breeders whose primary products are commodities for public consumption. Commercial dairy operations produce milk; commercial sheep operations produce lamb, mutton, and wool; commercial swine operations produce pork; commercial poultry operations produce eggs, chicken, and turkey; and commercial cattle operations produce beef.

purebred
Wholly of one breed or **line.**

line
A group of related animals within a breed.

Because commercial producers are themselves end users, the best animal for a commercial producer is the animal that best fits that producer's own operation. As a result, commercial producers are relatively unconstrained in their choices of breeding animals. Their livestock need not be **purebred** or registered with any breed association. Commercial producers are free to select among different breeds and to mate females of one breed to males of another.

[3]This is one definition of a breeding objective. See Chapter 14 for another, much more specific definition.

[4]Some argue that the end users of traditional food and fiber species are not commercial producers but retail consumers. The point is valid—breeders should be concerned about the needs of consumers. They should not ignore product characteristics when they set breeding objectives. I choose to identify commercial producers as the end users, however, because they have many requirements beyond product characteristics. They need animals that not only produce a desirable product, but do it efficiently. In any case, if the invisible hand is at work, if economic incentives are creating communication throughout an industry—a *big* if—the needs of consumers and middlemen should be reflected in the needs of commercial producers.

seedstock
Breeding stock, animals whose role is to be a parent or, in other words, to contribute genes to the next generation.

Commercial producers are indeed breeders—they mate animals to produce offspring. But they are not in the business of selling **seedstock.** Seedstock are animals whose role is to be a parent or, in other words, to contribute genes to the next generation. In traditional food and fiber producing species, seedstock breeders or seedstock producers are those breeders at the elite breeder and multiplier levels of the industry. Historically, seedstock producers have been breeders of purebreds. But seedstock need not be purebred, and in species like swine and beef cattle there are increasing numbers of nonpurebred seedstock. If the distinction between seedstock producers and commercial producers seems unclear, remember this: seedstock producers sell animals to other breeders—commercial producers sell animals and(or) animal products for consumption.

The job of seedstock breeders is to service the seedstock needs of commercial producers, typically through the sale of males, females, and semen. It is important, therefore, that seedstock producers define "best" to mean best for their commercial customers. This means that seedstock breeders should be producing animals that best fit the environments, management policies, and economic conditions of the commercial sector. The system that seedstock breeders ought to analyze in order to determine what animal is best is not their own enterprise, but rather the commercial enterprise that will ultimately use the seedstock.

Commercial producers sometimes need different kinds of seedstock, and the breeding objectives of seedstock breeders should reflect this. For example, commercial beef cattle producers often have a need for "heifer bulls," which is jargon for bulls that will cause few calving problems when bred to young heifers. These same producers may also have a need for bulls that will produce fast growing offspring with valuable carcass characteristics. By tailoring their breeding programs accordingly, seedstock breeders can produce specialty seedstock to fill these different market niches.

Recreational and Companion Animal Species

The breeding industries for recreational and companion animal species (horses, dogs, cats, etc.) differ somewhat in structure from traditional livestock industries. The pyramid arrangement is still present, and markets for specialized types of animals exist, but seedstock/commercial divisions are usually less clear, and the end users may not be breeders at all. Consider, for example, Labrador retrievers. The end users of Labs are hunters and pet owners. These persons may or may not choose to breed their animals, and the qualities that are important to them are those that contribute to retrieving ability, companionship, aesthetics, or some combination of these traits. Among Labrador breeders there are elite breeders and multipliers, but the term *commercial producer* does not really fit here because no consumable commodity like meat, eggs, milk, or wool is being produced. The various horse industries provide similar examples. End users of horses range from owners of the most valuable racing animals to casual riders to those who keep miniature horses as pets.

Factors That Distort Breeding Objectives

Regardless of species, the best animal should be the animal that is best for the end user. Sometimes, however, this concept seems to get lost in the effort to satisfy expectations that really have little to do with the end user. This usually happens at the level of the elite breeder. A typical example of distorted breeding objectives is the emphasis placed by breeders of meat and dairy animals on particular spotting patterns or shades of coat color. Surely coat color has little to do with production efficiency in these species. Nevertheless, it has somehow become important.

What causes distorted breeding objectives? Competition among breeders is one factor. In an effort to convince buyers that his animals are superior to those of his competitors, a breeder may find it profitable to emphasize qualities in his animals that set them apart, but may not be particularly important. For example, if a breeder's animals are especially large, he may be tempted to promote the value of increased size whether or not size is a valuable measure. And if his promotional efforts are successful, he will be rewarded for having large animals. He will then breed for and promote even larger animals, and the race is on. The competitive forum provided by livestock shows has played its part in fostering this sort of thing. More recently, the competition inherent in national sire summaries has had similar effects.

Another cause for distortions in breeding objectives is an undue reliance on the part of end users upon the opinions of breeders higher up in the pyramid. In a perfect world, end users would be able to objectively evaluate their needs and communicate them to the breeders whose job it is to meet those needs. In reality, however, objective information is often scarce, and end users make choices based upon the information that is easily obtainable—the opinions of breeders contained in promotional materials.

The way to avoid distortions in breeding objectives is simply to remember who the end user is in your industry. Try to understand the major interactions affecting the end user's system and define the best animal accordingly.

CHANGE VERSUS STASIS

If we are to improve animals genetically, it only stands to reason that we must change them in some way. We ordinarily take this to mean that we should change them in an established direction. For example, in dairy species we continually breed for increased milk production. Should every trait be changed in a particular direction, however? Do we always want more milk, faster speed, higher fertility?

The answer is clearly no, and for some traits it is easy to see why. Take, for example, the conformation trait called hock set. Animals whose rear legs are too straight are postlegged— acking sufficient angle at the hock—and run a risk of going lame. Animals with too much angle at the hock are sickle-hocked—they too may develop soundness problems. The optimum hock set is somewhere between these extremes. The best animal has enough angle at the hock to be athletic, but not so much that it moves awkwardly. Clearly it would be a mistake to breed animals forever for increased or decreased set at the hock. Once a desirable conformation has been reached, there is no reason for further change.

intermediate optimum
An intermediate level of performance that is optimal in terms of profitability and(or) function.

Hock set is an obvious example of a trait with an **intermediate optimum.** Other traits with intermediate optima are not always so obvious. Size in dogs is an example. So is milk production in beef cows. For traits like these, improvement does not necessarily mean directional change. Improvement could better be defined as an increase in the proportion of animals with optimum or near optimum performance. In other words, improvement could be an increase in uniformity.

EXERCISES

Study Questions

1.1 Define in your own words:

trait
phenotype
genotype
environmental effect
biological type
system
interaction
genotype by environment interaction
breed
germ plasm
breeding objective
commercial producer
purebred
line
seedstock
intermediate optimum

1.2 For a species of your choice, list five traits and at least two phenotypes for each trait.

1.3 What is wrong with the following statement: "Endurance is a common *trait* of this line of horses"? How should the statement be worded?

1.4 For a species of your choice, identify components of the system that is a ranch, farm, kennel, etc. by describing at least three of each of the following:

a. biological types
b. physical environments
c. fixed resource/management combinations
d. economic scenarios

1.5 Using your answers to Question 1.4, graph an example of each of the following interactions:

a. genotype by physical environment
b. genotype by management
c. genotype by economics

Label each graph carefully and explain how the interaction could affect selection objectives.

1.6 Describe a situation which at first appears to be a $G \times E$ interaction involving two distinct genotypes, but is not because there is really only one genotype. The two groups of animals are genetically similar, but one is environmentally adapted and the other is not. (See the Thoroughbred example under *Common Misconceptions about Interactions.*)

1.7 For the animal industry that interests you the most:

a. How is the industry structured?
b. Who are the end users?
c. Is there a clear seedstock/commercial division? If so, what do commercial producers sell?
d. Do breeders consistently keep the interests of end users in mind? If not, why not?

1.8 Describe a situation in which changing the mean performance of an animal population by breeding methods would be neither necessary nor advisable.

Problems

1.1 Listed in the following table are net profits ($) for equivalent beef cattle operations given three biological types: large, medium, and small mature size; and three economic scenarios: standard cost/price relationships, doubled cow herd feed costs, and doubled feedlot feed costs.

	Biological Type		
Economic Scenario	Large	Medium	Small
Standard cost/price relationships	24,510	18,825	15,990
Doubled cow herd feed costs	−4,973	−2,552	−20,157
Doubled feedlot feed costs	−15,819	−9,986	−11,336

Graph profit versus economic scenario for the different biological types.

a. Do genotype by economics interactions exist? If so, describe them.

b. How would you explain these interactions? (Background information: No cattle are sold before slaughter, and all young nonreplacement animals are fed in the feedlot to a constant degree of fatness. Cattle are managed in such a way that the operation can run 222 large cows (but no yearling steers), 204 medium-sized cows, and 284 small cows. Large, medium, and small cows produce a herdwide total of 111, 97, and 106 tons of beef, respectively.)

c. How should breeding objectives for operations like these change as economic scenarios change?

1.2 Listed in the table below are typical survival percentages for newborn calves varying in birth weight.

Birth Weight, lb	Survival, %
30	15
40	42
50	63
60	93
70	98
80	94
90	90
100	82
110	71
120	58
130	35
140	26

Plot survival versus birth weight.

a. What concept is illustrated here?

b. Describe logical selection objectives for birth weight if the average birth weight in a herd is:

i. 60 lb
ii. 75 lb
iii. 97 lb

CHAPTER 2

How Are Animal Populations Improved?

population
A group of intermating individuals. The term can refer to a breed, an entire species, a single herd or flock, or even a small group of animals within a herd.

The purpose of animal breeding is not to genetically improve individual animals—once an individual is conceived, it is a bit late for that—but to improve animal **populations,** to improve future generations of animals. To this task breeders bring two basic tools: selection and mating. Both involve decision making. In selection, we decide which individuals become parents, how many offspring they may produce, and how long they remain in the breeding population. In mating, we decide which of the males we have selected will be bred to which of the females we have selected. This chapter examines both kinds of decisions from a broad perspective.

There is little in this chapter that is not discussed in much more detail later. Selection and mating, the main topics here, are also the subjects of Chapters 6 to 19. The purpose of this chapter is to provide an overview of the principles, techniques, and language of animal breeding—to present the big picture before delving into the details. Any redundancy with later material is strictly intentional.

SELECTION

selection
The process that determines which individuals become parents, how many offspring they may produce, and how long they remain in the breeding population.

natural selection
Selection that occurs in nature independent of deliberate human control.

artificial selection
Selection that is under human control.

The method used by breeders to make long-term genetic change in animals is called **selection.** Selection is the process that determines which individuals become parents, how many offspring they may produce, and how long they remain in the breeding population. Most of us are familiar with the term **natural selection.** Natural selection is the great evolutionary force that fuels genetic change in all living things. The term conjures up visions of fossil records, species creation, gradual anatomical and physiological changes, and mass extinctions. We commonly think of natural selection as affecting wild animals and plants, but in fact it affects both wild and domestic species. All animals with lethal genetic defects, for example, are naturally selected against—they never live to become parents.

Animal breeders cannot ignore natural selection, but the kind of selection of primary interest to them is called **artificial selection;** selection that is under human control. Artificial selection has two aspects: **replacement selection** and **culling.** In replacement selection we decide which individuals will become parents for the first time. Replacement selection gets its name from the fact that we select new animals to *replace* parents that have been culled. These new animals are termed *replacements.*

We normally think of replacements as being young animals. When you choose the pups in a litter, the lambs in a flock, or the calves in a herd to be kept for

replacement selection The process that determines which individuals will become parents for the first time.

culling The process that determines which parents will no longer remain parents.

artificial insemination (A.I.) A reproductive technology in which semen is collected from males, then used in fresh or frozen form to breed females.

breeding value The value of an individual as a (genetic) parent.

breeding purposes, you practice replacement selection with young animals. Broadly speaking, however, replacement selection need not be confined to young animals. If you were a dairyman and you chose to use for the first time a well-known bull via **artificial insemination (A.I.),** you would still be practicing replacement selection. The bull is not young, nor will he be a parent for the first time, but he will be a parent for the first time in *your* herd.

When we cull animals, we decide which parents will no longer remain parents. Replacement selection and culling are really just different sides of the same coin. They involve different sets of animals, but their purposes are the same: to determine which animals reproduce. Both are integral parts of selection as a whole.

The idea behind selection is simply this: to let the individuals with the best sets of genes reproduce so that the next generation has, on average, more desirable genes than the current generation. The animals with the best sets of genes are said to have the best **breeding values.** They are the individuals with the greatest value as genetic parents. The term *genetic* parent is used here to differentiate between a parent in the sense of being a contributor of genes to offspring and a parent in the sense of caring for offspring. In selection, we try to choose those animals with the best breeding values: the animals that will contribute the best genes to the next generation. The result of successful selection is then to genetically improve future generations of a population by increasing over time the proportion of desirable genes in the population.

Phenotypic Selection

phenotypic selection Selection based solely on an individual's own phenotype(s).

To see how selection works, consider the simplest form of selection: **phenotypic selection.** In phenotypic selection, the only information used is the individual performance of each animal being considered for selection. No attention is paid to the pedigree of the animal, or the performance of its sibs (brothers and sisters) or of any progeny it may have produced. For example, if you were using phenotypic selection for weaning weight to determine whether a particular ewe lamb was to be kept for breeding, you would base your decision strictly on her own weaning weight. Other considerations, such as the genetic merit of her parents for weaning weight, would be ignored. In practice (meaning outside of scientific laboratories), phenotypic selection in its pure form is increasingly rare, but it makes a good example.

Figure 2.1 depicts phenotypic selection for increased body size in mice. The largest mice in each generation are chosen to become parents of the next generation, and the result over time is a general increase in body size.

The replacement mice in Figure 2.1 are selected on the basis of their phenotype for body size with the expectation that phenotype for size is a reasonable indicator of the genes affecting body size. It is the genes, after all, which are transmitted from parent to offspring. In other words, there is a tacit assumption that phenotype for body size in mice is somehow related to breeding value for body size. If that were not the case, phenotypic selection for this trait would be a waste of time. The relationship between phenotype and breeding value is therefore a very important one, and its measure is termed **heritability.** When heritability is high, phenotypes are generally good indicators of underlying breeding values, and phenotypic selection will be effective. When heritability is low, phenotypes reveal little about breeding values, and phenotypic selection will be ineffective.

heritability A measure of the strength of the relationship between breeding values and phenotypic values for a trait in a population.

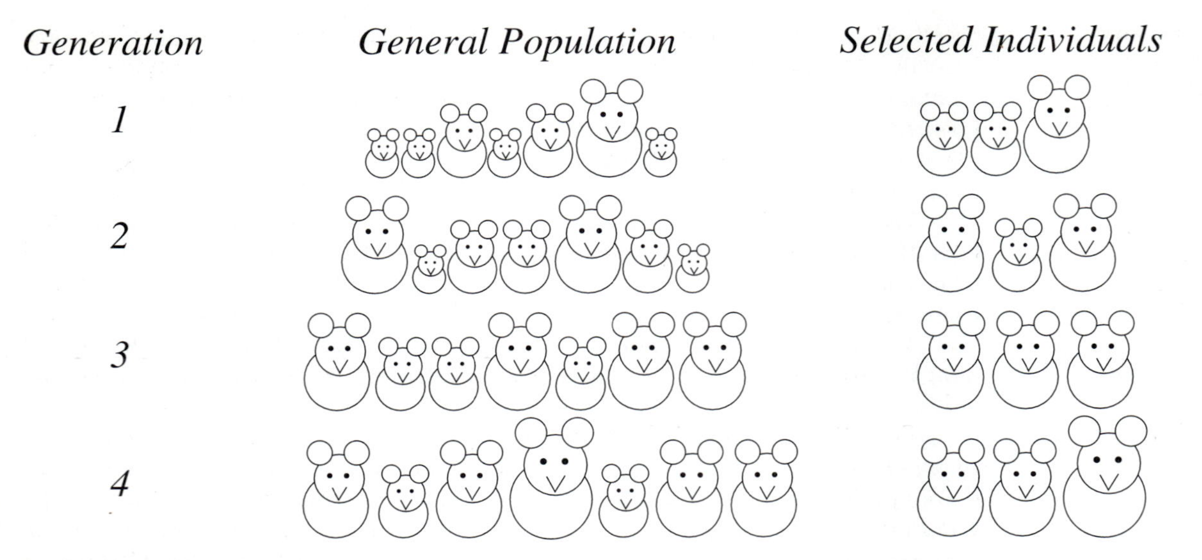

FIGURE 2.1 Phenotypic selection for increased body size in mice.

fertility
The ability (of a female) to conceive or (of a male) to impregnate.

Judging by the rapid increase in body size of the mice in Figure 2.1, body size must be quite heritable. Not all traits are as heritable. The heritability of **fertility** in mammals, for example, is generally quite low. Whether or not a female conceives during a breeding season typically has little to do with her breeding value for fertility. It is more a function of environmental effects.

Selection Using Information on Relatives

dam
A female parent.

sire
A male parent.

pedigree data
Information on the genotype or performance of ancestors and(or) **collateral relatives** of an individual.

collateral relatives
Relatives that are neither direct ancestors nor direct descendants of an individual, e.g., siblings, aunts, uncles, nieces, and nephews.

Most animal breeders are unlikely to limit themselves to individual performance information alone in making selection decisions. They will use information on relatives as well. For example, when a dog breeder purchases an eight-week-old puppy from another breeder, she probably does not base her choice on just the conformation and personality characteristics evident in such a young puppy. She wants to evaluate those same traits in the littermates, the **dam,** and the **sire.** She might want to see a copy of the puppy's extended pedigree to learn more about its ancestors. Similarly, when beef cattle breeders evaluate a sire to use via artificial insemination (A.I.), they look further than the sire's own performance for growth rate. They want to know something about the growth performance of his progeny.

The above examples illustrate the use of two different types of information (data) on relatives: **pedigree data** and **progeny data.** By examining the young puppy's parents, littermates, and extended pedigree, the dog breeder is using pedigree data. She is trying to learn something abut the genes made available to the puppy through its parents. Beef cattle breeders, on the other hand, are using progeny data. They are trying to learn something about an A.I. sire's genes by evaluating the performance of his offspring.

As the above examples should make clear, the information used to make selection decisions can be subjective, objective, or something in between. The pedigree data used by the dog breeder are, for the most part, subjective. The puppy's

progeny data
Information on the genotype or performance of descendants of an individual.

papers may include some semiobjective information on show championships won by ancestors, but the breeder's observations on conformation and personality are essentially subjective in nature. In contrast, the progeny data used by beef cattle breeders are relatively objective. They consist of carefully measured (we hope) weights of animals taken at specific ages.

Regardless of how subjective or objective the information used to make selection decisions, the purpose of that information is to help predict breeding values. When predictions of breeding values are derived from objective, numerical data—as in the case of beef cattle breeders—they will be expressed in objective, numerical terms. When predictions are derived from subjective data—as in the case of the dog breeder—they will be expressed in subjective terms, perhaps as simple perceptions. In either case, they are legitimate predictions of breeding value.

Sometimes we use information on relatives to predict breeding values because individual performance information is unavailable. Dairy sires, for example, do not produce milk, so we must rely on the records of their female relatives. More commonly, however, we use information on relatives because it increases the **accuracy** of our predictions. Accuracy measures the strength of the relationship between true values—often breeding values—and their predictions. When accuracy is high, predictions of breeding values will normally be good ones—they will closely reflect the true breeding values of the animals being evaluated. And because the predictions of breeding values are accurate, we can do a good job of selection.

accuracy
A measure of the strength of the relationship between true values and their predictions.

With phenotypic selection, accuracy of breeding value prediction is a function of heritability. When heritability is low, an animal's performance is generally not a good indicator of its breeding value. Accuracy is poor, and selection will not be very effective. Heritability also affects accuracy when breeding values are predicted from information on relatives—that information is, after all, nothing but phenotypic records—but heritability is not the limiting factor here that it is with phenotypic selection. In this case, accuracy depends not only on heritability, but on the *amount* of information as well. Most measures of fertility, for example, are lowly heritable. A female's own record for, say, interval from parturition to subsequent conception is not usually a very good indicator of her underlying breeding value for fertility. It is possible, however, to very accurately predict the breeding value of a sire for interval from parturition to conception based on large numbers of observations on daughters. In other words, the problem of low heritability can be overcome by using large amounts of information.

In most meat, dairy, and fiber producing species, **genetic prediction** technology has evolved to the point that objective predictions of breeding values and other related values are available. Predictions are based on performance information from large numbers of relatives of all kinds. In these species it is possible to make selection decisions on high-tech measures like EBVs (estimated breeding values), EPDs (expected progeny differences), PDs (predicted differences), ETAs (estimated transmitting abilities), MPPAs (most probable producing abilities), PCs (possible changes), and ACCs (accuracy values).[1] In beef cattle, dairy cattle, and swine you can use **sire summaries**—lists of genetic predictions, accuracy values, and other useful information—to help find the most outstanding sires in entire breeds. These technologies have caused revolutionary change in the way we select animals.

genetic prediction
The area of academic animal breeding concerned with measurement of data, statistical procedures, and computational techniques for predicting breeding values and related values.

sire summary
A list of genetic predictions, accuracy values, and other useful information about sires in a breed.

[1]For precise definitions of these terms, see Chapters 11 and 12.

In recreation and companion animal species like horses, dogs, and cats, high-tech genetic prediction technology of the kind used for meat, dairy, and fiber producing species is rare. Breeders rely on older, more traditional methods for selecting animals. In those species where there is sufficient economic incentive, more sophisticated prediction technology will probably be adopted. In some species it may never be used.

Selection for Simply-Inherited Traits

polygenic trait
A trait affected by many genes, no single gene having an overriding influence.

The traits mentioned in this chapter—weaning weight in sheep, body size in mice, fertility in mammals, conformation and personality in dogs, and growth rate in cattle—have all been **polygenic traits.** They are affected by many genes, and no single gene is thought to have an overriding influence. We know very little (if anything) about the specific genes affecting these traits—we just know there are lots of them. Because we cannot identify specific genes, we rely on phenotypic performance, predictions of breeding value, accuracy measures, etc. to help characterize the genotypes of animals.

simply-inherited trait
A trait affected by only a few genes.

polled
Naturally without horns.

Most economic traits in animals are polygenic in nature. Some traits, however, are **simply-inherited—**they are affected by only a few genes. A good example is the horned/polled character in cattle of European origin. (**Polled** means naturally without horns.) A single pair of genes determines whether a cow is horned or polled. Because simply-inherited traits are influenced by only a few genes, selection for simply-inherited traits is different from selection for typical polygenic traits. With simply-inherited traits, we do not deal with breeding values and their predictions, or even with concepts like heritability. Rather, we are interested only in knowing whether an individual possesses the specific gene or genes of interest, and we select animals based on that knowledge.

Sometimes an individual's genes affecting a particular simply-inherited trait are obvious. We know, for example, that a horned cow has two genes for the horned condition. In more complicated situations, we may not know by looking at an animal the specific genes it carries, but we can make an educated guess based on pedigree and progeny information.

The goal of selection is the same whether the trait under selection is polygenic or simply-inherited. In both cases we want to select the animals with the best sets of genes. The difference is that with polygenic traits, we predict breeding value in order to measure the overall effect of an individual's genes; with simply-inherited traits, we try to identify specific genes.

major gene
A gene that has a readily discernible effect on a trait.

Research in molecular genetics may soon blur the differences between simply-inherited and polygenic traits. Scientists may find **major genes—**genes that have readily discernible effects—for traits that were formerly thought to be polygenic. If that happens, breeders will be selecting replacements for both specific genes and desirable predictions of breeding value.[2]

Between-Breed Selection

between-breed selection
The process that determines the breed(s) from which parents are selected.

When we think of selection, we normally envision selection of individual animals. It is also possible to select groups of animals—even entire breeds. **Between-breed selection** provides a way of using breed differences to make very rapid genetic

[2]For a more complete discussion of the potential effects of biotechnology on animal breeding, see Chapter 20.

change. When commercial breeders choose the breeds they want to use in crossbreeding systems, they are practicing between-breed selection. Likewise, when seedstock breeders choose the component breeds they want in developing new breeds, they are practicing between-breed selection.

For many traits, breed differences can be very large. By taking advantage of such large differences, between-breed selection can produce genetic change much faster than the gradual change possible from selection within a breed. For example, production of cattle in many less-developed countries has increased dramatically in recent years—not through selection within native breeds, but through introduction of more productive breeds from Europe and North America.

Mating and Mating Systems

mating
The process that determines which (selected) males are bred to which (selected) females.

mating system
A set of rules for mating.

Selection is the first of the two basic tools used by animal breeders to make genetic change. The second tool is **mating.** Mating is the process that determines which (selected) males are bred to which (selected) females. It is distinctly different from selection. In selection, you choose the group of animals you want to be parents; in mating, you match males and females from the selected group.

There are many different methods for mating animals, and each method can be defined by a set of mating rules: a **mating system.** The mating system you choose depends on the kind of result you want. Listed in Table 2.1 are several examples of mating rules and the outcomes you could expect if you followed them.

The examples in Table 2.1 illustrate the three reasons breeders use mating systems: (1) to produce offspring with extreme breeding value in order to increase the rate of genetic change, (2) to make use of complementarity, and (3) to obtain hybrid vigor. By mating the largest males to the largest females (the first mating system in Table 2.1), you could, with a little luck, produce an offspring even larger than its parents. Assuming that larger animals are desired, this extreme individual then becomes a good candidate for selection. The purpose of this kind of mating system is, therefore, to speed genetic change caused by selection.

complementarity
An improvement in the overall performance of offspring resulting from mating individuals with different but complementary breeding values.

The next two mating systems in the table are designed to take advantage of **complementarity,** an improvement in the overall performance of offspring resulting from mating individuals with different but complementary breeding values. If an animal of intermediate size is desired, mating large animals to small animals is one way to produce it. The parental genotypes are quite different, and neither one is optimal, but the mating is complementary because the offspring is optimal. Similarly, if a palomino horse (tan with blonde mane and tail) is desired, one way to produce it is to mate a sorrel (all red) to a cremello (a particular kind of white horse). This too is a complementary mating.

TABLE 2.1 Examples of Mating Systems and Corresponding Outcomes

Mating Rule	Expected Outcome
Largest to largest	Produces an extreme
Large to small	Produces an intermediate
Sorrel to cremello	Produces a palomino
Charolais to Angus	Produces a hybrid
Half brother to half sister	Produces an inbred

crossbreeding
The mating of sires of one breed or breed combination to dams of another breed or breed combination.

crossbred
Having parents of different breeds or breed combinations.

hybrid vigor or **heterosis**
An increase in the performance of hybrids over that of purebreds, most noticeably in traits like fertility and survivability.

hybrid
An individual that is a combination of species, breeds within species, or lines within breeds.[3]

inbreeding
The mating of relatives.

inbreeding depression
The reverse of hybrid vigor—a decrease in the performance of inbreds, most noticeably in traits like fertility and survivability.

Because body size is influenced by many genes, mating a large animal to a small animal to obtain an animal of intermediate size is a complementary mating for a polygenic trait. Palominos differ from sorrels and cremellos by only one gene. Mating a sorrel to a cremello to obtain a palomino is, therefore, a complementary mating for a simply-inherited trait.

Mating a Charolais to an Angus is an example of **crossbreeding;** the mating of sires of one breed or breed combination to dams of another. Breeders often crossbreed to produce breed complementarity, and, in fact, the Charolais × Angus mating is a complementary one. Charolais are large French cattle known for fast growth and heavy muscling, Angus are smaller British cattle known for their maternal ability, and the **crossbred** offspring benefit from having both kinds of parents. Another reason for crossing these two breeds is to produce **hybrid vigor** or **heterosis.** Hybrid vigor is an increase in performance of crossbred or **hybrid** animals over that of purebreds. Hybrid vigor occurs to a greater or lesser degree in many traits, but it is most noticeable in traits like fertility and survivability—traits that are usually important economically.

Hybrid vigor is caused not by the presence of particular genes in an individual, but by the presence of particular gene *combinations.*[4] This suggests a fundamental, gene-level difference between selection and mating. While the purpose of selection is to increase the proportion of favorable genes in future generations of a population, the purpose of mating—at least when mating rules are designed to produce hybrid vigor—is to increase the proportion of favorable gene combinations in a population.

The last mating system listed in Table 2.1 involves **inbreeding** or the mating of relatives. We typically think of inbreeding in a negative way. We associate it with genetically defective individuals and **inbreeding depression,** which is the reverse of hybrid vigor—a decrease in the performance of inbreds, most noticeably in traits like fertility and survivability. But inbreeding can be very helpful. We can use it to create the breeds within species or lines within breeds that, when crossed, produce hybrid vigor.

Mating Systems and Industry Structure

The structure of a breeding industry and a breeder's place within that structure often influence the type of mating system he chooses. For example, in the sheep, beef cattle, and swine industries—all industries in which there is a fairly clear division between seedstock and commercial sectors—it is common to find breeders of purebred seedstock and increasingly less common to find breeders of purebred commercial animals. Furthermore, breeders of purebred seedstock will often practice inbreeding—*mild* inbreeding, anyway—whereas commercial producers will not. The reason for this has to do with the products marketed in each case. Commercial producers of these species sell lamb, wool, beef, and pork. They want to take advantage of all breeding methods that will increase the efficiency with which these things are produced, and because breed complementarity and hybrid vigor are important considerations for them, they crossbreed. Breeders of seedstock for these species, on the other hand, are typically not as concerned about complementarity and hybrid vigor.

[3]Some purists consider a hybrid the offspring of *purebreds* of two different species, lines, or breeds. I think the definition shown here is more useful.

[4]See Chapter 17 for a more detailed explanation.

There is relatively little complementarity to be found within a pure breed, purebred animals do not exhibit hybrid vigor themselves, and *inbred* purebreds may even show inbreeding depression. But because the seedstock breeder's product is an animal designed to be crossed with animals of another breed, these concerns are not important. The lack of complementarity and hybrid vigor in the purebred does not lessen its seedstock value to its buyer, the commercial producer.

SELECTION AND MATING TOGETHER

We normally think of selection preceding mating; first a pool of animals is selected, then mating decisions are made for the animals within the pool. The definition of mating given earlier in this chapter reinforced this idea by referring to the mating of "*selected* males" to "*selected* females." Indeed, animals must first be selected before they can be mated. In practice, however, selection and mating decisions are not always arrived at independently or in precise sequence.

corrective mating
A mating designed to correct in their progeny faults of one or both parents.

Suppose, for example, that you wanted animals of intermediate size, and many of the females in your breeding population were too small. One breeding strategy would be to make **corrective matings** by breeding the small females to especially large males. With this plan in mind, you select the males you need and then carry out the matings. In this example the first choice made was a choice of mating system—in this case mating for complementarity—followed by selection, followed by actual mating.

The interdependence of selection and mating decisions can also be seen in the initiation of a crossbreeding program. The first step is to decide on a crossbreeding system by weighing costs and management concerns against the potential benefits of hybrid vigor and complementarity. The next step is to decide which breeds to use—in other words, to practice between-breed selection. The final steps are to select individual animals and make individual matings. As in the previous example, mating and selection decisions are not independent of one another.

The relative gains to be made over time from selection (both between and within breeds) and mating (specifically crossbreeding) are depicted schematically in Figure 2.2. A large, onetime improvement can be realized simply by selecting

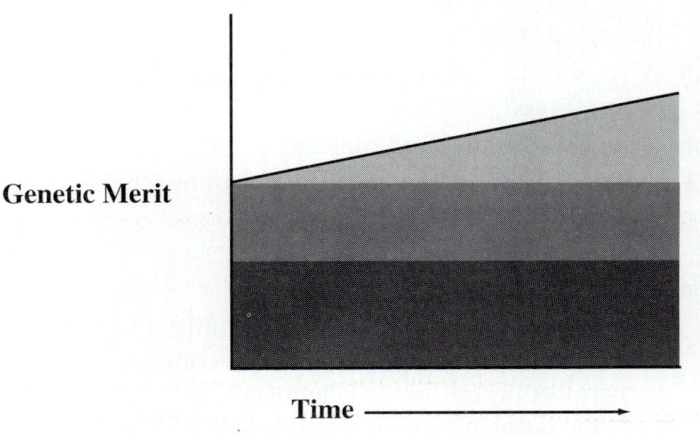

FIGURE 2.2
Schematic representation of genetic gain over time resulting from selection (both within and between breeds) and mating (specifically crossbreeding).

appropriate breeds. A further onetime gain comes from the complementarity and hybrid vigor associated with crossbreeding. Additional progress is made by selecting individuals within the parent breeds or within the crossbred population. Note that this last form of selection is the only breeding method that produces continuous improvement in the long term.

EXERCISES

Study Questions

2.1 Define in your own words:

population
selection
natural selection
artificial selection
replacement selection
culling
artificial insemination (A. I.)
breeding value
phenotypic selection
heritability
fertility
dam
sire
pedigree data
progeny data
accuracy
genetic prediction
sire summary
polygenic trait
simply-inherited trait
polled
major gene
between-breed selection
mating
mating system
complementarity
crossbreeding
crossbred
hybrid vigor or heterosis
hybrid
inbreeding
inbreeding depression
corrective mating

2.2 Explain how selection causes change in the performance of future generations of a population.

2.3 Why is selection generally more effective for highly heritable traits than for lowly heritable ones?

2.4 How does selection for polygenic traits differ from selection for simply-inherited traits?

2.5 Why should between-breed selection produce faster genetic change than within-breed selection?

2.6 Contrast selection and mating.

2.7 List three reasons to use mating systems.

2.8 Why are the mating systems used by seedstock breeders so often different from those used by commercial breeders?

2.9 For a species of your choice:

a. What polygenic traits are commonly selected for?

b. What simply-inherited traits are commonly selected for?

c. Is selection limited to phenotypic selection or are records of relatives used? If the latter, are pedigree data used? Progeny data? Both?

d. How high-tech are the criteria used in selection—i.e., do breeders select on looks alone, own performance, EPDs?

e. Are sire summaries available?

f. Is crossbreeding practiced? If so, why?
g. Describe the most common mating system.
h. Give an example of a typical complementary mating.
i. If there is a clear distinction between seedstock breeders and commercial producers, do the two kinds of breeders use different mating systems? Explain.
j. Are selection and mating decisions entirely independent? If not, describe the relationship between them.

PART II

Animal Breeding from the Bottom Up

If you read and studied the first two chapters of this book carefully, you already know a great deal about animal breeding. You now have an appreciation for terms like *breeding value*, *heritability*, and *hybrid vigor*—all extremely important concepts. You probably know as much about the principles of animal breeding as anyone living before the American Civil War. Many of the early breeders understood breeding value, heritability, and hybrid vigor too, although they probably used different names for them. Their understanding was limited, however, by a lack of information about the basic laws of inheritance. They did not know, for example, how genetic information is transmitted from parent to offspring. They were unaware of the genetic mechanisms that preserve variability in a population or that cause an outbred animal to be more vigorous than an inbred animal. Gregor Mendel and his successors changed this situation forever. Now we understand the fundamental mechanisms and rules of inheritance, and this knowledge allows us to better comprehend the important concepts of animal breeding.

The next three chapters review the basics of classical genetics. They introduce the terms that comprise the jargon (albeit *useful* jargon) of the discipline, and lay the groundwork for an understanding of animal breeding at the level of the gene—from the bottom up.

CHAPTER 3

Mendelian Inheritance

In the mid-nineteenth century, Gregor Mendel performed his now famous breeding experiments with peas. He knew nothing about the details of meiosis, chromosomes, or DNA, but he was perceptive enough to infer the basic rules of inheritance simply by observing the outcomes of his matings. Today we refer to Mendel's laws as Mendelian inheritance, the understanding of which is the basis for all genetic and animal breeding theory developed since Mendel's time.

In this chapter we will examine Mendelian inheritance from a modern perspective. We will add to Mendel's findings more recently discovered phenomena like chromosomes and DNA. The essentials of this chapter, however, are things that Mendel would have understood.

Genes, Chromosomes, and Genotypes

gene
The basic physical unit of heredity consisting of a DNA sequence at a specific location on a chromosome.

DNA
Deoxyribonucleic acid, the molecule that forms the genetic code.

chromosome
One of a number of long strands of DNA and associated proteins present in the nucleus of every cell.

homolog
One of a pair of chromosomes having corresponding loci.

locus
The specific location of a gene on a chromosome.

allele
An alternative form of a gene.

multiple alleles
More than two possible alleles at a locus.

The basic unit of inheritance is called a **gene.** Today we understand genes to be segments of deoxyribonucleic acid or **DNA,** the very complex molecule that forms the genetic code for all living things. Genes are relatively short sections of **chromosomes,** which are very long strands of DNA and associated proteins present in the nucleus of every cell of an organism. Chromosomes come in pairs, one chromosome of a pair inherited from an individual's sire and the other chromosome inherited from its dam. The number of chromosome pairs depends on the species. Humans, for example, have 23 pairs. Cattle have 30, dogs 39. A representative pair of chromosomes or **homologs** is depicted in Figure 3.1.

Shown in Figure 3.1 are a hypothetical "J" locus and "B" locus. **Locus** is the Latin word for location and denotes the site of a particular gene. At each locus is a pair of genes, one gene on the paternal chromosome and one on the maternal chromosome. The genes at a locus are denoted symbolically by a single letter or letter combination. For example, the two genes at the J locus in an individual might be designated *J* and *j*. If one gene is represented by an uppercase letter and the second gene by a lowercase letter (or some other variant), the implication is that there is a chemical and functional difference between them. *J* and *j* are called **alleles** (pronounced *uh-leels*), alternative forms of genes found at the J locus. If the two genes at the J locus were functionally alike, they would both have the same symbol.

Although there are only two genes at any particular locus in an individual, the two may be a subset of a larger series of alternative forms of a gene. In other words, there may be **multiple alleles.** In dogs, for example, there is a locus affecting coat color known as the E locus (for *extension* of pigmentation). There are three different

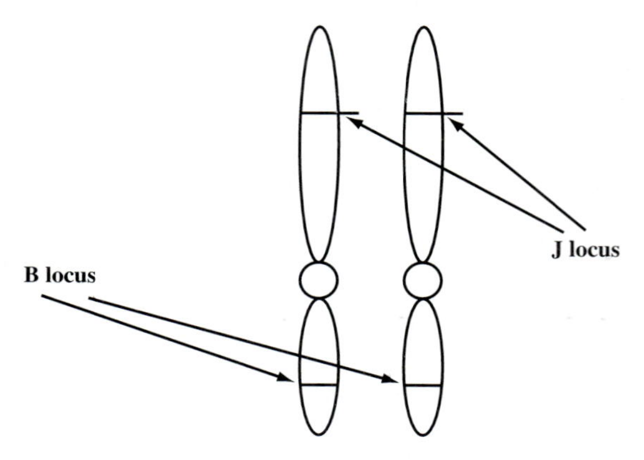

FIGURE 3.1
A representative pair of chromosomes.

genotype
The combination of genes at a single locus or at a number of loci. We speak of one-locus genotypes, two-locus genotypes, and so on.[2]

homozygote (homozygous genotype)
A one-locus genotype containing functionally identical genes.

heterozygote (heterozygous genotype)
A one-locus genotype containing functionally different genes.

E-locus alleles: *E,* which causes full extension of pigment, i.e., does not inhibit pigmentation; E^{br}, which causes brindling or tiger striping; and *e,* which inhibits pigmentation.[1] Any one dog can have a maximum of two of the three alleles in the E series. In the hypothetical example shown in Figure 3.1, if there were four *possible* alleles at the B locus, they might be represented by *B, b, b′,* and *b″.*

The combination of genes at a particular locus is referred to as a **genotype,** specifically a *one-locus* genotype. If *J* and *j* are the only possible alleles at the J locus, then three genotypes can occur: *JJ, Jj,* and *jj.* Because there are four possible alleles at the B locus, there is the potential for many more genotypes there: *BB, Bb, Bb′, Bb″, bb, bb′, bb″, b′b′, b′b″,* and *b″b″.* If we consider both the J and B loci (the plural of locus—pronounced *low-sigh*) together, then there is a much larger number of *two-locus* genotypes: *JJBB, JJBb, JJBb′, . . . jjb″b″*—30 two-locus genotypes in all.

A one-locus genotype is considered **homozygous** if both genes at that locus are functionally the same. The *JJ, jj, BB, bb, b′b′,* and *b″b″* genotypes are all examples of **homozygotes.** One-locus genotypes containing functionally different genes are considered **heterozygous.** The *Jj, Bb, Bb′, Bb″, bb′, bb″,* and *b′b″* genotypes are examples of **heterozygotes.**

Germ Cells and Their Formation

segregation
The separation of paired genes during germ cell formation.

germ cell or **gamete**
A sex cell—a sperm or egg.

meiosis
The process of germ cell formation.

Mendel's first law is known as the *law of* **segregation.** It states that in the formation of a **germ cell** or **gamete** (in the male, a sperm; in the female, an egg), the two genes at a locus in the parent cell are separated, only one gene being incorporated into each germ cell. Today we call the process that creates germ cells **meiosis.** Meiosis is quite complicated, involving a number of intricate steps during which not only genes but entire homologous chromosomes are separated (see Figure 3.2). Mendel knew nothing of the details, but he had it essentially right—gametes contain only one gene of a pair.

[1]More detailed discussion of the E locus can be found near the end of this chapter in the section on epistasis.

[2]Note the difference between this very precise definition of genotype and the much broader definition given in Chapter 1. Genotype can mean different things depending on the context.

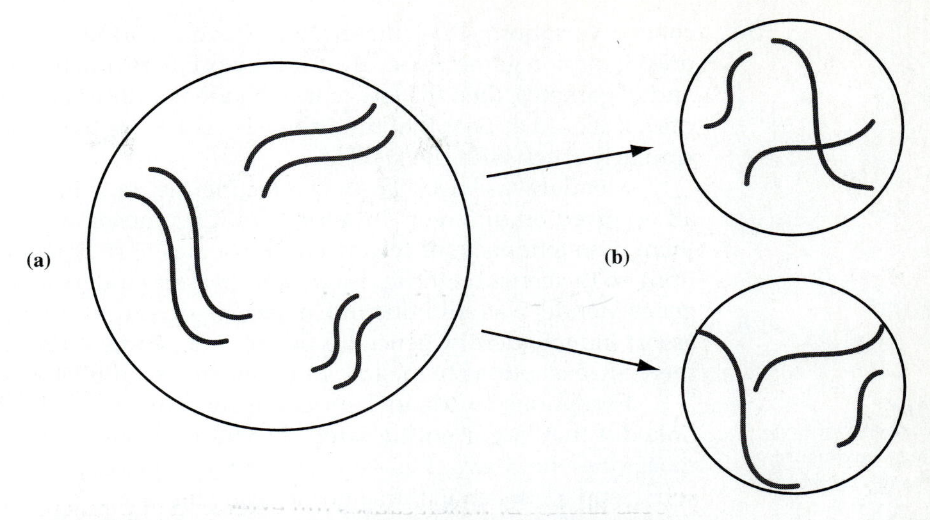

FIGURE 3.2 Schematic representation of the separation of homologous chromosomes in the formation of germ cells: (*a*), primordial sex cell containing both homologs; and (*b*), germ cells containing only one homolog from each pair.

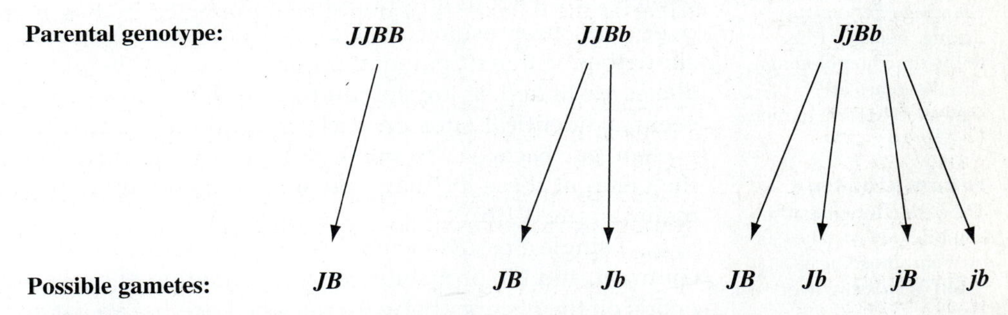

FIGURE 3.3 Two-locus genotypes and the gametes that can be produced from them.

The gametes that are obtainable from several two-locus genotypes are shown in Figure 3.3. Note that each gamete contains only one gene from each locus. While the original two-locus genotypes contained four genes altogether, each gamete has only two. As a rule, germ cells contain half the number of chromosomes and therefore half the number of genes of normal body cells.

The number of gametes that can be obtained from a parental genotype depends on how heterozygous the genotype is. The *JJBB* genotype, for example, is completely homozygous. It can produce only one kind of gamete: *JB*. Two kinds of gametes are obtainable from the partially heterozygous *JJBb* genotype, and the completely heterozygous *JjBb* genotype can produce four different kinds of gametes.

independent assortment The independent segregation of genes at different loci.

Figure 3.3 illustrates Mendel's second law, the *law of* **independent assortment.** Genes assort independently during meiosis if all possible gametes are formed in equal proportions. For this to happen, a given gene from one locus must have an equal probability of being present in the same germ cell with either of the two genes from some other locus. As an example, consider the *JjBb* genotype in Figure 3.3. *JjBb* individuals can produce four possible gametes: *JB*, *Jb*, *jB*, and *jb*. If all four gametes occur in equal proportions (allowing some leeway for chance variation), then these genes have assorted independently. If, however, only *JB* and *jb* gametes occur or if they occur at much higher frequencies than *Jb* and *jB* gametes, then the law of independent assortment has been violated. In this case, it would appear that the *J* allele is "stuck" with the *B* allele and the *j* allele is similarly stuck with the *b* allele.

Mendel was lucky. The loci affecting the traits he observed in his pea plants all occurred on different chromosomes. Chromosomes assort independently (i.e., there is no tendency for certain chromosomes to stick together in germ cell formation), so the genes on those chromosomes assort independently too. Because all the genes Mendel was studying did, in fact, assort independently, he believed *all* genes assort independently, hence his *law* of independent assortment. Today we know there are exceptions to the law, but they are exceptions, not the rule.

linkage
The occurrence of two or more loci of interest on the same chromosome.

crossing over
A reciprocal exchange of chromosome segments between homologs. Crossing over occurs during meiosis prior to the time homologous chromosomes are separated to form gametes.

recombination
The formation of a new combination of genes on a chromosome as a result of crossing over.

Exceptions to Mendel's second law are caused by **linkage.** Two loci are linked if they occur on the same chromosome. Because entire homologous chromosomes—not just genes—are separated at meiosis, genes on the same chromosome tend to end up in the same gamete. This is only a tendency, however, because of a phenomenon known as **crossing over.** Crossing over involves a reciprocal exchange of chromosome segments between homologs and occurs during meiosis prior to the time the chromosomes are separated to form gametes. Figure 3.4 depicts homologous chromosomes (a) before crossing over and (b) after crossing over. The chromosomes in Figure 3.4(a) have different background patterns to show their different parental origins. Note that before crossing over, the *J* and *B* alleles are linked, as are the *j* and *b* alleles. In the crossover process, mutual breaks occur at identical sites on each chromosome, and chromosome fragments are exchanged. Because the break in Figure 3.4 occurred between the J and B loci, the genes at these loci have **recombined** and are now linked in a new arrangement (Figure 3.4(b)).

A single crossover event is shown in Figure 3.4. Multiple crossover events are common, and the probability of recombination of genes at any two linked loci depends on the distance between the loci. Loci that are far apart (like the J and B loci) are likely to recombine often. For practical purposes, the genes at these loci will assort independently, just as they would if they had been on different chromosomes altogether. Recombination is much less likely for loci that are very close together because the probability of a break occurring between them is much less. These *closely linked* loci create exceptions to Mendel's second law. But in the species of

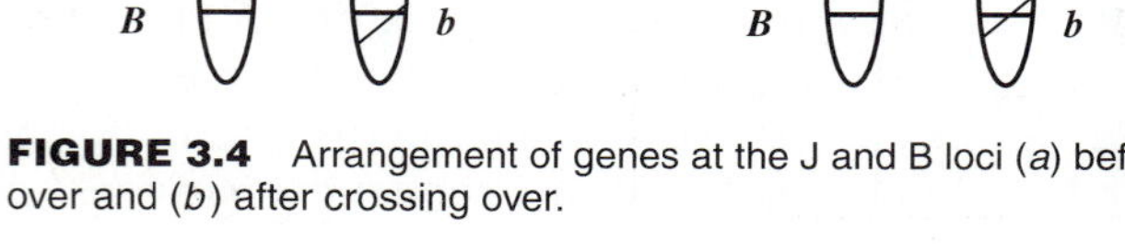

FIGURE 3.4 Arrangement of genes at the J and B loci (*a*) before crossing over and (*b*) after crossing over.

most interest to animal breeders, genes are distributed across a large number of chromosomes, and close linkage between two loci of interest is a relative rarity. In general then, we can assume independent assortment, and in the examples used in the remainder of this book, we will.

FORMATION OF THE EMBRYO

embryo
An organism in the early stages of development in the shell (bird) or uterus (mammal).

zygote
A cell formed from the union of male and female gametes. A zygote has a full complement of genes—half from the sperm and half from the egg.

gamete selection
The process that determines which egg matures and which sperm succeeds in fertilizing the egg.

Punnett square
A two-dimensional grid used to determine the possible zygotes obtainable from a mating.

When a male is successfully mated to a female, sperm and egg unite, and an **embryo** is formed. In genetic jargon, we say that gametes from the sire and dam combine to form a **zygote.** Zygotes are offspring. They have the normal number of genes and chromosomes, half from the gamete contributed by the sire, and half from the gamete contributed by the dam. The process that determines which egg matures (physiologically develops and is ovulated) and which sperm succeeds in fertilizing the egg is called **gamete selection.** Some gametes contain genetic defects that cause them to be nonviable. These gametes are naturally selected against. Aside from this form of natural selection, however, selection of gametes is essentially random. In other words, almost all gametes have an equal chance of contributing to a zygote.

A commonly used device for determining the possible zygotes obtainable from the mating of any two parental genotypes is the **Punnett square.** A Punnett square is a two-dimensional grid. Along the top of the grid are listed the possible gametes from one parent, and along the left side are listed the possible gametes from the other parent. Inside the cells of the grid are the zygotes that are possible from the mating. They are obtained by simply combining the gametes that head each row and column of the square. A two-locus example is shown in Figure 3.5. In this example, a *JjBb* male is mated to a *JjBb* female. Each parent can produce four distinct gametes: *JB, Jb, jB,* and *jb,* so there are four rows and four columns in the Punnett square, resulting in 16 cells. Not every cell contains a unique zygote,

♂ \ ♀	*JB*	*Jb*	*jB*	*jb*
JB	*JJBB* (1)	*JJBb* (2)	*JjBB* (3)	*JjBb* (4)
Jb	*JJBb* (2)	*JJbb* (5)	*JjBb* (4)	*Jjbb* (6)
jB	*JjBB* (3)	*JjBb* (4)	*jjBB* (7)	*jjBb* (8)
jb	*JjBb* (4)	*Jjbb* (6)	*jjBb* (8)	*jjbb* (9)

FIGURE 3.5
Punnett square showing the possible gametes (on the outside of the square) and possible zygotes (inside the square) from the mating of two individuals with the *JjBb* two-locus genotype. The small numbers in the corners of each cell identify unique offspring genotypes—nine in all.

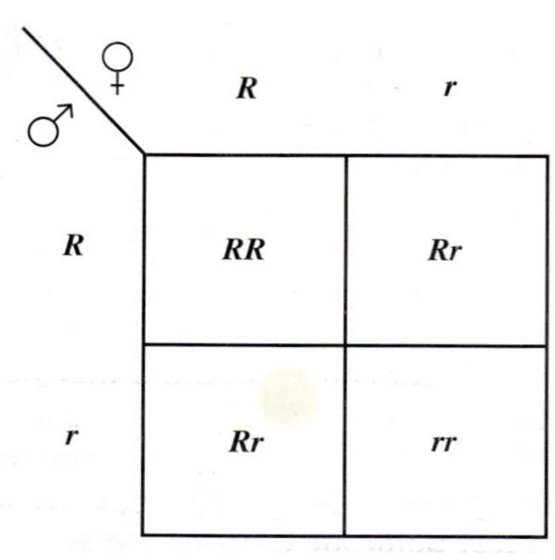

FIGURE 3.6
Punnett square representing the mating of roan Shorthorns.

however. Some cells of the square contain the same genotype. In this particular case, there are nine distinct zygotic types.

If each gamete listed along the top and side of a Punnett square occurs with equal frequency, each cell within the square should also occur with equal frequency. It is possible, therefore, to determine the likelihood of any particular offspring genotype by noting the frequency of the cells that contain that genotype. And if you know what phenotype is associated with each genotype—as is the case with simply-inherited, but not polygenic traits—you can also determine the expected proportions of offspring phenotypes.

Coat color in Shorthorn cattle provides a good example. Shorthorns come in three basic colors: red, white, and roan (a combination of both red and white hairs). These colors are controlled by the R locus. *RR* individuals are red, *rr* individuals white, and *Rr* individuals roan. The mating of two roan animals is illustrated by the Punnett square in Figure 3.6. As indicated by the frequency of the cells containing each genotype, the three offspring genotypes and phenotypes should occur in a 1:2:1 ratio—one red to two roans to one white. This ratio is an *expectation;* we cannot say that of every four calves from roan matings, one will be red, two roan, and one white. *On average,* however, the ratio will hold, and with large numbers of offspring from this mating, we would anticipate the coat colors of the calves to fit the 1:2:1 ratio quite closely.

Figure 3.6 is an example of a one-locus Punnett square. Figure 3.5 is a two-locus example. Punnett squares can be used to illustrate matings involving any number of loci. But for practical purposes, squares involving more than a few loci become unwieldy. A Punnett square showing the mating of individuals heterozygous at four loci would contain 256 cells.

The Randomness of Inheritance

The significance of Mendel's laws lies in their explanation of the particulate nature of inheritance, the "particles" being what we now call genes, and also in their explanation of how genetic variability is maintained in a population. Prior to Mendel's findings, the most widely accepted school of genetic thought involved a "blending" theory of inheritance in which hereditary information was contained in fluids, perhaps even in blood, and it was the mixing of parental fluids that determined the genetic makeup of an offspring. The blending theory was fatally flawed. It could never explain why, after many generations of mixing of fluids, all individuals in a population did not have a similar blend. In other words, it could not explain why there is so much genetic variation in most populations and why that variation does not diminish over time. Mendel's work provided the answer and disproved the blending theory forever, yet even today we use terms like "percent blood" to describe the ancestry of an animal.

To get a better feeling for the effect of Mendelian inheritance on the preservation of genetic variability, consider an individual that is heterozygous at 100 loci. Assuming segregation and independent assortment, this individual can produce over 1.2×10^{30} uniquely different gametes. And if this individual were mated to another individual just like it, over 5×10^{47} unique zygotes would be possible. That is *500 billion billion billion billion billion* zygotes—no two alike. These numbers are so large they are incomprehensible, yet they grossly underestimate the true number of possibilities. Most domestic animals are heterozygous at many more than 100 loci. A more realistic number of heterozygous loci might be in the thousands or tens of thousands. The resulting numbers for possible gametes and zygotes are staggering.

Calculating Numbers of Possible Gametes and Zygotes

The following examples show how to use an estimate of the number of loci at which an individual is heterozygous to determine mathematically the number of unique gametes the individual can produce. An individual with the genotype *AABBCC* has no heterozygous loci and produces just one kind of gamete: *ABC*. An individual heterozygous at one locus—say, *AaBBCC*—can produce two different gametes: *ABC* and *aBC*. (Note that in this example only the heterozygous A locus contributes to variation in gametes—the homozygous B and C loci do not.) Individuals heterozygous at two loci (*AaBbCC*) can produce four kinds of gametes:

ABC	*aBC*
AbC	*abC*

And individuals heterozygous at three loci (*AaBbCc*) can produce eight kinds of gametes:

ABC	*aBC*
ABc	*aBc*
AbC	*abC*
Abc	*abc*

There is a pattern here which can be summarized in the following formula:

$$\text{Number of unique gametes} = 2^n$$

where n is the number of loci at which an individual is heterozygous.

By similar reasoning, assuming only two possible alleles per locus,

$$\text{Number of unique zygotes} = 3^n \times 2^m$$

where n is the number of loci at which both parents are heterozygous, and m is the number of loci at which only one parent is heterozygous.

In the example of an individual heterozygous at 100 loci, the number of unique gametes possible is

$$2^n$$
$$= 2^{100}$$
$$1.27 \times 10^{30} \text{unique gametes}$$

And if that individual is mated to another individual *just like it,* the number of unique zygotes possible is

$$3^n \times 2^m$$
$$= 3^{100} \times 2^0$$
$$= 3^{100} \times 1$$
$$5.15 \times 10^{47} \text{unique zygotes}$$

The processes that ensure genetic variability are random (or *nearly* random) in nature. Independent assortment of genes during germ cell formation is almost entirely random; only close linkage prevents complete randomness. There is no way to predict what combination of genes will be present in a particular gamete. Some gametes will receive a favorable sample of genes; others will not. The process of gamete selection in the formation of the embryo is equally random. There is no way to predict the genetic makeup of the egg that is the next to mature or to predict the genetic makeup of the one sperm among millions that succeeds in fertilizing the egg. You can think of the random processes of independent assortment and gamete selection as two separate processes or pieces of a single process. Either way, the outcome is the same: the sample of genes that an offspring receives from its parents is a random one.

The randomness of inheritance is critically important from an evolutionary standpoint and, as we will see in Chapters 9 and 10, is also vitally important to the success of artificial selection. Nevertheless, it creates a problem for animal breeders—it reduces our ability to control the outcomes of matings. We can increase the probability of getting a superior offspring by mating parents we know to have superior breeding values, but we have no control over the **Mendelian sampling** of genes which determines the genetic makeup of the offspring. The fact that a sire and dam have together produced an outstanding offspring in the past is no guarantee that they will produce an equally outstanding one in the future. Likewise, just because the first mating of two individuals produced a less than desirable result does not mean that better results are not possible from this mating.

Mendelian sampling
The random sampling of parental genes caused by segregation and independent assortment of genes during germ cell formation and by random selection of gametes in the formation of the embryo.

Mendelian sampling in fish is illustrated in Figure 3.7. (I chose fish for this example because they are highly *fecund*—a single mating produces many offspring.) In (a) two individuals with inferior genetic merit for growth rate are mated. The offspring from this mating are not all the same because Mendelian sampling has

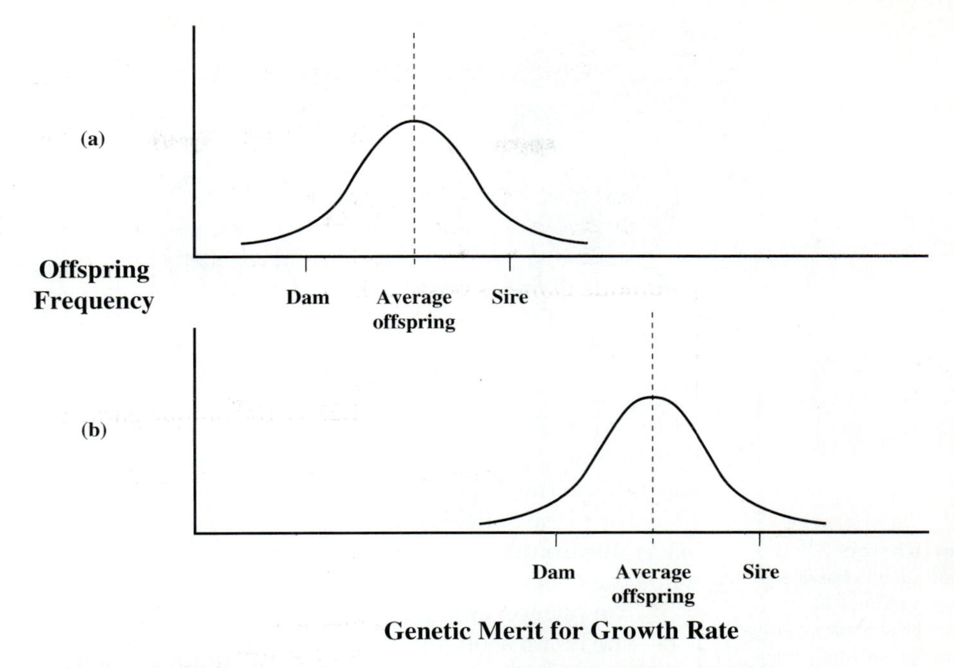

FIGURE 3.7 Distributions of progeny from a mating of (*a*) two fish with inferior genetic merit for growth rate and (*b*) two fish with superior genetic merit for growth rate.

caused them to receive different sets of genes from their parents. With respect to genetic merit for growth rate (the horizontal scale in Figure 3.7) the progeny appear to have a bell-shaped distribution. Most have genetic merit for growth rate close to the average merit of their parents, which, in this case, is inferior. Some (those at the extreme left of the distribution) are really poor, but a few (those at the extreme right of the distribution) are genetically capable of quite fast growth. In (b) two individuals with superior genetic merit for growth rate are mated. Again, Mendelian sampling causes variation in the offspring. In this case, most of the offspring are superior—some extremely good—and a few are inferior. Note that even though Mendelian sampling causes considerable variation in the progeny produced by any one mating, the *probability* of getting a superior offspring is much greater when superior parents are mated than when inferior parents are mated.

In some respects, offspring are dealt genes from their parents in much the same way you are dealt cards from a deck: sometimes you get a good hand, sometimes you get a poor one. This is an important practical point to remember and one that many breeders do not appreciate enough. Genetics, like a card game, involves chance and—to a degree, anyway—a certain amount of luck. When we think that we as breeders are entirely in control, we seriously overestimate our abilities.

Dominance and Epistasis

Mendel discovered that the expression of a gene at a locus depends on the other gene present at that locus. His pea plants were either tall or so short as to be considered dwarves. Dwarf plants were of the *tt* genotype, but tall plants were either

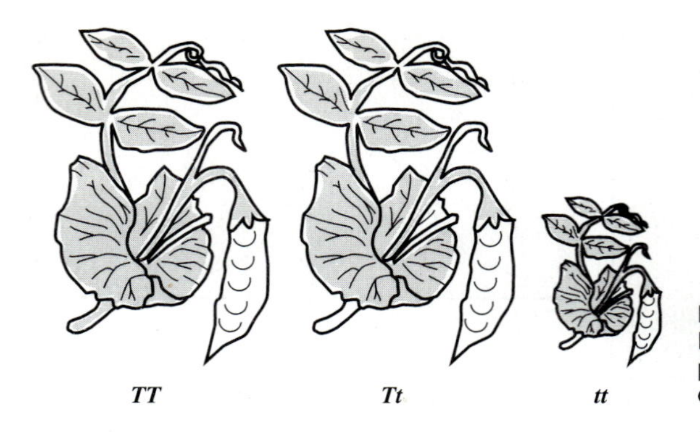

FIGURE 3.8
Dominance for height in Mendel's pea plants. The *T* (tall) allele is dominant over the *t* (dwarf) allele.

TT or *Tt* (see Figure 3.8). The gene for shortness *(t)* produced a dwarf when paired with another *t* gene. But when the *t* allele was paired with a tall *(T)* allele, the plant was not intermediate in size as you might expect. Rather, it was just as tall as the *TT* plants—the *t* allele appeared to have no effect at all. Today we say that the *T* allele is **dominant** over the *t* allele. In the heterozygote, the *T* allele expresses itself while the *t* allele does not. The *t* allele is then said to be **recessive.**

dominance
An interaction between genes at a single locus such that in heterozygotes one allele has more effect than the other. The allele with the greater effect is **dominant** over its **recessive** counterpart.

The phenomenon of dominance is important to animal breeding for two reasons. The first reason relates to simply-inherited traits like the ones Mendel studied in his peas. For these traits, dominance explains why we get various phenotypes in particular proportions when we make specific matings. Understanding the nature of dominance in these situations allows us to predict the outcomes of matings. This chapter contains a number of examples involving traits like coat color. The second reason involves polygenic traits. For these traits, dominance is the chief source of hybrid vigor and inbreeding depression.[3] (Epistasis, a related concept to be explained later in this chapter, is important to animal breeding for precisely the same reasons.)

Dominant alleles are usually represented by an uppercase letter and recessive alleles by a lowercase letter. At the J locus, then, the *JJ* genotype is called the *homozygous dominant genotype,* the *Jj* genotype is the *heterozygous genotype,* and the *jj* genotype is the *homozygous recessive genotype.* The letter or letter combination chosen to represent a locus is usually some form of abbreviation related to the characteristics of the dominant gene (hence the T locus for Mendel's tall versus dwarf plants). Unfortunately, the genetics literature is full of exceptions to this convention. Mendel studied loci affecting seed color and seed shape in pea plants, and these loci have been designated G and W even though green color and wrinkled shape are *recessive* conditions.

There are several forms of dominance possible at a locus. Really, these forms of dominance are not fundamentally different—they vary only in degree.

Complete Dominance

complete dominance
A form of dominance in which the expression of the heterozygote is identical to the expression of the homozygous dominant genotype.

In Mendel's peas, the mode of gene expression at the T locus was **complete dominance.** This is the classic form of dominance in which the expression of the heterozygous genotype is no different from the expression of the homozygous genotype having two dominant genes. *Tt* heterozygotes and *TT* homozygotes were equally tall; phenotypically they were indistinguishable from each other.

[3]See Chapters 7 and 17 for an explanation of the relationship between dominance, hybrid vigor, and inbreeding depression.

Complete dominance is common in a number of simply-inherited traits of animals. Typical examples in cattle are the polled trait (the *P* allele for polled is completely dominant over the *p* allele for horned) and black/red coat color (the *B* allele for black is completely dominant over the *b* allele for red).

Many lethal, semilethal, or otherwise deleterious conditions in animals involve complete dominance, and the problem gene is usually the recessive allele. An example is spider syndrome in sheep. The *S* allele is the normal gene at the S locus. The *s* allele is the recessive gene responsible for spider syndrome—often debilitatingly crooked legs in lambs. *SS* and *Ss* individuals are perfectly normal. Only the homozygous recessive *ss* lambs show the spider condition. Because deleterious, completely recessive genes can be carried and proliferated by apparently normal heterozygous animals, they are of particular concern to animal breeders.

Complete dominance is the one form of dominance in which heterozygous and homozygous dominant genotypes have the same phenotypic expression. Unlike the coat color example in Shorthorns in which each genotype (*RR*, *Rr*, or *rr*) is associated with a distinct phenotype (red, roan, or white), traits affected by complete dominance have more than one genotype for a phenotype. Polled cattle, for example, can be either *PP* or *Pp*. As a result, the classic 1:2:1 ratios expected from mating two heterozygotes will not occur with complete dominance.

To see how matings involving complete dominance turn out, consider crosses of Angus and Horned Hereford cattle. Angus cattle are polled, and purebreds are homozygous polled *(PP)*. Horned Herefords, on the other hand, are horned *(pp)*. Angus bulls bred to Horned Hereford cows (or vice versa) produce all heterozygous but phenotypically polled offspring as you can see from the following Punnett square.

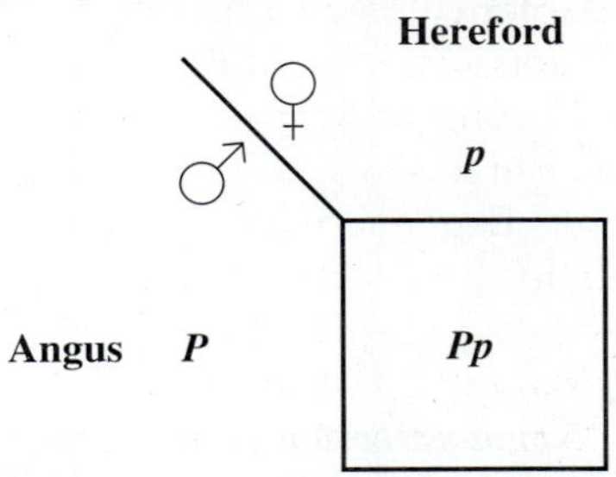

When Angus × Hereford crosses are mated inter se (among themselves), polled and horned offspring are produced with a ratio of three polled to one horned (see the next Punnett square). Two out of three polled types will be heterozygous; one out of three will be homozygous polled.

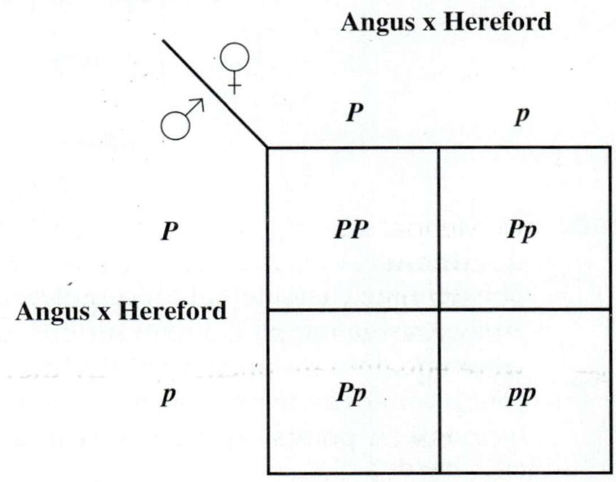

The polled character is not the only simply-inherited trait in which Angus and Herefords differ. Herefords are red. A large majority of Angus are black. Combining the P and B loci provides a more complex example of complete dominance. Assuming that the Angus bulls are homozygous for the black color gene, the mating of Angus bulls to Hereford cows produces all black, polled offspring.

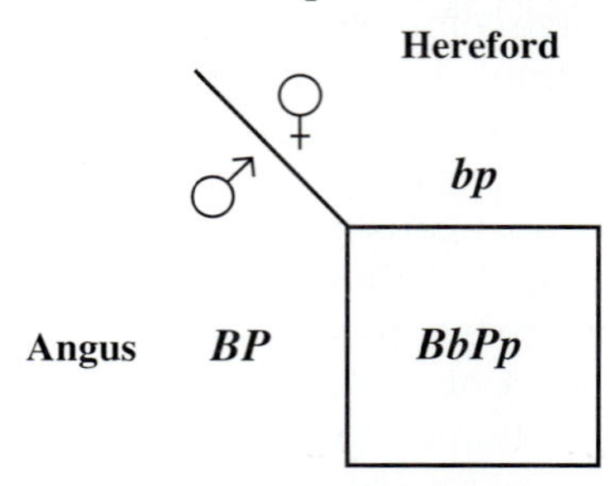

When these "black baldies" are mated to each other, nine distinct genotypes are produced, but because of complete dominance, only four phenotypes are recognizable: black/polled, black/horned, red/polled, and red/horned. These will occur with a ratio of approximately 9:3:3:1.

Angus x Hereford (♀ across top, ♂ down side)

♂ \ ♀	*BP*	*Bp*	*bP*	*bp*
BP	Black, polled *BBPP*	Black, polled *BBPp*	Black, polled *BbPP*	Black, polled *BbPp*
Bp	Black, polled *BBPp*	Black, horned *BBpp*	Black, polled *BbPp*	Black, horned *Bbpp*
bP	Black, polled *BbPP*	Black, polled *BbPp*	Red, polled *bbPP*	Red, polled *bbPp*
bp	Black, polled *BbPp*	Black, horned *Bbpp*	Red, polled *bbPp*	Red, horned *bbpp*

Angus x Hereford

Partial Dominance

The defining characteristic of complete dominance is that the expression of the heterozygous genotype is identical to the expression of the homozygous dominant genotype. This is shown graphically in Figure 3.9. The horizontal line represents a continuum of gene expression. If the J locus affected, say, height (like Mendel's T locus), then points on the line would signify greater and greater expression of

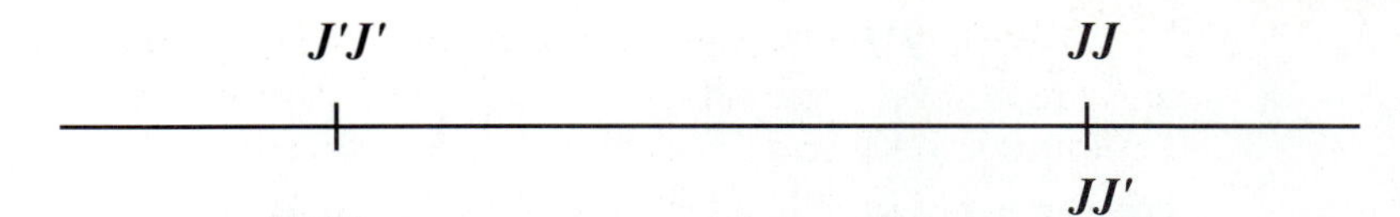

FIGURE 3.9 Schematic of complete dominance at the J locus. The *J* allele is dominant in this example.

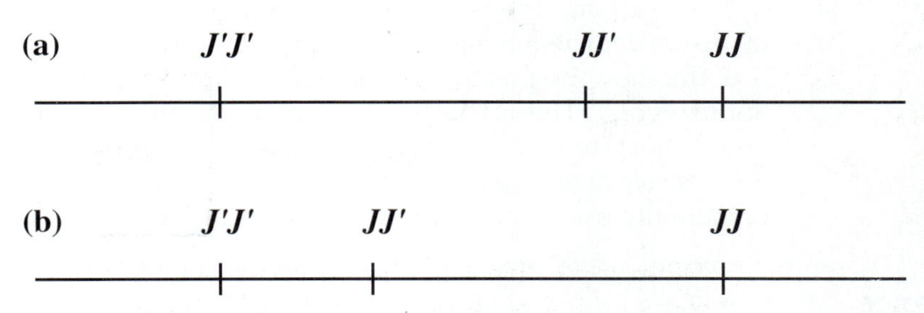

FIGURE 3.10 Partial dominance at the J locus. The dominant allele is *J* in (*a*) and *J'* in (*b*).

height as you go from left to right. Note that the point on the line for the expression of the heterozygote *(JJ')* is identical to the point representing the expression of the homozygous *JJ* genotype. Dominance is therefore complete, and the *J* allele is the dominant allele because it completely masks the expression of the *J'* allele in the heterozygote.

partial dominance A form of dominance in which the expression of the heterozygote is intermediate to the expressions of the homozygous genotypes and more closely resembles the expression of the homozygous dominant genotype.

Other forms of dominance can be shown with a line diagram like the one in Figure 3.9. **Partial dominance** is depicted in Figure 3.10. With partial dominance, the expression of the heterozygote is intermediate to the expressions of the homozygous genotypes and more closely resembles the expression of the homozygous dominant genotype. In Figure 3.10, the *JJ'* genotype lies somewhere between *J'J'* and *JJ*. In (a) the heterozygote more closely resembles the *JJ* homozygote. In this case the *J* allele is partially dominant over the *J'* allele because it has greater expression in the heterozygote. *J* is then the dominant allele. Partial dominance is also shown in (b), only in this case the heterozygote lies closer to *J'J'* than *JJ*, making *J'* the dominant allele.

A real-world example of partial dominance is the condition known as HYPP (hyperkalemic periodic paralysis) in horses. HYPP causes episodes of muscle tremors ranging from shaking or trembling to complete collapse. In some instances it can be fatal. The mutant gene causing HYPP is inherited as a partial dominant. Although clinical signs of HYPP vary considerably among individual horses, symptoms are generally more severe for homozygous HYPP animals than for heterozygotes.

HYPP is a particularly interesting case because it spread very rapidly among show and pleasure horses in the United States. Normally you would expect a deleterious dominant gene, even if it is only partially dominant, to be self-eliminating. After all, it is expressed in heterozygotes, not "hidden" in heterozygotes the way completely recessive genes are hidden. The HYPP gene was not quickly eliminated, however, because (1) it is not completely lethal, and (2) carriers often exhibit heavy muscling—a desirable characteristic in halter competition. The gene has persisted in horses because of the tendency of breeders to unknowingly select for it.

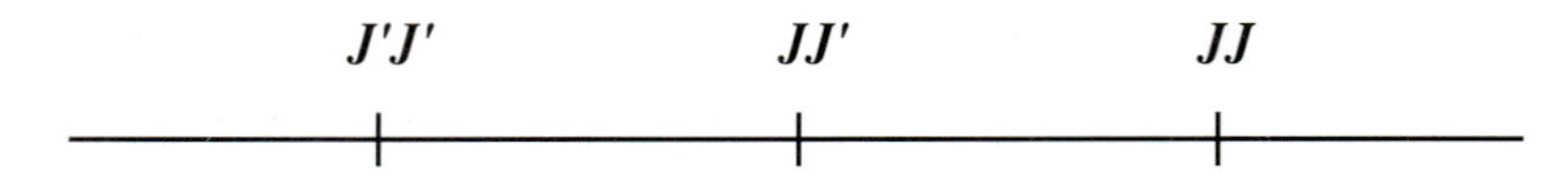

FIGURE 3.11 No dominance at the J locus.

If you compare complete dominance as shown in Figure 3.9 with partial dominance as depicted in Figure 3.10, you can see that the particular kind of dominance is defined by the position of the heterozygote relative to the positions of the two homozygotes. This is true for any form of dominance and leads to this general rule: *to determine the kind of dominance and the dominant allele, compare the expression of the heterozygote to the expressions of the homozygous genotypes.* An easy way to do this is to mentally use a line diagram like those in Figures 3.9 and 3.10.

No Dominance

no dominance
A form of dominance in which the expression of the heterozygote is exactly midway between the expressions of the homozygous genotypes.

No dominance exists if the expression of the heterozygote is *exactly* midway between the expressions of the homozygous genotypes. Neither allele is dominant in this case because both alleles appear to have equal expression in the heterozygote. No dominance is depicted in Figure 3.11.

For a hypothetical example of no dominance, consider resistance to a particular disease—say, tuberculosis. If, when exposed to the tuberculosis pathogen, animals with two copies of a tuberculosis-resistant gene (T^r) survive 100% of the time, animals with two copies of a tuberculosis-susceptible gene (T^s) survive only 40% of the time, and heterozygotes (T^rT^s) survive 70% of the time, then no dominance exists at this locus. The expression of the heterozygote is exactly midway between the expressions of the homozygous genotypes.

Overdominance

overdominance
A form of dominance in which the expression of the heterozygote is outside the range defined by the expressions of the homozygous genotypes and most closely resembles the expression of the homozygous dominant genotype.

The last form of dominance, **overdominance,** is illustrated in Figure 3.12. With overdominance, the expression of the heterozygote is outside the range defined by the expressions of the homozygous genotypes and most closely resembles the expression of the homozygous dominant genotype. Overdominance is often characterized as having a "superior heterozygote." "Superior" is probably not the best word—"extreme" might be more correct. In Figure 3.12(a) the heterozygote lies to the right of the *JJ* genotype. *J* is therefore the dominant allele. In (b) the heterozygote lies to the left of the *J′J′* genotype, making *J′* the dominant allele.

Survivability in wild rats provides an example of overdominance. The gene for resistance to the anticoagulant poison warfarin is inherited as a dominant with respect to resistance to the poison. Both homozygotes and heterozygotes are unaffected by warfarin. Unfortunately—at least from the rats' standpoint—homozygotes need a higher level of vitamin K than is available in normal diets. Thus, in places where warfarin is used, rats without the resistance gene succumb to warfarin poisoning, rats homozygous for the gene suffer from vitamin K deficiency, and heterozygotes remain healthy. With respect to *survivability,* the warfarin locus displays overdominance.

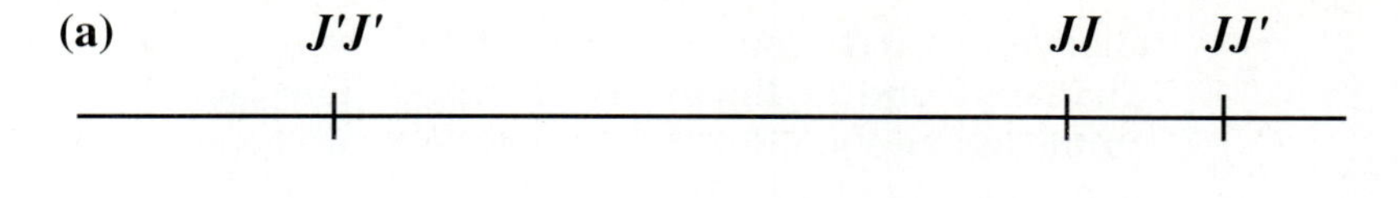

FIGURE 3.12 Overdominance at the J locus. The dominant allele is *J* in (*a*) and *J′* in (*b*).

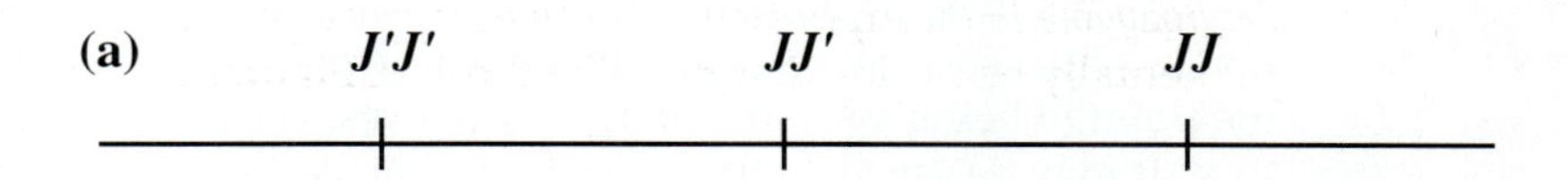

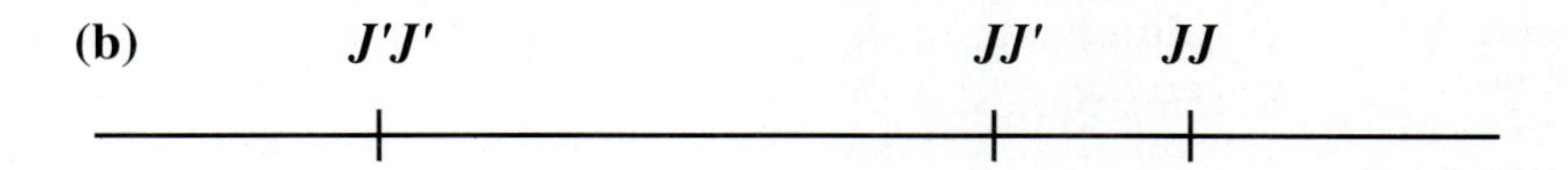

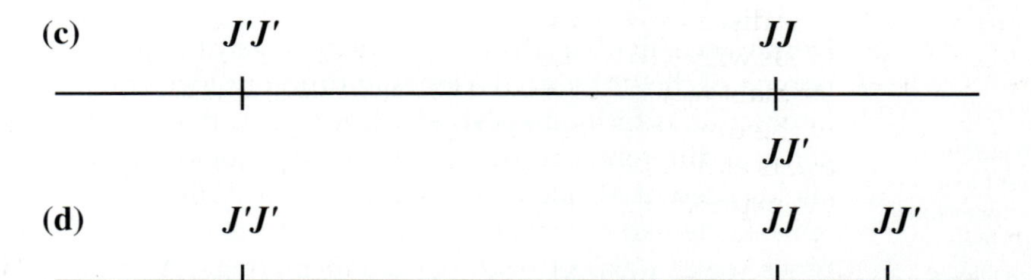

FIGURE 3.13 Degrees of dominance at the J locus: (*a*), no dominance; (*b*), partial dominance; (*c*), complete dominance; and (*d*), overdominance. Except for (*a*), the dominant allele is *J* in each case.

Overdominance is the most extreme form of dominance. If the four kinds of dominance are ordered by degree of dominance, they would progress from no dominance to partial dominance to complete dominance to overdominance (Figure 3.13).

Common Misconceptions about Dominance

The phenomenon of dominance is often misunderstood. Students often assume, for instance, that dominant genes are "good" and recessive genes are "bad." In many cases they are right. Lethal and semilethal recessive genes are clearly bad, and there is reason to believe that the ability of more favorable alleles to be dominant over less favorable alleles is something that has evolved over time. (Unfavorable dominants, after all, would tend to be eliminated by natural selection.) As we will see in Chapters 7 and 17, the theory of hybrid vigor and inbreeding depression is based on the assumption that dominant alleles are generally more favorable than recessive alleles. But there are exceptions to the rule. Clearly the HYPP gene in horses, a partial dominant, cannot be considered favorable with respect to animal health. Nor can the genes for red coat color or the presence of horns in cattle necessarily be considered unfavorable.

An even more pervasive misconception is that dominant genes are more common than recessive genes. Indeed, lethal recessive genes tend to be rare because they are self-eliminating, and any unfavorable recessives are likely to become less common as they are selected against over time. Again, however, exceptions abound. In the population of Horned Hereford cattle, for example, the dominant genes for polledness and black coat color are rare to the point of being nonexistent, and the recessive genes for horns and red coat color are common to the point of being the only alleles at their respective loci.

It may be *generally* true that dominant alleles are "better" and more common than recessive alleles. It is important to remember, however, that these characteristics of dominant and recessive alleles are not part of the definition of dominance. Dominance has to do with the relative expression of alleles in the heterozygote—nothing more.

Epistasis

epistasis
An interaction among genes at different loci such that the expression of genes at one locus depends on the alleles present at one or more other loci.

Dominance involves the *interaction* of genes at a single locus as they affect the phenotype of the individual. Genes at different loci can interact also, and this type of interaction is termed **epistasis.** Epistasis can be defined as an interaction among genes at different loci such that the expression of genes at one locus depends on the alleles present at one or more other loci. With respect to simply-inherited traits, epistasis is like dominance in that it affects the kinds and proportions of phenotypes we can expect from particular matings. And like dominance, epistasis is a source of hybrid vigor and inbreeding depression in polygenic traits.

A simply-inherited example of epistasis that is relatively easy to understand is coat color in Labrador retrievers. Labs come in three basic colors: black, chocolate, and yellow. These colors are determined by genes at two loci: the B (black) locus and E (extension of pigmentation) locus, as follows:

$B\_E \Rightarrow$ black
$bbE\_ \Rightarrow$ chocolate
$\_\_ee \Rightarrow$ yellow

The dashes in these genotypes indicate that either allele could be substituted without changing the phenotype. Black Labradors, for example, can be *BBEE, BBEe, BbEE,* or *BbEe.* Yellow Labs can be *BBee, Bbee,* or *bbee.*[4] Note that the expression of genes at the black locus depends on the alleles present at the extension locus. So long as there is at least one *E* allele at the extension locus, there appears to be straightforward, complete dominance at the black locus, with black being dominant to chocolate. However, if the genotype at the extension locus is *ee,* then genes at the black locus become irrelevant—all animals will be yellow.

[4]A third allele at the E locus, the E^{br} allele for brindling, occurs in a number of dog breeds, but not in Labradors.

A sampling of Labrador matings is shown with Punnett squares in Figure 3.14. The mating of two fully heterozygous black animals is depicted in (a). This mating produces a mixture of black, chocolate, and yellow puppies. Two chocolates *(bbEe)* are mated in (b). Because chocolate is a recessive condition, you would normally expect chocolates to **breed true** or produce only chocolates. However, due to the epistatic effect of genes at the E locus, this mating also produces yellow pups. Only yellow Labradors breed true (c)—a yellow mated to a yellow produces only yellows.

breed true
A phenotype for a simply-inherited trait is said to breed true if two parents with that phenotype produce offspring of that same phenotype exclusively.

In this chapter we have just scratched the surface of genetics. I have included here only the minimum information needed for a practical understanding of animal breeding, leaving out whole areas related to the biochemistry of

(a)

Black (*BbEe*) ♀ × Black (*BbEe*) ♂

♂ \ ♀	*BE*	*Be*	*bE*	*be*
BE	Black *BBEE*	Black *BBEe*	Black *BbEE*	Black *BbEe*
Be	Black *BBEe*	Yellow *BBee*	Black *BbEe*	Yellow *Bbee*
bE	Black *BbEE*	Black *BbEe*	Chocolate *bbEE*	Chocolate *bbEe*
be	Black *BbEe*	Yellow *Bbee*	Chocolate *bbEe*	Yellow *bbee*

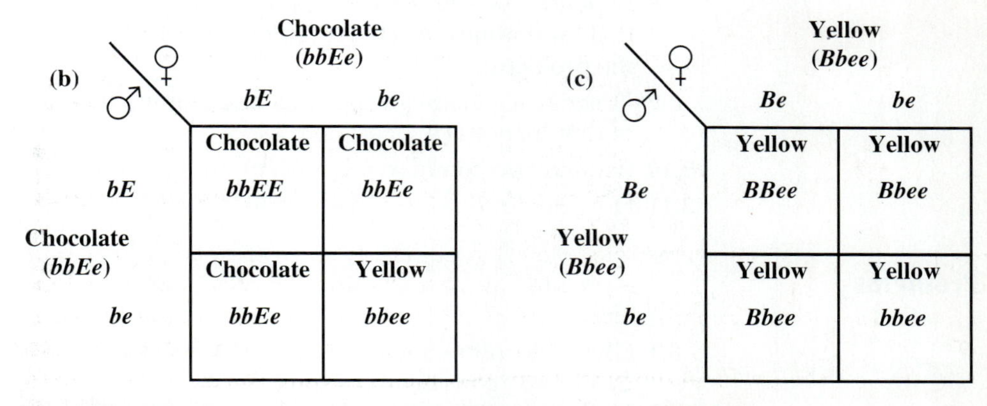

FIGURE 3.14 Punnett squares showing a sampling of Labrador matings: (*a*), black (*BbEe*) × black (*BbEe*); (*b*), chocolate (*bbEe*) × chocolate (*bbEe*); and (*c*), yellow (*Bbee*) × yellow (*Bbee*). Only yellow Labs breed true.

DNA and related molecules, cell physiology—even sex-related inheritance. For more complete information on these subjects, see any one of the many texts on classical genetics or molecular biology.

EXERCISES

Study Questions

3.1 Define in your own words:

gene
DNA
chromosome
homolog
locus
allele
multiple alleles
genotype
homozygote
heterozygote
segregation
germ cell or gamete
meiosis
independent assortment
linkage
crossing over
recombination
embryo
zygote
gamete selection
Punnett square
Mendelian sampling
dominance
recessive
complete dominance
partial dominance
no dominance
overdominance
epistasis
breed true

3.2 How do the concepts of *gene, allele,* and *locus* differ?

3.3 How does the definition of *genotype* given in this chapter differ from the definition given in Chapter 1?

3.4 Explain what occurs during meiosis.

3.5 Explain why there are exceptions to Mendel's law of independent assortment.

3.6 Why do we use Punnett squares? What are listed on the outside of a Punnett square? On the inside?

3.7 Discuss how the blending theory of inheritance fails to explain genetic variation and how Mendelian inheritance succeeds in explaining it.

3.8 Explain how Mendelian sampling leads to randomness in inheritance, why that randomness can be a problem for breeders, and how breeders get around the problem.

3.9 Describe a general protocol for determining the dominant allele and the kind of dominance at a locus.

3.10 Explain two common misconceptions about dominance.

3.11 How do dominance and epistasis differ?

Problems

3.1 Given the following alleles at the J locus: *J, J′, j,* and *j′*, list all possible:

a. homozygous combinations.
b. heterozygous combinations.

3.2 If you made 32 matings of roan Shorthorn cattle, how many calves would you *expect* to be red? Roan? White? If 32 calves result, will their coat colors match expectations? Why or why not?

3.3 A sire's five-locus genotype is *AaBBCcDdee*. A dam's genotype is *AABbCcDdEe*. Considering just these five loci:

a. How many unique gametes can the sire produce?
b. How many unique gametes can the dam produce?
c. How many unique zygotes can be produced from this mating?

3.4 Consider a hypothetical locus for tuberculosis resistance/susceptibility with alleles T^r (resistant) and T^s (susceptible). When exposed to the tuberculosis pathogen, T^rT^r individuals survive 90% of the time, and T^sT^s individuals survive 30% of the time. What is the value (range of values) for survival percentage of T^rT^s individuals if the locus exhibits:

a. complete dominance and T^r is the dominant allele?
b. complete dominance and T^s is the dominant allele?
c. no dominance?
d. partial dominance and T^r is the dominant allele?
e. partial dominance and T^s is the dominant allele?
f. overdominance and T^r is the dominant allele?
g. overdominance and T^s is the dominant allele?

3.5 A roan, heterozygous polled bull *(RrPp)* is mated to a roan, horned cow. What are the possible *phenotypes* and their expected proportions from this mating?

3.6 Your prize chocolate Labrador is from a mixed litter of black, yellow, and chocolate pups. He is by a yellow dog and out of a black bitch. What colors in what proportions would you expect if you mated your dog to his dam?

CHAPTER 4

Genes in Populations

Mendelian principles explain genetic mechanisms in individuals. As breeders, however, our task is not to change individuals, but populations. So we must take our knowledge of Mendelian inheritance and extend it from the level of the individual to the level of the population.

Gene and Genotypic Frequencies

In describing an individual for a simply-inherited trait, you might refer to the specific genes that the individual possesses, or you might describe its one-locus or two-locus genotype. For example, you might refer to a Blue Andalusian chicken as having both a black (*B*) and white (*b*) allele at a locus affecting feather color, or you might say that the chicken has a heterozygous *Bb* genotype at that locus. How do you genetically describe a population, however? How do you describe a whole flock of Andalusian chickens? The answer is to use gene and genotypic frequencies.

gene frequency or **allelic frequency** The relative frequency of a particular allele in a population.

A **gene frequency** or **allelic frequency** is the relative frequency of a particular allele in a population. It is a measure of how common that allele is relative to other alleles that occur at that locus. Relative frequencies range from zero to one. For example, if an allele does not exist in a population, its gene frequency is zero. If it is the *only* allele at its locus in the population, its gene frequency is one. If it comprises 35% of the genes at that locus in the population, its gene frequency is .35.

When there are just two possible alleles at a locus, the frequency of the "dominant" allele is commonly represented by the lowercase letter p and the frequency of the "recessive" allele by the lowercase letter q. (The terms *dominant* and *recessive* are set in quotes here because there are situations when neither allele is dominant. In cases of no dominance, the assignment of p or q to refer to the frequency of a particular allele is arbitrary.)

As an example, consider a flock of 100 Andalusians. Thirty-six are black (*BB*), 44 are blue (*Bb*—actually gray in color), and 20 are white (*bb*). At the locus affecting Andalusian feather color there is a total of 200 genes in this population—two genes for each of 100 individuals. The 36 black individuals each have two black genes, the 44 blues each have one black gene, and the whites have no black genes. The total number of black genes in the flock is therefore $2 \times 36 + 44 = 116$, and the gene frequency of the black allele is then 116 out of 200 or, in decimal form, .58. Likewise, there are no white genes in the black chickens, 44 white genes in the blues, and 40 white genes in the white individuals, for a total of 84 white genes in the flock. The gene frequency of the white allele is then 84 out of 200 or .42. In each case we have simply counted up the number of genes of a particular type and divided by the total number of genes at that locus in the population. The equations for the gene frequencies in our Andalusian example can be written as

$$p = \frac{2(36) + 44}{200} = \frac{116}{200} = .58$$

$$q = \frac{44 + 2(20)}{200} = \frac{84}{200} = .42$$

Note that $p + q = 1$. This will always be true if there are just two possible alleles at a locus. If there are multiple alleles at a locus, then the sum of the gene frequencies of each allele must equal one. For example, if possible alleles were *B*, *b*, and *b′*, we might call the gene frequencies of these alleles *p*, *q*, and *r*, respectively. Then $p + q + r = 1$.

genotypic frequency The relative frequency of a particular one-locus genotype in a population.

A **genotypic frequency** is the relative frequency of a particular one-locus genotype in a population. Uppercase letters are used to notate genotypic frequencies. With just two possible alleles at a locus, *P* refers to the genotypic frequency of the homozygous "dominant" genotype, *H* refers to the frequency of the heterozygous genotype, and *Q* refers to the frequency of the homozygous "recessive" genotype.

In our Andalusian flock, there are 36 *BB* genotypes, 44 *Bb* genotypes, and 20 *bb* genotypes out of a total of 100 individuals. Hence

$$P = \frac{36}{100} = .36$$

$$H = \frac{44}{100} = .44$$

$$Q = \frac{20}{100} = .20$$

To calculate genotypic frequencies, simply count up the number of individuals of a particular genotype and divide by the total number of individuals in the population. Note that $P + H + Q = 1$. This will always be true if there are just three possible genotypes at a locus. If there are more than three possible genotypes, then the sum of the genotypic frequencies of each genotype must equal one. For example, if possible alleles were *B*, *b*, and *b′*, with possible one-locus genotypes *BB*, *bb*, *b′b′*, *Bb*, *Bb′*, and *bb′*, we could call the genotypic frequencies *P*, *Q*, *R*, $H_{(Bb)}$, $H_{(Bb')}$, and $H_{(bb')}$, respectively. Then $P + Q + R + H_{(Bb)} + H_{(Bb')} + H_{(bb')} = 1$.

population genetics The study of factors affecting gene and genotypic frequencies in a population.

There are a number of factors that affect gene and genotypic frequencies in a population, not the least of which are the basic tools of animal breeding: selection and mating systems. The study of these factors comprises the branch of genetics known as **population genetics.**

THE EFFECT OF SELECTION ON GENE AND GENOTYPIC FREQUENCIES

Selection increases the gene frequency of favorable alleles. When we select replacement animals, we try to select those that have the best sets of genes and reject those with poorer sets of genes. As a result, members of the next generation should

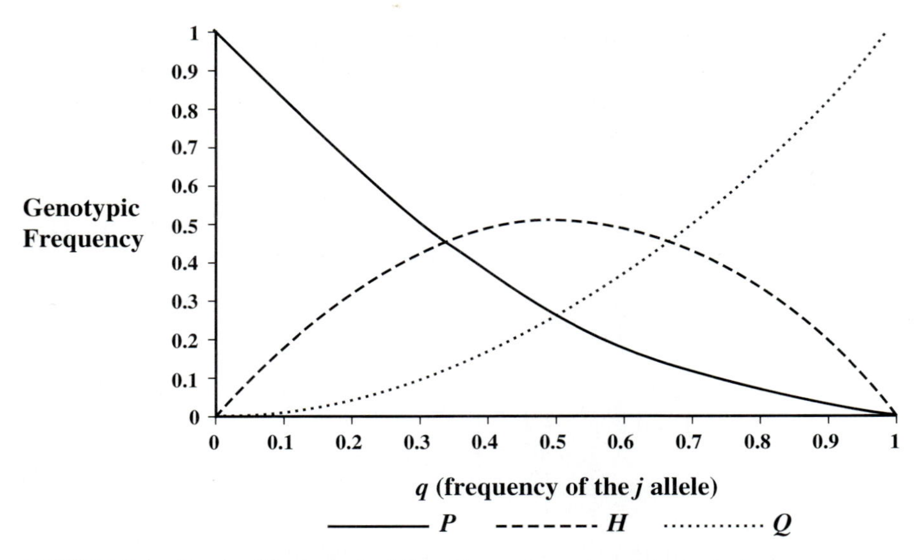

FIGURE 4.1 Graph of the typical relationships among gene and genotypic frequencies in a population. The frequency of the *JJ* genotype (*P*) is highest when *p*, the gene frequency of the *J* allele, is high (or, alternatively, when *q*, the frequency of the *j* allele, is low); the frequency of the *Jj* genotype (*H*) is highest when *p* is intermediate; and the frequency of the *jj* genotype (*Q*) is highest when *p* is low (*q* is high).[1]

have, on average, better sets of genes than members of the current generation. With continued selection, the following generation should have even better sets of genes, and so on. Over time, selection for better and better sets of genes causes the frequency of more favorable alleles in the population to increase and the frequency of less favorable alleles to decrease.

Another way of saying "better sets of genes" is to say better *breeding values.* When we select animals with better sets of genes generation after generation and increase the frequency of favorable alleles in the process, what we are really doing is increasing the average breeding value (and, therefore, the mean performance) of the population. Gene frequencies, mean breeding values, and mean performance, then, are inextricably tied. If we wish to increase the mean breeding value and performance of a population through selection, we necessarily want to change gene frequencies.

The immediate effect of selection is to change gene frequencies, but genotypic frequencies necessarily "tag along." The typical relationships among gene and genotypic frequencies are shown in Figure 4.1. The horizontal axis represents the gene frequency of the *j* allele. The vertical axis represents the genotypic frequencies of the three genotypes at the J locus. At the right side of the graph, the frequency of the *j* allele is high, so the frequency of the *J* allele is necessarily low

[1]In drawing these curves, I assume the existence of conditions necessary for *Hardy-Weinberg equilibrium,* a concept explained later in this chapter. Under other assumptions, the curves would look different, though not so different as to change the conclusions presented here.

(close to zero). Because there are very few *J* genes in the population, there are also very few *JJ* genotypes (*P* is low), relative few *Jj* genotypes (*H* is low too), and lots of *jj* genotypes (*Q* is high). If *J* is a favorable allele, then with selection, its frequency (*p*) will increase, and as we move from right to left on the graph, genotypic frequencies will change also—*JJ* genotypes will become more common (*P* increases), *jj* genotypes will become less common (*Q* decreases), and heterozygotes will become more common for a while and then diminish in number (*H* increases, then decreases). Ultimately, *p* may increase to the point that there are no other alleles at the J locus in the population except *J*. If this happens, we say the *J* allele is **fixed** or has reached **fixation.** Because the only genotype then possible is *JJ*, $P = 1$, and $H = Q = 0$.

fixation
The point at which a particular allele becomes the only allele at its locus in a population—the frequency of the allele becomes one.

To see how selection changes gene and genotypic frequencies in a population over time, consider the example of a completely recessive lethal gene. Let's call it the "killer" gene denoted by *k*. *K* is the normal allele at the K locus and is completely dominant to *k*. Thus *KK* and *Kk* individuals are perfectly normal, but *kk* individuals die at birth. Figure 4.2 illustrates the effects of natural selection against the killer gene over seven generations. The lines on the graph represent gene frequencies for the two alleles and genotypic frequencies for the *KK* and *Kk* genotypes in animals that survive to become parents. (The genotypic frequency of *kk* types is necessarily zero—none of them survive.) Natural selection causes an increase in the frequency of the normal allele and a decrease in the frequency of the killer allele. These changes in gene frequencies are accompanied by a corresponding increase in the frequency of homozygous normal genotypes and decrease in the frequency of heterozygotes.

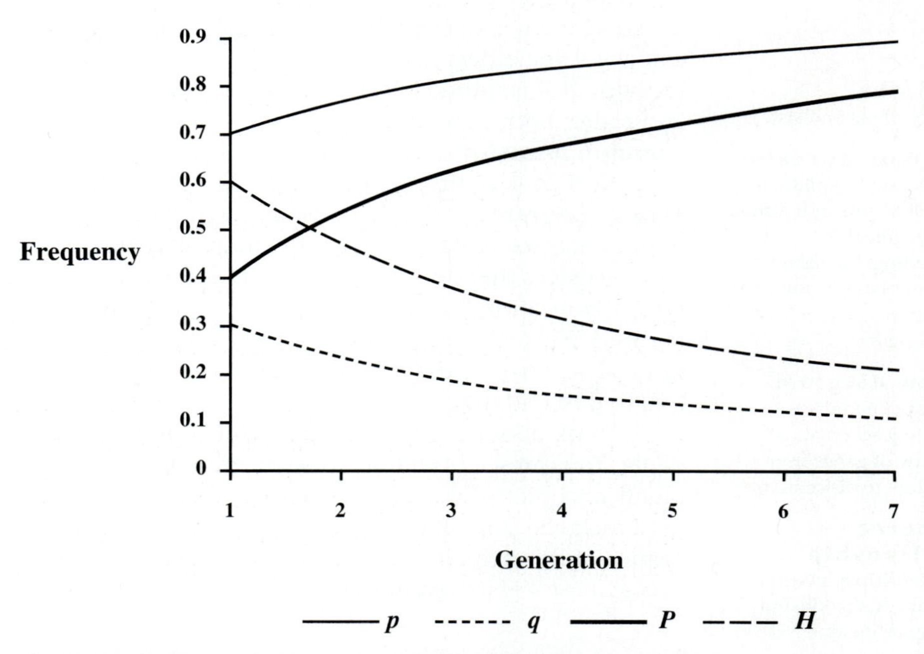

FIGURE 4.2 Gene and genotypic frequencies at the K locus over seven generations of natural selection against the completely recessive and lethal *k* ("killer") allele.

THE EFFECT OF MATING SYSTEMS ON GENE AND GENOTYPIC FREQUENCIES

Mating systems alone cannot change gene frequencies in a population, but occasionally a mating system, when combined with selection, can. For example, breeders interested in increasing body size typically select the largest replacements (male and female) available. From this pool of animals, they often mate the largest males to the largest females to produce especially large sons which they then select to be the sires of the next generation. When selection is combined with mating rules in this way, the frequencies of alleles that cause larger body size increase faster in future generations than would be the case if selection for size were combined with strictly random mating.[2]

The more common use of mating systems, however, is to change genotypic frequencies, specifically to increase either the number of homozygous gene combinations or the number of heterozygous combinations. Gene frequencies may or may not change as a result. Mating systems designed to affect homozygosity and heterozygosity fall under the general categories of inbreeding and outbreeding.

Inbreeding

Inbreeding, the mating of relatives, increases the frequency of homozygous genotypes. To see why, look at the pedigrees in Figure 4.3. The left-hand pedigree (a) is typical of animal pedigrees—the sire's pedigree comprises the upper half, the dam's pedigree the lower half, and increasingly younger generations appear further and further to the left. (In contrast, human pedigrees are typically oriented sideways, the oldest generations near the top and the youngest near the bottom.) Individual X is inbred because his sire (S) and dam (D) are **half sibs,** half brother and sister. Both parents have the same sire (A), and A is therefore considered a **common ancestor** to the parents of X.

half sibs
Half brothers and sisters.

common ancestor
An ancestor common to more than one individual. In the context of inbreeding, the term refers to an ancestor common to the parents of an inbred individual.

Now look at the right-hand pedigree (b). This type of pedigree is called an **arrow diagram,** and this particular arrow diagram corresponds to the traditional pedigree to its left. In arrow diagrams, individuals may appear only once, and ancestors that do not contribute to inbreeding or **pedigree relationship** are typically excluded. (I have included two noncontributing ancestors, B and C, in Figure 4.3(b) just to make the correspondence between the two types of pedigrees more clear.) Arrow diagrams depict schematically the flow of genes from ancestors to descendants.

arrow diagram
A form of pedigree depicting schematically the flow of genes from ancestors to descendants.

pedigree relationship
Relationship between animals due to kinship. Examples include full-sib, half-sib, and parent-offspring relationships.

Think about the flow of genes for a particular locus from A to his descendants. You can see from the arrow diagram that it is possible for S and D to inherit identical copies of the same gene from their common ancestor A, and for X to inherit the gene from both S and D. X would then be homozygous for that gene. The chance of this actually occurring is one in eight.[3] Considering all loci, we can expect at least ⅛ of inbred individual X's gene pairs to be homozygous because he in-

[2] This is an example of what is termed *positive assortative mating*. For more discussion, see Chapter 16.

[3] See Chapter 17 to learn how this probability is calculated.

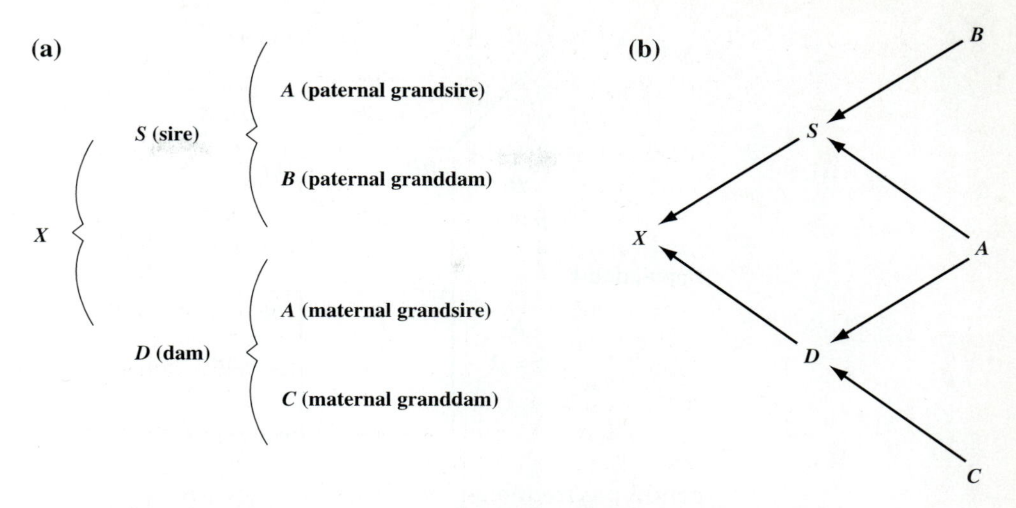

FIGURE 4.3 Pedigree (*a*) and arrow diagram (*b*) showing a half-sib mating.

herited identical genes from his parents' common ancestor *A*. (Actually, considerably more than ⅛ of *X*'s gene pairs should be homozygous. Many would have been homozygous even without inbreeding.) *The result of inbreeding, then, is an increase in homozygosity and a corresponding decrease in heterozygosity.*

Outbreeding (Crossbreeding)

outbreeding or **outcrossing**
The mating of unrelated individuals.

Outbreeding, the mating of unrelated individuals, has just the opposite effect of inbreeding. Outbreeding increases heterozygosity. The following rather long, drawn-out example will demonstrate the change in heterozygosity brought about by crossbreeding and, in the process, introduce one of the most basic concepts in population genetics: Hardy-Weinberg equilibrium.

Imagine two unrelated populations. The gene frequencies at the B locus in populations 1 and 2 are

$$p_1 = .8 \qquad p_2 = .1$$
$$q_1 = .2 \qquad q_2 = .9$$

Note that the frequencies are very different in the two populations. This is evidence that the two groups are indeed unrelated. Related populations can be expected to have similar gene frequencies.

F_1
Referring to the first generation of crosses between two unrelated (though not necessarily purebred) populations.[4]

Now let's cross populations 1 and 2 to create a new population of first-cross animals—an **F_1** generation. The results of this cross are shown with a Punnett square in Figure 4.4. Genotypic frequencies for the F_1s appear within each cell of the square. These were calculated by simply multiplying the appropriate gene frequencies from the parent populations. To see why, consider the

[4]The term "F_1" originally referred to the first generation of crosses between two unrelated *purebred* populations. The broader definition given here is, in my opinion, more useful.

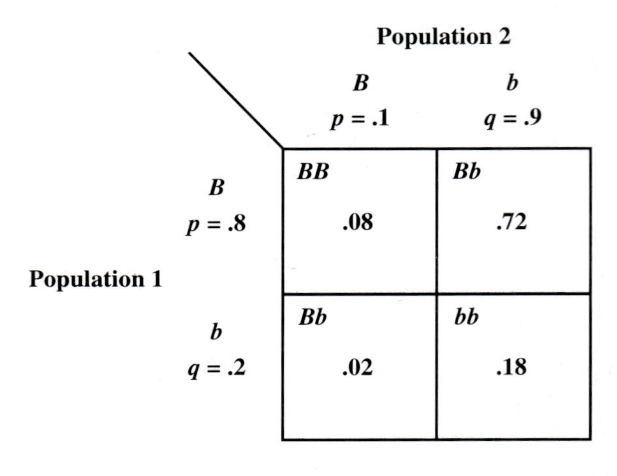

FIGURE 4.4
Punnett square showing genotypic frequencies at the B locus for F_1 offspring from a cross of two unrelated populations.

genotypic frequency of the homozygous *BB* genotype in F_1 animals (.08). The gene frequency of the *B* allele in population 1 is .8, and the gene frequency of the *B* allele in population 2 is .1. There is therefore an 80% chance that an offspring will inherit a *B* gene from population 1 and a 10% chance that it will inherit a similar *B* gene from population 2. The probability of inheriting two *B* genes is the product of these individual probabilities or $.8 \times .1 = .08$.

After combining the values in the two heterozygous cells of the Punnett square, genotypic frequencies in the F_1 generation are then

$$P_{F_1} = .08$$
$$H_{F_1} = .74 \qquad (.72 + .02)$$
$$Q_{F_1} = .18$$

The gene frequencies in the F_1 population can be determined from the genotypic frequencies. To make it easy, assume that the F_1 population contains 100 individuals. If eight of these carry two *B* genes and 74 carry one *B* gene, then out of a total of 200 genes at the B locus in the population, ninety $(2 \times 8 + 74)$ are *B* genes. The frequency of the *B* allele is then $90 \div 200 = .45$. The frequency of the *b* allele is $1 - .45 = .55$. Thus

$$p_{F_1} = .45$$
$$q_{F_1} = .55$$

(The reasoning used to derive gene frequencies from genotypic frequencies is summarized in the formulas $p = P + \frac{1}{2}H$ and $q = Q + \frac{1}{2}H$. You can use these formulas as a shortcut method.)

F_2
Referring to the generation of crosses produced by mating F_1 (first-cross) individuals among themselves.

If F_1 animals are mated inter se (among themselves), the resulting second cross or **F_2** generation will have the genotypic frequencies shown in Figure 4.5. Again, adding the values for the heterozygous cells, F_2 genotypic frequencies are

$$P_{F_2} = .2025$$
$$H_{F_2} = .495 \qquad (.2475 + .2475)$$
$$Q_{F_2} = .3025$$

F₁ individuals		*B* $p = .45$	*b* $q = .55$
F₁ individuals	*B* $p = .45$	*BB* .2025	*Bb* .2475
	b $q = .55$	*Bb* .2475	*bb* .3025

FIGURE 4.5
Punnett square showing genotypic frequencies at the B locus for F_2 offspring from inter se matings of F_1 individuals.

and F_2 gene frequencies are

$$p_{F_2} = P_{F_2} + \frac{1}{2}H_{F_2}$$
$$= .2025 + \frac{1}{2}(.495)$$
$$= .45$$

$$q_{F_2} = 1 - p_{F_2}$$
$$= 1 - .45$$
$$= .55$$

Note that gene frequencies did not change from the F_1 to the F_2 generation. They remained at .45 and .55. Because gene frequencies stay constant, and because genotypic frequencies in an offspring generation are a function of gene frequencies in the parent generation, we would expect that if F_2 animals were mated inter se, the resulting F_3 generation would have the same genotypic frequencies as the F_2 generation. In other words, if matings are made at random within a population, gene and genotypic frequencies do not change.

Hardy-Weinberg equilibrium
A state of constant gene and genotypic frequencies occurring in a population in the absence of forces that change those frequencies.

This conclusion reflects what is called **Hardy-Weinberg equilibrium.** The Hardy-Weinberg law—named after its codiscoverers—states:

> In a large, random mating population, in the absence of selection, mutation and migration, gene and genotypic frequencies remain constant from generation to generation, and genotypic frequencies are related to gene frequencies by the formulas

$$P = p^2$$
$$H = 2pq \qquad \text{and}$$
$$Q = q^2$$

The Hardy-Weinberg law essentially says that if there are no forces to change gene and genotypic frequencies in a population, those frequencies will not change. What are the forces that change gene and genotypic frequencies? The law specifically mentions selection, mutation, and migration. We know selection changes frequencies. In fact, the whole point of artificial selection is to

mutation (specifically **point mutation**) The process that alters DNA to create new alleles.

migration The movement of individuals into or out of a population.

random mating A mating system in which all matings are equally likely.

change gene frequencies. **Mutation,** the process that alters DNA to create new alleles, has some effect on gene and genotypic frequencies, but because mutations are rare events, the effect is small. **Migration** is the movement of individuals into or out of a population. Migration, particularly if it involves the introduction of a large number of genetically different individuals into a population, can have large effects on gene and genotypic frequencies. The Hardy-Weinberg law also states that a population in equilibrium must be large and randomly mated. Small populations become rather quickly inbred, and, as we saw earlier, inbreeding changes genotypic frequencies by increasing homozygosity. **Random mating** implies the absence of any systematic mating scheme. In our B-locus example, if we had not randomly mated, but had applied a rule that, say, *BB* genotypes could only mate with other *BB* genotypes, then we could expect rather different (and no longer static) genotypic frequencies. So random mating is required for Hardy-Weinberg equilibrium, and furthermore, as our example suggests, only *one generation* of random mating is needed to reach equilibrium. Random mating among F_1 animals created equilibrium in the F_2 generation.

Derivation of the Hardy-Weinberg formulas relating genotypic frequencies to gene frequencies is simple. If the gene frequencies at a particular locus in an equilibrium population are p and q, then the formulas are apparent from the following Punnett square.

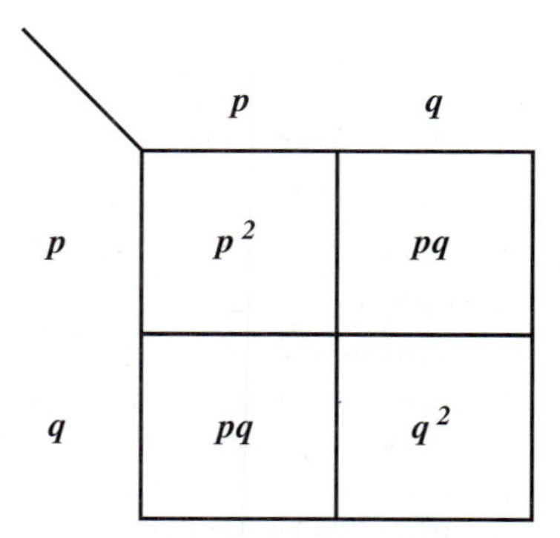

Be careful not to confuse the p^2, $2pq$, and q^2 of Hardy-Weinberg with the 1:2:1 ratio expected when mating two heterozygotes. (Students often do this.) Although the 1:2:1 ratio can be considered a special application of the Hardy-Weinberg law (when $p = q = .5$), the two concepts are used in very different contexts.

Do populations in Hardy-Weinberg equilibrium actually exist? Clearly no. No population is infinitely large, perfectly random mating, or free of natural selection and mutation. Some populations are close enough to equilibrium, however, that the Hardy-Weinberg formulas fit reasonably well and can be quite useful.

As an example of the utility of Hardy-Weinberg, let's go back to population 1, population 2, and their F_1 offspring. The point of making this cross (if you can remember back that far) was to show how outbreeding—crossbreeding in this case—increases heterozygosity. Recall from Figure 4.4 that the genotypic frequency of heterozygotes for the B locus in the F_1 population was .74. We determined this by multiplying gene frequencies from the parent populations. What we do not know are the frequencies of heterozygotes in those original populations. We can calculate them, however, assuming Hardy-Weinberg equilibrium.

The frequencies of heterozygotes in populations 1 and 2, as determined from the Hardy-Weinberg formula, are

$$\begin{aligned} H_1 &= 2p_1q_1 \\ &= 2(.8)(.2) \\ &= .32 \end{aligned}$$

$$\begin{aligned} H_2 &= 2p_2q_2 \\ &= 2(.1)(.9) \\ &= .18 \end{aligned}$$

The average heterozygote frequency in the parent populations is $(H_1 + H_2) \div 2$ or $(.32 + .18) \div 2 = .25$. Compare that figure with the frequency of heterozygotes in the F_1 crosses (.74). By crossing population 1 and population 2, heterozygosity was almost *tripled.* This is a dramatic example of how *crossbreeding increases heterozygosity.* Had the parent populations been more similar (i.e., had their B-locus gene frequencies not been so different), the increase in heterozygosity from crossing would have been less sensational but still evident.

How to Use the Hardy-Weinberg Formulas to Determine Gene and Genotypic Frequencies at a Locus Exhibiting Complete Dominance

In a population of Andalusian chickens (see the example given near the beginning of this chapter), it is relatively easy to calculate gene and genotypic frequencies for feather color because each color genotype is clearly identifiable: *BB* types are black, *Bb* types are blue, and *bb* types are white. To determine frequencies, you need only do some counting and a little arithmetic. But how do you proceed when the locus of interest exhibits complete dominance—i.e., when you cannot tell the difference between heterozygous and homozygous dominant genotypes by observing their phenotypes? For example, how would you determine gene and genotypic frequencies for coat color in a population of Hampshire swine when the *W* allele for white belt is completely dominant to the *w* allele for solid color? How do you know which of the belted pigs are *Ww* and which are *WW?*

The problem is a sticky one, but if (and *only* if) we can assume that the population is in (or nearly in) Hardy-Weinberg equilibrium, a solution is possible. For practical purposes, we can assume Hardy-Weinberg equilibrium if the population is reasonably large, if recent additions of animals from outside the population are negligible, and if, in the most recent generation of parents, selection and mating decisions are unrelated to the trait of interest.

Consider a population of 1,000 Hampshires of which 910 are belted and 90 are solid colored. At the outset, the only frequency we can be sure of is the frequency of the homozygous recessive genotype (Q), which is simply the proportion of solid colored pigs in the population. Mathematically,

$$Q = \frac{90}{1,000} = .09$$

We have no idea of the frequency of the heterozygous genotype (H) or the homozygous dominant genotype (P) because both genotypes look alike—they both have white belts. But if we assume Hardy-Weinberg equilibrium, then $Q = q^2$, so

$$\begin{aligned} q &= \sqrt{Q} \\ &= \sqrt{.09} \\ &= .3 \end{aligned}$$

and

$$\begin{aligned} p &= 1 - q \\ &= 1 - .3 \\ &= .7 \end{aligned}$$

therefore

$$\begin{aligned} P &= p^2 \\ &= (.7)^2 \\ &= .49 \end{aligned}$$

$$\begin{aligned} H &= 2pq \\ &= 2(.7)(.3) \\ &= .42 \end{aligned}$$

Of the 910 belted pigs in the population, we can expect approximately 420 (.42 × 1,000) to be heterozygotes and 490 to be homozygous for the W allele.

EXERCISES

Study Questions

4.1 Define in your own words:

gene frequency or allelic frequency
genotypic frequency
population genetics
fixation
half sibs
common ancestor
arrow diagram
pedigree relationship
outbreeding
F_1
F_2
Hardy-Weinberg equilibrium
mutation
migration
random mating

4.2 How is a gene frequency different from a genotypic frequency? How would you calculate gene and genotypic frequencies in a population for a simply-inherited trait affected by a single locus? (Assume all genotypes are easily identifiable.)

4.3 With respect to simply-inherited traits, we characterize *individuals* by their genotypes. We characterize *populations* by gene and genotypic frequencies. Explain.

4.4 How does selection affect gene frequencies? Genotypic frequencies?

4.5 Explain the following statement: A change in the average breeding value of a population necessarily means a change in gene frequencies.

4.6 Describe how inbreeding affects genotypic frequencies.

4.7 Describe how outbreeding affects genotypic frequencies.

4.8 The greater the difference in gene frequencies between two populations, the less related the populations. Explain.

4.9 Restate the Hardy-Weinberg law in your own words.

4.10 Think of an application of the Hardy-Weinberg law.

Problems

4.1 At the C locus in horses, chestnuts (sorrels) are CC, palominos are Cc^{cr}, and cremellos are $c^{cr}c^{cr}$. In a herd of 10 horses, there are 3 chestnuts, 6 palominos, and 1 cremello. What are the gene and genotypic frequencies at the C locus in this herd?

4.2 Skin color in Bohemian bullfrogs is determined by a single locus (S) such that:

$SS \Rightarrow$ solid
$Ss \Rightarrow$ striped
$ss \Rightarrow$ spotted

Of the Bohemians entered in the Backcounty Bullfrog Bounce, a regional frog jumping contest, 70% were solid colored, 20% were striped, and 10% were spotted. What were the gene and genotypic frequencies at the S locus for the frogs at the contest?

4.3 Construct a pedigree and arrow diagram for:
a. a sire × daughter mating.
b. a full-sib mating.
Identify the common ancestor(s) in each case.

4.4 Two large populations of horses are being systematically crossed (mares from one population bred to stallions of the other and vice versa). Coat color is *not* a factor in determining which animals are selected and which individual matings are made. Frequencies of coat color genes at the C locus are:

Population 1		Population 2	
C	c^{cr}	C	c^{cr}
.8	.2	.3	.7

a. What will be the gene and genotypic frequencies at the C locus in the offspring (F_1) population?
b. If the crossbred offspring are mated among themselves, what will be the gene and genotypic frequencies in the F_2 generation? (Assume random mating, no selection, etc.)
c. In the F_3 and subsequent generations?

4.5 In a population of Garden-digging Armadillos in Hardy-Weinberg equilibrium, the C allele for long claws is completely dominant to the *c* allele for clawlessness. Extensive sampling of this population showed 16% of the armadillos to be clawless. What are the gene and genotypic frequencies at the C locus for these varmints?

CHAPTER 5

Simply-Inherited and Polygenic Traits

The first two chapters in this book presented a broad overview of animal breeding. Most of the traits mentioned in those chapters—traits like milk production, weaning weight, body size, fertility, conformation, personality, and growth rate—were polygenic traits. The next two chapters dealt with the specifics of Mendelian inheritance and population genetics. In contrast to the earlier material, most of the traits used as examples in these chapters—coat color, presence of horns, warfarin resistance, and susceptibility to HYPP and other lethal and semilethal conditions—were simply-inherited traits. By now you may be seriously confused about the differences between polygenic and simply-inherited traits, about the importance of the two kinds of traits to practical animal breeding, and about the relevance of Mendelian principles to both kinds of traits. The purpose of this chapter is to clear up the confusion. This is important, not only because unresolved confusion about these issues will plague you for the rest of this book, but because animal breeders use very different approaches in breeding for simply-inherited versus polygenic traits.

SIMPLY-INHERITED AND POLYGENIC TRAITS

simply-inherited trait
A trait affected by only a few genes.

qualitative or **categorical trait**
A trait in which phenotypes are expressed in categories.

quantitative trait
A trait in which phenotypes show continuous (numerical) expression.

Simply-inherited traits were defined in Chapter 2 as traits affected by only a few genes. Coat color, presence of horns, and genetic defects like spider syndrome in sheep are all examples of simply-inherited traits. Only a single locus or, at most, a few loci are involved in their expression.

There are two common secondary characteristics of simply-inherited traits. First, phenotypes for these traits tend to be "either/or" or categorical (described by placing them in categories) in nature. A Labrador is either black, chocolate, or yellow; a cow is either horned or polled; and a lamb either has the spider condition or it does not. Coat color, presence of horns, and spider syndrome are also termed **qualitative** or **categorical traits** because of their "either/or" expression. It is possible (though rare) for a simply-inherited trait to be **quantitative;** to have phenotypes that are measured with numbers that range more or less continuously from large through intermediate through small values. An example is body weight when it is affected by a gene for dwarfism. The trait is measured in number of pounds or kilos, but is simply-inherited.

Second, simply-inherited traits are typically affected very little by environment. Yes, if a chocolate Lab spends a lot of time in the sun, its coat will bleach out, but there is still no mistaking it for a black or yellow Lab. Its phenotype is clearly chocolate.

polygenic trait
A trait affected by many genes, no single gene having an overriding influence.

In contrast, **polygenic traits** are affected by many genes, and no single gene is thought to have an overriding influence. Examples of polygenic traits include growth rate, milk production, and time to run a given distance. We know very little about the specific genes affecting these traits and can only conclude that there are many of them.

Like simply-inherited traits, polygenic traits commonly have similar secondary characteristics. Phenotypes for polygenic traits are usually described by numbers. We speak of 500-lb weaning weights, 30,000-lb lactation yields, and 20-second times in the quarter mile. Instead of being "either/or" in nature or falling into a few distinct categories as phenotypes for simply-inherited traits almost always do, phenotypes for polygenic traits are typically quantitative or continuous in their expression. Most (but not all) polygenic traits are therefore quantitative traits. Polygenic traits are clearly affected by environment. If cows, pigs, and sheep are fed less, they grow more slowly and produce less milk. If horses are not trained as well, they will not run as fast.

We must be careful, however, not to classify a trait as simply-inherited or polygenic on the basis of secondary characteristics alone. There are a number of simply-inherited traits that have a secondary characteristic of polygenic traits and vice versa. Body weight, when it is affected by a dwarfism gene, is one example. Another example is the trait called **dystocia** or difficulty in delivery at birth. Phenotypes for dystocia often fall into two categories: assisted or unassisted. Because of the either/or nature of these phenotypes, you might assume that dystocia is a simply-inherited trait. It is polygenic, though, because it is affected by many genes. Many genes influence the size of the fetus (or fetuses), the size of the dam's pelvic opening, and her perseverance in delivery. The critical thing to remember in deciding whether a trait is simply-inherited or polygenic is that secondary characteristics should not be the determining factor—it is the number of genes involved that counts.

dystocia
Difficulty in giving birth or being born.

New (and some not so new) students of animal breeding often confuse the terms "simply-inherited" with "qualitative," and "polygenic" with "quantitative." They are *not* synonymous. Most simply-inherited traits are qualitative, and most polygenic traits are quantitative. But body weight, when it is affected by a dwarfism gene, is an example of a simply-inherited, quantitative trait, and dystocia is an example of a polygenic, qualitative trait. Simply-inherited versus polygenic refers to how a trait is *inherited.* Qualitative versus quantitative refers to how it is *expressed.*

threshold trait
A polygenic trait in which phenotypes are expressed in categories.

Dystocia is an example of a special category of traits called **threshold traits.** Threshold traits are polygenic traits that exhibit categorical phenotypes. Other examples of threshold traits are fertility (as measured by success or failure to conceive) and natural gait (trotter or pacer). Threshold traits present unique problems for animal breeders and will be discussed further in Chapters 7 and 10.

Which are more important: simply-inherited or polygenic traits? As a rule, polygenic traits are. In food and fiber species, it is polygenic traits—traits like growth rate, fertility, milk production, etc.—that determine productivity and profitability. Polygenic traits are generally more important in recreational and companion species as well. For example, speed and endurance, important traits for racing animals, are polygenic—there is no single "speed gene" or "endurance gene."

There are instances where simply-inherited traits assume economic importance. Some markets are sensitive to coat or feather color. In cattle populations in which the polled allele is rare, polled animals may be particularly valuable. And animals with simply-inherited genetic defects are inevitably worth less. By and large, however, simply-inherited traits are less important than polygenic traits. Hence the emphasis on polygenic traits in this book.

COMMON CHARACTERISTICS OF SIMPLY-INHERITED AND POLYGENIC TRAITS

Simply-inherited and polygenic traits have a great deal in common. To begin with, the genes affecting both kinds of traits are subject to the same Mendelian mechanisms. Mendel's laws of segregation and independent assortment apply to genes that influence polygenic traits just as they do to genes that influence simply-inherited traits. Dominance and epistasis affect gene expression for both kinds of traits too. Admittedly, most of the practical examples used to illustrate Mendelian mechanisms involve simply-inherited traits. This is only because the genes affecting these traits are well understood—there are, after all, just a few of them. Because so many genes influence polygenic traits, and because the effect of each gene is so small, we know little to nothing about them. It is difficult, therefore, to use polygenic traits as examples of Mendelian inheritance.

Secondly, the basic tools of animal breeding—selection and mating—are the same for both simply-inherited and polygenic traits. When breeders select for either kind of trait, they are trying to increase the frequencies of favorable alleles. A breeder who selects only polled animals from a herd of both polled and horned cattle will increase the frequency of the polled allele in the herd. Likewise, a breeder who selects for increased rib eye area (a measure of muscling) in a herd of swine will increase the frequencies of the many genes—distributed among many loci—that favorably influence muscling. In the first example, the breeder selected for a simply-inherited trait, and in the second example, he selected for a polygenic trait, but the effect on gene frequencies was the same.

Mating systems affect gene combinations in the same way for both simply-inherited and polygenic traits. When a horse breeder crosses sorrels (*CC*) and cremellos ($c^{cr}c^{cr}$) to produce palominos (Cc^{cr}), the genotypic frequency of heterozygotes at the C locus increases. Likewise, when breeders cross unrelated lines or breeds to produce hybrid vigor, heterozygosity increases at many loci. Whether breeders make specific matings to affect a simply-inherited trait like coat color, or cross breeds to affect the whole array of polygenic traits that respond to hybrid vigor, they are using mating systems to create desirable gene combinations.

DIFFERENT BREEDING APPROACHES FOR SIMPLY-INHERITED VERSUS POLYGENIC TRAITS

Despite the fact that both simply-inherited and polygenic traits are subject to the same Mendelian rules, and that selection and mating systems are used to improve both kinds of traits, very different breeding approaches are taken in each case. This difference in approach is a function of the number of genes involved. The more

♂ \ ♀	*P*	*p*
P	Polled *PP*	Polled *Pp*
p	Polled *Pp*	Horned *pp*

FIGURE 5.1
The mating of two heterozygous polled cattle. A polled offspring from this mating would have a ⅔ probability of being *Pp* and a ⅓ probability of being *PP*.

genes affecting a trait, the more difficult it is to observe the effects of individual genes, and therefore the less specific information we have about those genes. The amount of available information affects the way we characterize genotypes and therefore determines the animal breeding technology we use.

Because few loci—often only one—influence simply-inherited traits, the effects of specific genes are typically well understood. It is therefore often possible to identify individual genotypes. For instance, horned cattle (with the exception of certain African types) are known to have the *pp* genotype at the polled/horned locus, and red cattle are known to have the *bb* genotype at the black/red color locus. Sometimes exact genotypes are not known, but a *probable* genotype is identifiable. Figure 5.1 depicts the mating of two polled cattle known to be carriers of the horned allele (*p*). If a polled calf were produced from this mating, its genotype might be either *PP* or *Pp,* but its most probable genotype would be *Pp* because a heterozygous genotype is twice as likely from this mating as a homozygous dominant genotype. To determine its genotype exactly, you could conduct **test matings,** matings designed to reveal the genotype of an individual for a small number of loci—a common technique used in breeding for simply-inherited traits.[1]

test mating or **test cross**
A mating designed to reveal the genotype of an individual for a small number of loci.

Regardless of whether genotypes for simply-inherited traits are known exactly or have probabilities associated with them, we characterize them by explicitly identifying genes and gene combinations. For example, we characterize red, horned cattle as *bbpp,* and we denote red, polled cattle as either *bbPP* or *bbPp.* In selecting and mating animals for simply-inherited traits like these, we consider the known or probable genotypes of the animals at the loci of interest.

Polygenic traits, on the other hand, are affected by so many genes that it is extremely difficult to observe the effects of specific loci and specific alleles at those loci. It is impossible, then, to explicitly identify an individual's many-locus genotype for a polygenic trait. Imagine, for example, writing out an animal's genotype for speed or growth rate. Where would you start?

Because identifying the actual genotype of an individual for a polygenic trait is out of the question, the logical alternative is to characterize the *net effect* of the individual's many genes influencing the trait—in other words, to quantify the individual's performance and breeding value (and related genetic values) for a trait. This requires the use of statistical tools including statistical concepts like heritability and accuracy. The technology and jargon associated with polygenic traits are

[1] For more on test matings, see Chapter 6.

therefore quite different from those used with simply-inherited traits. We move from the alphabet soup of specific genotypes for simply-inherited traits (Cc^{cr}, *BBPp*, and so on) to the alphabet soup of polygenic traits (EBVs, EPDs, ACCs, etc.).

A note of warning: In later chapters of this book there will be examples where, in an effort to explain polygenic concepts like breeding value or hybrid vigor, specific "genotypes" for polygenic traits are written out just as you might write out a genotype for a simply-inherited trait. Such genotypes are purely hypothetical and vastly oversimplified. They are used only for illustration. Do not be misled into thinking that genotypes like these could be identified in the real world of polygenic traits.

EXERCISES

Study Questions

5.1 Define in your own words:

simply-inherited trait	polygenic trait
qualitative or categorical trait	threshold trait
quantitative trait	test mating

5.2 List one or more traits that fit in each compartment of the following grid:

	Simply-inherited	**Polygenic**
Qualitative		
Quantitative		

Which compartment contains threshold traits?

5.3 Which tend to be more important economically: simply-inherited or polygenic traits?

5.4 List the characteristics that simply-inherited and polygenic traits have in common.

5.5 How do breeding approaches for simply-inherited and polygenic traits differ?

Selection

The next nine chapters in this book are devoted to the topic of selection. The sheer volume of material should give you some idea of the importance of selection to animal breeding. Selection is one of only two basic tools available to animal breeders for genetically improving populations—the other is mating—and is the classic means by which seedstock breeders make genetic change.

Chapter 6, the first chapter in this section, deals with selection for simply-inherited traits. The remaining chapters in the section cover selection for polygenic traits. Topics in these chapters range from fairly theoretical discussions of factors involved in genetic prediction to such practical concerns as how to read and interpret a sire summary. Along the way, you should acquire an understanding of breeding value, gene combination value, heritability, and a number of other basic, yet abstract concepts in quantitative genetics.

CHAPTER 6

Selection for Simply-Inherited Traits

Selection for simply-inherited traits is straightforward. You need only know how many loci are involved, how many alleles at each locus, how those alleles are expressed, and the genotypes or probable genotypes of potential parents. With this information, it is a relatively simple matter to select those individuals with the most favorable genotypes.

Unfortunately, it is not always possible to know an animal's exact genotype. As a result, much of breeding for simply-inherited traits involves methods for ascertaining the genotypes of individuals. A good deal of this chapter, therefore, is concerned with the determination of genotypes.

SIMPLE ONE-LOCUS CASE

As an example of selection for a simply-inherited trait, consider coat color in Shetland sheepdogs. Shetlands come in a number of colors including a pattern called merle. *MM* individuals are white merle—nearly all white with blue eyes; *Mm* heterozygotes are merle, having a dappled appearance; and *mm* individuals are "normal" (i.e., nonmerle). If you were a Shetland sheepdog breeder and were particularly fond of the white merle coloration, you could simply select for white merle, discarding dogs with any other color. In one generation you could have a population of purely white merle animals. (This would actually be unwise as white merle dogs are usually deaf.) If you wanted to avoid the merle allele altogether, you could select only animals with normal color and discard white merle and merle dogs. Again, in one generation you could rid the population of the merle gene. If you wanted the merle coloration, your job would be more difficult because, as shown in Figure 6.1, merle is a heterozygous condition—merle bred to merle produces all three genotypes. Whatever your color preference, however, selection for this simply-inherited trait is quite easy because there is only one locus involved, partial dominance is understood to be the mode of gene expression, and all three genotypes are easily identifiable.

Now consider black/red coat color in cattle. The black allele (*B*) is completely dominant to the red allele (*b*), so that black cattle are either *BB* or *Bb*, and red cattle are necessarily *bb*. Suppose your herd of cattle is a mixture of black and red animals and you want a uniform color. Selecting for red cattle would be easy enough—just keep the reds and discard the blacks as fast as you can afford to. But what if you

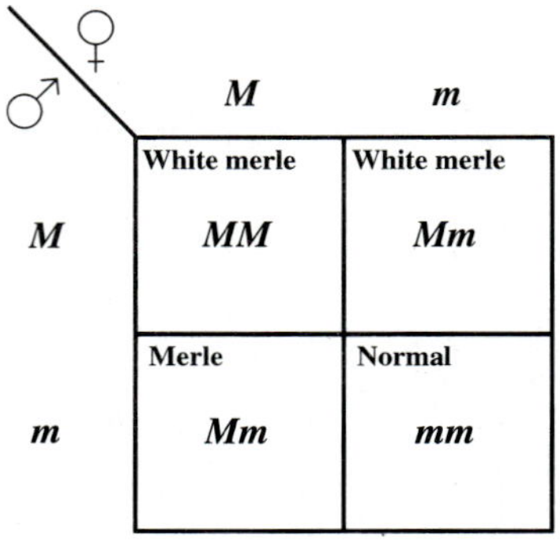

♂ \ ♀	*M*	*m*
M	White merle *MM*	White merle *Mm*
m	Merle *Mm*	Normal *mm*

FIGURE 6.1
The mating of two heterozygous merle sheepdogs. All three coat color phenotypes—white merle, merle, and normal—are possible in the offspring from this mating.

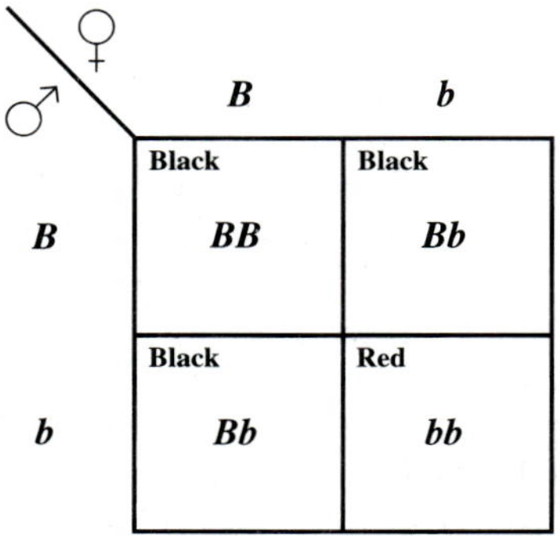

♂ \ ♀	*B*	*b*
B	Black *BB*	Black *Bb*
b	Black *Bb*	Red *bb*

FIGURE 6.2
The mating of two heterozygous black cattle. There is one chance in four that a red offspring will result.

want all black animals? Would it do to simply keep the blacks and discard the reds? Clearly not, because many blacks are likely to be carriers of the red gene, and matings of these heterozygotes can produce red offspring (see Figure 6.2). In this situation, breeding for a simply-inherited trait is more complicated because not all genotypes are known exactly.

Proving Parental Genotypes—Test Matings

test mating or **test cross**
A mating designed to reveal the genotype of an individual for a small number of loci.

When the genes affecting a simply-inherited trait show complete dominance, homozygous dominant and heterozygous individuals cannot be told apart by their phenotypes, and selection becomes more difficult. Selection would be less problematic if there were a way to determine which animals are carriers of the recessive allele. For example, if it were possible to perform some sort of laboratory test on black Angus cattle to see if they are carriers of the gene for red coat color, then elimination of the red factor from a herd would be much easier. Such tests are now possible for some genes—the HYPP gene in horses is an example—and advances in molecular biology may soon make tests of this sort feasible at coat color loci.[1] Until that time, however, the only way to prove the genotype of an animal is to use **test matings.**

To test a black bull to see if he carries the red allele, we could mate him to red cows. Figure 6.3 shows the possible outcomes of such test matings if (a) the bull is

[1]See Chapter 20 for a more detailed discussion of biotechnology and animal breeding.

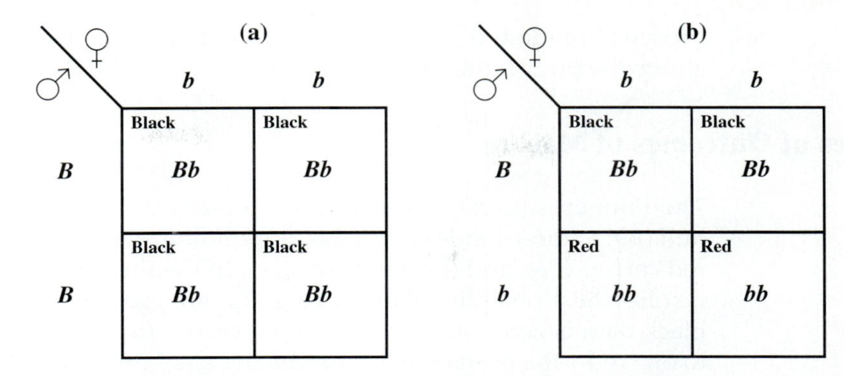

FIGURE 6.3 Punnett squares showing the possible outcomes of test matings of a black bull (*BB* or *Bb*) to red cows (*bb*). In (*a*) the bull is homozygous black, and all his offspring are black. In (*b*) he carries the red allele, and half his offspring (on average) are red.

homozygous black (*BB*) or (b) a heterozygote (*Bb*). A homozygous black bull produces all black calves from these matings. A heterozygote produces roughly half black and half red calves. If our bull produces at least one red calf, the test is conclusive—he carries the red allele. If he produces all black calves, the test is not conclusive—by pure chance he may not have transmitted the red allele to this set of calves. But if he produces many black calves and no red calves, we can be fairly sure that he does not carry the red allele.

How sure? That depends on how many matings we have made and to whom. Knowing these things, we can calculate the *probability* that our bull does not carry the red gene. And if that probability is high enough, we can safely assume that he is not a red carrier. Animal breeders sometimes (though not often) test for completely recessive genes by making multiple matings that will either produce a homozygous recessive offspring—and therefore prove an individual to be a carrier of a recessive gene—or produce no homozygous recessives, but reveal the probability that the individual is *not* a carrier.

Testing of this kind is typically done for males, partly because females are usually not valuable enough to justify a test, and partly because males can produce the required number of progeny and females commonly cannot. There is no theoretical difference between testing males and females, however, and with **embryo transfer (E.T.)** technology it is possible to produce enough offspring from a valuable female to have a conclusive test.

embryo transfer (E.T.)
A reproductive technology in which embryos from donor females are collected and transferred in fresh or frozen form to recipient females.

The probability that an individual does not carry a completely recessive allele depends on the number of matings made and the number of offspring born without a homozygous recessive showing up. Clearly, if our black bull had been mated to just one red cow and she produced a black calf, we have not learned much. The probability that the bull does not carry the red gene is not very high—at least based on such meager evidence. On the other hand, if he had been mated to a dozen red females and no red calves were born, we would be confident in concluding that he does not carry the red gene. The probability of his not being a carrier is high.

The probability that an individual does not carry a recessive allele also depends on the type of matings being made. Mating to 10 known carriers of a particular recessive is a very different test than mating to 10 daughters or 10 individuals chosen at random from the population at large. To understand why, we must examine the probabilities of outcomes from various kinds of matings.

Probabilities of Outcomes of Matings

The Punnett squares in Figure 6.4 represent the matings of a black, red carrier bull (*Bb*) to three kinds of females: (a), a homozygous black cow (*BB*); (b), a black, red carrier (*Bb*); and (c), a red cow (*bb*). In the first mating (a), all black offspring result—the probability of producing a homozygous red calf out of a homozygous black dam is zero. Mating to a known carrier (b) produces three black offspring to one red—the probability of producing a red offspring is .25. The probability of getting a red calf from the mating with the red cow (c) is one out of two or .5.

The outcomes of the above matings are simple enough to predict because the genotypes of the parents are known exactly. What happens when they are not known exactly? For example, what happens if we mate our black, red carrier bull to one of his own daughters or to a black daughter of a known carrier? We do not know for sure the genotypes of these mates. They could be either *BB* or *Bb*.

Recall from Chapter 4 that when populations with known gene frequencies are crossed, it is possible to use those frequencies in combination with a Punnett square to determine genotypic frequencies in the offspring generation. If we mate two individuals with known *probabilities* of having particular genotypes, we can use the same kind of technique to predict the probable proportions of genotypes resulting from that mating.

For simplicity, assume that the daughter's dam is a homozygous black (*BB*) cow (i.e., the daughter is the result of the mating depicted in Figure 6.4(a)). She is phenotypically black, but there is a 50% chance that she is homozygous black (*BB*) and a 50% chance that she is a red carrier (*Bb*).

Now, let p and q stand not for gene frequencies, but for the probabilities that an individual will contribute a gamete with a dominant or recessive allele, respectively. In this example, p is the probability that an individual will contribute a gamete with the *B* allele, and q is the probability that the individual will contribute a gamete with the *b* allele. The bull, being a heterozygote, has an equal chance of contributing either allele, so $p = q = .5$ in his case. Things are not so simple for his daughter. She is equally likely to be either *BB* or *Bb*. To determine the probability

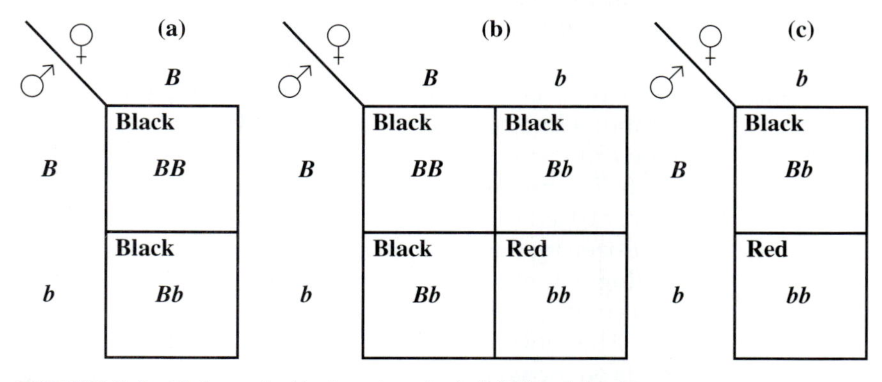

FIGURE 6.4 Matings of a black, red carrier bull (*Bb*) to (*a*), a homozygous black cow (*BB*); (*b*), a black, red carrier (*Bb*); and (*c*), a red cow (*bb*). The probabilities of producing a red calf are 0, .25, and .5, respectively.

that she will contribute a particular allele, think of her as just one of a large number of this bull's daughters. Half of these should be *BB* and half *Bb*. When gametes are sampled from this *pool* of daughters, three out of four gametes should contain the *B* allele, and one out of four the *b* allele. Thus, for any one of the bull's daughters, $p = .75$ and $q = .25$.

The sire × daughter mating is shown in Figure 6.5. By multiplying the probabilities of each parent contributing each type of gamete, we can predict the probable proportions of genotypes in the offspring. The probability of producing a red calf from the mating of the black, red carrier bull to a randomly chosen daughter is .125 or ⅛. The probability of producing a black calf from the same mating is .375 + .125 + .375 = .875 or ⅞. This procedure can be used to predict the probable proportions of genotypes resulting from any type of mating.

The results from Figures 6.4 and 6.5 (and more) are summarized in the first two columns of Table 6.1. Mating a heterozygote to a homozygous recessive individual has the greatest probability of producing a homozygous recessive offspring. That is because the homozygous recessive parent has the greatest likelihood of contributing a recessive allele. At the opposite extreme, mating a heterozygote to a randomly selected individual from a population in which the recessive allele is rare has the smallest probability of producing a homozygous recessive offspring. In that case, the mate is unlikely to even carry the recessive allele, much less transmit it. Matings to known carriers, daughters of the suspect individual, and daughters of known carriers result in outcome probabilities somewhere between these extremes.

Because the probability of producing a homozygous recessive offspring varies depending on the type of mating, tests that use these matings vary in their power to prove an individual free of a recessive gene. To achieve a given level of confidence in the result of a set of test matings, you must account for the type of mating when deciding how many matings to make. The numbers of exclusively normal offspring required to be 99% sure that a tested individual does not carry a particular recessive gene are listed in the last column of Table 6.1. In some cases (e.g., mating to homozygous recessive individuals), relatively few matings are necessary. In other cases (e.g., mating to individuals chosen at random from the population), many matings are required.

Given the information in Table 6.1, you might conclude that it is always best to test a suspect individual using homozygous recessive mates. Such matings provide the most powerful test and require the smallest number of matings for a given level of confidence in the result. The problem with this idea is that for many

♂ \ ♀	*B* $p = .75$	*b* $q = .25$
B $p = .5$	Black (*BB*) .375	Black (*Bb*) .125
b $q = .5$	Black (*Bb*) .375	Red (*bb*) .125

FIGURE 6.5
The mating of a black, red carrier bull (*Bb*) to his daughter (*BB* or *Bb*). The probability of producing a red calf from this mating is .125 or ⅛, and the probability of producing a black calf is .375 + .125 + .375 = .875 or ⅞.

TABLE 6.1 (1) Probability of Producing a Homozygous Recessive Offspring from a Single Mating of a Carrier Individual (Bb) to Different Types of Mates, and (2) Number of Exclusively Normal Offspring Required to Be 99% Sure That a Tested Individual Does Not Carry a Particular Recessive Gene.[a]

Type of Mate	(1) Probability	(2) Number
Homozygous dominant (*BB*)	0	∞
Known carrier (*Bb*)	.25	16
Homozygous recessive (*bb*)	.50	7
Daughter (*BB* or *Bb*)[b]	.125	35
Daughter of *any* known carrier (*BB* or *Bb*)[b]	.125	35
Mate chosen randomly from the population (*BB* or *Bb*)[cd]	.05[f]	90
Mate chosen randomly from the population (*BB* or *Bb*)[de]	.0125[f]	367

[a]One offspring per mating.
[b]The daughter's dam is assumed to be homozygous dominant (*BB*).
[c]Twenty percent of the population are assumed to be carriers (*Bb*) of the recessive allele.
[d]Homozygous recessive mates are assumed not viable or otherwise excluded from the population.
[e]Five percent of the population are assumed to be carriers (*Bb*) of the recessive allele.
[f]Formulas used to calculate these probabilities are presented in the boxed section that follows.

recessive conditions, homozygous recessive individuals are either infertile, culled from the population, or simply not viable. For example, you could not use mates homozygous for a lethal recessive gene. For recessive conditions that are just "undesirable," however, mating to homozygous recessives is the best alternative. To test a black bull for the red gene, mate him to red cows. To test a polled bull for the horned gene, mate him to horned cows.

Mating to known carriers requires relatively few matings but, like homozygous recessive mates, known carriers may be hard to find. Breeders often cull them as soon as these animals are proven to carry a deleterious recessive allele. Daughters of a known carrier are more easily found, but more are needed to achieve a given level of confidence in the test.

You might think of daughters of the suspect individual as an unlikely choice for use in a test. A relatively large number of daughters are needed to prove their sire "clean." Such a test is time consuming—the daughters must be conceived, be born, and reach breeding age—so the test cannot be used with young sires, and the progeny of sire × daughter matings are inbred and often less valuable. But the great advantage of breeding to daughters is that these matings test for *all* recessive genes, not just one specific gene. If a sire carries several deleterious recessive genes, his daughters are equally likely to inherit any one of them, so mating to daughters tests for all of them.

Tests involving mates chosen randomly from the general population require the largest number of matings and, as indicated in the bottom two rows of Table 6.1, that number is even greater when the recessive allele is rare in the population. Nevertheless, this type of test is probably the most common of all because it requires no special effort—just mate a suspect individual as you normally would (albeit to *many* females) and see what recessives show up. Like mating to daughters, mating to the general populations tests for all recessive genes, not just one. Artificial insemination studs routinely use this technique to test their sires.

Test Matings in More Depth

Listed in Table 6.2 are probabilities of detection of a recessive allele for different numbers of matings and types of mates. The numbers in the table are based on the assumption of one offspring per mating (typical of cattle and horses) and 100% mating success—all matings produce an offspring. In this context, "probability of detection" means the probability that, if the animal being tested is indeed a carrier of the recessive allele (i.e., a heterozygote) at least one homozygous recessive offspring will result from these matings. A useful way to interpret this probability is to consider it a measure of the confidence we can have that the individual being tested does *not* carry the recessive allele if the test matings have been made and no homozygous recessive offspring have been produced. For example, if the probability of detection is .64 and no homozygous recessives result, then we can be 64% confident that the animal being tested is not a carrier. That is not much confidence at all. On the other hand, if the probability of detection is .99+ and no homozygous recessives have been produced, we can be very confident that he is not a carrier. Note from Table 6.2 that our confidence in the test depends on the type of mate and increases as the number of matings increases.

The effect of number of matings is different for twin and litter bearing species like sheep, swine, dogs, cats, and poultry than it is for species that give birth to just one offspring. Listed in Table 6.3 are probabilities of detection (confidence in the test) associated with different numbers of matings (n), litter sizes (m), and types of mates. For a given type of mate and number of matings, probabilities of detection are higher with litters than with single offspring. This is because a litter provides multiple opportunities for a homozygous recessive offspring to be produced from one mating. The larger the litter size, the more of these opportunities and the greater the probability of detection.

TABLE 6.2 Probability of Detection of a Completely Recessive Allele for Different Numbers of Matings and Types of Mates[a]

Type of Mate	Number of Matings		
	5	20	100
Homozygous dominant (*BB*)	0	0	0
Known carrier (*Bb*)	.76	.99+	.99+
Homozygous recessive (*bb*)	.97	.99+	.99+
Daughter (*BB* or *Bb*)[b]	.49	.93	.99+
Daughter of *any* known carrier (*BB* or *Bb*)[b]	.49	.93	.99+
Mate chosen randomly from the population (*BB* or *Bb*)[cd]	.23	.64	.99
Mate chosen randomly from the population (*BB* or *Bb*)[de]	.06	.22	.72

[a]One offspring per mating.
[b]The daughter's dam is assumed to be homozygous dominant (*BB*).
[c]Twenty percent of the population are assumed to be carriers (*Bb*) of the recessive allele.
[d]Homozygous recessive mates are assumed not viable or otherwise excluded from the population.
[e]Five percent of the population are assumed to be carriers (*Bb*) of the recessive allele.

TABLE 6.3 Probability of Detection of a Completely Recessive Allele for Different Numbers of Matings (n), Litter Sizes (m), and Types of Mates

Type of Mate	$m = 5$			$m = 10$		
	$n = 5$	20	100	5	20	100
Homozygous dominant (*BB*)	0	0	0	0	0	0
Known carrier (*Bb*)	.99+	.99+	.99+	.99+	.99+	.99+
Homozygous recessive (*bb*)	.99+	.99+	.99+	.99+	.99+	.99+
Daughter (*BB* or *Bb*)[a]	.91	.99+	.99+	.96	.99+	.99+
Daughter of *any* known carrier (*BB* or *Bb*)[a]	.91	.99+	.99+	.96	.99+	.99+
Mate chosen randomly from the population (*BB* or *Bb*)[bc]	.56	.96	.99+	.65	.98	.99+
Mate chosen randomly from the population (*BB* or *Bb*)[cd]	.18	.54	.98	.21	.62	.99

[a]The daughter's dam is assumed to be homozygous dominant (*BB*).
[b]Twenty percent of the population are assumed to be carriers (*Bb*) of the recessive allele.
[c]Homozygous recessive mates are assumed not viable or otherwise excluded from the population.
[d]Five percent of the population are assumed to be carriers (*Bb*) of the recessive allele.

TABLE 6.4 Number of Matings Required for 95% and 99% Probabilities of Detection ($P[D_n]$) of a Completely Recessive Allele for Different Litter Sizes (m) and Types of Mates

Type of Mate	$m = 1$		$m = 5$		$m = 10$	
	95%	99%	95%	99%	95%	99%
Homozygous dominant (*BB*)	∞	∞	∞	∞	∞	∞
Known carrier (*Bb*)	11	16	3	4	2	2
Homozygous recessive (*bb*)	5	7	1	2	1	1
Daughter (*BB* or *Bb*)[a]	23	35	10	16	6	8
Daughter of *any* known carrier (*BB* or *Bb*)[a]	23	35	10	16	6	8
Mate chosen randomly from the population (*BB* or *Bb*)[bc]	59	90	19	28	15	23
Mate chosen randomly from the population (*BB* or *Bb*)[cd]	239	367	78	119	62	96

[a]The daughter's dam is assumed to be homozygous dominant (*BB*).
[b]Twenty percent of the population are assumed to be carriers (*Bb*) of the recessive allele.
[c]Homozygous recessive mates are assumed not viable or otherwise excluded from the population.
[d]Five percent of the population are assumed to be carriers (*Bb*) of the recessive allele.

Table 6.4 shows the number of matings required for 95% and 99% probabilities of detection of a recessive allele for different litter sizes (m) and types of mates. You can think of each value in the table as the number of matings that need to have been made—without any homozygous recessive offspring resulting—in order to have 95% or 99% confidence that the individual being tested does not carry the recessive allele. (These confidence levels, corresponding to 1 in 20 and 1 in 100 chances of being wrong, are commonly used in statistics.) Again, 100% mating success is assumed. If you design a test for a recessive allele, you should plan more matings than Table 6.4 suggests in order to account for less than perfect conception and fetal survival rates.

You can draw several conclusions from Table 6.4. First, the number of matings required varies a great deal depending on the type of mate. Secondly, fewer matings are needed when litters are produced than when single offspring are produced. The larger the litter size, the smaller the number of required matings. Lastly, if you desire a higher level of confidence in the test, you have to make more matings.

Implicit in Tables 6.2 to 6.4 is the assumption that all the mates involved in testing for a recessive allele are of one type (e.g., known carriers *or* daughters of the suspect individual *or* randomly chosen individuals). This is often not the case. A test might include some known carriers, some daughters of the individual being tested, and some mates about which nothing is known. It is mathematically possible to determine confidence levels for tests involving any combination of mate types.

Calculating Confidence Levels and Required Numbers of Test Matings

I. One offspring per mating and one uniform group of mates. The next two formulas can be used to determine confidence levels and required numbers of test matings for species that typically have just one offspring per mating (e.g., cows and horses) when all mates belong to one uniform group. In other words, all mates should have the same probability of having a particular genotype at the locus of interest. In practice, this means that all mates are daughters of the sire being tested, or all mates are known carriers, or all mates are randomly selected from the general population, etc.

Let n = the number of "successful" matings—successful in the sense that an offspring results.

$P[D_n]$ = probability of *detection* in n matings—i.e., the probability that at least one homozygous recessive offspring will be born given n matings. This is our level of confidence in the test.

P_{BB} = probability that a mate is homozygous dominant at the locus of interest.

P_{Bb} = probability that a mate is heterozygous at the locus of interest.

P_{bb} = probability that a mate is homozygous recessive at the locus of interest.

Then

$$P[D_n] = 1 - \left(P_{BB} + \frac{3}{4}P_{Bb} + \frac{1}{2}P_{bb}\right)^n$$

and

$$n = \frac{\log(1 - P[D_n])}{\log\left(P_{BB} + \frac{3}{4}P_{Bb} + \frac{1}{2}P_{bb}\right)}$$

Example

Suppose we wish to test a stallion to see if he is heterozygous for a particular recessive condition, and we have 10 known carrier (heterozygous) mares available for the test. Then

$$P[D_n] = 1 - \left(P_{bb} + \frac{3}{4}P_{Bb} + \frac{1}{2}P_{bb}\right)^n$$

Because the mates are known carriers, $P_{Bb} = 1$ and $P_{BB} = P_{bb} = 0$. Then

$$\begin{aligned} P[D_n] &= 1 - \left(0 + \frac{3}{4}(1) + \frac{1}{2}(0)\right)^{10} \\ &= 1 - \left(\frac{3}{4}\right)^{10} \\ &\approx .94 \end{aligned}$$

If all 10 foals are normal, we can be 94% confident that the stallion does not carry the recessive allele in question. (If one or more foals shows the homozygous recessive condition, we *know* the stallion carries the recessive allele.)

Suppose the 10 mates are daughters of the stallion instead of known carriers. Assuming the stallion is a carrier, we expect half his daughters to be *BB* and half to be *Bb*. So

$$\begin{aligned} P[D_n] &= 1 - \left(P_{BB} + \frac{3}{4}P_{Bb} + \frac{1}{2}P_{bb}\right)^n \\ &= 1 - \left(\frac{1}{2} + \frac{3}{4}\left(\frac{1}{2}\right) + \frac{1}{2}(0)\right)^{10} \\ &= 1 - \left(\frac{7}{8}\right)^{10} \\ &\approx .74 \end{aligned}$$

If all 10 foals are normal, we can be 74% confident that the stallion does not carry the recessive gene in question—considerably less confident than if we had used known carriers.

If we want 94% confidence in test results using daughters—the same level possible using 10 known carriers—we will need

$$\begin{aligned} n &= \frac{\log(1 - P[D_n])}{\log\left(P_{BB} + \frac{3}{4}P_{Bb} + \frac{1}{2}P_{bb}\right)} \\ &= \frac{\log(1 - .94)}{\log\left(\frac{1}{2} + \frac{3}{4}\left(\frac{1}{2}\right) + \frac{1}{2}(0)\right)} \end{aligned}$$

$$= \frac{\log(.06)}{\log\left(\frac{7}{8}\right)}$$

$$= \frac{-1.2218}{-.0580}$$

$$\approx 21 \text{ matings}$$

We need 21 sire × daughter matings to achieve the same level of confidence obtainable from 10 matings with known carriers.

II. One offspring per mating and multiple groups of mates. When there is more than one group of mates—for example, a group of known carriers, a group of daughters, etc.—use the following formula to calculate the level of confidence in the test:

$$P[D_n] = 1 - \prod_{i=1}^{k}\left(P_{BB_i} + \frac{3}{4}P_{Bb_i} + \frac{1}{2}P_{bb_i}\right)^{n_i}$$

where i = a counter referencing different groups of mates.
k = the number of groups of mates.
$\prod_{i=1}^{k}$ = the symbol for a *product* of computations for each group of mates.

Because there is more than one group of mates—more than one n—there is no simple formula for determining the number of matings required for a given level of confidence in the test. The same level of confidence can be achieved with different combinations of group sizes.

Example

Say we had 20 mares for the test, five of which are known carriers and 15 of which are daughters of the stallion being tested. Then

$$P[D_n] = 1 - \prod_{i=1}^{k}\left(P_{BB_i} + \frac{3}{4}P_{Bb_i} + \frac{1}{2}P_{bb_i}\right)^{n_i}$$

$$= 1 - \left(0 + \frac{3}{4}(1) + \frac{1}{2}(0)\right)^5\left(\frac{1}{2} + \frac{3}{4}\left(\frac{1}{2}\right) + \frac{1}{2}(0)\right)^{15}$$

$$= 1 - \left(\frac{3}{4}\right)^5\left(\frac{7}{8}\right)^{15}$$

$$\approx .97$$

III. Multiple offspring per mating and one uniform group of mates. For twinning or litter bearing species that average m offspring per mating,

$$P[D_n^m] = 1 - \left(P_{BB} + \left(\frac{3}{4}\right)^m P_{Bb} + \left(\frac{1}{2}\right)^m P_{bb}\right)^n$$

and

$$n = \frac{\log(1 - P[D_n^m])}{\log\left(P_{BB} + \left(\frac{3}{4}\right)^m P_{Bb} + \left(\frac{1}{2}\right)^m P_{bb}\right)}$$

Example

We want to test a boar for a particular recessive allele by mating him to three daughters that average 9.6 pigs per litter.

$$P[D_n^m] = 1 - \left(P_{BB} + \left(\frac{3}{4}\right)^m P_{Bb} + \left(\frac{1}{2}\right)^m P_{bb}\right)^n$$

$$= 1 - \left(\frac{1}{2} + \left(\frac{3}{4}\right)^{9.6}\left(\frac{1}{2}\right) + \left(\frac{1}{2}\right)^{9.6}(0)\right)^3$$

$$= 1 - \left(\frac{1}{2}\left(1 + \left(\frac{3}{4}\right)^{9.6}\right)\right)^3$$

$$\approx .85$$

How many matings of this kind are needed to have 95% confidence in the test?

$$n = \frac{\log(1 - P[D_n^m])}{\log\left(P_{BB} + \left(\frac{3}{4}\right)^m P_{Bb} + \left(\frac{1}{2}\right)^m P_{bb}\right)}$$

$$= \frac{\log(1 - .95)}{\log\left(\frac{1}{2} + \left(\frac{3}{4}\right)^{9.6}\left(\frac{1}{2}\right) + \left(\frac{1}{2}\right)^{9.6}(0)\right)}$$

$$= \frac{\log(.05)}{\log\left(\frac{1}{2}\left(1 + \left(\frac{3}{4}\right)^{9.6}\right)\right)}$$

$$= \frac{-1.3010}{-.2744}$$

$$\cong 4.74 \text{ or } 5 \text{ matings}$$

IV. Multiple offspring per mating and multiple groups of mates. Use the following formula to calculate the level of confidence in the test when there are both multiple offspring per mating and multiple groups of mates. This particular formula is generalized—you can use it for any set of test matings.

$$P[D_n^m] = 1 - \prod_{i=1}^{k}\left(P_{BB_i} + \left(\frac{3}{4}\right)^{m_i} P_{Bb_i} + \left(\frac{1}{2}\right)^{m_i} P_{bb_i}\right)^{n_i}$$

Example

Let's assume that no more daughters are available for testing our boar, but a known carrier of the recessive allele has been located. If she produces a litter of eight, then, combining that information with the data from the three sire × daughters matings, we have

$$
\begin{aligned}
P[D_n^m] &= 1 - \prod_{i=1}^{k}\left(P_{BB_i} + \left(\frac{3}{4}\right)^{m_i} P_{Bb_i} + \left(\frac{1}{2}\right)^{m_i} P_{bb_i}\right)^{n_i} \\
&= 1 - \left(\frac{1}{2} + \left(\frac{3}{4}\right)^{9.6}\left(\frac{1}{2}\right) + \left(\frac{1}{2}\right)^{9.6}(0)\right)^{3}\left(0 + \left(\frac{3}{4}\right)^{8}(1) + \left(\frac{1}{2}\right)^{8}(0)\right)^{1} \\
&= 1 - \left(\frac{1}{2}\left(1 + \left(\frac{3}{4}\right)^{9.6}\right)\right)^{3\,+\left(\frac{3}{4}\right)^{8}} \\
&\cong .98
\end{aligned}
$$

The addition of a single litter out of a known carrier increases our level of confidence in the test considerably.

FACTORS INFLUENCING THE EFFECTIVENES OF SELECTION

Selection increases the frequency of favored alleles, and the easiest way to judge the effectiveness of selection for simply-inherited traits is to look at how fast gene frequencies change. There are several factors that influence selection's effect on the rate of change of gene frequency. The first is the current or *initial* gene frequency.

Figure 6.6 illustrates the change in gene frequency caused by selection against a less desirable allele at a hypothetical J locus containing two possible alleles, J_1 and J_2. The vertical axis represents initial gene frequency, and the horizontal axis represents time in generations. For the moment, ignore the broken line on the graph and focus on the solid line. The slope of this line represents the rate of change in gene frequency of the less desirable allele (J_2)—the steeper the fall of the line, the faster selection is eliminating J_2 genes. Note that the fastest change in gene frequency occurs at intermediate gene frequencies. With few exceptions, this is the rule in selection. At high frequencies of the less desirable allele, there are relatively few of the more desirable genes in the population to favor by selection, and at low frequencies of the less desirable allele, there are relatively few less desirable genes to select against. Only when alleles occur at intermediate frequencies are there enough "good" genes to favor and enough "bad" genes to select against that change in gene frequencies can be fast.

fitness
The ability of an individual and its corresponding phenotype and genotype to contribute offspring to the next generation. The term refers to the number of offspring an individual produces—not just its ability to be selected.

A second factor influencing the effectiveness of selection is the degree to which the various genotypes in a population differ in **fitness.** An individual's fitness is the ability of the individual and its corresponding phenotype and genotype to contribute offspring to the next generation. The fittest animals are those that contribute the most offspring, so fitness refers not just to an individual's ability to be selected, but to the number of offspring it produces.

The two lines in Figure 6.6 depict contrasting differences in fitness among genotypes. The solid line represents larger differences—relative to J_1J_1 individuals, J_1J_2 individuals produce ¾ as many offspring, and J_2J_2 individuals produce ½ as many offspring. The broken line represents smaller differences in fitness—relative to J_1J_1 individuals, J_1J_2 individuals produce ⅞ as many offspring, and J_2J_2 individuals produce ¾ as many offspring. The faster change in gene frequency is shown by the solid line, indicating that larger differences in fitness allow selection to be more effective.

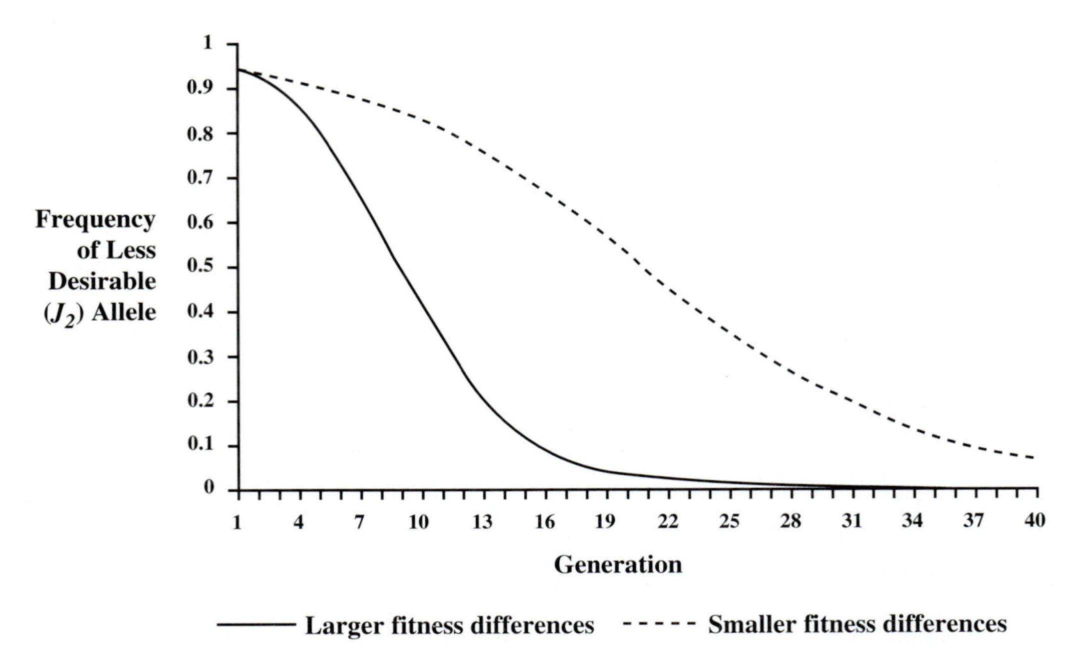

FIGURE 6.6 Change in gene frequency of the less desirable (J_2) allele over 40 generations of selection. In this example there is no dominance with respect to fitness at the J locus. The solid line represents larger differences in fitness among genotypes—relative to J_1J_1 individuals, J_1J_2 individuals produce ¾ as many offspring, and J_2J_2 individuals produce ½ as many offspring. The broken line represents smaller differences in fitness—relative to J_1J_1 individuals, J_1J_2 individuals produce ⅞ as many offspring, and J_2J_2 individuals produce ¾ as many offspring.

This only makes sense. For example, consider the case of natural selection against a recessive lethal gene versus natural selection against a gene that reduces fertility to some small degree. The difference in fitness between normal animals and those with two copies of the lethal gene is very large—normal individuals survive and reproduce; homozygous recessive individuals die. As a result, natural selection against the lethal allele is quite effective (at intermediate gene frequencies, anyway—see the following discussion). But in the case of the gene with a marginal negative influence on fertility, differences in fitness among genotypes are much smaller, and natural selection is not nearly as effective. Many individuals carrying the gene survive and reproduce, and the gene remains in the population.

A third factor affecting the rate of change in gene frequency at a particular locus is the degree of dominance expressed at that locus with respect to fitness. To understand "degree of dominance with respect to fitness," think of fitness as a trait like any other. When dominance is complete, homozygous dominant individuals and heterozygous individuals are equally fit—they contribute, on average, the same number of offspring. Likewise, when no dominance exists, the number of offspring contributed by heterozygotes is exactly midway between the number contributed by homozygous dominant types and the number contributed by homozygous recessive types. You can use the same logic to explain partial dominance and overdominance with respect to fitness.

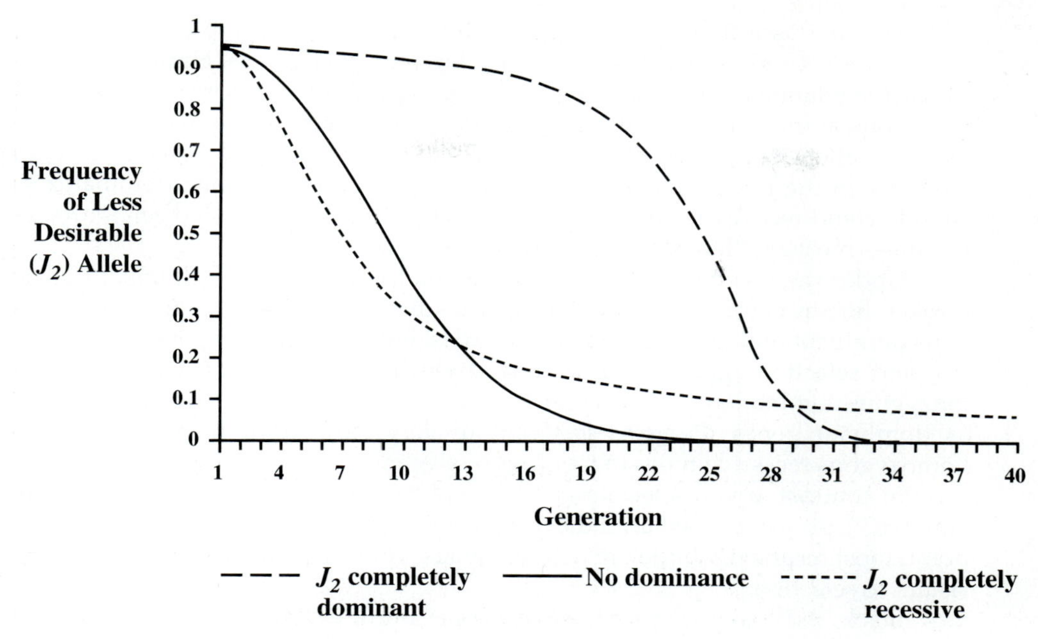

FIGURE 6.7 Change in gene frequency of the less desirable (J_2) allele over 40 generations of selection. Differences in fitness among genotypes are fairly large. When J_2 is completely dominant with respect to fitness (broken line), relative to J_1J_1 individuals, J_1J_2 and J_2J_2 individuals produce ½ as many offspring. With no dominance (solid line), J_1J_2 individuals produce ¾ as many offspring, and J_2J_2 individuals produce ½ as many offspring. When J_2 is completely recessive (dotted line), J_1J_2 individuals produce equally as many offspring, and J_2J_2 individuals produce ½ as many offspring.

Figure 6.7 illustrates change in the frequency of a less desirable allele (J_2) given three scenarios: selection against a gene showing no dominance for fitness (solid line), selection against a completely dominant gene (broken line), and selection against a completely recessive gene (dotted line). When no dominance exists, the pattern of change in gene frequency is the same as in Figure 6.6—the fastest change occurs at intermediate gene frequencies, and the less desirable allele is eliminated relatively quickly (if you consider 30 or so generations "quick"). But the patterns of change in gene frequency are quite different for alleles that are completely dominant or completely recessive. In the first case, progress is very slow when the unwanted allele is at high frequencies, but eventually becomes rapid. In the second case, progress is initially fast, but then slows, and the undesirable allele stubbornly persists in the population.

Why is this? Two facts, taken together, provide the answer:

1. When dominance is complete (or nearly complete), the presence of a recessive allele is virtually undetectable in heterozygous individuals.
2. At low frequencies of an allele, a higher proportion of genes of that allelic type reside in heterozygous individuals than in homozygous individuals, and the lower the frequency of the allele, the more exaggerated this imbalance becomes.

The first assertion is simply part of the definition of complete dominance. Proof of the second assertion can be seen in Figure 6.8, a repeat of Figure 4.1, which shows the relationships among gene and genotypic frequencies in a population.

Look at the left side of the graph in Figure 6.8 where q, the frequency of the recessive allele, is .1 or smaller. Note how few homozygous recessive genotypes there are in the population relative to heterozygous types. When dominance is complete and recessive gene frequencies are this low, few recessive genes are detectable—most are "hidden" in heterozygotes.

Under these conditions, if you select against a completely dominant allele (broken line in Figure 6.7), dominant genes are recognizable enough in homozygous dominant and heterozygous individuals, but few recessive genes are apparent, and selection against heterozygotes (which are indistinguishable from homozygous dominants) results in selection against "hidden" recessive genes as well as dominant genes. Progress is therefore very slow until there are enough homozygous recessive types to favor with selection.

In contrast, if you select *against* a completely recessive allele (dotted line in Figure 6.7), progress is fast until the recessive allele becomes relatively rare. At that point, most recessives "hide" in heterozygotes. They remain there undetected, resistant to selection.

Black/red coat color provides a classic example of how difficult it is to select against a completely recessive gene. For many years, Angus breeders in the United States considered red coat color to be a genetic defect and culled red animals from their herds. Still, they were never successful in eliminating the red gene, and it remains at low frequency in the Black Angus population to this day. Figures 6.7 and 6.8 explain why. They also explain why so many recessive genes for true genetic

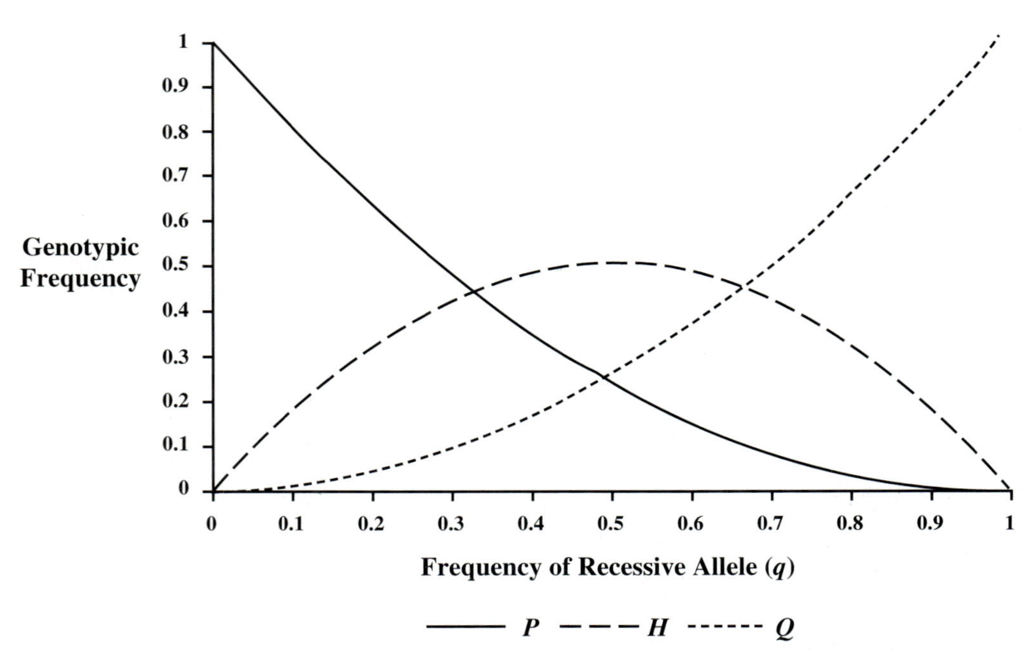

FIGURE 6.8 Relationships among gene and genotypic frequencies before selection (assuming random mating).

defects, despite strong natural and artificial selection against them, still persist in animal populations.

Degree of dominance affects the rate at which an undesirable allele is eliminated from a population, but, given enough time, the allele is either eliminated or reduced to trivial frequency regardless of degree of dominance—*with one exception.* The exception occurs when a locus displays overdominance for fitness and the heterozygote is the fittest genotype. In this case, selection for heterozygotes favors both alleles and causes gene frequencies to approach an intermediate equilibrium value. This value is a function of the relative fitness of the two homozygous genotypes.

An example of selection at a locus exhibiting overdominance is shown in Figure 6.9. Relative to heterozygotes, J_1J_1 individuals produce ¾ as many offspring, and J_2J_2 individuals produce ½ as many offspring. Because J_2 is the less favorable allele with respect to fitness, the equilibrium frequency of J_2 is less than .5 (.33 in this case). It does not matter whether the frequency of J_2 is initially high (solid line) or low (broken line), selection and overdominance cause it to approach the same intermediate value.

Examples of selection for heterozygotes resulting in equilibrium gene frequencies are not common in nature, but neither are they unheard of. Selection for warfarin resistance in rats is one example. Selection for malaria resistance in humans (and consequent selection for the allele causing sickle cell anemia) is another.

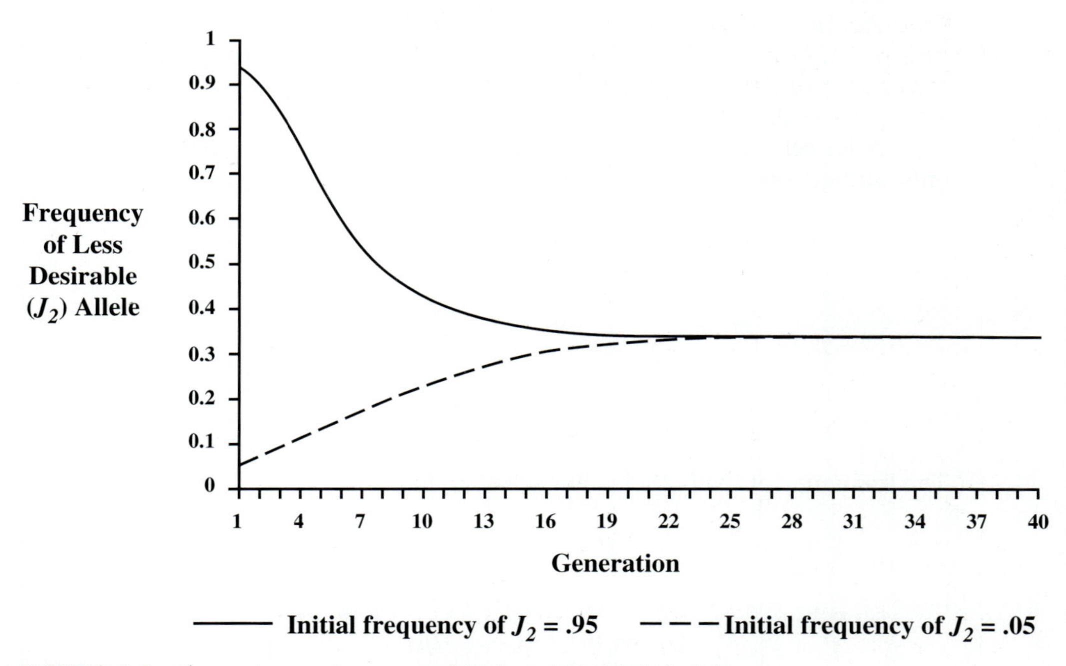

FIGURE 6.9 Change in gene frequency of the less desirable (J_2) allele over 40 generations of selection when overdominance exists at the J locus. In this example, the heterozygote is superior, and differences in fitness among genotypes are of the same magnitude as in Figure 6.7. Relative to J_1J_2 individuals, J_1J_1 individuals produce ¾ as many offspring, and J_2J_2 individuals produce ½ as many offspring. Whether the initial frequency of J_2 is high (solid line) or low (broken line), selection and overdominance cause the frequency of J_2 to approach an intermediate equilibrium value (.33 in this case).

Calculating Changes in Gene and Genotypic Frequencies Caused by Selection

To mathematically determine the effects of selection on gene and genotypic frequencies given different initial gene frequencies and degrees of dominance, consider a hypothetical J locus with alleles J_1 and J_2. Let q be the frequency of the J_2 allele (regardless of whether J_2 is dominant or recessive). Also let p and q represent gene frequencies, and P, H, and Q represent genotypic frequencies in a population consisting of a generation of offspring *before any of them have been selected to become parents.* If we assume that their (selected) parents were randomly mated, then the offspring population should be in Hardy-Weinberg equilibrium, and

$$P = p^2$$

$$H = 2pq$$

$$Q = q^2$$

Now let's select a portion of this population to become parents of the next generation. To accomplish this, let *relative* fitness values for the J_1J_1, J_1J_2, and J_2J_2 genotypes be $(1 - s_1)$, $(1 - s_2)$, and $(1-s_3)$, respectively. By "relative," we mean relative to the most fit genotype. For example, if the J_1J_1 genotype produces the most offspring (is the most fit), then $s_1 = 0$, and the relative fitness of J_1J_1 individuals is $1 - s_1 = 1-0 = 1$. If J_1J_2 individuals produce 25% fewer offspring than J_1J_1 types, then the relative fitness of J_1J_2 individuals is $1 - s_2 = 1 - .25 = .75$. And so on.

After selection, the offspring population becomes a population of parents, and genotypic frequencies in this new population are

$$P_1 = \frac{(1 - s_1)p^2}{(1 - s_1)p^2 + (1 - s_2)2pq + (1 - s_3)q^2}$$

$$H_1 = \frac{(1 - s_2)2pq}{(1 - s_1)p^2 + (1 - s_2)2pq + (1 - s_3)q^2}$$

$$Q_1 = \frac{(1 - s_3)q^2}{(1 - s_1)p^2 + (1 - s_2)2pq + (1 - s_3)q^2}$$

The frequency of the J_2 allele after selection is then

$$q_1 = Q_1 + \frac{1}{2}H_1$$

$$= \frac{(1 - s_3)q^2 + (1 - s_2)pq}{(1 - s_1)p^2 + (1 - s_2)2pq + (1 - s_3)q^2}$$

Rearranging terms,

$$q_1 = \frac{(1 - s_2)q + (s_2 - s_3)q^2}{1 - s_1 + 2(s_1 - s_2)q + (2s_2 - s_1 - s_3)q^2}$$

This formula can be used to calculate the new gene frequency given any initial gene frequency, any degree of dominance, and any differnce in fitness values among genotypes.

Example

Consider the different dominance scenarios illustrated in Figure 6.7. Assume that the J_2 allele is the less desirable allele at the J locus and that q, the initial gene frequency of J_2 before selection, is .3. When J_2 is completely dominant with respect to fitness, relative to J_1J_1 individuals, J_1J_2 and J_2J_2 individuals produce half as many offspring. Then

$$s_1 = 0$$
$$s_2 = .5$$
$$s_3 = .5$$

and

$$\begin{aligned} q_1 &= \frac{(1 - s_2)q + (s_2 - s_3)q^2}{1 - s_1 + 2(s_1 - s_2)q + (2s_2 - s_1 - s_3)q^2} \\ &= \frac{(1 - .5)(.3) + (.5 - .5)(.3)^2}{1 - 0 + 2(0 - .5)(.3) + (2(.5) - 0 - .5)(.3)^2} \\ &= \frac{.5(.3)}{1 - (.3) + .5(.3)^2} \\ &= .20 \end{aligned}$$

With no dominance, J_1J_2 individuals produce 25% fewer offspring than J_1J_1 types, and J_2J_2 individuals produce half as many. Then

$$s_1 = 0$$
$$s_2 = .25$$
$$s_3 = .5$$

and

$$\begin{aligned} q_1 &= \frac{(1 - s_2)q + (s_2 - s_3)q^2}{1 - s_1 + 2(s_1 - s_2)q + (2s_2 - s_1 - s_3)q^2} \\ &= \frac{(1 - .25)(.3) + (.25 - .5)(.3)^2}{1 - 0 + 2(0 - .25)(.3) + (2(.25) - 0 - .5)(0.3)^2} \\ &= \frac{.75(.3) - .25(.3)^2}{1 - .5(.3)} \\ &= .24 \end{aligned}$$

When J_2 is completely recessive, J_1J_2 individuals produce equally as many offspring as J_1J_1 types, and J_2J_2 individuals produce ½ as many. Then

$$s_1 = 0$$
$$s_2 = 0$$
$$s_3 = .5$$

and

$$q_1 = \frac{(1 - s_2)q + (s_2 - s_3)q^2}{1 - s_1 + 2(s_1 - s_2)q + (2s_2 - s_1 - s_3)q^2}$$

$$= \frac{(1 - 0)(.3) + (0 - .5)(.3)^2}{1 - 0 + 2(0 - 0)(.3) + (2(0) - 0 - .5)(.3)^2}$$

$$= \frac{.3 - .5(.3)^2}{1 - .5(.3)^2}$$

$$= .27$$

In one generation of selection against the J_2 allele, the frequency of J_2 decreases considerably (from .3 to .2) if J_2 is completely dominant, decreases somewhat less (from .3 to .24) if there is no dominance at the J locus, and decreases relatively little (from .3 to .27) if J_2 is completely recessive. You can see the same result graphically in Figure 6.7.

EXERCISES

Study Questions

6.1 Define in your own words: test mating
fitness

6.2 What kind of information is needed to do a good job of selecting for simply-inherited traits?

6.3 Why is it easier to select for simply-inherited traits controlled by loci that exhibit no dominance or partial dominance than to select for comparable traits affected by completely dominant alleles?

6.4 What is meant by the term "*probable* genotype"? What does it mean to have a high level of confidence in an animal's probable genotype?

6.5 What factors determine the level of confidence to be placed in the result of a set of test matings?

6.6 In testing a sire for a particular recessive allele, offspring out of certain types of mates provide more information than equal numbers of offspring out of other types. Rank the following types of mates for the amount of information their offspring provide:

Daughters of the sire
Homozygous recessives
Females chosen at random from the general population
Daughters of known carriers
Known carriers

6.7 What particular advantage do sire × daughter matings have over other types of test matings?

6.8 a. If you repeatedly bred a dog to a single daughter, and she had several very large litters without any abnormal puppies, would you be confident that the dog does not carry any deleterious recessive alleles?
b. If you repeatedly bred the dog to a bitch known to carry a particular deleterious recessive allele, and she had several very large litters without any abnormal puppies, would you be confident that the dog does not carry that allele?
c. If your answers to (*a*) and (*b*) differ, why do they differ?

6.9 What three factors affect the rate of change in gene frequency caused by selection?

6.10 Why is it generally true that selection causes the fastest change in gene frequency when the gene being selected for or against is at intermediate frequency?

6.11 Provide hypothetical fitness values for *JJ*, *Jj*, and *jj* genotypes when, with respect to fitness, the form of dominance at the J locus is:
a. complete dominance
b. partial dominance
c. no dominance
d. overdominance
Assume *J* is the dominant allele (except when there is no dominance).

6.12 Explain why the three curves in Figure 6.7 are shaped the way they are.

6.13 Why are completely recessive alleles so difficult to eliminate from a population?

6.14 In long-term selection experiments, what happens to gene frequencies when the most fit genotype is heterozygous? How does this differ from scenarios in which the most fit genotype is homozygous?

Problems

6.1 A Labrador breeder analyzed the pedigrees of two of her dogs and determined that the black male has a 50% chance of having the genotype *BBEe* and a 50% chance of having the genotype *BbEe.* The yellow female has a 75% chance of having the genotype *BBee* and a 25% chance of having the genotype *Bbee.* For matings of animals with probable genotypes like these, what proportion of puppies is expected to be chocolate? (See Chapter 3 for an explanation of coat color in Labs.)

6.2 A large artificial insemination stud has just purchased a promising bull. Management is concerned, however, that the bull might be a carrier of osteopetrosis (marble bone disease), a recessive lethal condition. Five percent of all cows are thought to be carriers of the osteopetrosis allele.
a. A.I. matings to randomly selected cows have already produced 100 normal calves (and no homozygous recessive calves). What is management's level of confidence that the bull is not a carrier of the gene for osteopetrosis?
b. How many successful A.I. matings to randomly selected cows are required to be 99% sure he does not carry the gene?
c. If the bull sires a calf with marble bone disease next year, what will be management's level of confidence that the bull does not carry the osteopetrosis gene? What will be their level of confidence that he *does* carry the gene?

6.3 A ram was bred to eight of his daughters to see if he carries any undesirable recessive genes. Four daughters produced twins, three produced singles, and 1 produced triplets. All lambs were normal.

a. How confident are we that the ram does not carry any undesirable recessives?

b. The ram was also bred to three known carriers of the recessive allele for spider syndrome. Each of these ewes produced a set of normal twin lambs. How confident are we that the ram does not carry the spider allele?

6.4 Consider a herd of 100 Hampshire sows in Hardy-Weinberg equilibrium.

a. What is the expected ratio of heterozygotes (white belted carriers of the allele for solid color) to homozygous recessive (solid colored) animals if:

i. twenty-five sows are solid colored.

ii. four sows are solid colored.

iii. one sow is solid colored.

b. What is the ratio of recessive genes found in heterozygotes to recessive genes found in homozygotes for (i), (ii), and (iii) above?

c. relationship is evident here?

6.5 There are two alleles at the J locus: J_1 and J_2. J_2 is the less desirable allele—in homozygous form it is lethal—and its frequency in the current generation is .2. What will be the frequency of J_2 in the next generation if:

a. with respect to fitness, J_1 is completely dominant to J_2?

b. J_2 is partially dominant such that J_1J_2 individuals produce 70% fewer offspring than J_1J_1 types?

c. overdominance for fitness exists at the J locus such that J_1J_1 individuals produce 25% fewer offspring than J_1J_2 types?

d. Compare your results for (*a*), (*b*), and (*c*) above and explain why they differ.

CHAPTER 7

The Genetic Model for Quantitative Traits

In selecting for simply-inherited traits, the breeder's task is to identify genotypes of individuals for loci of interest and select those individuals with the most favorable genotypes. The breeder's job in selecting for polygenic traits is much the same, except that identifying genotypes is out of the question. Instead, the breeder tries to identify *breeding values* of individuals for traits of importance and to select those individuals with the best breeding values.

An individual's genotype for a simply-inherited trait, even though we cannot see or feel it, is a relatively concrete thing. We know, for example, that if a black Labrador sires chocolate and yellow puppies, his genotype for coat type is *BbEe*. An individual's breeding value, on the other hand, is an abstract, mathematical idea. It can never be measured directly and, being a relative concept, its numerical value depends upon the breeding values of all the other individuals in the population.

To fully understand breeding values and other related notions, we need a mathematical model—a conceptual framework on which to hang definitions in some kind of logical, consistent way. That model, the genetic model for quantitative traits, is the subject of this chapter.

As its name suggests, the genetic model for quantitative traits is designed to be used with *quantitative* traits: traits in which phenotypes show continuous (numerical) expression. But in practice the model is used with polygenic traits in general. Threshold traits—*qualitative,* polygenic traits—are not continuously expressed and, at first glance, do not appear to be good candidates for a quantitative model. But we apply the model to them anyway. I discuss how that is done in the last section of this chapter.

quantitative genetics
The branch of genetics concerned with influences on, measurement of, relationships among, genetic prediction for, and rate of change in traits that are or can be treated as quantitative.

The branch of genetics dealing with the genetic model for quantitative traits and its applications is called **quantitative genetics.** It is concerned with influences on, measurement of, relationships among, genetic prediction for, and rate of change in traits that are or can be treated as quantitative.

THE BASIC MODEL

The basic genetic model for quantitative traits is represented by the following equation:

$$P = \mu + G + E$$

phenotypic value (P)
A measure of performance for a trait in an individual—a performance record.

population mean (μ)
The average phenotypic value of all individuals in a population.

genotypic value (G)
The effect of an individual's genes (singly and in combination) on its performance for a trait.

environmental effect (E)
The effect that external (nongenetic) factors have on animal performance.

where P = the **phenotypic value** or performance of an individual animal for a trait,
μ = (the Greek letter *mu*) the **population mean** or average phenotypic value for the trait for all animals in the population,
G = the **genotypic value** of the individual for the trait, and
E = the **environmental effect** on the individual's performance for the trait.

A phenotypic value is an individual performance record. It is the measure of an animal's own performance for a specific trait. Genotypic value refers to the effect of the individual's genes (singly and in combination) on its performance for the trait. Unlike phenotypic value, it is not directly measurable. The environmental effect is comprised of all nongenetic factors influencing an individual's performance for a trait.

The basic model presented here is slightly different from the better known model introduced in Chapter 1. That model did not include a population mean (μ). The reason for adding the mean is to emphasize that in animal breeding, genotypic values, environmental effects, and all the other elements of the genetic model discussed in this chapter are *relative*—relative to the population being considered. They are not absolutes. Their numerical values depend on the average performance of the population, and they are therefore expressed as *deviations* from the population mean.

Examples of the basic model for quantitative traits are illustrated schematically in Figure 7.1. Weaning weights (phenotypic values) for three calves are signified by the solid black columns in the figure. These columns extend from a line representing the average weaning weight in the population—a population mean of 500 lb. The black column extending above the line denotes an above average weaning weight, and the black columns extending below the line denote below average weaning weights.

The gray and white columns in the "background" represent the contributions of genotype and environment to each performance record. They signify genotypic values and environmental effects, respectively. (This is a hypothetical example used for illustration. In reality, of course, we cannot know an individual's genotypic value or environmental effect. All that we can measure directly is its phenotypic value.) Note that some of the columns representing genotypic values and environmental effects are located above the line and some below, and that those above the line have positive values and those below have negative values. This is because these values are expressed as deviations from the population mean. A positive deviation means greater than average; a negative deviation means less than average.

Calf (a), for example, weighs 600 lb—an above average phenotypic value. His 100-lb weaning weight advantage over the typical calf is partially due to a higher than average genotypic value. He is genetically 30 lb above average for weaning weight. He also experienced a better than average environmental effect—some 70 lb worth. Perhaps his dam was a particularly good milker.

Calves (b) and (c) both weigh 450 lb—50 lb below average. Calf (b) has a lower than average genetic value and experienced a worse than average environmental effect. Calf (c), on the other hand, is better than average genetically, but experienced a very poor environment. Perhaps he got sick or his dam had very little milk.

The basic model for quantitative traits is nothing more than a mathematical representation of how performance (P) is affected by both nature (G) and

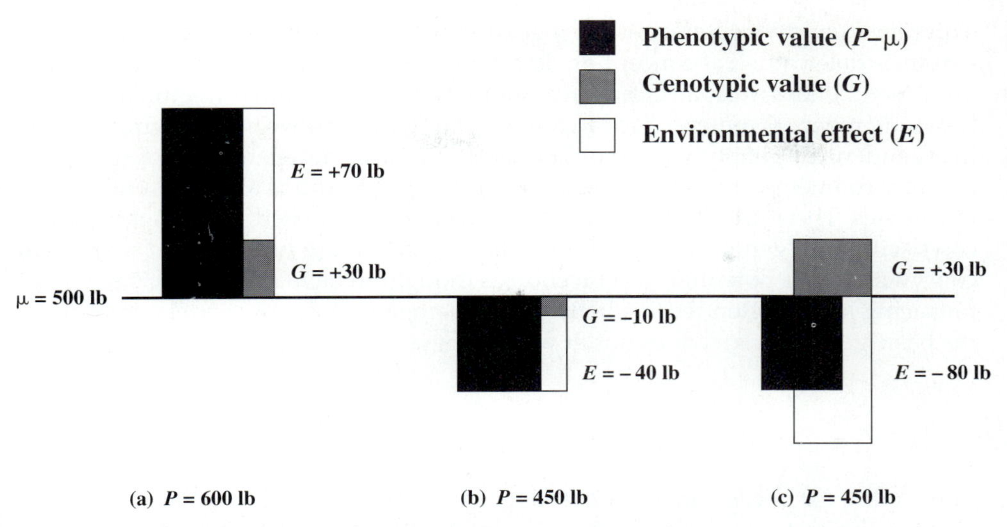

FIGURE 7.1[1] Schematic representation of genetic and environmental contributions to the weaning weights of three calves. Calf (*a*) weighs 600 lb (100 lb above average), has a higher than average genotypic value (*G*), and experienced a better than average environmental effect (*E*). Calf (*b*) weighs 450 lb (50 lb below average), has a lower than average genotypic value, and experienced a worse than average environmental effect. Calf (*c*) weighs 450 lb also. His genotypic value for weaning weight is higher than average, but his actual performance is below average due to a very poor environment.

nurture (E). There are some characteristics of the model that are important to remember, however. First, the model represents the genetic and environmental contributions to a *single* performance record on *one* animal. For every performance record, there exist precise—though not necessarily observable—numerical values for P, G, and E. The calf weaning weights in Figure 7.1 are examples. P, G, and E are called **values,** which in an animal breeding context means that they apply to *individuals* within a population. We speak of a particular calf's phenotypic value for weaning weight or a particular sow's genotypic value for litter size.

value
(in animal breeding) Any measure applied to an individual as opposed to a population. Examples are phenotypic value, genotypic value, breeding value, and environmental effect.

Secondly, these values are trait specific. The sow's genotypic value is not generic in any sense—it is her genotypic value for the specific trait of litter size.

Thirdly, because G and E are expressed as deviations from a mean, the *average* of genotypic values and the *average* of environmental effects across an entire population are zero. (When deviations are added in the process of averaging, the negative deviations cancel out the positive ones, and the sum and average are zero.) In statistical notation—a bar above a variable denotes a mean—then

$$\overline{G} = \overline{E} = 0$$

Finally, G and E are considered independent. This means that the genotype of the individual has no influence on the environmental effect that the individual

[1]Diagrams similar in format to Figure 7.1 appear repeatedly in the remainder of this book. If you take the time to develop a clear understanding of the format of Figure 7.1, subsequent diagrams will be much easier to follow.

experiences and vice versa. A calf's genotypic value for weaning weight, for example, is determined at conception. It is completely unaffected by preweaning environment. Likewise, if all calves in a population receive similar treatment regardless of their genetic potential for weaning weight, then preweaning environment is independent of genotypic value. The assumption of independence of genotype and environment is needed to keep the model simple and is a safe assumption in most cases. There are situations, however, where it is violated. For example, dairy cows with high genetic potential for milk production are typically fed more than cows with lower potential, and racehorses thought to have great genetic potential for racing ability often receive better than average training. A model as simple as the basic genetic model does not fit well in these cases.

BREEDING VALUE

breeding value (BV)
The value of an individual as a (genetic) parent.

In selecting for polygenic traits, breeders try to choose as parents those individuals with the best sets of genes—those individuals with the best **breeding values.** Breeding value (BV) is defined here as parental value—the value of an individual as a contributor of genes to the next generation. But breeding value does not appear in the basic model for quantitative traits ($P = \mu + G + E$). In fact, the only truly genetic component in the model is genotypic value (G). You might then ask whether breeders should choose as replacements those individuals with the best *genotypic* values. In other words, you might ask whether breeding value and genotypic value are the same thing.

The answer is no. While genotypic value represents the overall effect of an individual's genes, *breeding value represents only that part of genotypic value that can be transmitted from parent to offspring.* To get a better understanding of how breeding value and genotypic value differ, let's look at a hypothetical one-locus example.

independent gene effect
The effect of a gene independent of the effect of the other gene at the same locus (dominance) and the effects of genes at other loci (epistasis).

Assume that the B locus is one of many loci that affect mature weight. There are two possible alleles at the B locus: *B* and *b*. The average effect of each *B* gene is to increase mature weight by 10 grams, and the average effect of each *b* gene is to decrease mature weight by 10 grams. (These amounts may seem tiny, but remember that mature weight is a polygenic trait and individual gene effects on polygenic traits are thought to be small.) These 10-gram gene effects are known as **independent gene effects.** They reflect the value of each gene *independent* of the effects of the other gene at the same locus (dominance) and the effects of genes at other loci (epistasis). In other words, each independent gene effect reflects the inherent value of a gene as we would measure it if we would consider the gene in isolation.

An animal's breeding value for mature weight is simply the sum of the independent effects of that animal's genes at the B locus and at all other loci affecting mature weight. Here is why: Parents transmit a sample half of their genes to their offspring—one gene from each pair of genes at a locus. They do not transmit both genes from one locus, nor do they (as a rule) transmit intact combinations of genes from different loci. The Mendelian processes of segregation and independent assortment of genes prevent the inheritance of the particular gene combinations present in the parent. Because breeding value is the value of an individual as a contributor of genes to its offspring, and because gene combinations are not transmitted, then breeding value should reflect only the *independent* effects of genes, not effects due to gene combinations. The breeding value of an animal is then just the sum of the independent effects of all that animals' genes on a trait.

In our mature weight example, let's assume for simplicity that genotypes at all loci affecting mature weight except the B locus are identical for all animals. Then the breeding values for each of the three B-locus genotypes (*BB*, *Bb*, and *bb*) are

$$BV_{BB} = 10 + 10 = 20 \text{ g}$$
$$BV_{Bb} = 10 + (-10) = 0 \text{ g and}$$
$$BV_{bb} = -10 + (-10) = -20 \text{ g}$$

Animals with the *BB* genotype have the highest breeding value for mature weight. They can contribute only a *B* gene to their offspring, and *B* genes have a positive effect on mature weight. Animals with the *bb* genotype have the lowest breeding value for mature weight. They can contribute only a *b* gene to their offspring, and *b* genes have a negative effect on mature weight. Animals with the *Bb* genotype, having one gene of each kind, are intermediate in breeding value. Half of the time they will contribute a *B* gene, positively influencing the mature weight of the offspring, and half of the time they will contribute a *b* gene, negatively influencing offspring mature weight.

Now let's assume complete dominance at the B locus. In other words, the effect of the heterozygous combination of genes at the B locus (*Bb*) on mature weight is exactly the same as that of the homozygous dominant combination (*BB*). The *overall* effect that the genes of a *Bb* individual have on that individual's own mature weight—an effect that includes both independent gene effects and the effect of dominance at the B locus—is then no different than the overall effect that the genes of a *BB* individual have on its mature weight. Both individuals have the same *genotypic* value. As shown previously, however, they do not have the same *breeding* value.

For simplicity, let's assume that for homozygous genotypes, genotypic values are equal to breeding values.[2] We can then construct the following table.

Genotype	Breeding Value (*BV*)	Genotypic Value (*G*)
BB	20 g	20 g
Bb	0 g	20 g
bb	−20 g	−20 g

BB and *Bb* individuals are likely to have similar mature weights because both gene combinations have the same effect on an individual's own mature weight. They will not produce offspring with the same mature weights, however. The progeny of *BB* individuals will, on average, be heavier. Homozygous *bb* individuals will themselves be the lightest at maturity and will also product offspring with the lowest mature weights.

breeding value (*BV*)
The part of an individual's genotypic value that is due to independent and therefore transmittable gene effects.

To summarize, genotypic value represents the overall effect of an individual's genes (singly and in combination) on that individual's own performance for a trait. Not all of genotypic value is heritable, however. **Breeding value** is the

[2]Strictly speaking, this statement is not true. Genotypic values and breeding values are functions of gene frequencies in a population, and as such, the relationship between them is more complicated than I have presented here. I choose to oversimplify in this and later examples because the value gained from being technically correct is far outweighed by the clarity lost in the process.

part of an individual's genotypic value that is due to independent gene effects that can be transmitted from parent to offspring. An alternative (and useful) way to understand the difference between genotypic value and breeding value is to think of genotypic value as the value of an individual's genes to its own performance, and breeding value as the value of an individual's genes to its progeny's performance.

estimated breeding value (EBV)
A prediction of a breeding value.

Just as genotypic values are not directly measurable, neither are breeding values. We can predict them, however, using performance data.[3] A prediction of a breeding value is known as an **estimated breeding value** or **EBV.**

Progeny Difference

A parent passes on a sample half of its genes and therefore a sample half of the independent effects of those genes to its offspring. Because breeding value is the sum of the independent effects of all of an individual's genes affecting a trait, a parent passes on, *on average,* half its breeding value to its offspring. Half the parent's breeding value for a trait is our expectation of what is inherited from the parent and is called **progeny difference (*PD*)**[4] or **transmitting ability (*TA*).** Another way to understand progeny difference is to consider it the expected breeding value of a gamete produced by an individual. By either definition,

progeny difference (*PD*) or transmitting ability (*TA*)
Half an individual's breeding value—the expected difference between the mean performance of the individual's progeny and the mean performance of all progeny (assuming randomly chosen mates).

$$PD = \tfrac{1}{2}BV$$

Progeny difference is a very practical concept. Think of it as the expected difference between the mean performance of an individual's progeny and the mean performance of all progeny (assuming randomly chosen mates). For example, if a particular ram has a progeny difference of +1.2 lb for grease fleece weight, and we are careful to choose mates randomly (i.e., we are careful to mate him to a cross section of ewes, not just those with especially heavy or especially light grease fleece weights), we can expect the fleeces of his progeny to average 1.2 lb heavier than the average fleece.

expected progeny difference (EPD), predicted difference (PD), or estimated transmitting ability (ETA)
A prediction of a progeny difference.

Like breeding values, progeny differences are not directly measurable, but can be predicted from performance data. Such predictions are called **expected progeny differences (EPDs), predicted differences (PDs),** or **estimated transmitting abilities (ETAs)** and are commonly used to make genetic comparisons among animals.

It is important to understand than an individual does not transmit its progeny difference (exactly half its breeding value) to every offspring. A parent always passes on half its genes, but the genes transmitted constitute a random sample of the parent's genes. Some samples are better than others, Figure 7.2 depicts the distribution of genetic merit likely to be transmitted from an individual to its progeny. *On average,* half its breeding value (its progeny difference) is transmitted. Typically, however, Mendelian sampling causes the sample half of a parent's genes that are transmitted to an offspring to vary from half the true merit (*BV*) of the parent's genes. And it is impossible to control or predict whether a particular offspring will inherit a good, average, or mediocre sample.

[3]See Chapter 11.

[4]Not to be confused with *PD* in dairy literature. *PD* in a dairy context stands for "predicted difference," which is almost identical to progeny difference except that predicted difference is a *prediction* of a progeny difference, and progeny difference is the real thing—a true (but unknown) value.

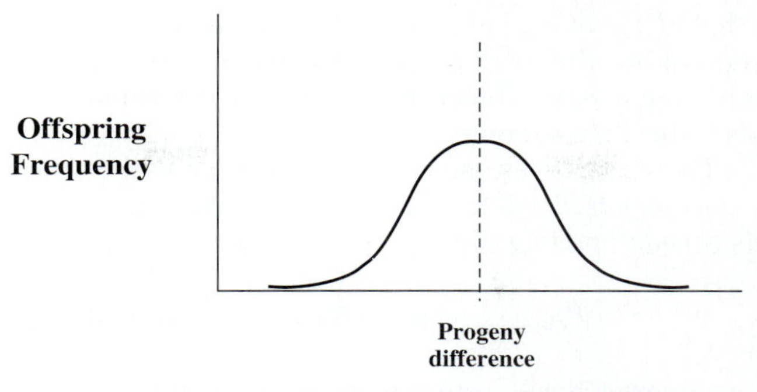

FIGURE 7.2
Distribution of genetic merit transmitted from a parent to its offspring. *On average*, a parent passes on its progeny difference or half its breeding value. Typically, however, it passes on more or less than that amount.

Additive Properties of Breeding Values and Progeny Differences

An animal's breeding value or progeny difference for a trait is a function of the independent effects of the animal's genes influencing the trait. "Independent," in this context, means independent of the effects of other genes. Because of their independence, these gene effects are *additive* in the sense that the animal's breeding value, the heritable part of its genotypic value for a trait, is simply the sum of the independent effects of all the animal's genes influencing the trait. For example, if a trait were affected by five (hypothetical) loci, and an individual's independent gene effects for the 10 genes at those loci were +3.0, −.6, +.2, +4.2, −1.4, −2.3, +.4, −.1, +.9, and +1.5 units, then the individual's breeding value for the trait would be 3.0 + (−.6) + .2 + 4.2 + (−1.4) + (−2.3) + .4 + (−.1) + .9 + 1.5 = +5.8 units. Because of their additivity, independent gene effects are often referred to as **additive gene effects.** Similarly, breeding value is often referred to as **additive genetic value** or simply **additive value.**

additive gene effect
Independent gene effect.

additive genetic value or **additive value**
Breeding value.

If independent gene effects are truly additive, then it stands to reason that the breeding value of an offspring is just the sum of the independent effects of the genes inherited from its sire and the independent effects of the genes inherited from its dam. Because the independent gene effects inherited from the sire comprise, on average, half the sire's breeding value, and the independent gene effects inherited from the dam comprise, on average, half the dam's breeding value, then

$$\overline{BV}_{Offspring} = \frac{1}{2}BV_{Sire} + \frac{1}{2}BV_{Dam}$$

In other words, *an offspring's breeding value for a trait will be, on average, the average of its parents' breeding values for the trait.* This relationship, as simple as it may seem, is a key one because it allows us to predict the breeding values of members of the next generation based upon our predictions of the breeding values of their parents. For example, if the estimated breeding value of a quarter horse sire for time to run a quarter mile is −3 seconds (three seconds faster than average), and the estimated breeding value of a quarter horse dam is −1 seconds, then

$$\begin{aligned}\overline{\hat{BV}}_{Offspring} &= \frac{1}{2}\hat{BV}_{Sire} = \frac{1}{2}\hat{BV}_{Dam} \\ &= \frac{1}{2}(-3) + \frac{1}{2}(-1) \\ &= -2 \text{ seconds}\end{aligned}$$

(A "hat" (^) above a value indicates a prediction of that value—in this case a prediction of breeding value or an EBV.) We would expect the average offspring from this mating to have a breeding value for quarter-mile time of −2 seconds—two seconds faster than average.

There is an even more useful extension to this concept: *an offspring's own performance for a trait will be, on average, the mean performance for the trait plus the average of its parents' breeding values for the trait,* or

$$\overline{P}_{Offspring} = \mu + \overline{BV}_{Offspring} = \mu + \frac{1}{2}BV_{Sire} + \frac{1}{2}BV_{Dam}$$

In our quarter horse example, if the population mean for quarter-mile time were 20 seconds, then

$$\begin{aligned}\hat{\overline{P}}_{Offspring} &= \mu + \frac{1}{2}\hat{BV}_{Sire} + \frac{1}{2}\hat{BV}_{Dam} \\ &= 20 + \frac{1}{2}(-3) + \frac{1}{2}(-1) \\ &= 20 + (-2) \\ &= 18 \text{ seconds}\end{aligned}$$

We would expect the average offspring from this mating not only to have a breeding value two seconds faster than average, but also to run two seconds faster than average.

You might ask, What about environmental effects? Don't they affect the performance of offspring? Shouldn't they be included somehow in this last equation? The answer is yes for an individual offspring, but no for the average of many offspring. This is because some offspring from a given mating will experience negative environmental effects and some will experience positive environmental effects, but the *average* environmental effect will be zero—the positives and negatives will cancel each other out.

Again, it is important to understand that the average of parental breeding values does not determine the breeding value or performance of *every* offspring from a mating—just the *average* offspring. Mendelian sampling causes variation in the breeding values of progeny, and differences in environmental effects cause additional variation in progeny performance. As shown in Figure 7.3, however, the average of parental breeding values can be used to predict offspring breeding value and performance. Although there is considerable variation in the progeny from any one mating, the likelihood of getting offspring with higher breeding values and performance is much greater when parents with superior breeding values are mated (b) than when parents with inferior breeding values are mated (a).

Breeding Value and Selection

The breeding values of parents are the key ingredient determining progeny breeding value and performance. Clearly, if we could define the "best" animal in terms of breeding values and could know exactly the breeding values of potential parents, then selection would be easy—we would simply pick those individuals with the best breeding values. In real life, however, true breeding values are unknown, and we must use *predictions* of breeding values instead. Because the effectiveness

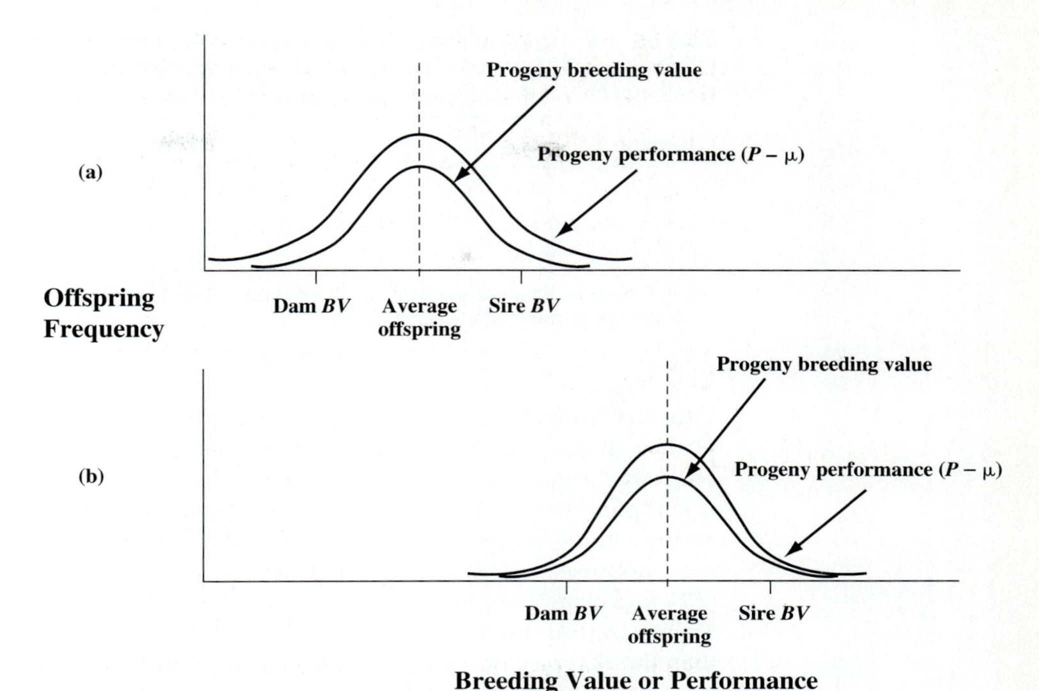

FIGURE 7.3 Distributions of progeny breeding values and performance from matings of (*a*) inferior parents and (*b*) superior parents.

of selection depends on the quality of such predictions, much of animal breeding technology involves the prediction of breeding values (or progeny differences). Chapter 11 in this book is devoted to that subject. To put all this in perspective, consider the fact that seedstock producers sell not only animals, but also information about those animals—namely, predictions of breeding values. The more and better the information, the more valuable the product.

GENE COMBINATION VALUE

gene combination value (GCV)
The part of an individual's genotypic value that is due to the effects of gene combinations (dominance and epistasis) and cannot, therefore, be transmitted from parent to offspring.

gene combination effect
The effect of a combination of genes, i.e., a dominance or epistatic effect.

Breeding value is the part of genotypic value that can be transmitted from parent to offspring. The remaining portion of genotypic value is called **gene combination value (GCV).** Gene combination value is the part of an individual's genotypic value for a trait that is due to **gene combination effects**—dominance and epistasis. Because individual genes and not gene combinations survive segregation and independent assortment during meiosis, gene combination value cannot be transmitted from parent to offspring.

An animal's breeding value and gene combination value together constitute its genotypic value for a trait. In model form,

$$G = BV + GCV$$

For a numerical example of how independent gene effects and gene combination effects relate to this equation, consider the T locus, a hypothetical locus affecting

TABLE 7.1 Contributions of Independent Gene Effects and Gene Combination Effects to Breeding Value (*BV*), Genotypic Value (*G*), and Gene Combination Value (*GCV*) at a Hypothetical T Locus[a]

Genotype	*BV*	*G*	*GCV* (*G* − *BV*)
TT	.1 + .1 = .2	.2	0
Tt	.1 + (−.1) = 0	.2	.2
tt	.1 + (−.1) = −.2	−.2	0

[a]As in the earlier example involving the B locus and mature weight, simplifying assumptions are made here for clarity.

litter size in swine. For simplicity, assume that the animals in this example are identical at all the other loci influencing litter size. Assume also that the independent effect of a *T* gene is +.1 pigs, the independent effect of a *t* gene is −.1 pigs, and *T* is completely dominant to *t*. Breeding values, genotypic values, and gene combination values for the three T-locus genotypes are listed in Table 7.1.

Because of complete dominance at the T locus, homozygous dominant types (*TT*) and heterozygous types (*Tt*) have the same genotypic value for litter size. Their breeding values are different, however, because the *TT* type is more likely than the *Tt* type to transmit the favorable *T* allele to its offspring. The difference between genotypic value and breeding value in the heterozygous animal (*G* − *BV*) is that animal's gene combination value, in this case .2 pigs.

In the T-locus example, the gene combination effect is due entirely to dominance. A similar effect due to epistasis could be shown with a multilocus example. Although dominance is generally believed to be the major cause of gene combination effects, gene combination value is probably the result of quite complicated dominance and epistatic relationships among genes.

Gene combination effects are not additive in the same way that independent gene effects are. You cannot simply sum gene combination effects to determine an individual's gene combination value. This is because gene combination effects at one or more loci may depend on genes present at yet other loci. Likewise, gene combination value is not additive in the same way that breeding value is. You cannot average the gene combination values of parents to predict the gene combination value of an offspring. Mathematically,

$$\overline{GCV}_{Offspring} \neq \frac{1}{2}GCV_{Sire} + \frac{1}{2}GCV_{Dam}$$

nonadditive gene effects Gene combination effects.

nonadditive genetic value or **nonadditive value** Gene combination value.

Because of their lack of additivity, gene combination effects are commonly called **nonadditive gene effects,** and gene combination value is typically referred to as **nonadditive genetic value** or simply **nonadditive value.**

A common misconception among students of animal breeding is the notion of "additive and nonadditive genes." Genes should not be considered one or the other. If a gene has *any effect at all* on a polygenic trait, it has an independent (additive) effect. If the expression of the gene is also influenced by dominance and(or) epistasis, the gene contributes to a gene combination (nonadditive) effect. It is possible for some genes to be entirely additive in their influence on a trait, but no gene can be entirely nonadditive in its influence. It is best not to think of genes as being additive or nonadditive. Think instead of genes as having independent effects and, in many cases, contributing to gene combination effects.

Unlike breeding values, gene combination values of individuals are rarely, if ever, predicted. This is partly because prediction of gene combination values is difficult, but mostly because a prediction of gene combination value has little practical use. Gene combination value is not, after all, transmitted from parent to offspring. Knowledge of breeding values helps us greatly in selection. Knowledge of gene combination values does not.[5]

This is not to say that gene combination value is not important. To the contrary, an individual's gene combination value for a trait can have a great influence on its own performance. As we will see in Chapter 17, hybrid vigor and inbreeding depression are just alternative names for favorable and unfavorable gene combination value. The particular importance of gene combination value depends on whether you are a seedstock producer or commercial producer. Seed stock producers market breeding potential, so breeding value is of primary concern to them. Commercial producers market performance, and to the degree that gene combination value contributes to performance, gene combination value is of concern to them.

A New Model

The basic genetic model for quantitative traits can now be expanded to include breeding value (*BV*) and gene combination value (*GCV*). Mathematically,

$$P = \mu + BV + GCV + E$$

The weaning weights of the three calves depicted in Figure 7.1 are shown again in Figure 7.4, this time with their (hypothetical) breeding values and gene combination values included. Calf (a) has the heaviest weaning weight, but note that most of his superior performance is due to either environment or genetic effects that cannot be transmitted to offspring. If heavier weaning weights are desirable, the best breeding animal should be calf (c). Despite his mediocre performance, he has the highest breeding value.

The new model for quantitative traits retains many of the characteristics of the basic model. Like genotypic value, breeding value and gene combination value are expressed as positive and negative deviations from a population mean. The average of breeding values and the average of gene combination values across an entire population are therefore zero. Thus

$$\overline{BV} = \overline{GCV} = \overline{G} = \overline{E} = 0$$

Furthermore, breeding value and gene combination value are considered independent of environment and of each other. It is very possible to have highly inbred animals that suffer serious inbreeding depression (unfavorable gene combination value) and yet are superior breeding animals (favorable breeding value). Likewise, many outcross animals perform well due to hybrid vigor (favorable gene combination value), but are not particularly good breeding animals (unfavorable breeding value).

[5]If cloning of animals becomes commonplace, gene combination value may become an important selection criterion. See Chapter 20 for more explanation.

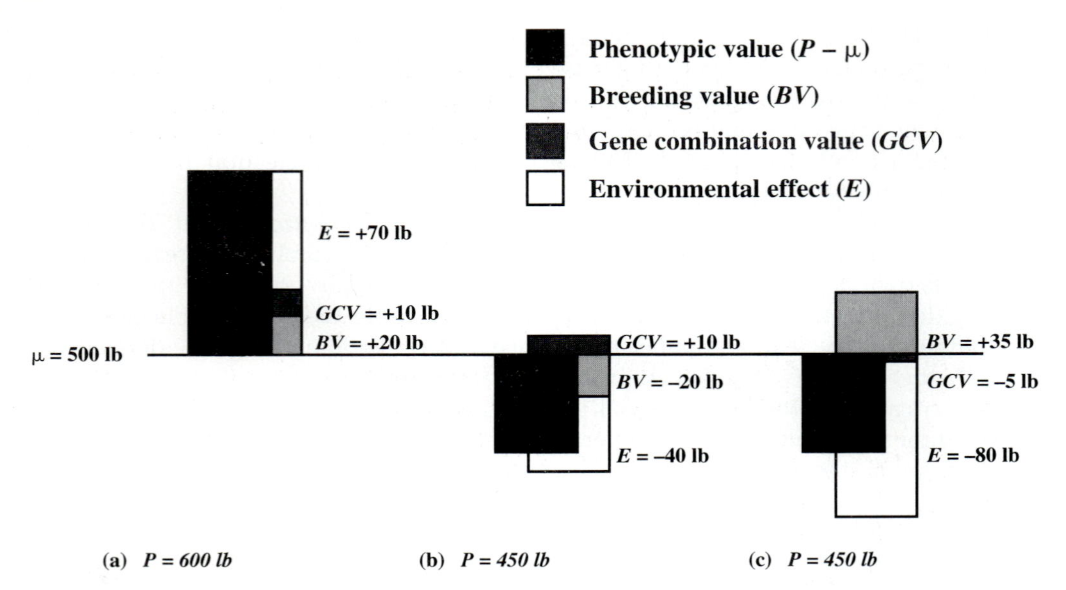

FIGURE 7.4 Schematic representation of the contributions of breeding value, gene combination value, and environmental effect to the weaning weights of the three calves depicted earlier in Figure 7.1. Calf (*a*) weaned the heaviest, but much of his superiority is due to factors that cannot be transmitted to offspring. If heavier weaning weights are desirable, the best breeding animal should be calf (*c*).

An Application of the New Model: Proof that PD = ½BV

The following proof illustrates the additive properties of breeding values; the independence of *BV*, *GCV*, and *E*; and the canceling of deviations when many records are averaged. Consider a sire with breeding value BV_s and progeny difference PD_s. If the sire is mated to a large number of females chosen at random from the general population, the average performance of his offspring ($\overline{P}_o$) should be

$$\overline{P}_o = \mu + \overline{BV}_o + \overline{NAV}_o + \overline{E}_o$$

$$= \mu + \frac{1}{2}BV_s + \frac{1}{2}\overline{BV}_d + \overline{NAV}_o + \overline{E}_o$$

Gene combination values and environmental effects are independent of breeding values, and, in a large population of offspring, positive deviations for these values should be balanced by negative deviations. Thus

$$\overline{P}_o = \mu + \frac{1}{2}BV_s + \frac{1}{2}\overline{BV}_d + 0 + 0$$

$$= \mu + \frac{1}{2}BV_s + \frac{1}{2}\overline{BV}_d$$

If dams are truly chosen at random from the general population, those with above average breeding values should be balanced by those with below average breeding values. The average breeding value for dams is therefore 0. Then

$$\overline{P}_o = \mu + \frac{1}{2}BV_s + 0$$

$$= \mu + \frac{1}{2}BV_s$$

The sire's progeny difference is simply the difference between the average performance of his progeny and the overall performance average. So

$$PD_s = \overline{P}_o - \mu$$

$$= \frac{1}{2}BV_s$$

And, in general,

$$PD = \frac{1}{2}BV$$

PRODUCING ABILITY

repeated trait
A trait for which individuals commonly have more than one performance record.

producing ability (PA)
The performance potential of an individual for a repeated trait.

Breeding value is important to selection, and gene combination value is important for its contribution to animal performance. For **repeated traits—**traits for which individuals commonly have more than one performance record—another value known as **producing ability (PA)** is important too. Producing ability refers to the performance potential of an individual for a repeated trait.

The following traits are repeated traits: milk production in dairy species, wool production in sheep, twinning in sheep (considered as a trait of the ewe, not the lamb), weaning weight in beef cattle (considered as a trait of the cow, not the calf), and racing ability in horses. In each case, the individual cow, sheep, or horse commonly has more than one record. Producing ability for each of these traits represents the potential of an animal to produce milk, wool, twins, calf weaning weight, or racing results, respectively. The term "producing ability" was first applied to milk production in dairy species. It referred to the ability of an animal to manufacture a product—milk. The term sometimes seems less appropriate for other traits in other species—for example, racing ability in horses. The "product" in this case is not immediately clear. Nevertheless, the concept of producing ability is still applicable to horses; it represents the racing potential of a horse.

Permanent and Temporary Environmental Effects

Producing ability is a function of all those factors that *permanently* affect an individual's performance potential. Genotypic value and its components, breeding value and gene combination value, are determined at conception. They are therefore permanent effects. Some environmental influences can also permanently affect performance potential. For example, calfhood nutrition is known to have a permanent effect on the ability of dairy and beef cows to produce milk. Training of young horses can have a permanent effect on their racing ability in later years.

permanent environmental effect (E_p) An environmental effect that permanently influences an individual's performance for a repeated trait.

Environmental effects that permanently influence an animal's performance for a repeated trait are known as **permanent environmental effects *(Ep)***. Producing ability is then

$$PA = G + E_p$$

and because

$$G = BV + GCV$$

therefore

$$PA = BV + GCV + E_p$$

Note that producing ability is neither a purely genetic value nor a purely environmental one; it is a combination of both.

temporary environmental effect (E_t) An environmental effect that influences a single performance record of an individual, but does not permanently affect the individual's performance potential for a repeated trait.

Many environmental effects do not permanently influence an animal's performance potential. These effects are temporary, influencing the performance of the animal a single time, but not every time. For the beef cow, forage digestibility is an example of such a **temporary environmental effect (E_t)**. High forage digestibility in good years allows the cow to produce plentiful milk and wean a heavier calf. Low digestibility in bad years has the opposite effect. The effect of forage digestibility is temporary, however. It does not *permanently* influence the cow's ability to produce calf weaning weight. An analogous example for racehorses is track condition. The condition of the track on race day (dry, muddy, etc.) affects a horse's performance in that day's race, but does not have a permanent effect on the horse's racing ability. Temporary environmental effects like these are therefore not a part of producing ability.

The Genetic Model for Repeated Traits

The genetic model for a single performance record of an individual in a trait for which individuals typically have multiple records can be written:

$$P = \mu + G + E_p + E_t$$

$$\text{or} \quad P = \mu + BV + GVC + E_p + E_t$$

Figure 7.5 illustrates the genetic model for repeated traits. Shown schematically are two records apiece on two dairy cows, Bessie and Flossie, for 305-day lactation yield (measured in total pounds of milk produced). Each black column in the foreground represents a single phenotypic record. The columns in the background represent the contributions of elements of the genetic model for repeated traits—including producing ability—to each record.

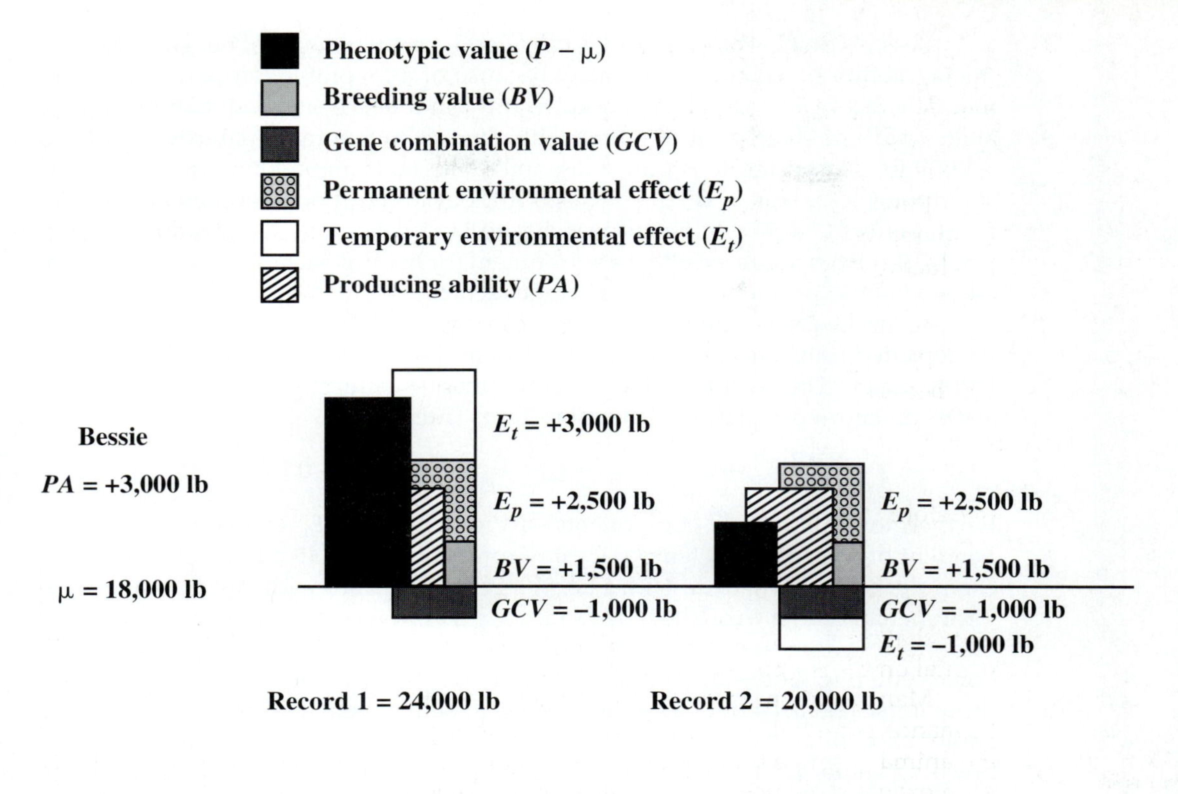

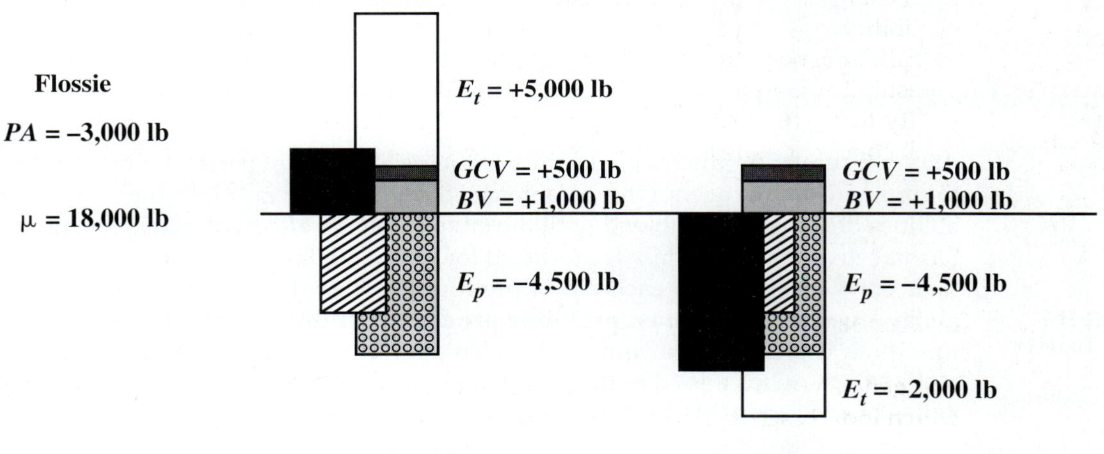

FIGURE 7.5 Schematic representation of the contributions of components of the genetic model for repeated traits—including producing ability—to two records apiece on two dairy cows (Bessie and Flossie) for 305-day lactation yield.

Bessie's first lactation record is very good, partly because of her superior producing ability (+3,000 lb) and partly because of a favorable temporary environmental effect (+3,000 lb). Her second record is not nearly as good. Her producing ability did not change—it is permanent—but an unfavorable environment (E_t = −1,000 lb) caused her to produce less milk. Flossie is a good cow from a genetic standpoint ($G = BV + GVC$ = +1,500 lb), but a very poor permanent environmental effect (−4,500 lb) has caused her to be a below average producer (PA = −3,000 lb). Even so, an excellent environment for her first lactation (E_t = +5,000 lb) allowed her to post a better than average record.

As in the earlier models we have seen, the components of the genetic model for repeated traits are expressed as positive and negative deviations from a population mean. The average of each component—including producing ability—across an entire population is therefore zero. Thus

$$\overline{BV} = \overline{GVC} = \overline{G} = \overline{E}_p = \overline{E}_t = \overline{PA} = 0$$

Furthermore, all the *basic* components—BV, GVC, E_p, and E_t—are considered independent of each other. A temporary environmental effect like track condition, for example, has nothing to do with a horse's genetic merit or with any permanent environmental influence on the horse's racing performance. Likewise, permanent environment is assumed independent of genetic merit, although in the case of race horses you might question this assumption—the best training may be reserved for those horses perceived to have the greatest genetic potential. Note that producing ability should be considered independent of temporary environment, but not of breeding value, gene combination value, or permanent environment. Because producing ability is the sum of these values, it cannot be independent of them.

The Importance of Producing Ability

As a measure of productive capacity, producing ability is of particular importance to commercial producers. On a commercial dairy farm, for example, the producing abilities of the cows combined with the environmental effects of management determine just how much milk is produced for sale. And dairy farmers typically feed their cows according to each cow's producing ability. Predictions of producing ability—usually called **most probable producing abilities (MPPAs)**—are therefore quite useful and are commonly calculated from performance data. An animal's MPPA, when added to the population mean for a trait, is a prediction of the animal's *next record*. Mathematically,

most probable producing ability (MPPA)
A prediction of producing ability.

$$\hat{P} = \mu + \text{MPPA}$$

where $\hat{P}$ indicates a *prediction* of performance. We cannot predict an individual's next record exactly because we never know what the temporary environmental effect on the record will be. We can often come close, however, with a prediction of producing ability.

Producing ability is generally less important to seedstock breeders than to commercial producers. The chief concern of seedstock breeders is, as always, breeding value. It is not unusual, however, for a seedstock breeder to select replacements on the basis of genetic potential (EBVs or EPDs) and cull poor producing mature animals on the basis of MPPA.

Racehorses provide a good example of how predictions of breeding value and producing ability are used differently. If you were deciding which horse to bet on, the most useful information to have would be the MPPA for racing ability of each horse in the race. The horse with the best MPPA should have the best producing ability (i.e., the best likelihood of winning). You would be betting not just on the horse's genetic merit, but also on its permanent environment—its training. On the other hand, if you were an owner and had just retired your horse to stud, the most useful piece of information would be the horse's EBV or EPD for racing ability. The stallion's breeding value, after all, will determine its ultimate success as a sire.

THE GENETIC MODEL AND THRESHOLD TRAITS

Threshold traits were defined in Chapter 5 as polygenic traits that are not continuous in their expression, but exhibit categorical phenotypes. An example of a threshold trait is fertility (as measured by success or failure to conceive). Fertility is believed to be influenced by many genes and is therefore a polygenic trait. But with only two phenotypes (pregnant or nonpregnant), fertility is no typical polygenic trait—it is a threshold trait.

We have every reason to believe that genotypic values, breeding values, gene combination values, and environmental effects for threshold traits like fertility are continuously distributed (*i.e.*, animals can differ in these values by very small increments). Threshold traits are no different from quantitative polygenic traits in this respect. But phenotypes for threshold traits are not expressed on a continuous scale, and that creates a problem: How can P be equal to $\mu + BV + GVC + E$ when P can take on just two or a few values? How can we apply the genetic model for quantitative traits to threshold traits?

The answer is to think of a threshold trait as having a continuous but unobservable underlying scale. The underlying scale for fertility is depicted in Figure 7.6. It is technically incorrect to call this a phenotypic scale because phenotypes for threshold traits are not continuous. This scale is often called a scale of "liability," and you can think of an animal's liability for a threshold trait as the

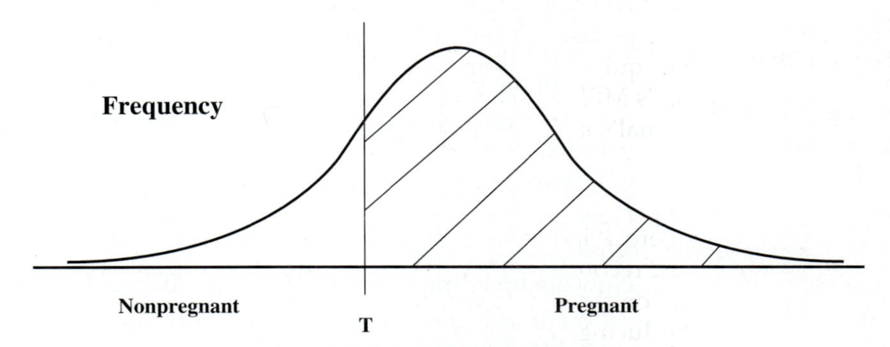

FIGURE 7.6 Distribution of liability for a threshold trait—in this case, fertility as measured by breeding success or failure. The underlying liability scale is continuous but unobservable. Animals to the left of the threshold (marked **T** on the liability scale) are nonpregnant. Those to the right are pregnant.

sum of its genetic values for the trait (continuously distributed values) and its environmental effect for the trait (also continuous). The **threshold** is the point on the liability scale above which animals exhibit one phenotype and below which they exhibit another. In Figure 7.6, all animals to the left of the threshold (**T**) fail to conceive, and all those to the right of the threshold conceive.

threshold
A point on the continuous liability scale for a threshold trait above which animals exhibit one phenotype and below which they exhibit another.

With the sleight of hand provided by the continuous liability scale, we can now apply the genetic model for quantitative traits to threshold traits. We can express EBVs, EPDs, MPPAs, etc. for these traits on the underlying liability scale itself or on related continuous scales that may be easier to understand.

EXERCISES

Study Questions

7.1 Define in your own words:
quantitative genetics
phenotypic value (*P*)
population mean (μ)
genotypic value (*G*)
environmental effect (*E*)
value
breeding value (*BV*) or additive genetic value or additive value
independent gene effect or additive gene effect
estimated breeding value (EBV)
progeny difference (*PD*) or transmitting ability (*TA*)
expected progeny difference (EPD) or predicted difference (PD) or estimated transmitting ability (ETA)
gene combination value (GCV) or nonadditive genetic value or nonadditive value
gene combination effect or nonadditive gene effect
repeated trait
producing ability (PA)
permanent environmental effect (E_p)
temporary environmental effect (E_t)
most probable producing ability (MPPA)
threshold

7.2 What is the purpose of a mathematical model for quantitative traits?

7.3 For a trait of your choice, construct several hypothetical records illustrating the basic genetic model for quantitative traits.

7.4 Contrast the various genetic models described in this chapter. What advantages do the more complex models have over simpler models?

7.5 The elements on the right-hand side of the genetic model are usually defined to be *independent*. What does independent mean in this context? Describe a situation in which the assumption of independence is violated.

7.6 How are genotypic value, breeding value, and gene combination value related?

7.7 How are breeding value and progeny difference related?

7.8 Why are independent gene effects and not gene combination effects transmitted from parent to offspring?

7.9 Why are independent gene effects sometimes called *additive* gene effects (i.e., what is additive about them)?

7.10 Why are gene combination effects sometimes called *nonadditive* gene effects (i.e., what is nonadditive about them)?

7.11 What is wrong with the notion of "additive and nonadditive genes"?

7.12 Why is gene combination value more important to commercial producers than to seedstock producers?

7.13 For a species of your choice, list the most important traits and state which of them are *repeated* traits.

7.14 Contrast breeding value and producing ability. For a repeated trait of your choice, describe how each value might best be used.

7.15 For a repeated trait of your choice, list permanent and temporary environmental effects.

7.16 Some breeders select animals on the basis of EBVs, and cull animals on the basis of MPPAs. What are their reasons for doing so?

7.17 How are threshold traits—polygenic *qualitative* traits—reconceptualized so that the genetic model for *quantitative* traits can be used to describe them?

Problems

7.1 **a.** Construct to scale a diagram like Figure 7.1 showing the following sample of records for milk production in dairy cows (μ = 13,600 lb):

Cow #	*P*	*G*	*E*
1	12,100	−300	−1,200
2	14,600	+1,200	−200
3	14,600	−400	+1,400

b. Cows 2 and 3 have similar records, but for very different reasons. Explain.

7.2 A famous draft horse was mated to a large number of *randomly selected* mares. On average, offspring of these matings pull 200 lb more than the average horse in major contests.

a. What is the sire's progeny difference for pulling power?

b. What is his breeding value for pulling power?

c. The horse was later mated to a large number of mares handpicked for pulling power. Offspring of these matings pull 300 lb more than the average horse. What is the mean breeding value of their *dams* for pulling power?

7.3 Consider a hypothetical quantitative trait (a weight of some kind) affected by five loci. Assume the following:

Complete dominance at all loci. No epistasis.

The independent effect of each dominant gene is +10 lb.

The independent effect of each recessive gene is −4 lb.

For homozygous combinations, genotypic values are equal to breeding values.

μ = 600 lb.

a. Fill in the following table:

Genotype	*BV*	*G*	*GCV*	*E*	*P*
(1) *AaBbCcDdEe*				+17	
(2) *AAbbCCddEE*				−21	

b. Which individual is the heaviest? Explain.
c. Which would product the heaviest offspring (on average)? Explain.

7.4 Consider the Thoroughbred stallions Raise a Ruckus and Presidium. Raise a Ruckus's breeding value for racing time is −8 seconds. He was particularly well trained, having a permanent environmental effect of −6 seconds. Presidium's breeding value is −12 seconds, but his permanent environmental effect is +2 seconds. Assuming both horses have gene combination values of 0,
a. Calculate progeny difference for each horse.
b. Calculate producing ability for each horse.
c. Which horse would you bet on in a race? Why?
d. Which horse would you breed mares to? Why?

7.5 Calving difficulty in beef cattle is a threshold trait, and breeders often record just two categories of calving difficulty scores: assisted and unassisted. Using Figure 7.6 as a guide, show the distributions of liability for calving difficulty if:
a. about 90% of the cows in a population calve unassisted.
b. only 50% of the cows in a population calve unassisted.

Statistics and Their Application to Quantitative Traits

How heritable is this trait? How variable is animal performance for the trait? If I select for it, will any other traits be affected? How do I predict my animals' breeding values and producing abilities for the trait? These are all questions you might ask about a quantitative trait. Unfortunately—or fortunately, depending on your point of view—all but the most simplistic answers to these questions are typically couched in statistical language. Quantitative traits (and threshold traits, which can be treated as quantitative) are described using terms like *mean, standard deviation,* and *correlation.* For a better understanding of quantitative traits we need to know some very basic statistics. The purpose of this chapter is to present fundamental statistical concepts as th ey apply to quantitative genetics and animal breeding.

INDIVIDUAL VALUES AND POPULATION MEASURES

value (in animal breeding)
Any measure applied to an individual as opposed to a population. Examples are phenotypic value, genotypic value, breeding value, and environmental effect.

population measure
Any measure applied to a population as opposed to an individual.

With the exception of the population mean (μ), all the elements of the genetic model for quantitative traits are considered **values.** As explained in the last chapter, the term *value,* in an animal breeding context, refers to any measure applied to an individual as opposed to a population. Phenotypic value, genotypic value, breeding value, producing ability—all are values that describe a characteristic of an individual animal for a specific trait. Animal breeders are concerned not only with values, however, but also with the *distributions* of and *relationships* between values in a population. They might want to know, for example, the average rate of growth in a breed of chickens and whether growth rate is highly variable or quite uniform across the breed. They might want to know whether beef cattle with very high breeding values for yearling weight tend to produce calves with heavy birth weights and increased calving difficulty. They might want to know whether horses with great racing performance typically have high breeding values for the trait, i.e., produce offspring with superior racing performance. (This is the same as asking whether racing performance is highly heritable.)

I call measures that provide information on distributions of and relationships between values in a population—measures that apply to a population instead of an individual—**population measures,** and I devote much of this

population parameter
A true (as opposed to estimated) population measure. Examples are true population means, variances, and standard deviations; true correlations between traits; and true heritabilities.

sample statistic
An estimate of a population parameter.

chapter to defining various population measures. There are two types. True (as opposed to estimated) measures are termed **population parameters.** Examples include true population means, variances, and standard deviations; true correlations between traits; and true heritabilities. **Sample statistics** are estimates of population parameters.[1] They are derived from sample data taken from a population. Examples include heritability estimates, estimates of variance, etc.

Statisticians are usually very careful about the distinction between parameters and statistics. Animal breeders, even academic animal breeders, are typically less finicky. We do not know exact values for population parameters—we only have access to sample statistics—yet we readily substitute statistics for parameters and often speak of both as if they were the same. For example, a horse breeder might refer to the mean wither height for a breed as 15 hands, as though 15 hands were a population parameter. In fact it is a sample statistic (or maybe just a guess). We even go so far as to apply the mathematical notation meant for parameters to statistics. Sigma squared (σ^2) is the proper notation for a population variance, and *s* squared (s^2) or $\hat{\sigma}^2$ is the proper notation for a sample variance, yet it is standard procedure in animal breeding to equate the symbol σ^2 with a numerical value clearly derived from a sample. In this book I follow animal breeding tradition and do the same. Only in actual calculations of sample statistics am I more fastidious.

Students often confuse individual values with population measures. They may speak, for example, of a certain sire's "heritability" for a trait. As will be explained in detail in Chapter 9, heritability is a population measure. We refer to the heritability of a given trait in a given population, but an individual animal cannot have a heritability of its own. In other words, heritability is not a value. To avoid problems later on, it is important at this point to clearly understand the distinction between values and population measures.

THE NORMAL DISTRIBUTION

Values of individuals, when viewed across an entire population, tend to follow a certain pattern or *distribution.* For example, most of the phenotypic values for weaning weight in beef cattle tend to be within 50 lb of the average weaning weight for the population. Only about a third of all weaning weights are more than 50 lb heavier or lighter than the population mean, and weaning weights more than 100 lb heavier or lighter than average are quite rare. The pattern exhibited by weaning weights (Figure 8.1) is typical of quantitative traits in general and is referred to as a **normal distribution.**

normal distribution
The statistical distribution that appears graphically as a symmetric, bell-shaped curve.

The normal distribution appears graphically as a symmetric, bell-shaped curve. The horizontal axis represents levels of some value—for example, phenotypic value (as in Figure 8.1), breeding value, or producing ability. The vertical axis represents the frequency of different levels of the value in the population. The area between the curve and the horizontal axis and bounded on each side by a given interval of values represents the proportion of observations in the population likely to be within that interval. In Figure 8.1, for example, about 14% of weaning weights are between 400 lb and 450 lb. (The hashed area in Figure 8.1 comprises about 14%

[1]This definition of a statistic is probably too restrictive for many statisticians (and some quantitative geneticists). I think it is the best definition for readers of this book.

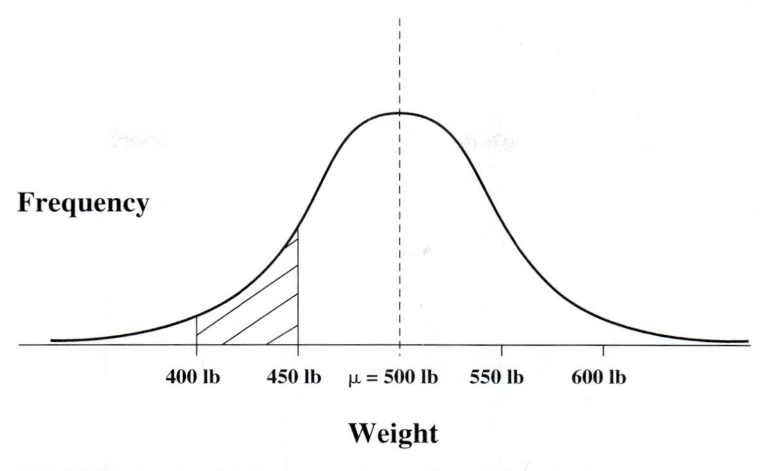

FIGURE 8.1 Normal distribution of weaning weight in beef cattle.

of the total area under the curve.) For normally distrib-uted traits, most observations are near the mean (dashed line in Figure 8.1), and relatively few observations occur at the *tails* of the distribution, i.e., far from the mean.

The great majority of quantitative traits are normally or near-normally distributed. That is because they are affected by many genes. Figure 8.2 shows the distribution of genotypic values in a population for traits affected by (a) one locus, (b) two loci, and (c) four loci. The horizontal axis represents (from left to right) increasing genotypic values, and the vertical axis represents the frequency of those values in the population. With just one locus and two alleles influencing a trait, there are only three possible genotypes and therefore only three possible genotypic values, each noticeably different from the others. As the number of loci affecting the trait increases, the number of genotypes and the number of possible genotypic values increase, and the differences between adjacent genotypic values generally become smaller. Some simplifying assumptions have been made in constructing the graphs, but it is clear that the greater the number of loci influencing a trait, the more nearly the distribution of genotypic values resembles a normal distribution.

A six-locus example is illustrated in Figure 8.3 with a normal curve superimposed. The continuous curve suggests what the distribution would look like if the trait were affected by many more than six loci or if environmental effects were included that would cause variation in phenotypic values to appear continuous, thus "smoothing" out the distribution.

Not all traits are normally distributed. Threshold traits—polygenic qualitative traits—have only a small number of possible phenotypes, so phenotypic values for these traits cannot follow any continuous distribution.

It is easy to envision measures of actual performance like weaning weights being normally distributed. Normal distributions for less tangible values such as breeding values, gene combination values, or environmental effects are intuitively less obvious. It is important to understand, however, that these values are distributed across a population in much the same fashion as phenotypic values. Figure 8.4 depicts the distributions of phenotypic values (P), breeding values (BV), gene combination values (GCV), and environmental effects (E) for a typical quantitative trait. The distributions are not identical, but they are all normal.

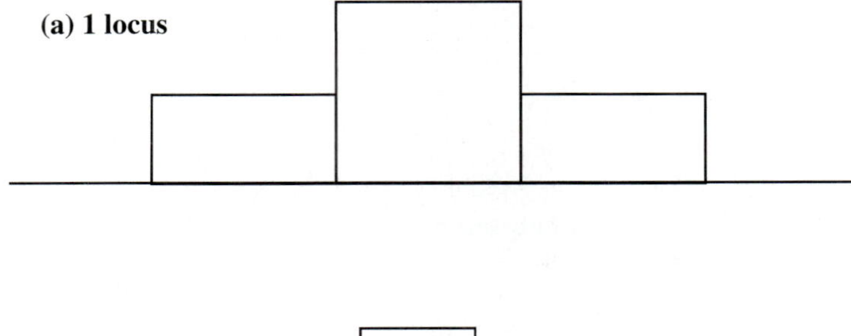

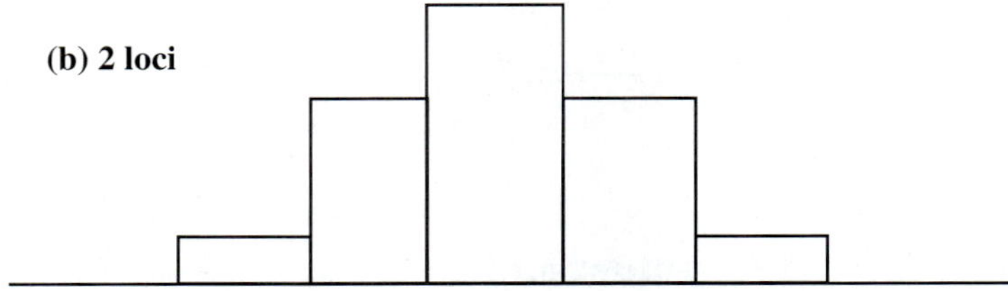

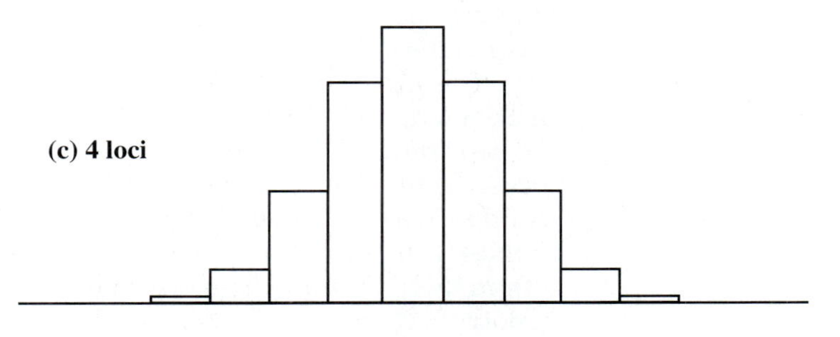

FIGURE 8.2 Distribution of genotypic values in a population for traits affected by (*a*) one, (*b*) two, and (*c*) four loci. Genotypic values increase from left to right. The area within each rectangular column indicates the proportion of individuals in the population having a particular genotypic value. (Assumptions: no dominance or epistasis, two possible alleles per locus, equal gene effects at all loci, all gene frequencies = .5.)

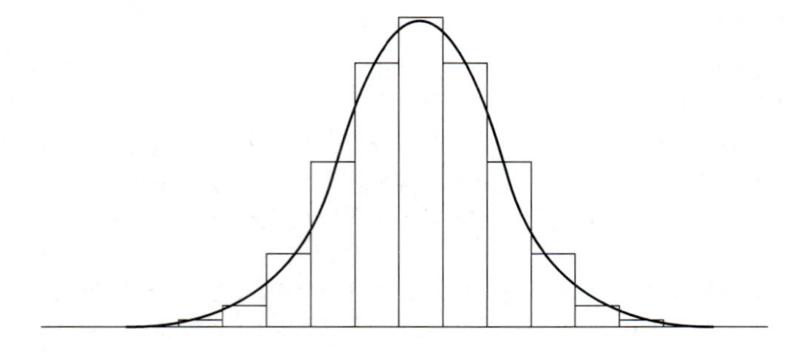

FIGURE 8.3
Distribution of genotypic values in a population for a trait affected by six loci. A continuous normal curve is superimposed to suggest how this distribution would appear with a large number of loci or after adding the "smoothing" influence of environmental effects (same assumptions as in Figure 8.2).

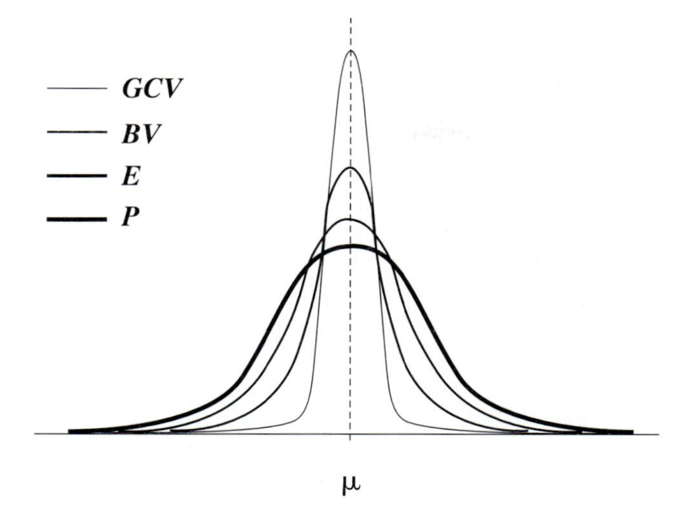

FIGURE 8.4
Normal distributions of phenotypic values (*P*), breeding values (*BV*), gene combination values (*GCV*), and environmental effects (*E*) for a typical quantitative trait.

THE MEAN

mean
An arithmetic average.

The **mean** is a population measure. It is simply an arithmetic average. You have already been introduced to the mean phenotypic value for an entire population (μ). For normally distributed traits, the mean determines the center of the distribution—the point on the horizontal axis where the bell-shaped curve is the highest. In Figure 8.1, for example, the normal distribution of weaning weight has its center at the mean value of 500 lb.

Examples of phenotypic means for a number of traits and species are listed in Table 8.1. Means vary greatly with breed, management, and physical environment, and selection causes them to change over time. The values listed in the table are therefore not universally representative. They do, however, offer some rough perspective on typical performance in various traits.

Means are calculated by simply adding up values from a population or from a sample taken from a population and dividing by the number of values. In mathematical notation, the formula for a mean phenotypic value is

$$\overline{P} = \frac{1}{n}\sum_{i=1}^{n} P_i$$

$$\text{or} \quad \overline{P} = \frac{1}{n}(P_1 + P_2 + P_3 + \cdots + P_n)$$

where $\overline{P}$ (p-bar) represents the mean phenotypic value (a bar over a variable indicates a sample mean)
P_i is the ith phenotypic value
and n is the number of phenotypic values in the sample

The upper formula uses *summation notation*—mathematical, shorthand denoting a sum. The sigma (Σ) in the formula indicates that the elements to the right of the sigma are to be added together. The lower formula shows the same thing in more explicit fashion.

Means are used in many ways in animal breeding. We refer to the overall mean performance for a trait in a population (μ) in the genetic model for quantitative

TABLE 8.1 Typical Phenotypic Means (μ) for a Number of Traits and Species

Species	Trait	μ
Cattle (beef)	Calving interval	380 days
	Birth weight	80 lb
	Weaning weight	475 lb
	Yearling weight (bulls)	950 lb
	Mature weight	1,100 lb
	Feed conversion (feed per gain)	7 lb per lb
	Scrotal circumference	34.5 cm
	Backfat thickness (steers)	.4 in
Cattle (dairy)	Days dry	83 days
	Calving interval	404 days
	Milk yield	13,000 lb
	% fat	4.4%
	% protein	3.5%
Horses	Wither height	60 in
	Cannon bone circumference	7.7 in
	Mature weight	1,180 lb
	Time to trot one mile	130 sec
	Time to run ¼ mile	20 sec
	Time to run one mile	96 sec
	Weight started (draft)	2,000 lb
	Cutting score	209 points
Swine	Litter size (number born alive)	9.8 pigs
	Litter size (number weaned)	7.3 pigs
	Weaning weight	14 lb
	21-day litter weight	100 lb
	Days to 230 lb	175 days
	Feed conversion (feed per gain)	3.8 lb per lb
	Loin eye area	4.3 in^2
	Backfat thickness	1.3 in
Poultry	No. of eggs in first year (layers)	300
	Egg weight (layers)	58 g
	Hatchability (chickens)	90%
	Feed conversion ratio (broilers)	2.45 kg/kg
	Hot carcass weight (broilers)	1.5 kg
	Mature body weight (broilers)	2.4 kg
	Shank length (turkeys)	103 mm
	Breast weight (broilers)	290 g
Sheep	Number born	1.3 lambs
	Birth weight	9 lb
	60-day weaning weight	45 lb
	Yearling weight	150 lb
	Loin eye area	2.1 in^2
	Grease fleece weight	8 lb
	Staple length	2.5 in

traits. We also speak of the mean performance of a management group—a group of animals that have been managed similarly. We deal with means of other values besides phenotypic values. Recall that across an entire population, means of breeding values, gene combination values, producing abilities, environmental effects, etc. are defined to be zero. In some sires summaries, the mean EPD of animals born in a specified year is used as a reference point for the published EPDs of all animals.

VARIATION

variation (in most animal breeding applications): Differences among individuals within a population.

The mean indicates the population average, but it reveals nothing about how individuals deviate from the average. It tells us nothing about the uniformity of the population, or, put differently, it tells us nothing about **variation** in the population. In an animal breeding context, variation usually refers to differences among individuals within a population. For almost all traits there exists variation in performance, breeding values, producing abilities, environmental effects—variation in *values* of any kind and in their predictions as well.

Figure 8.5 depicts variation in a sample of 16 animals taken from a larger population. The horizontal line represents mean performance (μ) and each column (in the top diagram, each pair of columns) represents an individual animal's deviation from the mean. Columns extending above the mean indicate positive deviations, and those extending below the mean indicate negative deviations. The top diagram shows actual performance or phenotypic value (P, black columns) with each individual's underlying breeding value (BV), gene combination value (GCV), and environmental effect (E) in the background. The lower diagrams represent the same 16 animals, but breeding values, gene combination values, and environmental effects are broken out separately so that variation in these component values is easier to see.

If you look closely at Figure 8.5, you can see that of the four values shown, phenotypic value is the most variable—phenotypic deviations tend to be the largest. In this population, gene combination values are the least variable, and breeding values and environmental effects are intermediate in variability.

The Importance of Variation

Variation is the source of genetic change. If there is little genetic variation in a trait—more specifically, if there is little variation in breeding values for the trait—selection will be difficult because no individual is much better than any other as a genetic parent. From a genetic standpoint, it is important that a population be variable. At the same time, there is often economic value associated with reduced variation. A more uniform product is more impressive and therefore more easily marketed than a variable one. Breeders like to tout the "peas in a pod" appearance and performance of their animals. Maintaining genetic variation in today's animals in order to make genetic change possible in tomorrow's animals and at the same time fostering uniformity is a constant challenge for animal breeders.

It is important not only to have variation in a population, but also to have some way to measure it. Without some yardstick of variability, it would be impossible, for example, to know whether an animal's performance is extreme or just slightly above average. Measures of variation provide reference points in this respect.

Measures of variation can also be used to compare the importance of different components of the genetic model. In the top diagram in Figure 8.5, variation in performance can be seen to result from variation in breeding values, gene combination values, and environmental effects. When you look at the lower diagrams in the figure, it should be clear that variation in environmental effects and variation in breeding values are major contributors to phenotypic variation in this case, and variation in gene combination values is a relatively minor contributor. In other

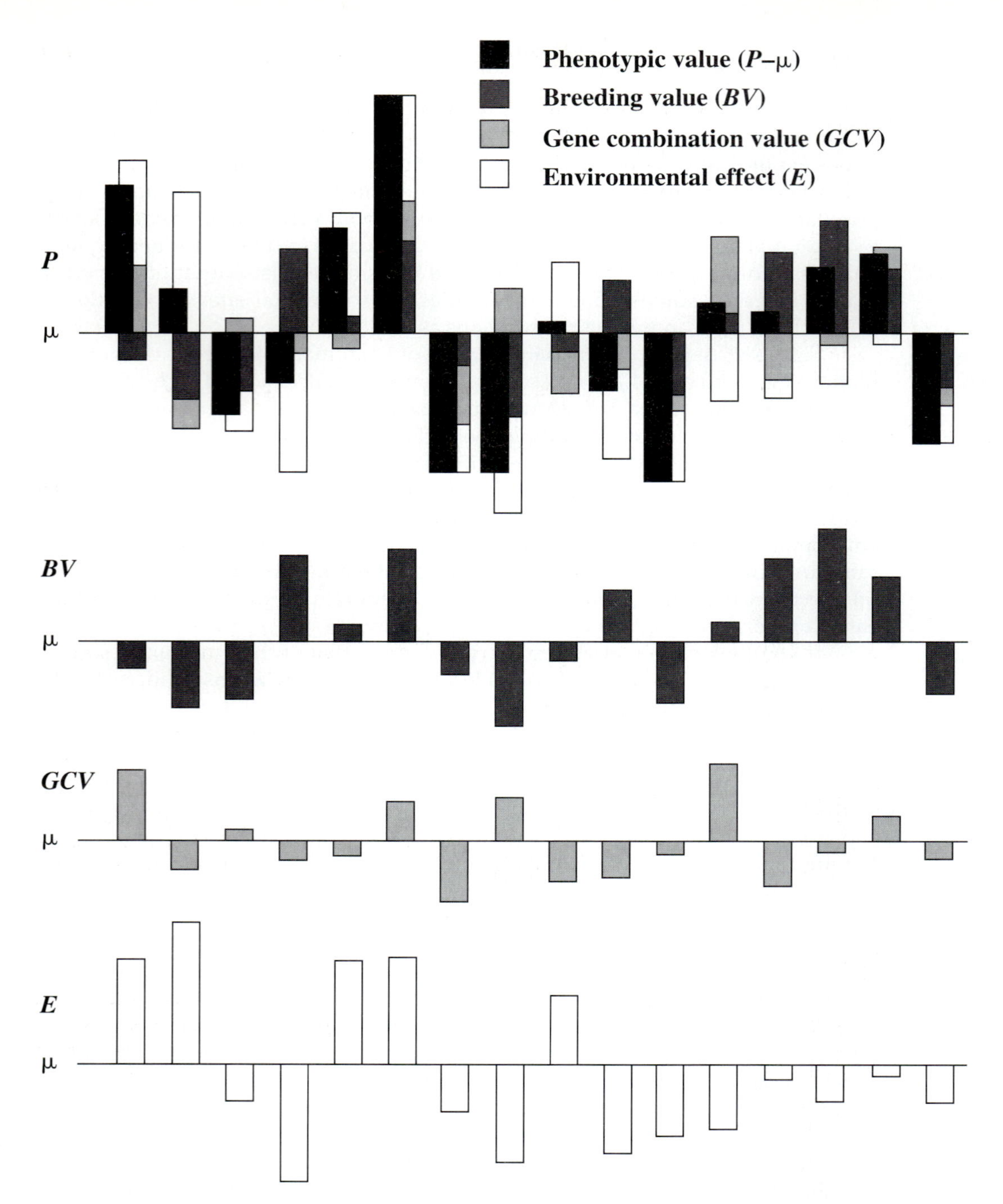

FIGURE 8.5 Schematic representation of variation in a sample of 16 animals from a larger population. The top diagram shows phenotypic variation (*P*, black columns) with variation of component values in the background. The lower three diagrams separate variation in breeding values (*BV*), gene combination values (*GCV*), and environmental effects (*E*).[2]

[2]Because this figure represents a *sample* of animals from a larger population—not the entire population—the average deviation from the population mean (μ) for sample animals is somewhat different from zero. Had the entire population been included in the figure, the average deviation would have been exactly zero. This apparent discrepancy is noticeable in all similar figures that involve samples in this and later chapters.

words, environmental effects and breeding values play important roles in the expression of the hypothetical trait represented in Figure 8.5, and the role of gene combination values is much less important.

Using measures of variation to quantify the importance of model components for a trait has great practical value. As we will see in the next chapter, the relative importance of breeding value is of particular interest. It determines *heritability,* a measure that determines the degree to which offspring resemble their parents in performance, and a measure that is needed for predicting animals' breeding values, progeny differences, and producing abilities. In short, the ability to measure variation is important to selection.

Measures of Variation

variance (σ^2) A mathematical measure of variation.

standard deviation (σ) A mathematical measure of variation that can be thought of as an average deviation from the mean. The square root of the variance.

The most commonly used mathematical measures of variation are the **variance** and the square root of the variance or the **standard deviation.** Both are, of course, population measures. The symbol for a variance is σ^2 (sigma squared), and the symbol for a standard deviation is σ (sigma). Subscripts are often used to specify exactly what kind of variance or standard deviation is implied. A number of examples are given in Table 8.2. The first column of the table indicates the type of value—in traditional mathematical terms, the type of **variable—**being considered. The second and third columns show the appropriate notation for the variance and standard deviation.

Variances are expressed in squared units. For example, the variance of calf weaning weight used to construct Figure 8.1 is 2,500 *pounds squared.* While a phenotypic mean for weaning weight of 500 lb is easy to imagine, a 2,500 lb^2 variance is difficult to conceptualize. Variances have very useful mathematical properties, but little intuitive appeal. The standard deviation is better in this respect. A standard deviation is just what its name implies—a "standard" deviation from the

TABLE 8.2 Notation for Variances and Standard Deviations of Variables of Interest to Animal Breeders

Variable	Variance	Standard Deviation
Phenotypic value (P)	σ^2_P	σ_P
Breeding value (BV)	σ^2_{BV}	σ_{BV}
Progeny difference (PD)	σ^2_{PD}	σ_{PD}
Gene combination value (GCV)	σ^2_{GCV}	σ_{GCV}
Permanent environmental effect (E_p)	$\sigma^2_{E_p}$	σ_{E_p}
Temporary environmental effect (E_t)	$\sigma^2_{E_t}$	σ_{E_t}
Producing ability (PA)	σ^2_{PA}	σ_{PA}
Estimated breeding value (EBV)[a]	$\sigma^2_{\hat{BV}}$	$\sigma_{\hat{BV}}$
Expected progeny difference (EPD)[a]	$\sigma^2_{\hat{PD}}$	$\sigma_{\hat{PD}}$
Most probable producing ability (MPPA)[a]	$\sigma^2_{\hat{PA}}$	$\sigma_{\hat{PA}}$

[a] A "hat" (ˆ) above a variable indicates an estimate or a prediction.

TABLE 8.3 Typical Phenotypic Standard Deviations (σ_P) for a Number of Traits and Species

Species	Trait	σ_P
Cattle (beef)	Calving interval	20 days
	Birth weight	10 lb
	Weaning weight	50 lb
	Yearling weight (bulls)	60 lb
	Mature weight	85 lb
	Feed conversion (feed per gain)	.5 lb per lb
	Scrotal circumference	2 cm
	Backfat thickness (steers)	.1 in
Cattle (dairy)	Days open	90 days
	Calving interval	75 days
	Milk yield	560 lb
	% fat	.5%
	% protein	.4 %
Horses	Wither height	1.8 in
	Cannon bone circumference	.23 in
	Mature weight	110 lb
	Time to trot one mile	3.5 sec
	Time to run ¼ mile	.6 sec
	Time to run one mile	1.3 sec
	Cutting score	10.3 points
Swine	Litter size (number born alive)	2.8 pigs
	Litter size (number weaned)	2.8 pigs
	Weaning weight	1.5 lb
	21-day litter weight	15 lb
	Days to 230 lb	12 days
	Feed conversion (feed per gain)	.2 lb per lb
	Loin eye area	.25 in^2
	Backfat thickness	.15 in
Poultry	No. of eggs in first year (layers)	3 eggs
	Egg weight (layers)	4.6 g
	Hatchability (layers)	2.2%
	Feed conversion ratio (broilers)	.4 kg/kg
	Hot carcass weight (broilers)	10.5 g
	Mature body weight (broilers)	.9 kg
	Shank length (turkeys)	.5 mm
	Breast weight (broilers)	3 g
Sheep	Number born	.3 lambs
	Birth weight	3 lb
	60-day weaning weight	8 lb
	Yearling weight	30 lb
	Loin eye area	.1 in^2
	Grease fleece weight	1.1 lb
	Staple length	.5 in

mean. You can think of the standard deviation as the *average* deviation from the mean (though technically it is not). For example, the standard deviation of weaning weight in Figure 8.1 is $\sqrt{2{,}500} = 50$ lb. The "average" deviation from the mean weaning weight in this population is therefore ± 50 lb. Estimates of phenotypic standard deviations for a number of traits and species are listed in Table 8.3.

Variances and standard deviations are population measures, not individual values. We speak of the variance of loin eye area in swine or the standard deviation of producing ability in dairy goats. In both cases, we are referring to variation in some value for a trait in a population. It would be incorrect to speak of a particular pig's variance for loin eye area or a particular goat's standard deviation for producing ability.

Variation and the Normal Distribution

The shape of the normal curve for a particular value indicates the amount of variation of that value in the population. A relatively flat, broad distribution indicates a high degree of variability. A tall, narrow distribution, on the other hand, indicates a high degree of uniformity. Figure 8.6 illustrates the distributions of weaning weight and birth weight in beef cattle. Note that weaning weight is much more variable ($\sigma_P = 50$ lb) than birth weight ($\sigma_P = 10$ lb). Almost all birth weights lie within a 60-lb range (55 to 115 lb) while the range of weaning weights is over 300 lb (350 to 650 lb).

The shape of a normal distribution and the standard deviation of the distribution are closely related. On the graph of a normal distribution, the standard deviation appears as the distance between the mean and the *point of inflection*[3] of the normal curve. A generic example is shown in Figure 8.7.

Knowing the mean and standard deviation for a value, you can make certain generalizations about the distribution of that value in a population. That is because for normally distributed values, approximately ⅔ or 68% of all observations lie within one standard deviation either side of the mean (hashed area in Figure 8.7). Ninety-five percent of all observations lie at a distance less than two standard deviations from the mean, and virtually all observations (99%) are less than three standard deviations from the mean. Thus, for the weaning weight example in Figure 8.6, we can expect 68% of calf weaning weights to be between 450 and 550 lb ($\mu - \sigma_P = 500 - 50 = 450$, $\mu + \sigma_P = 500 + 50 = 550$). For the birth weight example in the same figure, we can expect 95% of calf birth weights to be between 65 and 105 lb ($\mu - 2\sigma_P = 85 - 2(10) = 65$, $\mu + 2\sigma_P = 85 + 2(10) = 105$).

COVARIATION

covariation
How two traits or values vary together in a population.

Animal breeders are often concerned with how two traits or two values *vary together.* They are concerned, in a word, with **covariation.** For example, a swine breeder might want to know if daily weight gain is related to feed conversion. A cattle breeder might want to know if breeding value for scrotal circumference is associated with breeding value for age at puberty. A dog breeder might want to know

[3]If you start at the top of the curve and proceed to the right, the curve becomes increasingly steep until at some point it begins to flatten out. That point is the point of inflection. A similar inflection point exists on the left side of the curve.

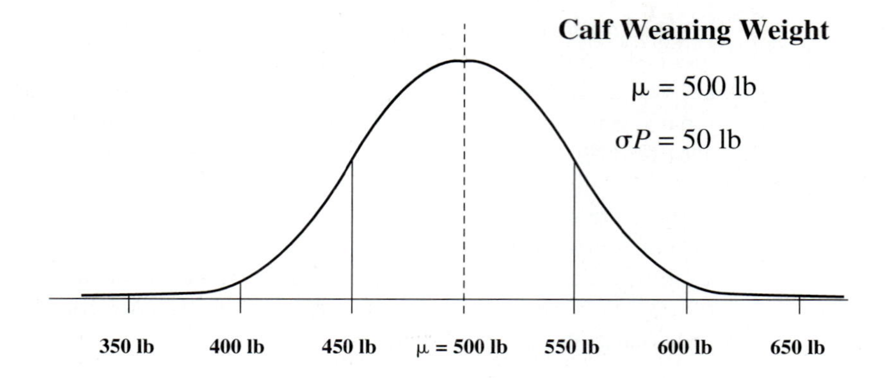

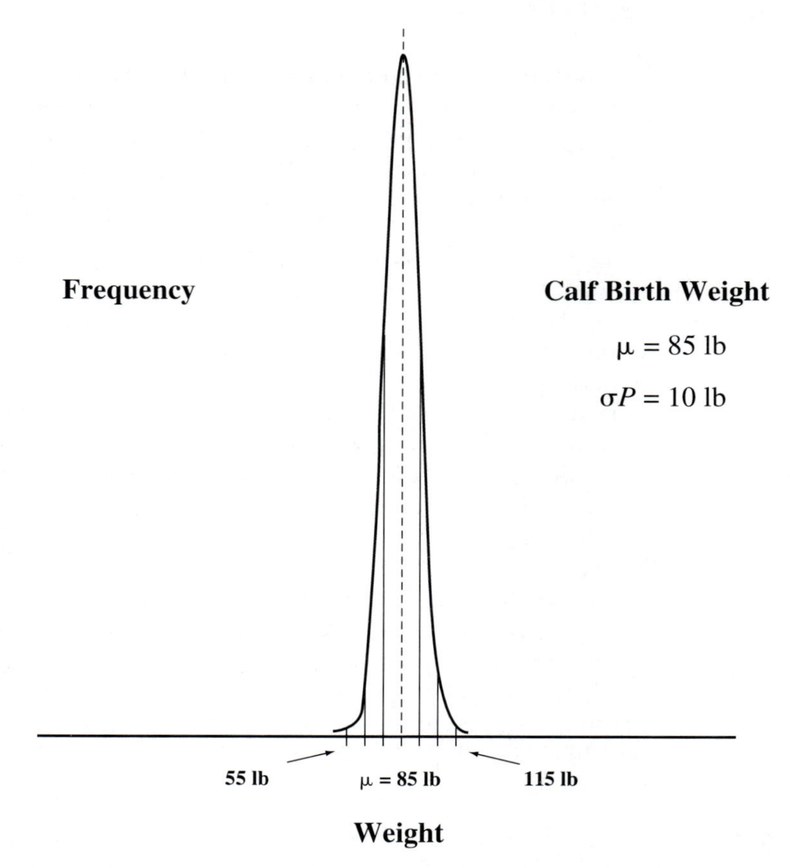

FIGURE 8.6 Normal distributions of weaning weight (*top*) and birth weight (*bottom*) in beef cattle. Note the large difference in the variability of these two traits.

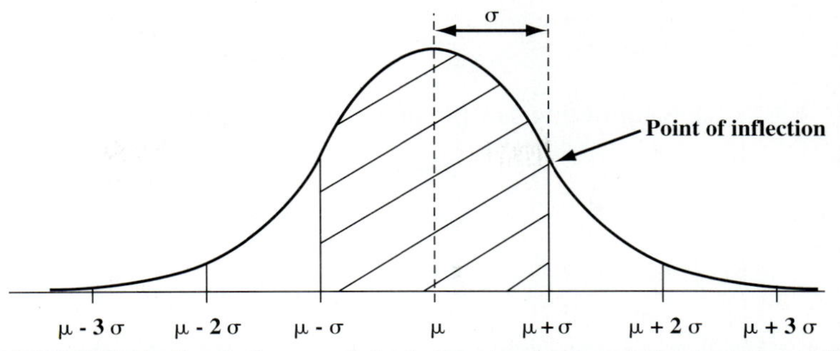

FIGURE 8.7 Graph of a normal distribution showing the relationship between the standard deviation (σ) and the normal curve. Approximately 68% of all observations lie within one standard deviation either side of the mean (hashed area on the graph).

Calculating Variances and Standard Deviations

A true measure of variance (the population parameter) can be defined with the following *word formula:*

Variance = the average squared deviation from the mean

Mathematically, the variance of some variable X is then

$$\sigma_X^2 = \frac{\sum_{i=1}^{n}(X_i - \mu_X)^2}{n}$$

where each X_i represents an observation for the variable X. For example, if X is weaning weight performance, then each X_i represents a weaning weight (phenotypic value) for an individual animal. The denominator (n) represents the number of observations. In expanded form, the formula appears as:

$$\sigma_X^2 = \frac{(X_1 - \mu_X)^2 + (X_2 - \mu_X)^2 + \cdots + (X_n - \mu_X)^2}{n}$$

The formulas for an *estimate* of a variance—the formulas you should use to estimate a variance from a sample of data—are slightly different:

$$\hat{\sigma}_X^2 = \frac{\sum_{i=1}^{n}(X_i - \hat{\mu}_X)^2}{n-1}$$

or, in expanded form,

$$\hat{\sigma}_X^2 = \frac{(X_1 - \hat{\mu}_X)^2 + (X_2 - \hat{\mu}_X)^2 + \cdots + (X_n - \hat{\mu}_X)^2}{n-1}$$

The standard deviation is just the square root of the variance. Thus

$$\sigma_X = \sqrt{\sigma_X^2}$$

and

$$\hat{\sigma}_X = \sqrt{\hat{\sigma}_X^2}$$

Example

The following sample of weaning weights was reported for a set of beef calves:

Calf#	Weaning Weight, lb
1	515
2	430
3	475
4	565
5	630
6	510
7	495
8	480
9	555
10	505
11	470
12	445

To calculate the variance and standard deviation of these weights, we must first determine the sample mean.

$$\hat{\mu}_{P_{WW}} = \overline{P}_{WW}$$
$$= \frac{1}{n}\sum_{i=1}^{n} P_{WW_i}$$
$$= \frac{1}{12}(515 + 430 + \cdots + 445)$$
$$= 506.25 \text{ lb}$$

Because this number represents the mean of phenotypic values for weaning weight, I use the notation P (for phenotypic value) and WW (for weaning weight). Now, our estimate of the phenotypic variance for weaning weight is

$$\hat{\sigma}^2_{P_{WW}} = \frac{\sum_{i=1}^{n}(P_{WW_i} - \hat{\mu}_{P_{WW}})^2}{n - 1}$$
$$= \frac{(515 - 506.25)^2 + (430 - 506.25)^2 + \cdots + (445 - 506.25)^2}{12 - 1}$$
$$= \frac{8.75^2 + (-76.25)^2 + \cdots + (-61.26)^2}{11}$$
$$= 3{,}082.4 \text{ lb}^2$$

and our estimate of the phenotypic standard deviation for weaning weight is

$$\hat{\sigma}_{P_{WW}} = \sqrt{\hat{\sigma}^2_{P_{WW}}}$$
$$= \sqrt{3{,}082.4}$$
$$= 55.5 \text{ lb}$$

if observed temperament is related to breeding value for temperament. In each case, the breeder wants to know how two traits or values—in traditional mathematical terms, two variables—*covary.*

Figure 8.8 depicts covariation for samples of 16 animals each from three larger populations. X and Y represent attributes of these animals of some kind. X and Y could be phenotypic values for different traits, breeding values for different traits, environmental effects for different traits, phenotypic values and breeding values for the same trait, or whatever values are of interest. Each pair of black and white columns represents the X and Y attributes for a single animal expressed as deviations from the overall mean (μ).

In the sample from the first population (a), X and Y show a strong relationship with each other. Positive deviations for X are quite consistently associated with positive deviations for Y. Likewise, negative deviations for X are quite consistently associated with negative deviations for Y. Furthermore, larger deviations tend to be paired with larger deviations, and smaller deviations tend to be paired with smaller deviations. There are exceptions to the rule. The fifth animal from the left, for example, shows a positive deviation for X and a negative deviation for Y, and the animal on the far left shows a very small deviation for X, but a fairly large deviation for Y. Given the general pattern, however, we say that in this population, X and Y exhibit *strong, positive covariation.*

In the sample from the second population (b), X and Y are closely related also. But in this population, *positive* deviations for X are quite consistently associated with *negative* deviations for Y and *vice versa.* Again, there are exceptions to the rule—the individuals on the far left and far right, for example. But because positive deviations in one attribute show a strong tendency to be paired with negative deviations in the other attribute, and because larger deviations tend to be paired with larger deviations and smaller deviations tend to be paired with smaller deviations, we say that in this population, X and Y exhibit *strong, negative covariation.*

In the sample from the third population (c), there is no clear pattern to the relationship between X and Y. Sometimes positive deviations are paired with positive deviations, sometimes negatives with negatives, sometimes positives with negatives, and there seems to be no consistency in the size of deviations within a pair. In this population, there appears to be little, if any, covariation between X and Y.

Figure 8.8 is not the easiest figure to understand. You need to look closely to see how covariation differs in the three populations. But the effort will pay off because the graphical method used here shows—more clearly than any other method I can think of—how pairs of deviations contribute to covariation. I use the same method again to illustrate the concepts of heritability, repeatability, and various types of correlations.

The Importance of Covariation

There are three basic aspects of covariation. The first has to do with the *direction,* or, in mathematical terms, the *sign* of the relationship between two variables. In other words, this aspect has to do with whether the relationship is positive, negative, or nonexistent. An example of each type is shown in Figure 8.8(a), (b), and (c), respectively. When positive deviations in one variable tend to be paired with positive deviations in another variable, and when negative deviations tend

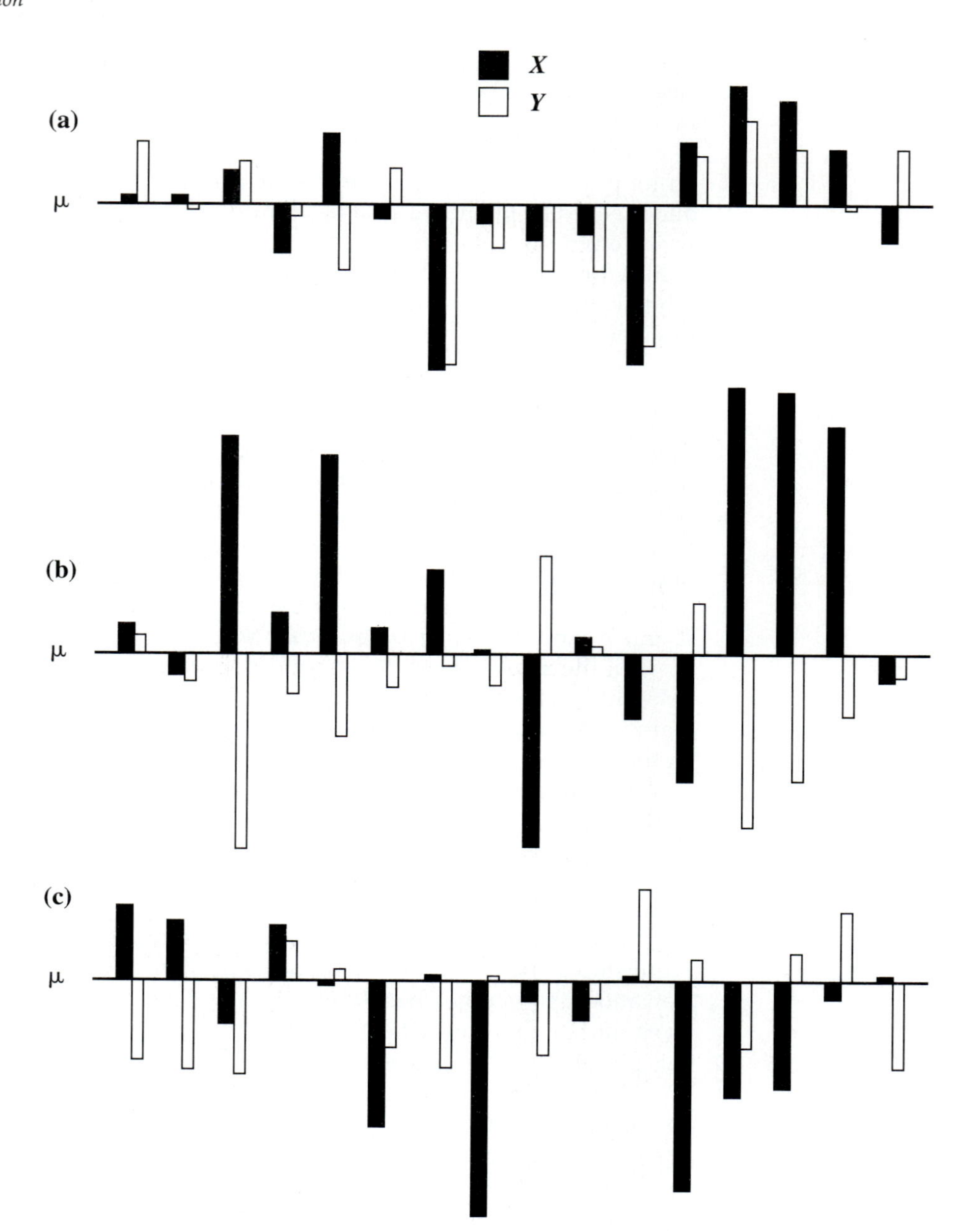

FIGURE 8.8 Schematic representation of covariation in samples from three populations. *X* and *Y* represent two animal attributes (e.g., phenotypic values for two traits, etc.—see text). In population (*a*), *X* and *Y* exhibit strong, positive covariation; in population (*b*), *X* and *Y* exhibit strong negative covariation; and in population (*c*), *X* and *Y* do not appear to covary.

to be paired with negative deviations (Figure 8.8(a)), covariation is positive. When positives tend to be paired with negatives (b), covariation is negative. And when there is no pattern to the pairing (c), covariation is zero or near-zero.

The sign of the covariation between two variables has very practical value. It tells us the direction of change of one variable that is expected with change in the other variable. For example, birth weight and yearling weight in cattle exhibit positive covariation. We can expect, therefore, that calves with heavier birth weights will likely have heavier yearling weights. Furthermore, we can expect that selection for heavier yearling weights will lead to heavier birth weights.

The second aspect of covariation has to do with the *strength* of the relationship between two variables. A strong relationship can be described as being consistent or reliable. In the populations represented in Figure 8.8(a) and (b), the variables X and Y exhibit strong, consistent, reliable relationships. In population (a), positive deviations are almost always paired with positive deviations, and negatives are almost always paired with negatives. In population (b), positives are almost always paired with negatives. In both cases, covariation between X and Y is quite consistent and reliable. In population (c), however, there is no consistency to the relationship between X and Y. Covariation between X and Y in population (c) is weak if it exists at all.

It is often important to know how strong the relationship between two variables is. For example, suppose you were a beef cattle breeder and were concerned about heavy birth weights and the calving problems associated with them. The relationship between yearling weight performance and breeding value for birth weight in cattle is quite strong. Animals with heavy yearling weights are fairly consistent in having high breeding values—and thus high progeny differences—for birth weight. Knowing the strength of this relationship, you would probably be reluctant to use a young bull with extremely high yearling weight performance for fear that he might sire heavy calves that will be hard to deliver. If the relationship between yearling weight and breeding value for birth weight were not so strong, you would be more likely to take a chance on a bull with exceptional growth performance.

The third aspect of covariation has to do with the *amount of change in one variable that can be expected for a given amount of change in another variable.* In Figure 8.8(a), for example, large negative deviations for variable X are associated with large negative deviations for variable Y, deviations for X that are close to zero are associated with similarly small deviations for Y, and large positive deviations for X are associated with large positive deviations for Y. It appears that as X increases, Y increases as well. This third aspect of covariation is concerned with just how much change in Y occurs with changes in X.

Knowing how much change to expect in one variable per unit change in another is useful for prediction. Consider our example of the bull with the extremely high yearling weight. If you knew how much change to expect in progeny difference for birth weight per pound of change in yearling weight, you could predict the average birth weight of the bull's calves based on his yearling weight. This would help you decide whether or not to use the bull. If the predicted birth weights were alarmingly high, you would probably avoid using him. If they were not especially high, you would be more likely to use him. The most common kinds of predictions in animal breeding are predictions of breeding values, progeny differences, and producing abilities. The ability to predict these values is an important aspect of selection, and to make such predictions possible, we need to know how much these values change with changes in animal performance.[4]

[4]For more information on prediction, see (1) the section in this chapter on prediction, and (2) Chapter 11.

Measures of Covariation

Just as variation is measured by the variance and standard deviation, covariation has its measures as well: covariance, correlation, and regression. Like variances and standard deviations, covariances, correlations, and regressions are mathematically related. And like variances and standard deviations, these measures of covariation are not individual values. Rather, they are population measures; they provide information about the relationship between two traits or two values in a population. Each of these measures tells us something about one or more of the aspects of covariation discussed earlier.

Covariance

covariance (cov(X,Y)) The basic measure of covariation.

The **covariance** is the basic measure of covariation. The symbol I choose to use for the covariance of variables X and Y is $\text{cov}(X,Y)$. Subscripts are often used to specify the type of covariance under consideration. Several examples are shown in Table 8.4. As you can see from the table, many different kinds of covariances are conceivable. The covariances listed in the table are particularly useful ones for animal breeding.

Covariances are to covariation as variances are to variation. Variation and covariation are phenomena; variances and covariances are measures of these phenomena. Covariances are very useful from a mathematical standpoint but, like variances, have little intuitive appeal. That is because the numerical values of covariances are often very large or very small and offer no point of reference for evaluating the relationship between two variables. Furthermore, the units of covariances are difficult to deal with conceptually. An estimate of the covariance between phenotypic values for 60-day weaning weight and grease fleece weight in lambs, for example, is 2.6 lb^2. An estimate of the covariance between breeding values for scrotal circumference and age at puberty in cattle is -17 cm·days—that's *centimeter·days*. As you can see from these examples, covariances by themselves are not very revealing.

The one piece of information that is clearly indicated by a covariance, however, is the *direction* or *sign* of the covariation between two variables. Weaning weight and grease fleece weight in lambs exhibit positive covariation. Greater than average weaning weights tend to be associated with greater than average fleece

TABLE 8.4 Examples of Notation for Covariances, Correlations, and Regressions Involving Variables of Interest to Animal Breeders

Variables	Covariance	Correlation	Regression
Phenotypic value (P) for trait X and phenotypic value for trait Y	$\text{cov}(P_X,P_Y)$	r_{P_X,P_Y}	$b_{P_Y \cdot P_X}$
Breeding value (BV) for trait X and breeding value for trait Y	$\text{cov}(BV_X,BV_Y)$	r_{BV_X,BV_Y}	$b_{BV_Y \cdot BV_X}$
Environmental effect (E) for trait X and environmental effect for trait Y	$\text{cov}(E_X,E_Y)$	r_{E_X,E_Y}	$b_{E_Y \cdot E_X}$
Phenotypic value (P) and breeding value (BV) for the same trait	$\text{cov}(P,BV)$	$r_{P,BV}$	$b_{BV \cdot P}$
Phenotypic value (P) and producing ability (PA) for the same trait	$\text{cov}(P,PA)$	$r_{P,PA}$	$b_{PA \cdot P}$
Breeding value (BV) and estimated breeding value (EBV) for the same trait	$\text{cov}(BV,\widehat{BV})$	$r_{BV,\widehat{BV}}$	$b_{\widehat{BV} \cdot BV}$

weights, and less than average weaning weights tend to be associated with less than average fleece weights. Breeding values for scrotal circumference and age at puberty in cattle, on the other hand, exhibit negative covariation. Greater than average breeding values for scrotal circumference tend to be associated with younger than average (less than average) breeding values for age at puberty, and less than average breeding values for scrotal circumference tend to be associated with older than average (greater than average) breeding values for age at puberty.

Calculating Covariances

A true measure of covariance (the population parameter) can be defined with the following *word formula:*

Covariance = the average product of deviations from the means of two variables

Mathematically, the covariance of variables X and Y is then

$$\text{cov}(X,Y) = \frac{\sum_{i=1}^{n}(X_i - \mu_X)(Y_i - \mu_Y)}{n}$$

where each X_i,Y_i pair represents two attributes of some entity. In animal breeding, that entity is usually an individual animal. Each X_i,Y_i pair might represent an individual's phenotypic values for two traits, an individual's breeding values for two traits, an individual's breeding value and phenotypic value for the same trait, etc. The denominator (n) is the number of X_i,Y_i pairs. In expanded form, the formula appears as:

$$\text{cov}(X,Y) = \frac{(X_1 - \mu_X)(Y_1 - \mu_Y) + (X_2 - \mu_X)(Y_2 - \mu_Y) + \cdots + (X_n - \mu_X)(Y_n - \mu_Y)}{n}$$

The formulas for an *estimate* of a covariance—the formulas you should use to estimate a covariance from a sample of data—are slightly different:

$$\widehat{\text{cov}}(X,Y) = \frac{\sum_{i=1}^{n}(X_i - \hat{\mu}_X)(Y_i - \hat{\mu}_Y)}{n - 1}$$

or, in expanded form,

$$\text{cov}(X,Y) = \frac{(X_1 - \hat{\mu}_X)(Y_1 - \hat{\mu}_Y) + (X_2 - \hat{\mu}_Y)(Y_2 - \hat{\mu}_Y) + \cdots + (X_n - \hat{\mu}_X)(Y_n - \hat{\mu}_Y)}{n - 1}$$

Note that the order of variables in these formulas is unimportant. We could reverse the order of deviations in computing each product of deviations, and the result would be the same. Mathematically,

$$(X_i - \hat{\mu}_X)(Y_i - \hat{\mu}_Y) = (Y_i - \hat{\mu}_Y)(X_i - \hat{\mu}_X)$$

and therefore

$$\hat{\text{cov}}(X,Y) = \hat{\text{cov}}(Y,X)$$

Example

Listed in the following table are weaning weights for the same 12 calves from the previous example plus their birth weights.

Calf#	Weaning Weight, lb	Birth Weight, lb
1	515	62
2	430	74
3	475	72
4	565	98
5	630	88
6	510	80
7	495	78
8	480	72
9	555	75
10	505	86
11	470	86
12	445	78

From earlier calculations, we know that

$$\hat{\mu}_{P_{WW}} = 506.25 \text{ lb}$$
$$\hat{\sigma}_{P_{WW}} = 55.5 \text{ lb}$$

Similar calculations for birth weight result in:

$$\hat{\mu}_{P_{BW}} = 79.08 \text{ lb}$$
$$\hat{\sigma}_{P_{BW}} = 9.39 \text{ lb}$$

The next step is to compute deviations from weaning weight and birth weight means and associated *cross products* (products of deviations). These are listed in the next table.

Calf#	WW Deviation, lb	BW deviation, lb	Cross Product, lb^2
1	+8.75	−17.08	−149.45
2	−76.25	−5.08	+387.35
3	−31.25	−7.08	+221.25
4	+58.75	+18.92	+1,111.55
5	+123.75	+8.92	+1,103.85
6	+3.75	+.92	+3.45
7	−11.25	−1.08	+12.15
8	−26.25	−7.08	+185.85
9	+48.75	−4.08	−198.90
10	−1.25	+6.92	−8.65
11	−36.25	+6.92	−250.85
12	−61.25	−1.08	+66.15

Most of the cross products in this example are positive (i.e., most calves are either lighter than average for both birth and weaning weight or heavier than average for both traits). This suggests positive covariation, and the calculated covariance confirms it.

$$\begin{aligned}\hat{\text{cov}}(P_{WW},P_{BW}) &= \frac{\sum_{i=1}^{n}(P_{WW_i} - \hat{\mu}_{P_{WW}})(P_{BW_i} - \hat{\mu}_{P_{BW}})}{n-1} \\ &= \frac{-149.45 + 387.35 + \cdots + 66.15}{12-1} \\ &= \frac{2{,}483.75}{11} \\ &= 225.8 \text{ lb}^2\end{aligned}$$

The unit of measure for this particular covariance is lb^2. That is because each deviation is measured in lb, and when deviations are multiplied together, the unit of measure for the product is lb · lb or lb^2.

Covariances and the Genetic Model

Consider the genetic model for a single individual:

$$P = \mu + BV + GCV + E$$

In a population, each element of this model varies (except μ, which is a constant) and therefore has a variance associated with it. We have σ_P^2, σ_{BV}^2, σ_{GCV}^2, and σ_E^2. How are these variances related? Is the phenotypic variance just the sum of the variances of *BV*, *GCV*, and *E*?

To answer these questions, let's determine σ_P^2. There is a set of rules for dealing with variances and covariances, and the rule we need here is this: The variance of a sum of random variables is the sum of the variances of each random variable plus two times the covariance of each pair of random variables.[5] Applying this rule to the genetic model, we have:

$$\begin{aligned}\text{var}(P) &= \text{var}(BV + GCV + E) \\ &= \sigma_{BV}^2 + \sigma_{GCV}^2 + \sigma_E^2 + 2\text{cov}(BV,GCV) + 2\text{cov}(BV,E) + 2\text{cov}(GCV,E)\end{aligned}$$

Recall, however, that the variables *BV*, *GCV*, and *E* are defined to be *independent* (i.e., covariances between them are zero). Then

$$\sigma_P^2 = \sigma_{BV}^2 + \sigma_{GCV}^2 + \sigma_E^2$$

Similar relationships occur whenever the variables being summed are independent of each other. Some examples:

$$\sigma_P^2 = \sigma_G^2 + \sigma_E^2$$

$$\sigma_G^2 = \sigma_{BV}^2 + \sigma_{GCV}^2$$

$$\sigma_{PA}^2 = \sigma_{BV}^2 + \sigma_{GCV}^2 + \sigma_{E_p}^2$$

[5]See the Appendix for a complete set of rules for the algebra of variances and covariances.

Correlation

correlation or **correlation coefficient** ($r_{X,Y}$) A measure of the strength (consistency, reliability) of the relationship between two variables.

The **correlation** or **correlation coefficient** is a measure of the *strength* (consistency, reliability) of the relationship between two variables. The notation for the correlation between variables X and Y is $r_{X,Y}$. Like covariances, correlations are used to describe the relationship between two traits in a population or between two values for the same trait in a population. And like covariances, correlations are population measures—not individual values. For example, we speak of the correlation between daily weight gain and feed conversion in swine, the correlation between breeding values for scrotal circumference and age at puberty in cattle, and the correlation between observed temperament and breeding value for temperament in dogs. Several examples of specific kinds of correlations and appropriate notation are listed in Table 8.4.

If a covariance is analogous to a variance, then a correlation is analogous to a standard deviation. Variances and covariances have uninformative numerical values and units that make them difficult to interpret. Standard deviations and correlations, on the other hand, have sensible numerical values and units that make them easy to interpret. Numerical values for a correlation coefficient range from -1 to $+1$. A correlation near -1 indicates very *strong, negative* covariation, a correlation near $+1$ indicates very *strong, positive* covariation, and a zero correlation indicates *no covariation* at all. The populations sampled in Figure 8.8 illustrate different correlations. The values for population (a) were generated using a correlation between X and Y of $+.8$. The correlation used for population (b) was $-.8$. Values for population (c) were generated using a zero correlation.

A correlation is essentially a "standardized" covariance. This means that a correlation is easily interpretable regardless of the variability or units of the traits involved. A correlation of $+.8$ always suggests strong, positive covariation (strong in an animal breeding context, anyway), and a correlation of $-.1$ always suggests weak, negative covariation—regardless of the traits or values involved. Correlations are, in fact, unitless. You need not worry about trait units when interpreting a correlation.

Another useful property of correlations—a property true for covariances as well—is that the order of variables is unimportant. In other words, the correlation between X and Y is the same as the correlation between Y and X, or

$$r_{XY} = r_{YX}$$

To get an idea of just how easy correlations are to work with, consider the correlation between weaning weight and grease fleece weight in lambs and the correlation between breeding values for scrotal circumference and age at puberty in cattle. Recall that covariances were 2.6 lb^2 in the lamb example and -17 cm·days in the cattle example—not very telling numbers. The corresponding correlations turn out to be $+.3$ and $-.9$. There is a moderately consistent, positive relationship between phenotypic values for weaning weight and fleece weight in lambs, and a very reliable, negative relationship between breeding values for scrotal circumference and age at puberty in cattle. You can see from these examples that correlations provide an easily interpretable measure of both the sign and the strength of the relationship between two variables.

Column charts like the ones in Figure 8.8 represent one way of illustrating correlations. A more common way to visually show correlations is to use a *scatter*

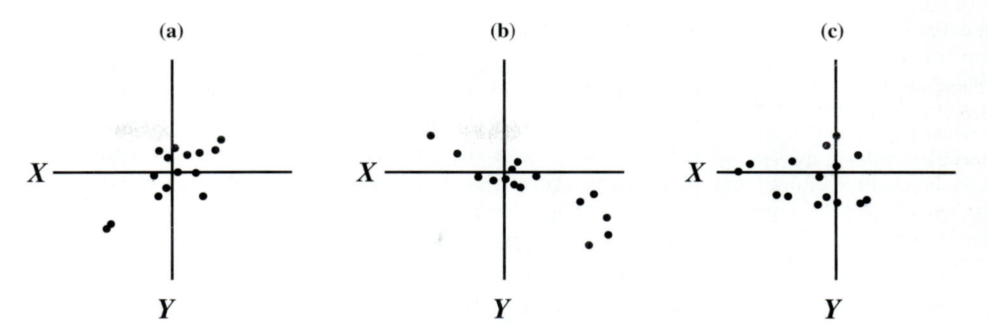

FIGURE 8.9 Scatter plots of the samples of three populations from Figure 8.8. Correlations ($r_{X,Y}$) used to generate values were (*a*) +.8, (*b*) −.8, and (*c*) zero.

plot. In an *XY* scatter plot, each pair of values for *X* and *Y* is represented as a coordinate on a two-dimensional plane. Scatter plots of the values used in Figure 8.8 are shown in Figure 8.9. Note that the points plotted for the sample from population (a) show an upward trend from left to right. This indicates a positive correlation between *X* and *Y*. The points plotted for the sample from population (b) show a downward trend from left to right, indicating a negative correlation between *X* and *Y*. It is hard to see any trend in the points plotted for the sample from population (c). The correlation between *X* and *Y* in this population appears to be near zero.

In a scatter plot like those in Figure 8.9, the *sign* of the correlation is determined by whether the trend in the points is upward or downward (from left to right). The *magnitude* of the correlation is determined by the degree to which the points are scattered. If X and Y are highly correlated, i.e., if the relationship between X and Y is strong (consistent, reliable), the points will tend to cluster tightly about an imaginary line. Clustering of this kind is marginally evident in plot (a) and clearly evident in plot (b). If the correlation between X and Y is low (i.e., if their relationship is weak), any kind of clustering about a line will be difficult to see. Such is the case in plot (c). It is important to remember that the *slope* of any trend in the points—the slope of the imaginary line—has nothing to do with the magnitude of the correlation. The size of the correlation and therefore the strength of the relationship are determined by how tightly the points are clustered about a line.[6]

phenotypic correlation (r_{P_X,P_Y}) A measure of the strength (consistency, reliability) of the relationship between performance in one trait and performance in another trait.

Some of the most common and useful correlations in animal breeding are correlations between traits. **Phenotypic correlations (r_{P_X,P_Y})** measure the strength of the relationship between performance (phenotypic value) in one trait and performance in another trait. Phenotypic correlations are helpful because they give us a sense of the observable relationship between traits. For example, the phenotypic correlation between weaning weight and grease fleece weight in lambs (.3) suggests that heavier lambs at weaning will tend to have heavier fleece weights, but there will likely be a number of exceptions—a number of light lambs with heavy fleece weights and vice versa.

[6]Technically, the correlation between two variables should be defined as a measure of the strength of their *linear* relationship because it measures how closely points on a graph cluster about a *line.* Conceivably, points could cluster tightly about a *curve,* producing a strong *curvilinear* relationship. That relationship, however, would not be reflected by a correlation coefficient.

genetic correlation (r_{BV_X,BV_Y})
A measure of the strength (consistency, reliability) of the relationship between breeding values for one trait and breeding values for another trait.

Perhaps even more important than phenotypic correlations are **genetic correlations (r_{BV_X,BV_Y})**. They measure the strength of the relationship between breeding values for one trait and breeding values for another trait. The reason why genetic correlations are so important is that if two traits are genetically correlated, selection for one will cause genetic change in the other. Furthermore, performance in one trait can be used to help predict breeding value in a genetically correlated trait. Consider, for example, scrotal circumference and age at puberty in cattle. The genetic correlation between these traits is very strong (−.9). If breeders select for larger scrotal circumference in bulls, they can expect younger age at puberty in both male and female offspring. And a bull's performance for scrotal circumference—an inexpensive, easy trait to measure—can also be used to predict his breeding value for age at puberty, a much more difficult trait to measure.

environmental correlation (r_{E_X,E_Y})
A measure of the strength (consistency, reliability) of the relationship between environmental effects on one trait and environmental effects on another trait.

An **environmental correlation** is a measure of the strength of the relationship between environmental effects on one trait and environmental effects on another trait. Environmental correlations are often useful for management purposes. For example, the environmental correlation between average daily gain and backfat thickness in swine has been estimated at +.4. This suggests that environments conducive to rapid weight gains tend to produce fatter pigs. If you were feeding pigs and wanted a lean product, you might, therefore, consider feeding your animals to grow at less than the maximum rate.[7]

Other useful correlations for animal breeding are correlations between different values for the same trait. Two of the more important correlations of this kind are the correlation between phenotypic values and breeding values ($r_{P,BV}$) and the correlation between phenotypic values and producing abilities ($r_{P,PA}$). An example of $r_{P,BV}$ is the correlation between observed temperament and breeding value for temperament in dogs. An estimate of this correlation in some dog populations is .45. This figure indicates that temperament in these populations is moderately heritable. An example of $r_{P,PA}$ is the correlation between a single lactation record and producing ability for milk production in dairy cattle. A typical estimate of this correlation is .7. This indicates that milk production is moderately to highly repeatable.[8]

Another useful correlation is the correlation between a value and a prediction of that value. Such a correlation is termed *accuracy of prediction* or simply *accuracy*, and it is easy to see why. When underlying values—breeding values, for example—are highly correlated with their predictions, we can assume that the predictions are accurate. If the correlation between values and their predictions is weak, we cannot put much faith in the predictions. Accuracy figures are commonly reported for predictions of breeding values, progeny differences, and producing abilities. These correlations suggest how much confidence we should place in any particular prediction.[9]

[7]Phenotypic, genetic, and environmental correlations are explained in much more detail in Chapter 13.

[8]The connection between $r_{P,BV}$ and $r_{P,PA}$ and the concepts of heritability and repeatability is probably not clear at this point. See Chapter 9 for a complete explanation. I introduce these correlations here simply to show that correlations between traits are not the only correlations of interest to animal breeders.

[9]See Chapter 11 for more on the concept of accuracy.

Calculating Correlation Coefficients

A correlation between two variables is a simple function of the covariance of the variables and their standard deviations. For variables X and Y,

$$r_{X,Y} = \frac{\text{cov}(X,Y)}{\sigma_X \sigma_Y}$$

Because the standard deviations in the denominator of the formula are necessarily positive, the sign of the correlation is determined by the sign of the numerator—the covariance. If the covariance is positive, the correlation is positive. If the covariance is negative, the correlation, is negative.

A correlation is nothing more than a "standardized" covariance—a covariance with the units removed. To show this, let's try standardizing a covariance. The formula for a covariance is:

$$\text{cov}(X,Y) = \frac{\sum_{i=1}^{n}(X_i - \mu_X)(Y_i - \mu_Y)}{n}$$

If we standardize each deviation (i.e., divide it by the appropriate standard deviation), the standardized covariance ($\text{cov}(X,Y)^*$) is then

$$\text{cov}(X,Y)^* = \frac{\sum_{i=1}^{n}\left(\frac{X_i - \mu_X}{\sigma_X}\right)\left(\frac{Y_i - \mu_Y}{\sigma_Y}\right)}{n}$$

By doing this, we divide deviations that are expressed in trait units by standard deviations that are expressed in the same units, and the units cancel. Because σ_X and σ_Y are constants, we can factor them out of the summation. Thus

$$\begin{aligned}\text{cov}(X,Y)^* &= \left(\frac{1}{\sigma_X}\right)\left(\frac{1}{\sigma_Y}\right)\frac{\sum_{i=1}^{n}(X_i - \mu_X)(Y_i - \mu_Y)}{n} \\ &= \frac{\text{cov}(X,Y)}{\sigma_X \sigma_Y} \\ &= r_{X,Y}\end{aligned}$$

The order of variables in a correlation is unimportant. Because

$$\text{cov}(X,Y) = \text{cov}(Y,X)$$

and

$$\sigma_X \sigma_Y = \sigma_Y \sigma_X$$

then

$$r_{X,Y} = r_{Y,X}$$

Example Calculation

Let's estimate the phenotypic correlation between birth weight and weaning weight in beef cattle using the data from the last example. We already know the following:

$$\hat{\sigma}_{P_{BW}} = 9.39 \text{ lb}$$

$$\hat{\sigma}_{P_{WW}} = 55.5 \text{ lb}$$

$$\hat{\text{cov}}(P_{BW}, P_{WW}) = 225.8 \text{ lb}^2$$

Therefore

$$\hat{r}_{P_{BW}, P_{WW}} = \frac{\hat{\text{cov}}(P_{BW}, P_{WW})}{\hat{\sigma}_{P_{BW}} \hat{\sigma}_{P_{WW}}} = \frac{225.8}{9.39(55.5)} = .43$$

There appears to be a moderate to strong, positive correlation between birth weight and weaning weight (if this small data set is any indicator).

Regression

regression or **regression coefficient** ($b_{Y \cdot X}$) The expected or average change in one variable (Y) per unit change in another (X).

The third aspect of covariation, the amount of change in one variable that can be expected for a given amount of change in another variable, is measured by a **regression** or **regression coefficient** ($b_{Y \cdot X}$). Like covariances and correlations, regressions are population measures—not individual values—and can involve all kinds of values in a population. Example notation for a number of different types of regressions is shown in Table 8.4.

Following are two practical examples of regressions. An estimate of the regression of grease fleece weight on 60-day weaning weight in lambs is .04 lb per lb. In other words, for every 1-lb increase in weaning weight, we can expect a .04-lb increase in fleece weight. Heavier lambs at weaning should, on average, have .04 lb more fleece for each additional lb of weaning weight. An estimate of the regression of breeding value for age at puberty on breeding value for scrotal circumference in cattle is −8.5 days per centimeter. Bulls with breeding values for larger scrotal circumference should have breeding values for age at puberty that are, on average, 8.5 days younger for each additional centimeter of breeding value for scrotal circumference. If you select for increased scrotal circumference, you can expect age at puberty to decrease 8.5 days for each centimenter of progress you make in scrotal circumference.

Regressions are used to help predict a value based on some other piece of information. For example, we could use the regression of fleece weight on weaning weight to help predict a lamb's fleece weight based on its weaning weight.

Alternatively, we could use the regression of breeding value for a trait on phenotypic value for the same trait to help predict an animal's breeding value based on its own performance. More on prediction in the next section.

Regressions appear graphically as the *slope* of the imaginary line about which points cluster in a scatter plot of two correlated variables. Figure 8.10 is identical to Figure 8.9 except that *regression lines* are superimposed. Note the positive slope of the line for population (a), the negative slope for population (b), and the zero slope for population (c). In population (a) we expect variable Y to increase .533 units for each unit increase in variable X, in population (b) we expect Y to *decrease* .533 units for each unit increase in X, and in population (c) we expect no change in Y as X changes.

Regressions, like covariances and correlations, reveal the *sign* or *direction* of covariation. Positive regressions—like the regression of fleece weight on weaning weight—indicate an increase in one variable as the other variable increases. Negative regressions—like the regression of breeding value for age at puberty on breeding value for scrotal circumference—indicate a decrease in one variable as the other variable increases.

Unlike covariances and correlations, the *order* of variables is important in regressions. Mathematically,

$$b_{Y \cdot X} \neq b_{X \cdot Y}$$

This is only sensible. While we might expect fleece weight to increase .04 lb per lb increase in weaning weight, we certainly would not expect weaning weight to increase .04 lb per lb increase in fleece weight. (And we *absolutely* would not expect breeding value for scrotal circumference to decrease 8.5 cm per day increase in breeding value for age at puberty—ouch!) We speak of the regression of Y *on* X, meaning the expected change in Y per unit change in X. That is very different from the expected change in X per unit change in Y.

Covariances and correlations tell us something about covariation, but they give no hint as to why two variables covary. In other words, they say nothing about cause and effect. Regressions, on the other hand, often imply cause. It seems intuitively sensible to refer to the regression of fleece weight on weaning weight

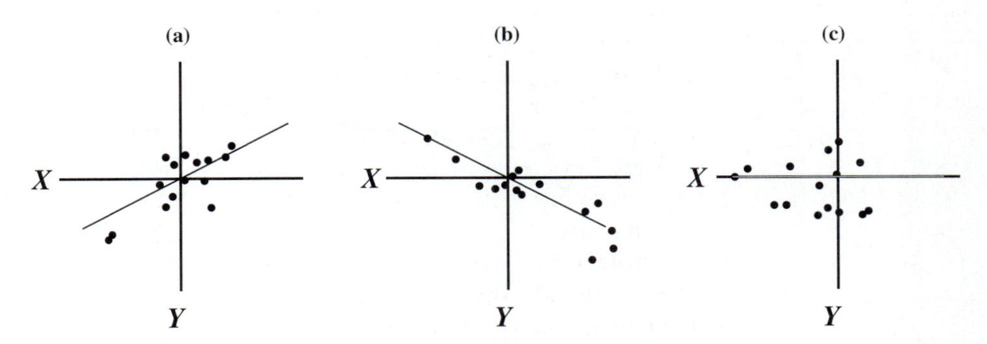

FIGURE 8.10 Scatter plots of the samples of three populations from Figure 8.8 with regression lines superimposed. Regressions ($b_{Y \cdot X}$) implied in the generation of values were (*a*) +.533, (*b*) −.533, and (*c*) zero.

because we expect fleece weight to change *as a result* of changes in weaning weight. Heavier lambs presumably have larger surface areas and consequently heavier fleece weights. It would make less sense to talk about regression of weaning weight on fleece weight since changes in fleece weight are unlikely to cause much change in weaning weight. Regressions do not necessarily imply cause, however. For example, the regression of breeding value for temperament on observed temperament is a useful population measure for predicting a dog's breeding value for temperament. Observed temperament (phenotypic value) does not *cause* breeding value for temperament, however. On the contrary, breeding value has a causal effect on phenotypic value.

Regressions are really just rates of change, and the units of a regression reflect that fact. The speed of a car, for example, is the change in distance per unit of time and is measured in miles per hour. In our lamb example, we refer to the change in fleece weight per unit change in weaning weight and measure that rate in pounds per pound. Likewise, a regression of age at puberty on scrotal circumference would be measured in days per centimeter.

Correlations and regressions are mathematically related, but the relationship between the two is not what you might think. In particular, a strong correlation does *not necessarily* imply a large regression coefficient. The scatter plot in Figure 8.11 is an example of a very strong correlation ($r_{X,Y}$ = .95) and a small regression ($b_{X \cdot Y}$ = .1). The points on the graph cluster very tightly about the regression line, but the slope of the line is quite flat. In other words, as X increases, Y increases in a very consistent, very reliable way, but the amount of change in Y per unit change in X is small. Without any additional information, the most we can say about the relationship between the correlation between two variables and the regression of one variable on the other is that they both have the same sign—if one is positive, the other will be positive; if one is negative, the other will be negative; and if one is zero, both will be zero.

FIGURE 8.11
Example of a very strong correlation ($r_{X,Y}$ = .95) and small regression ($b_{Y \cdot X}$ = .1). The points on the graph cluster very tightly about the regression line, but the slope of the line is quite flat.

Calculating Regression Coefficients

The regression of variable Y on variable X is a simple function of the covariance of the variables and the variance of X.

$$b_{Y \cdot X} = \frac{\text{cov}(X,Y)}{\sigma_X^2}$$

Because the variance in the denominator of the formula is necessarily positive, the sign of the regression, like the sign of the correlation, is determined by the sign of the numerator—the covariance. If the covariance is positive, the regression is positive. If the covariance is negative, the regression is negative.

Unlike a correlation, the order of variables in a regression *is* important. The regression of Y on X is *not* the same as the regression of X on Y. For proof, consider that

$$b_{Y \cdot X} = \frac{\text{cov}(X,Y)}{\sigma_X^2} \quad \text{and} \quad b_{X \cdot Y} = \frac{\text{cov}(X,Y)}{\sigma_Y^2}$$

but, unless $\sigma_X^2 = \sigma_Y^2$,

$$\frac{\text{cov}(X,Y)}{\sigma_X^2} \neq \frac{\text{cov}(X,Y)}{\sigma_Y^2}$$

so

$$b_{Y \cdot X} \neq b_{X \cdot Y}$$

Nor is one regression the reciprocal of the other.[10] That is,

$$b_{X \cdot Y} \neq \frac{1}{b_{Y \cdot X}}$$

To see where the units of measure of regression coefficients come from, consider the regression of age at puberty (AAP) on scrotal circumference (SC).

$$b_{AAP \cdot SC} = \frac{\text{cov}(AAP,SC)}{\sigma_{SC}^2}$$

The units in this formula are

$$\frac{\text{cm} \cdot \text{days}}{\text{cm}^2} \quad \text{or} \quad \frac{\text{days}}{\text{cm}}$$

[10]There is one exception to this statement. If X and Y are perfectly correlated, i.e., if $r_{X,Y} = 1$ or $r_{X,Y} = -1$, then $b_{X \cdot Y} = \dfrac{1}{b_{Y \cdot X}}$.

Let's estimate the phenotypic regression of weaning weight on birth weight from our sample of data. We already know the following:

$$\hat{\sigma}_{P_{BW}} = 9.39 \text{ lb}$$

$$\hat{\text{cov}}(P_{BW}, P_{WW}) = 225.8 \text{ lb}^2$$

Therefore

$$\hat{b}_{P_{WW} \cdot P_{BW}} = \frac{\hat{\text{cov}}(P_{WW}, P_{BW})}{\hat{\sigma}^2_{P_{BW}}} = \frac{225.8}{(9.39)^2} = 2.55 \text{ lb per lb}$$

For every 1-lb increase in birth weight, weaning weight increases an average of 2.55 lb.

Mathematical Relationship between Correlation and Regression

You can convert a correlation to a regression and vice versa using the following formulas:

$$b_{Y \cdot X} = r_{X,Y}\left(\frac{\sigma_Y}{\sigma_X}\right)$$

and

$$r_{X,Y} = b_{Y \cdot X}\left(\frac{\sigma_X}{\sigma_Y}\right)$$

Proof

$$b_{Y \cdot X} = \frac{\text{cov}(X,Y)}{\sigma^2_X} = \frac{\text{cov}(X,Y)}{\sigma^2_X}\left(\frac{\sigma_Y}{\sigma_Y}\right) = \frac{\text{cov}(X,Y)}{\sigma_X \sigma_Y}\left(\frac{\sigma_Y}{\sigma_X}\right) = r_{X,Y}\left(\frac{\sigma_Y}{\sigma_X}\right)$$

and, rearranging:

$$r_{X,Y} = b_{Y \cdot X}\left(\frac{\sigma_X}{\sigma_Y}\right)$$

Example

For birth weight and weaning weight, if

$$\sigma_{P_{BW}} = 9.39 \text{ lb}$$

$$\sigma_{P_{WW}} = 55.5 \text{ lb}$$

$$r_{P_{BW},P_{WW}} = .43$$

$$b_{P_{WW} \cdot P_{BW}} = 2.55 \text{ lb per lb}$$

then to convert the correlation to a regression,

$$\begin{aligned} b_{P_{WW} \cdot P_{BW}} &= r_{P_{BW},P_{WW}}\left(\frac{\sigma_{P_{WW}}}{\sigma_{P_{BW}}}\right) \\ &= .43\left(\frac{55.5}{9.39}\right) \\ &= 2.55 \text{ lb per lb} \end{aligned}$$

To convert the regression to a correlation,

$$\begin{aligned} r_{P_{BW},P_{WW}} &= b_{P_{BW} \cdot P_{WW}}\left(\frac{\sigma_{P_{BW}}}{\sigma_{P_{WW}}}\right) \\ &= 2.55\left(\frac{9.39}{55.5}\right) \\ &= .43 \end{aligned}$$

PREDICTION

True and Predicted Values

Consider the following genetic model for a quantitative trait:

$$P = \mu + BV + GCV + E$$

true value
An unknown, underlying attribute that affects animal performance. Examples include breeding value (BV), progeny difference (PD), gene combination value (GCV), producing ability (PA), environmental effect (E), etc.

The values in the model are called **true values,** underlying attributes that affect an animal's performance for a trait. Examples of true values include not only those listed in the genetic model above, but also progeny difference (PD), producing ability (PA), and permanent (E_p) and temporary (E_t) environmental effects. The difficulty with true values is that, with the exception of phenotypic value, they are not directly measurable and therefore never known precisely. We can never know an individual's true breeding value, for example. That is unfortunate because knowledge of true breeding values would make selection much easier.

predicted value
A prediction of a true value. The most common predicted values are estimated breeding value (EBV or $\hat{BV}$), expected progeny difference (EPD or $\hat{PD}$), and most probable producing ability (MPPA or $\hat{PA}$).

In the absence of known true values, we must deal instead with predictions of true values, i.e., **predicted values.** Predicted values are calculated from performance data using statistical techniques. The most common predicted values are estimated breeding value (EBV or $\hat{BV}$), expected progeny difference (EPD or $\hat{PD}$), and most probable producing ability (MPPA or $\hat{PA}$). (Note the use of a "hat" (ˆ) over a true value to indicate a prediction of that value.)

It is important to understand the distinction between true and predicted values. If you were to speak of an animal's EPD, for example, when what you really meant was his true progeny difference, you might easily confuse your audience. True progeny differences and EPDs can be quite different depending on the amount and quality of information used to calculate the EPD.

Prediction Equations

prediction equation
A mathematical equation used to calculate a predicted value based (usually) on phenotypic data.

Predicted values are calculated using either a single **prediction equation** or a set of prediction equations. In its most simple form, a prediction equation looks like the following:

$$\text{Predicted value} = \text{regression coefficient} \times \text{"evidence"}$$

The predicted value can be a prediction of any true value of interest. We typically predict breeding values, progeny differences, and producing abilities, but other values—for example, an as yet unmeasured phenotypic value—can be predicted as well. The "evidence" in this context usually refers to a phenotypic measure of some kind. It could be a single performance record for the individual whose true value we are trying to predict, or it could be an average of the performance records of his progeny, half sibs, etc. The regression coefficient in the equation is then the regression of true value on the evidence. In other words, it measures the expected change in true value per unit change in the evidence.

Suppose, for example, that you wanted to predict the future grease fleece weight of a lamb based upon its weaning weight. In this case, the true value to be predicted is phenotypic value for fleece weight, the evidence is the lamb's phenotypic value for weaning weight, and the appropriate regression is the regression of phenotypic value for fleece weight on phenotypic value for weaning weight (estimated to be .04 lb per lb). If the lamb in question weighed 10 lb more than average at weaning, then we would expect its grease fleece weight to be .04 lb heavier for each 1-lb advantage in weaning weight or $.04 \times 10 = .4$ lb heavier than average. If the lamb weighed 5 lb less than average at weaning, we would expect its fleece weight to be $.04 \times (-5) = -.2$ lb different from average or .2 lb *lighter* than average.

Now for a more interesting example, suppose you raised a boar whose loin eye area (a measure of lean meat yield) was ultrasonically measured at .6 square inches (in^2) above average, and you want to predict his breeding value for loin eye area. In this case, the true value to be predicted is breeding value for loin eye area, the evidence is the boar's phenotypic value for loin eye area, and the appropriate regression coefficient is the regression of breeding value for loin eye area on phenotypic value for the trait. Published estimates of this regression average near .5 in^2 per in^2. The boar's EBV for loin eye area would then be $.5 \times .6 = .3$ in^2 above average or—since the average breeding value is defined to be zero—simply $+.3$ in^2.

It is important to understand that predictions are—well—just predictions. Some predictions are very accurate and some are wildly inaccurate. Neither of the predictions in the above examples is likely to be very accurate. That is because the evidence in both cases is simply too meager. A single performance record for weaning weight or loin eye area is not a lot of information. Accurate predictions come from large amounts of high quality data and, for this reason, the collection of data and the statistical techniques used in making genetic predictions from those data are important parts of animal breeding.[11]

The Prediction Equation in Mathematical Form

Mathematically, a simple prediction equation appears as:

$$\hat{Y}_i = \hat{\mu}_Y + b_{Y \cdot X}(X_i - \hat{\mu}_X)$$

where
$\hat{Y}_i$ = a predicted value for animal *i*.
$\hat{\mu}_Y$ = the expected mean of predictions for animals in the population
$b_{Y \cdot X}$ = the regression of values being predicted (Y) on the evidence (X)
$X_i - \hat{\mu}_X$ = the evidence for animal *i*, expressed as a deviation from the populaton mean

Example

Let's predict the weaning weight of a young calf based on its birth weight. We need the phenotypic regression of weaning weight on birth weight because we are predicting a phenotypic value—weaning weight—using phenotypic evidence, the calf's birth weight. Assume the following:

$\hat{\mu}_{P_{WW}}$, our best guess at what the average weaning weight in the population will be, = 500 lb

$b_{P_{WW} \cdot P_{BW}}$, the phenotypic regression of weaning weight on birth weight (calculated earlier), = 2.55 lb per lb

P_{BW_i}, the calf's birth weight, = 85 lb

$\hat{\mu}_{P_{BW}}$, the average birth weight in the population, = 79 lb

Then

$$\hat{P}_{WW_i} = \mu_{P_{WW}} + b_{P_{WW} \cdot P_{BW}}(P_{BW_i} - \hat{\mu}_{P_{BW}})$$
$$= 500 + 2.55(85 - 79)$$
$$= 515 \text{ lb}$$

We predict a 515-lb weaning weight for this calf.

[11]See Chapter 11 for a much more detailed discussion of genetic prediction.

A Summary Example

This has been a long chapter laden with statistical concepts that can often be confusing. Table 8.5 serves as a chapter summary. It lists the population measures discussed in this chapter along with definitions, uses in animal breeding, and numerical examples taken from an actual beef cattle population. The traits used in the examples are birth weight (BW) and yearling weight (YW).

TABLE 8.5 A Summary of Population Measures, Their Definitions and Uses, Plus Examples Involving Birth Weight (BW) and Yearling Weight (YW) in Beef Cattle

Parameter	Definition, Uses, and Examples
Mean (μ)	*Definition* An arithmetic average. *Uses* Means are used wherever the notion of an average is informative (e.g., the average performance for a trait in a population, the average performance of a sire's progeny, or the average EPD of all animals born in a given year). Means mark the center of the distribution for normally distributed variables. Values (except phenotypic values) and their predictions are expressed as deviations from a mean. Means are required for the calculation of variances and covariances. *Examples* $\mu_{BW} = 74$ lb $\mu_{YW} = 772$ lb In this particular population of beef cattle, the mean birth weight is 74 lb and the mean yearling weight is 772 lb.
Variance (σ^2)	*Definition* A mathematical measure of variation. A measure (in most animal breeding applications) of differences among individuals within a population. *Uses* Variances are useful for comparing the variability of different traits and for comparing the variability of different values (P, BV, etc.) for a single trait. They are necessary for the calculation of such genetic parameters as heritability and repeatability and for the calculation of correlations and regressions of any kind. The typically uninformative numerical values of variances and the fact that they are expressed in squared units makes them difficult to deal with conceptually. *Examples* $\sigma^2_{P_{BW}} = 90$ lb^2 $\sigma^2_{P_{YW}} = 3{,}249$ lb^2 $\sigma^2_{BV_{BW}} = 31.5$ lb^2 $\sigma^2_{BV_{YW}} = 2{,}144$ lb^2 $\sigma^2_{E_{BW}} = 58.5$ lb^2 $\sigma^2_{E_{YW}} = 1{,}105$ lb^2 The variances of phenotypic values for birth weight and yearling weight in this population are 90 and 3,249 lb^2 (pounds squared), respectively. Clearly, yearling weight is the more variable trait. In this population, variation in breeding values for yearling weight is largely responsible for variation in observed yearling weights (2,144 is 66% of 3,249). Heritable factors are less important for birth weight (31.5 is 35% of 90).

Parameter	Definition, Uses, and Examples
Standard deviation (σ)	*Definition* A mathematical measure of variation that can be thought of as an average deviation from the mean. The square root of the variance. *Uses* Standard deviations are used for much the same purpose as variances. Standard deviations are generally easier to conceptualize than variances because they are measured in trait units and typically have meaningful numerical values that are more easily remembered. They provide practical information about the distribution of variables. On the graph of a normal distribution, the standard deviation appears as the distance between the mean and the point of inflection of the normal curve. Sixty-eight percent of observations occur within one standard deviation either side of the mean, and the entire range of observations is often no greater than five or six standard deviations. *Examples* $\sigma_{P_{BW}} = 9.5$ lb $\sigma_{P_{YW}} = 57$ lb $\sigma_{BV_{BW}} = 5.6$ lb $\sigma_{BV_{YW}} = 46$ lb $\sigma_{E_{BW}} = 7.6$ lb $\sigma_{E_{YW}} = 33$ lb Sixty-eight percent of birth weights in this population are between 64 and 84 lb ($\mu - \sigma = 74 - 9.5 \approx 64$, $\mu + \sigma = 74 + 9.5 \approx 84$) and assuming a 5–standard deviation range, birth weights run from approximately 50 lb to 98 lb. Breeding values for birth weight are less variable. Sixty-eight percent of them are between −6 and +6 lb with a range from −14 lb to +14 lb. Likewise, 68% of yearling weights are between 715 lb and 829 lb with a range from 629 lb to 915 lb. Sixty-eight percent of breeding values for yearling weight are between −46 lb and +46 lb with a range from −115 lb to +115 lb.
Covariance ($\mathrm{cov}(X,Y)$)	*Definition* The basic measure of covariation. A measure of how two traits or values vary together in a population. *Uses* Covariances reveal the direction or mathematical sign of the relationship between two variables and, to a limited degree, the strength of the relationship as well. Like variances, their units and uninformative numerical values make them difficult to deal with conceptually. Covariances are needed for the calculation of correlations and regressions. *Examples* $\mathrm{cov}(P_{BW},P_{YW}) = 216$ lb^2 $\mathrm{cov}(BV_{BW},BV_{YW}) = 143$ lb^2 $\mathrm{cov}(E_{BW},E_{YW}) = 73$ lb^2 In this population, the phenotypic, genetic, and environmental relationships between birth weight and yearling weight are all mathematically positive. Animals with heavier birth weights tend to have heavier yearling weights, animals with higher breeding values for birth weight tend to have higher breeding values for yearling weight, and environmental factors that cause heavier birth weights seem to be associated with environmental factors causing heavier yearling weights.
Correlation ($r_{X,Y}$)	*Definition* A measure of the strength (consistency, reliability) of the relationship between two variables. *Uses* Correlations reveal both the sign and the strength of the relationship between two traits or two values. They are easy to interpret because they are unitless and range from

table continues

TABLE 8.5 continued

Parameter	Definition, Uses, and Examples
	−1 to +1. Common correlations in animal breeding are phenotypic, genetic, and environmental correlations between traits; correlations between breeding values and phenotypic values (indications of heritability); and accuracy figures—correlations between true values and their predictions. *Examples* $r_{P_{BW},P_{YW}} = .40$ $r_{BV_{BW},BV_{YW}} = .55$ $r_{E_{BW},E_{YW}} = .29$ $r_{P_{BW},BV_{BW}} = .59$ $r_{P_{YW},BV_{YW}} = .81$ In this population, there is a strong positive relationship between birth weight and yearling weight performance, and an even stronger positive relationship between breeding values for the two traits. While environmental factors that cause heavier birth weights are associated with environmental factors causing heavier yearling weights, the relationship between environmental effects is not particularly strong. Both birth weight and yearling weight appear to be quite heritable; phenotypic values and breeding values for these traits are moderately to highly correlated.
Regression ($b_{Y \cdot X}$)	*Definition* The expected or average change in one variable (*Y*) per unit change in another (*X*). *Uses* Regressions are used to predict one variable based upon another. In animal breeding, regressions are used most commonly to predict breeding values, progeny differences, and producing abilities based on performance information. *Examples* $b_{P_{YW} \cdot P_{BW}} = 2.4$ lb per lb $b_{P_{BW} \cdot P_{YW}} = .066$ lb per lb $b_{BV_{BW} \cdot P_{BW}} = .35$ lb per lb $b_{PD_{YW} \cdot P_{YW}} = .33$ lb per lb In this population, for every 1-lb increase in birth weight, there is an average increase in yearling weight of 2.4 lb. Conversely, for every 1-lb increase in yearling weight, birth weight increases (on average) .066 lb. We expect the breeding values of cattle for birth weight to increase .35 lb per lb increase in actual birth weight, and for every 1-lb increase in the yearling weight of individuals, we expect a .33-lb increase in the yearling weight of their progeny.

EXERCISES

Study Questions

8.1 Define in your own words:

value
population measure
population parameter
sample statistic
normal distribution
mean (μ)
variation
variance (σ^2)
standard deviation (σ)
variable
covariation
covariance ($\text{cov}(X,Y)$)
correlation or correlation coefficient ($r_{X,Y}$)
phenotypic correlation (r_{P_X,P_Y})
genetic correlation (r_{BV_X,BV_Y})
environmental correlation (r_{E_X,E_Y})
regression or regression coefficient ($b_{Y \cdot X}$)
true value
predicted value
prediction equation

8.2 Explain the difference between a value and a population measure. Provide examples.

8.3 Explain the difference between a population parameter and a sample statistic. Provide examples.

8.4 Why are most quantitative traits normally distributed?

8.5 For a trait of your choice:

a. Create a relatively *uniform* five-record data set. Calculate the sample mean, variance, and standard deviation of these data.

b. Create a relatively *nonuniform* five-record data set and calculate the sample mean, variance, and standard deviation.

c. Construct column charts for each data set. (Use Figure 8.5 as a guide. Illustrate phenotypic values only.)

d. Draw the normal distributions implied by our parameter estimates. Use the same horizontal scale for each. (Guess at the height of the distributions.)

8.6 Why is it important to have variation in a population and be able to measure it?

8.7 For a pair of traits of your choice:

a. Create a data set consisting of five pairs of records that show *positive* covariation. Calculate the covariance of the two traits.

b. Create a data set consisting of five pairs of records that show *negative* covariation. Calculate the covariance of these.

c. Construct column charts for each data set. (Use Figure 8.8 as a guide.)

d. Construct scatter plots for each data set. (Use Figure 8.9 as a guide.)

8.8 List the three aspects of covariation and discuss the applications of each.

8.9 Calculate estimates of phenotypic correlations from the two data sets created for Question 8.7. Characterize each correlation for sign and strength.

8.10 Calculate all four possible phenotypic regressions from the two data sets created for Question 8.7. Describe (in words, using the appropriate units of measure) the meaning of each regression.

8.11 **a.** Contrast correlation and regression.
b. If a correlation is strong, are the corresponding regressions large?
c. Use measures computed for Questions 8.9 and 8.10 to show that

$$b_{Y \cdot X} = r_{X,Y}\left(\frac{\sigma_Y}{\sigma_X}\right)$$

and

$$r_{X,Y} = b_{Y \cdot X}\left(\frac{\sigma_X}{\sigma_Y}\right)$$

8.12 What do we use regression coefficients for?

8.13 Describe the elements of a simple prediction equation.

Problems

8.1 Given the following set of data on days to 230 lb (D230) and backfat thickness (BF) in pigs:

Pig#	Days to 230 lb, days	Backfat Thickness, in
1	164	1.1
2	181	1.2
3	158	1.3
4	160	1.5
5	198	1.3
6	172	1.4
7	187	1.2
8	180	1.4
9	178	1.4
10	186	1.0

a. Calculate:
i. $\hat{\mu}_{P_{D230}}$
ii. $\hat{\mu}_{P_{BF}}$
iii. $\hat{\sigma}^2_{P_{D230}}$
iv. $\hat{\sigma}^2_{P_{BF}}$
v. $\hat{\sigma}_{P_{D230}}$
vi. $\hat{\sigma}_{P_{BF}}$

b. Calculate $\text{cov}(P_{D230}, P_{BF})$. What is implied by the *sign* of the covariance? What will be the signs of $\hat{r}_{P_{D230}, P_{BF}}$ and $\hat{b}_{P_{BF} \cdot P_{D230}}$?
c. Calculate $\hat{r}_{P_{D230}, P_{BF}}$. Characterize this correlation.
d. Calculate $\hat{b}_{P_{BF} \cdot P_{D230}}$. Interpret this regression in your own words.
e. Using the means and regression coefficient you calculated already, predict the backfat thickness of a pig that reached 230 lb in
i. 156 days.
ii. 200 days.

CHAPTER 9

Heritability and Repeatability

Most everyone has heard of heritability. Heritability tells us to what extent the differences we observe in animal performance are due to inheritance. But what does that really mean? What causes some traits to be more heritable than others? The purpose of this chapter is to explain heritability and its sister concept repeatability in some detail and to show why these two population measures are so critically important in animal breeding.

HERITABILITY

heritability (h^2) A measure of the strength of the relationship between performance (phenotypic values) and breeding values for a trait in a population.

Probably the most commonly understood definition of **heritability** is that it measures the degree to which offspring resemble their parents in performance for a trait. If a trait is highly heritable, animals with high performance themselves tend to produce high performing offspring, and animals with low performance tend to produce low performing offspring. On the other hand, if a trait is not very heritable, performance records of parents reveal little about progeny performance.

This definition of heritability is perfectly correct and reasonably satisfying. However, it is not the best definition on which to build an understanding of how heritability can be used in animal breeding. The following definition will take some getting used to, but, in the end, will extend the meaning of heritability well beyond the resemblance between parents and offspring. Heritability is a measure of the strength (consistency, reliability) of the relationship between performance (phenotypic values) and breeding values for a trait in a population.

The concept of heritability is illustrated graphically in Figure 9.1. The upper diagram (a) depicts a sample of performance records for a highly heritable trait. Phenotypic values are represented by the black columns extending above and below the population mean (μ). The contributions of breeding values and environmental effects to each performance record are shown in the background.[1] Note that with high heritability, breeding values generally have a large influence on

[1]Note that gene combination values appear to be missing from this diagram. They have been lumped together with environmental effects. Mathematically,

$$"E" = GCV + E$$

Admittedly this makes little theoretical sense. Gene combination values are genetic, not environmental in origin. But as a practical matter, gene combination values and environmental effects are commonly pooled because (1) there is rarely any way to distinguish between them and (2) gene combination values are not inherited and are therefore of little interest from a selection standpoint. In the diagrams like Figure 9.1 that appear in the remaining chapters on selection (Chapters 9 through 14), gene combination values are assumed to be contained within environmental effects (E for nonrepeated traits, E_P for repeated traits).

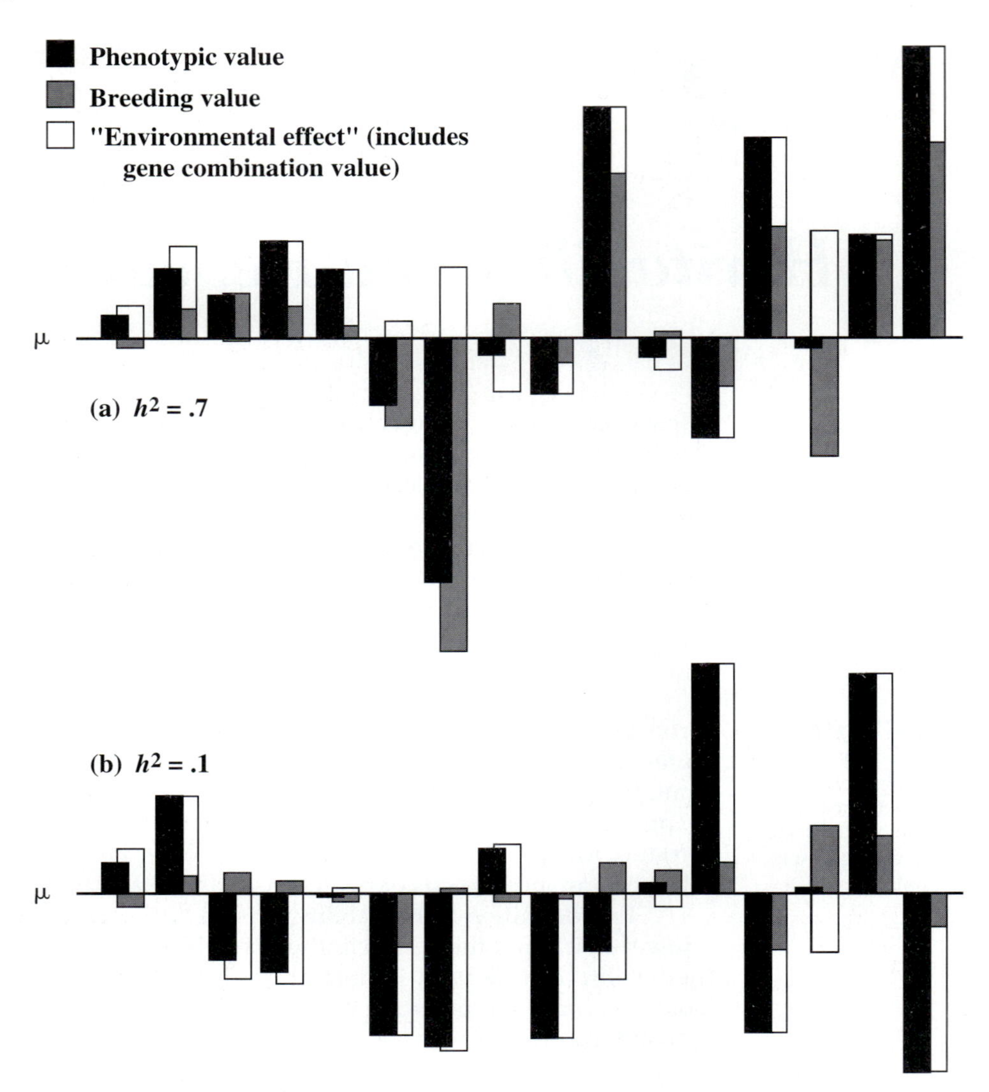

FIGURE 9.1 Schematic representation of animal performance for two traits that differ in heritability. Records for a sample of 16 animals are shown for each trait. Contributions of breeding values and "environmental effects" (environmental effects and gene combination values combined) are shown in the background. Heritabilities (h^2) for the traits depicted in the upper (*a*) and lower (*b*) diagrams are .7 and .1, respectively.

phenotypic values. Positive breeding values tend to be associated with above average performance, negative breeding values tend to be associated with below average performance, and the larger the breeding values (positive or negative), the larger the phenotypic deviations from the mean. There is a strong (consistent, reliable) relationship between breeding values and animal performance.

Another way of interpreting Figure 9.1(a) is to say that when heritability is high, performance is, on average, a good indicator of breeding value. With few exceptions, the black columns in the figure (phenotypic values) are reasonably good indicators of the gray columns (underlying breeding values). From a practical standpoint, this simply means that when a trait is highly heritable, the performance of animals reveals a lot about their breeding values. Animals with better performance themselves typically have better breeding values and therefore produce better performing offspring. Animals with poorer performance will typically have worse breeding values and produce poorer performing offspring.

Contrast Figure 9.1(a) with the lower diagram (b). Heritability is much lower for the trait represented in this second sample of records. There is little consistency to the relationship between phenotypic values and breeding values. In quite a few cases, negative breeding values result in better than average performance and vice versa, and extreme performance does not reflect particularly extreme breeding value. If there is a strong relationship in (b), it is not the relationship between performance and breeding values, but rather the relationship between performance and environmental effects. As a practical matter, we can conclude that when heritability is low, an animal's own performance is not likely to be a good indicator of its breeding value. The offspring of high performing parents will probably not perform much differently than the offspring of low performing parents.

As it has been described here—"a measure of the strength (consistency, reliability) of the relationship . . ."—heritability certainly sounds like a correlation. In fact, heritability (denoted as h^2) is *not* the correlation between phenotypic values and breeding values. It is the *square* of the correlation between phenotypic values and breeding values. Mathematically,

$$h^2 = r^2_{P,BV}$$

The square of a correlation is numerically somewhat different from the correlation itself, but from an interpretative standpoint, there is almost no difference. So think of heritability as you would any other correlation.

Strictly speaking, heritability as we have defined it is heritability in the *narrow sense.* A related notion, **heritability in the broad sense (H^2),** is a measure of the strength of the relationship between performance (phenotypic values) and *genotypic* values for a trait in a population. Mathematically,

heritability (in the broad sense, H^2)
A measure of the strength of the relationship between performance (phenotypic values) and genotypic values for a trait in a population.

$$H^2 = r^2_{P,G}$$

Broad sense heritability measures the total influence of genetics on the expression of a trait because it includes the contributions of both breeding value and gene combination value. It is not a particularly useful concept, however. Because gene combination values cannot be inherited, broad sense heritability does not reflect the relationship between the performance of animals and their potential as parents. From a selection perspective, therefore, it is not a very helpful measure.[2]

As a mathematical measure, heritability is always positive, ranging from zero to one or, in percentage terms, 0% to 100%. Traits with heritabilities near zero are barely heritable, and traits with heritabilities near one are extremely heritable. (Heritabilities above .7 are rare.)

Typical heritability estimates for a number of traits and species are listed in Table 9.1. As a rule, traits with heritabilities below .2 are considered lowly heritable, traits with heritabilities between .2 and .4 are considered moderately heritable, and traits with heritabilities above .4 are considered highly heritable. If you study

[2]If cloning of animals becomes widespread, H^2 could be very important. See Chapter 20.

TABLE 9.1 Typical Heritability Estimates for a Number of Traits and Species

Species	Trait	h^2
Cattle (beef)	Calving interval	.05
	Birth weight	.40
	Weaning weight	.30
	Yearling weight	.40
	Mature weight	.65
	Feed conversion	.40
	Scrotal circumference	.50
	Backfat thickness	.40
Cattle (dairy)	Calving interval	.10
	Milk yield	.25
	% fat	.55
	% protein	.50
	Udder support	.20
	Teat placement	.30
	Rear leg set	.15
	Stature	.50
Horses	Wither height	.40
	Cannon bone circumference	.45
	Temperament	.25
	Walking speed	.40
	Time to trot one mile	.45
	Time to run one mile	.35
	Pulling power	.25
	Cutting ability	.12
Swine	Litter size (number born alive)	.15
	Litter size (number weaned)	.10
	Weaning weight	.10
	21-day litter weight	.15
	Days to 230 lb	.25
	Feed conversion	.35
	Loin eye area	.50
	Backfat thickness	.50
Poultry	500-day egg production	.25
	Egg size	.45
	Shell thickness	.45
	Hatchability	.10
	Viability	.10
	Body weight	.45
	Shank length	.50
	Breast width	.25
Sheep	Number born	.15
	Birth weight	.30
	60-day weaning weight	.20
	Yearling weight	.40
	Loin eye area	.45
	Grease fleece weight	.40
	Fleece grade	.35
	Staple length	.50

Table 9.1, you will see that traits related to fertility and survivability tend to be lowly heritable. "Production traits" (traits like milk production and growth rate) tend to be moderately heritable. The most highly heritable traits are typically carcass or "product traits" and traits that are related to skeletal dimensions (e.g., structural size and mature body weight).

Common Misconceptions about Heritability

If a trait is genetically determined, we naturally assume that it is heritable—that is, the heritability of the trait is something greater than zero and less than one. But that is not always the case. When we say that a trait is heritable, what we really mean is that *differences* in performance for the trait are heritable. Some traits show no phenotypic differences, and so are not heritable even though they may be completely genetically determined. Totally confused? Consider the trait *number of legs in dogs.* (I am being only half facetious.) Dogs have four legs and, ruling out rare congenital deformities, we can say that *all* dogs are born with four legs. Number of legs is certainly genetically determined; it is coded for somewhere in every dog's DNA. But because *differences* in leg number do not exist, the trait is not heritable.[3]

Contrast leg number with leg *length.* There exist both genetic and phenotypic differences among dogs for leg length, and we can measure the strength of the relationship between breeding values and phenotypic values for the trait. Leg length is heritable in dogs, probably highly heritable.

Students sometimes assume that if the heritability of a trait is high, breeding values for the trait will be high also. Not so. High heritability indicates only that there is a strong relationship—a strong correlation—between phenotypic values and breeding values for a trait. Regardless of the magnitude of the heritability of a trait (so long as h^2 is not zero), there will be high breeding values, average breeding values, and low breeding values in a population.

We speak of the heritability of a trait in a population. Some examples are the heritability of speed in Thoroughbred horses, the heritability of egg production in layers, and the heritability of litter size in swine. Heritability is therefore a *population measure,* not a value to be associated with an individual animal. It would be an incorrect use of the term to speak of a certain animal's heritability for a trait. A particular stallion, for example, cannot have a high heritability for speed. What is probably meant by such a statement is that the horse's own performance and breeding value match closely. He was either fast or slow and his progeny perform much like him. This says nothing, however, about the heritability of speed—the strength of the relationship between observed speed and breeding values for speed—in horses. It is simply information about a particular horse's performance and breeding value for the trait.

Finally, the heritability of a trait is not fixed. It varies from population to population and from environment to environment. For example, the heritability of preweaning growth rate in beef cattle is different in different breeds. It tends to be greater in breeds that have high milk production relative to the growth potential

[3]Some might argue that the heritability of traits like number of legs is not zero. Rather, it is undefined. Fair enough. In animal breeding, anyway, we do not consider such traits heritable.

of calves. These breeds provide more favorable and more uniform nutrition for calf growth. Likewise, preweaning growth rate tends to be more heritable in good environments than in poor ones. This is because calves can more easily express inherent differences in growth potential in good environments. As we will see in the last section of this chapter, there are management techniques and mathematical procedures that breeders can use to increase the heritability of traits.

Heritability and Resemblance Among Relatives

When a trait is highly heritable, relatives tend to "resemble" each other in the trait (i.e., they tend to have similar performance). Conversely, when a trait is lowly heritable, there is little resemblance among relatives. How does heritability as it is defined here—the strength of the relationship between phenotypic values and breeding values for a trait—lead to resemblance (or lack of it) among relatives?

The answer is straightforward but follows a rather precise train of logic. To begin with, relatives share many of the same genes because they inherited them from common ancestors. Close relatives—full sibs, half sibs, parents and their progeny—share a large proportion of their genes (50%, 25%, and 50%, respectively), and more distant relatives share a smaller proportion. When individuals share genes, they also share the *independent effects* of those genes. As a result, there is a similarity in breeding values. In other words, the breeding values of relatives are correlated. This correlation has nothing to do with heritability (so long as h^2 is not zero). It is strictly a function of pedigree relationship. As you would expect, the correlation between breeding values of close relatives is higher than the correlation between breeding values of more distant relatives.

According to the definition of heritability, when heritability is high, there is a strong relationship between observed performance and breeding values; phenotypic values and breeding values are highly correlated. This means that when heritability is high, the similarity in breeding values of relatives will show up as similarity in phenotypic values as well. Relatives will exhibit similar performance, and the more closely related individuals are, the more similar their performance is likely to be.

You can use the same logic to explain why relatives show little resemblance in lowly heritable traits. As before, their breeding values are similar because of pedigree relationship. However, when heritability is low, breeding values and phenotypic values have little bearing on each other, and the similarity in breeding values of relatives will not show up as similarity in performance.

If you follow this line of thinking, you can see that the common understanding of heritability as the degree to which offspring resemble their parents in performance reflects just one manifestation of heritability. Heritability affects the resemblance among relatives of all kinds, not simply the resemblance between parents and progeny.

The various procedures used to estimate heritability are beyond the scope of this book. But they all involve measurement of the resemblance among relatives. In general, when relatives exhibit similar performance in a trait, the trait is quite heritable. When there is little more similarity in the performance of relatives than in the performance of individuals that have been randomly chosen from the population, the heritability of the trait is low.

Alternative Definitions of Heritability

In my opinion, the most instructive definition of heritability for most people is the one presented at the beginning of this chapter: the strength of the relationship between performance (phenotypic values) and breeding values for a trait in a population. But there are other, equally correct definitions of heritability that are useful in their own way.

Heritability can be thought of as *the change in breeding value expected per unit change in phenotypic value.* Mathematically, heritability is the regression of breeding value on phenotypic value, or

$$h^2 = b_{BV \cdot P}$$

Figure 9.2 contains scatter plots of the same samples of data shown in Figure 9.1. The slope of each regression line ($b_{BV \cdot P}$) indicates heritability. Regressions

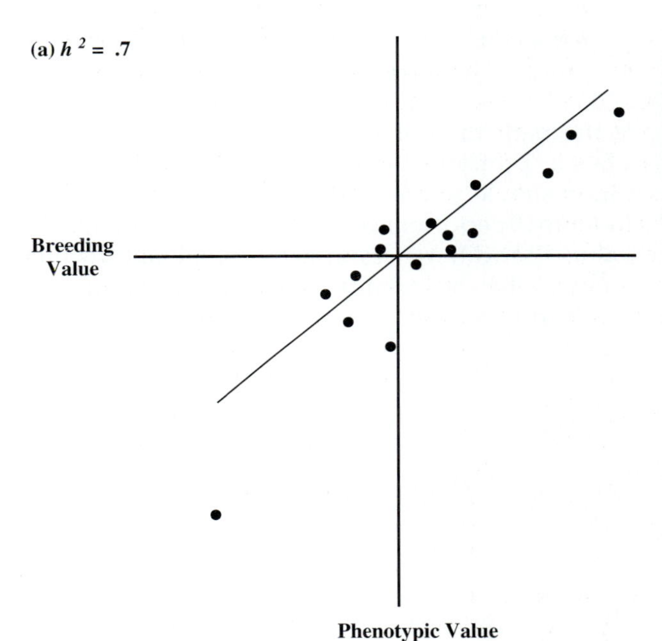

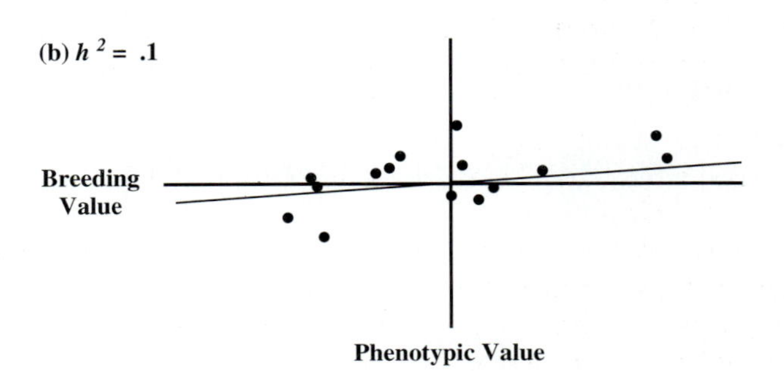

FIGURE 9.2
Scatter plots of the same samples of data shown in Figure 9.1. The slope of each regression line indicates heritability. Regressions implied in the generation of values were (*a*) $b_{BV \cdot P}$ = .7 and (*b*) $b_{BV \cdot P}$ = .1.

implied in the generation of values were (a) $b_{BV \cdot P} = .7$ and (b) $b_{BV \cdot P} = .1$. Heritability is higher in the upper plot, so there is greater change in breeding value per unit change in phenotypic value in the upper plot than in the lower plot.

As a regression coefficient, heritability can be used to predict an individual's breeding value from its phenotypic value. See the example in the next boxed section.

Heritability can also be thought of as a ratio of variances. It is the ratio of the variance of breeding value to the variance of phenotypic value, or

$$h^2 = \frac{\sigma^2_{BV}}{\sigma^2_P}$$

This is the most commonly used mathematical expression for heritability and is the most useful computationally. If you were to translate this expression into a *word formula*, it would read:

Heritability = the proportion of differences in performance for a trait that are attributable to differences in breeding value for the trait

In the mathematical expression, "differences" are represented by variances. Thus a "proportion of differences" becomes a ratio of variances. This definition should reinforce the idea that when heritability is high, differences in animal performance are largely attributable to differences in breeding value—not differences in gene combination value and(or) environmental effects. When heritability is low, differences in performance are determined less by differences in breeding value and more by differences in these other factors.

Example

Variances for mature body weight (MW) in broilers are:

$$\sigma^2_{P_{MW}} = .80 \text{ kg}^2$$

and

$$\sigma^2_{BV_{MW}} = .36 \text{ kg}^2$$

Therefore

$$h^2_{MW} = \frac{\sigma^2_{BV_{MW}}}{\sigma^2_{P_{MW}}} = \frac{.36}{.80} = .45$$

The Importance of Heritability

Heritability and Selection

Heritability is critically important to selection for polygenic traits. The object of selection is to choose those animals with the best breeding values to become parents of the next generation. To do a good job of this, we need good information about the candidates for selection. Because the only information available is phenotypic information, the strength of the relationship between phenotypic values and breeding values (i.e., heritability) is very meaningful.

Consider the simplest form of selection—*phenotypic selection.* In phenotypic selection, the only information used to determine whether an individual is selected or not is that individual's own performance. Pedigree and progeny data are disregarded. An example would be selection (or rejection) of animals on the basis of physical soundness. When heritability is low, phenotypic values generally reveal little about underlying breeding values, and it is difficult to determine which animals have the best breeding values and are therefore the best potential parents. *Accuracy of selection* or, more precisely, *accuracy of breeding value prediction* is poor and, as a result, the rate of genetic change is expected to be slow. When heritability is high, just the opposite is true. An animal's performance is generally a good indicator of its breeding value. Accuracy of selection is therefore good, and genetic change should be rapid.

The situation is somewhat different when the information used to make selection decisions is not limited to an individual's own performance. The advantage of using pedigree and(or) progeny information is that low heritability does not necessarily lead to poor selection accuracy.[4] However, whether information comes from an animal's own performance or the performance of relatives, that information still consists of phenotypic values, and the relationship between phenotypic values and breeding values is as important as ever. Given equal amounts of information, accuracy of selection will always be better for a more heritable trait than for a less heritable one. And if genetic variability is the same in both traits, the rate of genetic change due to selection will be faster for the more heritable trait.

Heritability and Prediction

Heritability plays an important role in the prediction of breeding values, progeny differences, and producing abilities. The equations used in prediction of these values are almost always functions of heritability. That is because heritability indicates how conservative a prediction should be.

For example, suppose you want to predict a sow's breeding value for number of pigs weaned based on the size of her first litter, which happened to be very small. The heritability of this trait is low, approximately .10. In other words, the connection between performance for number of pigs weaned and breeding value for the trait is a tenuous one. Knowing this, you would be reluctant to assign a highly negative estimated breeding value to this sow. Her EBV should be below average, but because of the trait's low heritability, it would be unfair to assume that

[4]See Chapter 11 for a detailed explanation.

her EBV is much below average based on just this one phenotypic record. The prudent approach to prediction in this case is a conservative one.

For a contrasting example, suppose you want to predict the breeding value for percent butterfat of a dairy cow whose first lactation milk tested very low in fat. The heritability of percent fat is quite high, approximately .55. Knowing that performance for percent fat is generally a fairly good indicator of underlying breeding value for the trait, you would want to take a less conservative approach than was taken in the litter size example. The cow's EBV for percent fat should be considerably below average.

Mathematical Examples of the Use of Heritability in Prediction

Suppose we want to predict a sow's breeding value for number of pigs weaned (NW) based on the size of her first litter. The basic prediction equation takes the form:

$$\hat{Y}_i = \hat{\mu}_Y + b_{Y \cdot X}(X_i - \hat{\mu}_X)$$

In the sow's case, the predicted value ($\hat{Y}_i$) is a prediction of her breeding value for number of pigs weaned ($\hat{BV}_{NW_i}$), and the evidence (X_i) is her first-litter performance (P_{NW_i}). The prediction equation is therefore:

$$\hat{BV}_{NW_i} = \hat{\mu}_{BV_{NW}} + b_{BV_{NW} \cdot P_{NW}}(P_{NW_i} - \hat{\mu}_{P_{NW}})$$

The average breeding value in the population ($\hat{\mu}_{BV_{NW}}$) is defined to be zero, and the regression of breeding value on phenotypic value ($b_{BV_{NW} \cdot P_{NW}}$) is simply heritability. So

$$\hat{BV}_{NW_i} = 0 + h^2_{NW}(P_{NW_i} - \hat{\mu}_{P_{NW}})$$

$$= h^2_{NW}(P_{NW_i} - \hat{\mu}_{P_{NW}})$$

If we want to predict an individual's breeding value using evidence other than a single record on the individual, the prediction equation is more complex than simply heritability times a phenotypic deviation. But heritability is still a factor. For example, to predict a dairy sire's breeding value for percent butterfat (PF) based on his daughters' first lactation records, we could use the equation:

$$\hat{BV}_{PF_i} = \frac{2ph^2_{PF}}{4 + (p - 1)h^2_{PF}}(\bar{P}_{BF_i} - \hat{\mu}_{P_{BF}})$$

In this equation, the expression $\frac{2ph^2_{PF}}{4 + (p - 1)h^2_{PF}}$ represents the regression of breeding value on the mean of single performance records of offspring. The regression is a function of both heritability and number of progeny records (p).[5]

[5]For more on prediction of breeding values, see Chapter 11. For a proof of this equation, see the Appendix.

Heritability and Management

Heritability indicates the extent to which differences in animal performance for a trait are determined by heritable factors as opposed to environmental effects. For highly heritable traits, differences in breeding values of animals have large effects on performance, and differences in environments are less important. Just the opposite is true for lowly heritable traits. As a rule, then, producers tend to select for more highly heritable traits knowing they can make significant genetic change. Because selection is less effective for lowly heritable traits, producers often choose not to change these traits genetically through selection, but rather to improve performance through management.

Growth traits, for example, tend to be quite heritable. Such traits are easy to improve through selection, so breeders do just that. Fertility traits, on the other hand, are usually lowly heritable. Breeders will typically put less emphasis on genetically improving fertility traits and instead will manage for good fertility by providing good nutrition.

The idea of using selection as the primary tool for improving performance in more heritable traits and using management as the primary tool for improving performance in less heritable traits is generally a sound one. But you should be careful not to follow this rule blindly. Some traits are so important economically that they deserve to be selected for despite low heritability. Fertility and survivability traits are typical examples in most species. And even though we know environmental effects are important determinants of performance in lowly heritable traits, that does not mean that we can identify and manipulate those effects. Embryonic loss, for example, is lowly heritable, but we know little about how to manage animals to prevent it. Finally, low heritability does not mean that improvement through selection is hopeless. As we will see in Chapter 11, modern genetic prediction technology allows us to improve even lowly heritable traits.

REPEATABILITY

repeatability (r) A measure of the strength of the relationship between repeated records (repeated phenotypic values) for a trait in a population.

Repeatability is a measure of the strength (consistency, reliability) of the relationship between repeated records (repeated phenotypic values) for a trait in a population.[6] Repeatability can be determined for any trait in which individuals commonly have more than one performance record. Examples of repeated traits include milk yield in dairy animals, racing and show performance in horses, litter size in swine, and fleece weight in sheep. The notation for repeatability is simply r.

The concept of repeatability is illustrated graphically in Figure 9.3. In both the upper (a) and lower (b) diagrams, a sample containing 10 pairs of records is displayed. (For clarity, pairs are separated by vertical lines.) Each pair represents two performance records made by the same individual for the same trait, so that the records of 10 individuals appear in each diagram. The contributions of producing

[6]In dairy publications, repeatability means something quite different. Repeatability in a dairy context refers to accuracy of prediction.

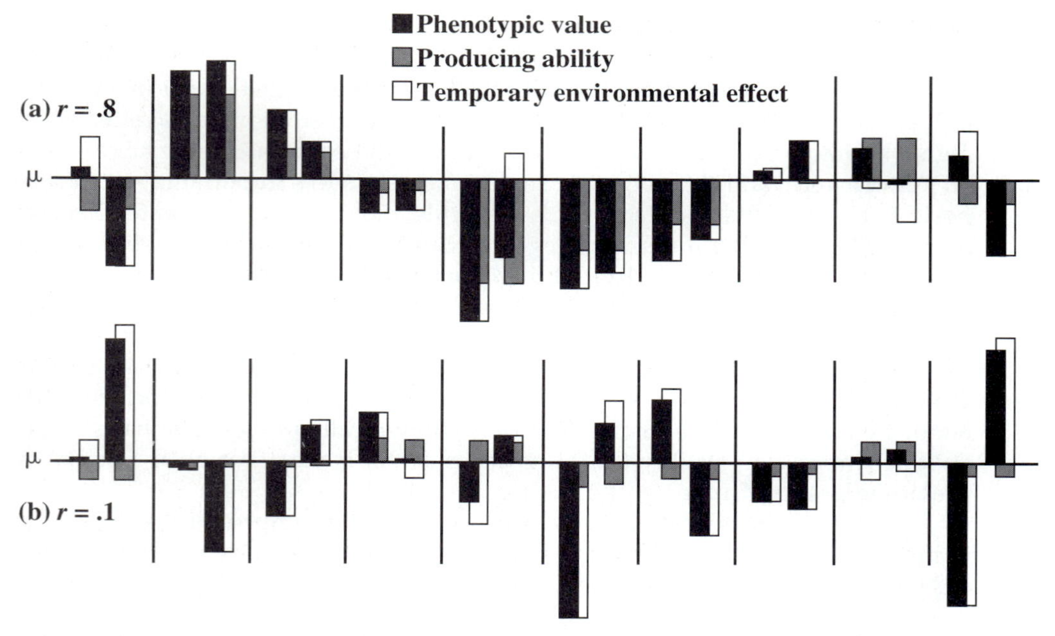

FIGURE 9.3 Schematic representation of animal performance for two traits that differ in repeatability. For each trait, pairs of repeated records from a sample 10 animals are illustrated. (For clarity, vertical lines separate each pair.) Contributions of producing abilities and temporary environmental effects are shown in the background. Repeatabilities (r) for the traits depicted in the upper (a) and lower (b) diagrams are .8 and .1, respectively.

ability and temporary environmental effect to each record are shown in the background. Note that only the temporary environmental effect changes from record to record on the same individual. Producing ability, as you recall, is made up of strictly *permanent* influences: breeding value, gene combination value, and permanent environmental effect.

The trait depicted in the upper diagram of Figure 9.3 is highly repeatable. If an animal's first record is above average, its second record is typically above average; if the first record is below average, the second is typically below average; and records on the same animal tend to be of similar magnitude.

In contrast, repeatability for the trait shown in the bottom diagram is low. Above average first records are not especially likely to be followed by above average second records, nor are below average first records especially likely to be followed by below average second records. There appears to be little relationship between the first and second records of individuals.

Another interpretation of Figure 9.3 is that when repeatability is high as in (a), the first record on an animal is, on average, a good indicator of that animal's second record. When repeatability is low as in (b), the first record is typically not a good indicator of the second.

Here is a second and equally useful definition of repeatability: Repeatability is a measure of the strength (consistency, reliability) of the relationship between single performance records (phenotypic values) and producing abilities for a trait in a population. In the upper diagram in Figure 9.3, above average performance records are quite consistently associated with above average producing abilities; below average performance records are quite consistently associated with below average producing abilities; and the magnitude of an animal's performance records tends to match the magnitude of its producing ability. Repeatability for the trait depicted in the upper diagram is therefore high. In the lower diagram, there appears to be little relationship between phenotypic records and producing abilities. Repeatability for this trait is therefore low.

When repeatability is high, we can say that a single record of performance on an animal is, on average, a good indicator of that animal's producing ability. When repeatability is low, a single phenotypic value tells us very little about producing ability.

Both definitions of repeatability—(1) the strength of the relationship between repeated records and (2) the strength of the relationship between single performance records and producing abilities—suggest that, in mathematical terms, repeatability is a correlation of some kind. And indeed it is. Repeatability is the correlation between repeated records for a trait in a population. In most cases, it is also the square of the correlation between single performance records (phenotypic values) and producing abilities for a trait in a population.[7] Mathematically, the two definitions of repeatability are then

$$r = r_{P_1,P_2}$$

and (usually)

$$r = r^2_{P,PA}$$

where the subscripts 1 and 2 refer to two different records made by the same individual for the same trait.

As a correlation, repeatability ranges from -1 to $+1$, though only rarely are repeatabilities negative. A repeatability near one indicates that a trait is extremely repeatable, and a repeatability near zero indicates that a trait is hardly repeatable at all. The repeatabilities used to generate the upper and lower diagrams in Figure 9.3 were .8 and .1, respectively.

Listed in Table 9.2 are typical repeatability estimates for a number of traits and species. Rules of thumb for repeatabilities are similar to those for heritabilities. Traits with repeatabilities below .2 are considered lowly repeatable, traits with repeatabilities between .2 and .4 are considered moderately repeatable, and traits with repeatabilities above .4 are considered highly repeatable.

[7]Repeatability does not equal $r^2_{P,PA}$ in those instances in which temporary environmental effects (E_t) for repeated records are correlated. Such cases are relatively rare, but do occur. For example, ewes that lamb late in one year, then early in the following year, record a short lambing interval. But if the starting date of the breeding season is fixed, the *next* lambing interval tends to be average or longer because, having lambed early in the previous year, these ewes have no opportunity to lamb even earlier, recording another short interval. The environmental influence of a fixed breeding season causes a tendency for shorter lambing intervals to be followed by longer ones, to be followed by shorter ones, and so on. In other words, repeatability of lambing interval can actually be *negative*.

TABLE 9.2 Typical Repeatability Estimates for a Number of Traits and Species

Species	Trait	*r*
Cattle (beef)	Calving date (trait of the dam)	.35
	Birth weight (trait of the dam)	.20
	Weaning weight (trait of the dam)	.40
	Body measurements	.80
Cattle (dairy)	Services per conception	.15
	Calving interval	.15
	Milk yield	.50
	% fat	.60
	Udder support	.50
	Teat placement	.55
	Rear leg set	.30
	Stature	.75
Horses	¼-mile time	.32
	1-mile time (flat races)	.57
	1-mile time (trotters)	.39
	1-mile time (pacers)	.45
	Cutting score	.22
Swine	Litter size (number born alive)	.15
	Litter size (number weaned)	.10
	Birth weight	.30
	Weaning weight	.15
	21-day litter weight	.15
Poultry	Egg weight	.90
	Egg shape	.95
	Shell thickness	.65
	Shell weight	.70
Sheep	Number born	.15
	Birth weight (trait of the dam)	.35
	60-day weaning weight (trait of the dam)	.25
	Grease fleece weight	.40
	Fleece grade	.60
	Staple length	.60

Common Misconceptions about Repeatability

Like heritability, repeatability is a *population measure,* a characteristic of a trait in a population. It is not a value to be associated with an individual animal. We speak of the repeatability of racing performance in horses or the repeatability of show placing in dogs, but it is an incorrect use of the term to speak of a particular horse's repeatability for racing performance or a particular dog's repeatability for placing.

Like heritability, repeatability is not fixed. It varies from population to population and from environment to environment. Factors that affect heritability tend to affect repeatability in a similar fashion. The last section in this chapter discusses management techniques and mathematical procedures that breeders can use to increase both the heritability and repeatability of traits.

Alternative Definitions of Repeatability

Just as there are alternative definitions of heritability, there are alternative definitions of repeatability.[8] Each definition is useful in its own way.

Repeatability can be thought of as *the change in producing ability expected per unit change in phenotypic value.* Mathematically, repeatability is the regression of producing ability on phenotypic value, or

$$r = b_{PA \cdot P}$$

Figure 9.4 contains scatter plots of the same samples of data shown in Figure 9.3. (Only the first performance record of a pair is plotted.) The slope of each regression line ($b_{PA \cdot P}$) indicates repeatability. Regressions implied in the generation of values were (a) $b_{PA \cdot P} = .8$ and (b) $b_{PA \cdot P} = .1$. Repeatability is higher in the upper plot, so there is greater change in producing ability per unit change in phenotypic value in the upper plot than in the lower plot.

As a regression coefficient, repeatability can be used to predict an individual's producing ability from its phenotypic value. See the example in the next boxed section.

Repeatability can also be thought of as a ratio of variances. It is the ratio of the variance of producing ability to the variance of phenotypic value, or

$$r = \frac{\sigma^2_{PA}}{\sigma^2_P}$$

If you were to translate this expression into a word formula, it would read:

Repeatability = the proportion of differences in performance for a trait that are attributable to differences in producing ability for the trait

Again, "differences" are represented by variances, and a "proportion of differences" becomes a ratio of variances. This definition should reinforce the idea that when repeatability is high, differences in animal performance are largely attributable to differences in producing ability—not differences in temporary environmental effects. When repeatability is low, differences in performance are determined less by differences in producing ability and more by differences in temporary environmental effects.

Example

Variances for cutting scores (CS) in horses are:

$$\sigma^2_{P_{CS}} = 106 \text{ points}^2$$

[8]The definitions of repeatability that follow are *usually* reliable. In most situations, repeatability $= r^2_{P,PA} = b_{PA \cdot P} = \frac{\sigma^2_{PA}}{\sigma^2_P}$. Only when temporary environmental effects are correlated (see the previous footnote) are these definitions inappropriate. It is always true that $r = r_{P_1,P_2}$.

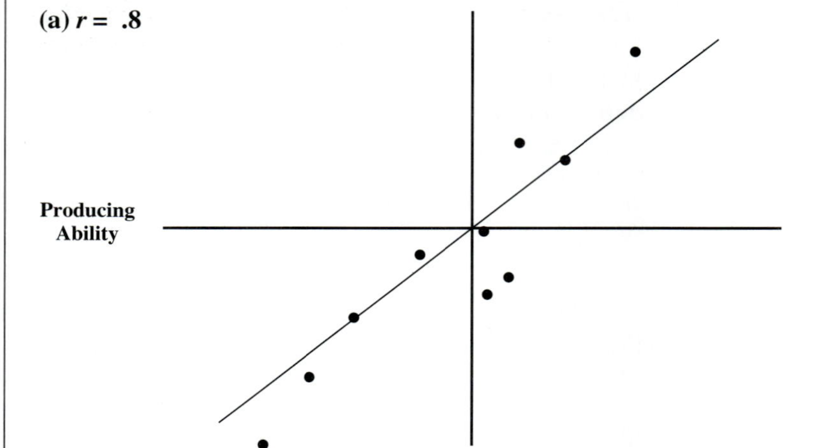

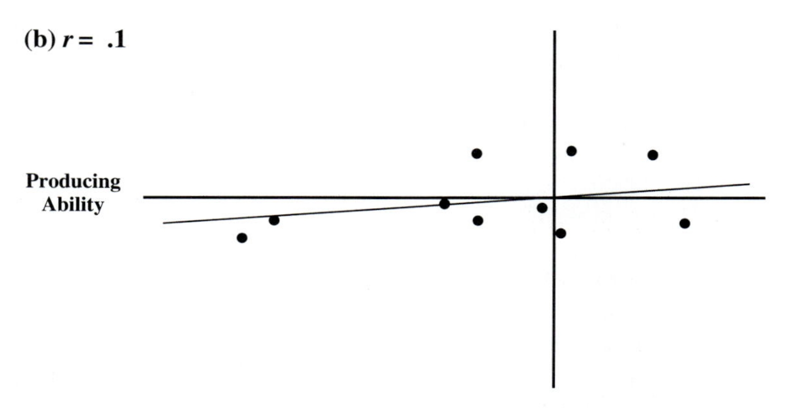

FIGURE 9.4 Scatter plots of the same samples of data shown in Figure 9.3. (Only the first performance record of a pair is plotted here.) The slope of each regression line indicates repeatability. Regressions implied in the generation of values were (*a*) $b_{PA \cdot P} = .8$ and (*b*) $b_{PA \cdot P} = .1$.

and

$$\sigma^2_{PA_{CS}} = 23 \text{ points}^2$$

Therefore

$$\begin{aligned} r_{CS} &= \frac{\sigma^2_{PA_{CS}}}{\sigma^2_{P_{CS}}} \\ &= \frac{23}{106} \\ &= .22 \end{aligned}$$

When heritability and repeatability are expressed as ratios of variances, you can see the relationship between the two parameters.

$$h^2 = \frac{\sigma^2_{BV}}{\sigma^2_P}$$

and

$$r = \frac{\sigma^2_{PA}}{\sigma^2_P}$$

$$= \frac{\sigma^2_{BV} + \sigma^2_{GCV} + \sigma^2_{E_p}}{\sigma^2_P}$$

$$= h^2 + \frac{\sigma^2_{GCV} + \sigma^2_{E_p}}{\sigma^2_P}$$

Repeatability is (in almost all cases) at least as large as heritability, and is often considered an upper limit for heritability.

The Importance of Repeatability

Repeatability and Culling

An awareness of the repeatability of a trait can be useful in making culling decisions. Suppose, for example, that you raise dairy cattle and want to cull lower producing cows. After examining the records of cows that have just completed their first lactation, you compile a list of potential culls. The first cow on the list—let's call her Ruby—is a decent milker, but she was slow to breed back, which means her second lactation will be delayed. You are tempted to cull her for this reason, but you note that calving interval and services per conception—both measures of ability to breed back—are not very repeatable ($r \approx .15$). In other words, Ruby's long interval between first and second lactations is not strong evidence that she will have long dry (nonlactating) periods in the future. Her poor first record is not a strong indication that her producing ability for calving interval is below average. So you give her the benefit of the doubt and keep her. If Ruby is slow to breed the next time, you will probably be tougher on her.

Emerald, Ruby's contemporary, is also on the short list for culling. Emerald bred back early, but her problem was poor milk yield. Noting that the repeatability of yield is quite high ($r \approx .5$), you choose to cull Emerald. Given such a high repeatability, Emerald's poor first-lactation record indicates that her future production records will also be poor and that her producing ability for yield is well below average.

The examples of Ruby and Emerald illustrate how repeatability should be used in culling. When repeatability is high, cull poor producing individuals on the basis of their first record. When repeatability is low, wait for more records before making a culling decision on an animal.

Repeatability and Prediction

In Ruby's and Emerald's cases, culling decisions were made intelligently, but still rather subjectively. A more objective approach would have been to calculate predictions of producing ability for calving interval and milk yield for each cow and cull on the basis of MPPA. (See the following boxed section or Chapter 11 for explanations and examples.) The MPPA calculation accounts for the repeatability of each trait. Thus Ruby's MPPA for calving interval, calculated under the assumption of low repeatability, would be longer than average, but not *much* longer than average—not long enough to justify culling her. On the other hand, Emerald's MPPA for milk yield, calculated under the assumption of high repeatability, would be quite low and would suggest that she be culled.

Just as heritability is necessary for predicting breeding values, repeatability is necessary for predicting producing abilities. In fact, repeatability is needed for *any* prediction calculation in which repeated records are involved. The reason for this is that when an individual has repeated records that are correlated (i.e., $r > 0$), each record is not really an independent piece of evidence. The higher the repeatability, the less the predictive value of each additional record on an individual. Repeatability is needed, therefore, to properly weight the contributions of repeated records.

Mathematical Examples of the Use of Repeatability in Prediction

Suppose we want to predict Emerald's producing ability for milk yield (MY) based on her first-lactation performance. The basic prediction equation takes the form:

$$\hat{Y}_i = \hat{\mu}_Y + b_{Y \cdot X}(X_i - \hat{\mu}_X)$$

In Emerald's case, the predicted value ($\hat{Y}_i$) is a prediction of her producing ability for milk yield ($\hat{PA}_{MY_i}$), and the evidence (X_i) is her first-lactation performance (P_{MY_i}). The prediction equation is therefore:

$$\hat{PA}_{MY_i} = \hat{\mu}_{PA_{MY}} + b_{PA_{MY} \cdot P_{MY}}(P_{MY_i} - \hat{\mu}_{P_{MY}})$$

The average producing ability in the population ($\hat{\mu}_{PA_{MY}}$) is defined to be zero, and the regression of producing ability on phenotypic value ($b_{PA_{MY} \cdot P_{MY}}$) is simply repeatability. So

$$\begin{aligned} \hat{PA}_{MY_i} &= 0 + r_{MY}(P_{MY_i} - \hat{\mu}_{P_{MY}}) \\ &= r_{MY}(P_{MY_i} - \hat{\mu}_{P_{MY}}) \end{aligned}$$

If we want to predict Emerald's producing ability for milk yield using more than one of her lactation records, then the prediction equation becomes:

$$\hat{PA}_{MY_i} = \frac{nr_{MY}}{1 + (n - 1)r_{MY}}(\bar{P}_{MY_i} - \hat{\mu}_{P_{MY}})$$

The expression $\frac{nr_{MY}}{1 + (n - 1)r_{MY}}$ represents the regression of producing ability on the mean of an individual's own repeated records, and is a function of both repeatability and number of records (n).[9]

To predict Emerald's *breeding value* for milk yield from her own lactation records, we could use the equation;

$$\hat{BV}_{MY_i} = \frac{nh^2_{MY}}{1 + (n - 1)r_{MY}}(\bar{P}_{MY_i} - \hat{\mu}_{P_{MY}})$$

Note that even though we are predicting breeding value (as opposed to producing ability) with this equation, repeatability is a factor because we are dealing with repeated records on an individual.

[9]For a proof of this equation, see the Appendix.

Ways to Improve Heritability and Repeatability

The higher the heritability of a trait, the better any one performance record is as an indicator of an animal's underlying breeding value. Likewise, the higher the repeatability of a trait, the better a single record is as an indicator of underlying producing ability. When heritability is high, prediction of breeding values will be more accurate, fewer mistakes will be made in replacement selection, and genetic progress will be faster. When repeatability is high, prediction of producing abilities will be more accurate and fewer mistakes will be made in culling. Given a choice, we would prefer heritability to be as high as possible and repeatability to be high as well.[10]

Breeders often assume that heritability and repeatability are immutable characteristics of a trait. In fact, they are not. It is possible to increase them, at least to an extent. In the remainder of this chapter we discuss the management strategies and mathematical techniques that are used to increase heritability and repeatability.

Environmental Uniformity

One important way to increase heritability and repeatability is to make the environment as uniform as possible. This means, in other words, to manage animals in such a way that environmental effects on the performance of different animals are as similar as possible.

[10]For predicting breeding values, it is actually best for heritability to be high and for repeatability to be high but no higher than heritability. This way, each performance record is likely to be a good indicator of breeding value (h^2 is high), and additional, repeated records on individuals provide as much independent evidence as possible (r is not *too* high).

Figure 9.5 illustrates the effect of a more uniform environment on heritability. The upper diagram (a) is identical to Figure 9.1(b). Heritability is low (h^2 = .1), so the relationship between animal performance and underlying breeding value is weak. The same animals are depicted in the lower diagram (b), but, in this case, each environmental effect has been reduced 75%—presumably through more uniform management. Because the environmental effects are now smaller, variation in environmental effects is less, and variation in performance is reduced as well. Note, however, that in the lower diagram, performance is a much better indicator of breeding value than it was in the upper diagram. In other words, heritability has increased.

For a less theoretical example, consider a group of Thoroughbred horses. Half of them are chosen at random and then trained so that they are very fit, seasoned runners. The other half receive no training whatsoever. They are inexperienced and out of shape. Now put all these horses together in the same race and record the outcome. Chances are the well-trained horses will outperform the others. Is it not likely, however, that some of the untrained horses have good breeding values for racing ability even though their performance is unexceptional due to lack of training? And is it not likely that some of the trained horses are genetically mediocre and only perform well because of their training? Is racing per-

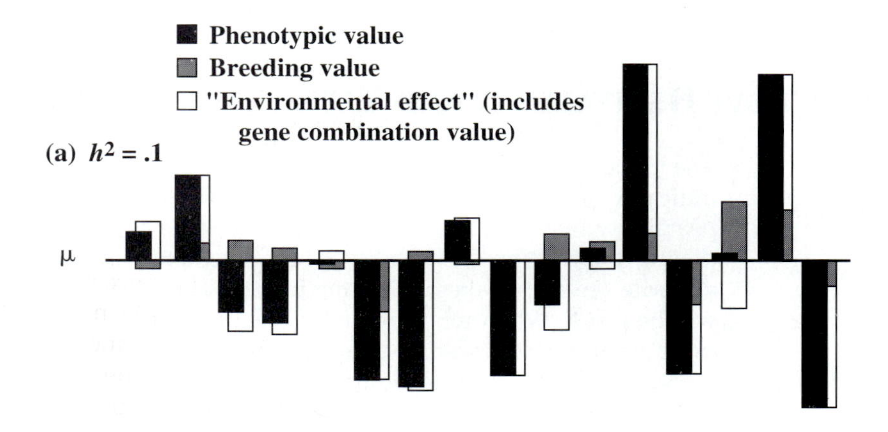

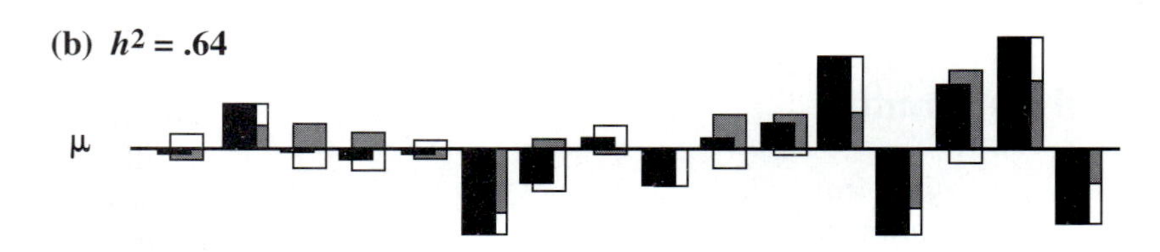

FIGURE 9.5 Schematic representation of the effect of a more uniform environment on heritability. The sample of records shown in (*a*) is identical to the sample depicted in Figure 9.1(b). Heritability is low (h^2 = .1). Records for the same animals are shown again in (*b*). This time, however, more uniform management has reduced each environmental effect by 75%. Phenotypic variation decreases, and heritability increases dramatically (h^2 = .64).

formance in this case a good indicator of breeding value for racing ability? In other words, is the heritability of racing ability high in this small population? The answer is clearly no. Performance is not a good indicator of underlying breeding value because differences in training—*environmental* differences—have biased performance.

Now imagine what would happen if all the horses in the group received similar training and then competed against each other. In the absence of any advantages or disadvantages with respect to training, the horses with the better breeding values are likely to outperform the horses with the poorer breeding values. In other words, the relationship between racing performance and breeding value for racing ability is stronger than before. Heritability is increased due to the more uniform training environment.

How do you manage animals so that environmental effects are as consistent as possible? The answer is to minimize the environmental advantages that some animals have over others. In the Thoroughbred example, the solution was to provide similar training for all horses. For other species and other traits, management practices will differ. In dairy animals, for example, one way to minimize environmental differences affecting milk production would be to provide all individuals with similar feed. In beef cattle, environmental differences in weaning weight can be minimized by making sure that all animals graze pastures of similar quality and receive the same vaccinations.

Note that minimizing environmental differences does not mean making the environment *better.* Rather, it means making the environment more *uniform.* It is not necessary that all the horses in our Thoroughbred example receive the very best training available. Heritability will be increased if they receive a similar level of training, whatever that level may be.

Accurate Measurement

The more accurate the measurement of performance in a trait, the higher the trait's heritability. Consider the weaning weight example in beef cattle. Weaning weights are measured most accurately if the scale is tested and frequently balanced, and all calves are equally full (i.e., all have about the same amount of water and feed in their digestive tracts). If these conditions are met, then measurement error is minimized, and differences among animals in recorded weights are more likely to represent underlying genetic differences. If manure is allowed to collect on the scale and the scale is not rebalanced, or if the first calves through the chute have had no access to water, but later ones have filled up at the trough, then measurement error is introduced. Differences in recorded weights are less likely to represent underlying genetic differences, and heritability is reduced.

Rebalancing the scale and ensuring similar fill levels increase measurement accuracy and heritability by creating a more uniform environment in which animals are measured. Better measurement *precision* increases heritability too. To use the weaning weight example, more *precise* measurements can be made by using a scale that can be read in 5-lb increments than by using one that can be read in 20-lb increments. Weaning weights recorded on the more sensitive scale should be more heritable (if only slightly) than weaning weights recorded on the less precise scale.

Mathematical Adjustments for Known Environmental Effects

Most environmental influences (training regimen, level of feed, pasture quality, etc.) are difficult to quantify. There is no straightforward way to adjust a horse's race record to account for the ability of the horse's trainer. Likewise, there is no good way to adjust a calf's weaning weight to account for pasture quality. Environmental effects like these are considered *unknown*, meaning that we cannot mathematically adjust animal performance to account for them. There are such things as *known* environmental effects, however. These are influences that are so consistent that researchers have developed mathematical adjustment factors or adjustment procedures to account for them. Listed in Table 9.3 are examples of environmental effects for which mathematical adjustments are available.

TABLE 9.3 Examples of Environmental Effects for which Mathematical Adjustment Procedures and(or) Adjustment Factors Are Available

Species	Trait	Environmental Effect
Cattle (beef)	Birth weight	Age of dam
	"	Sex of calf
	Weaning weight	Age of calf
	"	Age of dam
	Yearling weight	Days on test
	Scrotal circumference	Age of bull
	"	Age of dam
	Frame score	Age
	"	Sex
Cattle (dairy)	Milk yield	Length of lactation
	"	Milkings per day
	"	Age at calving
	Fat yield	Length of lactation
	"	Milkings per day
	"	Age at calving
	Fat corrected milk yield	Fat yield
Horses	Earnings	Year
Swine	Litter size (number born alive)	Parity (no. of litters)
	Litter size (number weaned)	Parity
	21-day litter weight	Age of pigs
	"	Parity
	"	Number of pigs nursed
	"	Number of pigs weaned
	Days to 230 lb	Age
	Backfat thickness	Weight
	Loin eye area	Weight
Sheep	Weights (30-day, 90-day, 120-day, etc.)	Age of lamb
	"	Type of birth/type of rearing (e.g., twin raised as twin, etc.)
	"	Age of dam
	"	Sex
	Wool traits	Age of lamb
	"	Type of birth/type of rearing
	"	Age of dam
	"	Sex

Weaning weight in beef cattle provides a typical example of a mathematical adjustment procedure to account for known environmental effects. Weaning weights are normally adjusted for age of the calf and age of the calf's dam. The age-of-calf adjustment removes the weight advantage of older calves (or the weight disadvantage of younger calves). The age-of-dam adjustment accounts for the increased milk production and therefore better nutritional environment provided by older dams. In the United States, weaning weights are adjusted to a 205-day, mature dam equivalent by the following formula:

$$\text{Adjusted weaning wt.} = \left(\frac{\text{Actual wt.} - \text{birth wt.}}{\text{Age at weighing}}\right) \times 205 + \text{birth wt.} + \text{age-of-dam factor}$$

The part of the formula in parentheses represents the growth rate of the calf from birth until weighing. Multiplying this value by 205 and adding back the calf's birth weight provides an estimate of the calf's weight had the calf been exactly 205 days old when weighed. The age-of-dam factor is taken from a table similar to Table 9.4.

Adjustment factors like those in Table 9.4 could be calculated by enterprising breeders. Typically, however, they are determined by animal breeding researchers using large amounts of performance data from a number of research populations and(or) private herds or flocks. Because each adjustment factor represents the *average* influence of an environmental effect, it should work reasonably well in most situations, but may not be appropriate in certain herds or flocks in some years.

Adjustment factors can be either *additive* or *multiplicative*. The age-of-dam factors in Table 9.4 are additive because they are simply added on to a calf's weaning weight. As you would expect, multiplicative factors are used as multipliers. For example, 21-day litter weight in swine is adjusted for age of pigs by multiplying litter weight by a factor that ranges from 1.3 for pigs that are just 14 days old at weighing to .73 for pigs that are 35 days old.

Adjustment procedures and adjustment factors often vary for different populations within a species. Many breeds have their own sets of adjustments. Differences in adjustment procedures more often than not represent real biological differences between breeds.

TABLE 9.4 Age-of-Dam Correction Factors for Weaning Weight in Beef Cattle

	Adjustment (lb)	
Age of Dam (yr)	Males	Females
2	60	54
3	40	36
4	20	18
5 to 10	0	0
11+	20	18

A Mathematical Perspective: How Reducing Environmental Effects Increases Heritability and Repeatability

You can see how reducing the size of environmental effects mathematically increases heritability and repeatability if you express heritability and repeatability as ratios of variances. For example,

$$h^2 = \frac{\sigma^2_{BV}}{\sigma^2_P}$$

Expanding the denominator,

$$h^2 = \frac{\sigma^2_{BV}}{\sigma^2_{BV} + \sigma^2_{GCV} + \sigma^2_E}$$

If we reduce the size of environmental effects—either by making the environment more uniform or by mathematically adjusting for known environmental effects—we reduce σ^2_E and make the denominator smaller. The entire expression (i.e., heritability) increases as a result.

Example

The data used to build Figure 9.5 provide a useful example. Before reducing the size of environmental effects, $\sigma^2_{BV} = .1$ and $\sigma^2_{E*_1} = .9$. (Gene combination value and environmental effects have been lumped together, so $\sigma^2_{E*} = \sigma^2_{GCV} + \sigma^2_E$.) Then

$$\begin{aligned} h_1^2 &= \frac{\sigma^2_{BV}}{\sigma^2_{BV} + \sigma^2_{E*_1}} \\ &= \frac{.1}{.1 + .9} \\ &= .1 \end{aligned}$$

After reducing environmental effects by 75%,

$$\sigma_{E*_2} = .25\sigma_{E*_1}$$

and

$$\begin{aligned} \sigma^2_{E*_2} &= (.25)^2\sigma^2_{E*_2} \\ &= .0625(.9) \\ &= .05625 \end{aligned}$$

so

$$h_2^2 = \frac{\sigma_{BV}^2}{\sigma_{BV}^2 + \sigma_{E^*_2}^2}$$

$$= \frac{.1}{.1 + .05625}$$

$$= .64$$

Heritability increases from .1 to .64.

By expressing repeatability as a ratio of variances, you can see how reducing environmental effects (*temporary* environmental effects, anyway) mathematically increases repeatability in the same way it increases heritability.

$$r = \frac{\sigma_{PA}^2}{\sigma_P^2}$$

Expanding numerator and denominator,

$$r = \frac{\sigma_{BV}^2 + \sigma_{GCV}^2 + \sigma_{E_p}^2}{\sigma_{BV}^2 + \sigma_{GCV}^2 + \sigma_{E_p}^2 + \sigma_{E_t}^2}$$

Any reduction in $\sigma_{E_t}^2$ will make the denominator of this expression smaller and increase repeatability.

Contemporary Groups

We can increase the heritability and repeatability of traits by mathematically adjusting for known environmental effects and by managing animals to minimize differences in animal performance caused by other environmental effects. But what can be done in situations where all animals cannot be managed the same? For example, what should a breeder of grazing animals do when some animals in a herd or flock have access to high quality pasture, but others must be confined to a poor quality pasture? And what can be done to make it possible for the performance of one breeder's animals to be fairly compared with the performance of another breeder's animals when management and physical environment differ between breeding operations?

contemporary group
A group of animals that have experienced a similar environment with respect to the expression of a trait. Contemporaries typically perform in the same location, are of the same sex, are of similar age, and have been managed alike.

One answer is to express animal performance not in absolute terms, but rather as a deviation from a contemporary group mean. A **contemporary group** is a group of animals that have experienced a similar environment with respect to the expression of a trait. A "similar environment" in this context typically means that animals in the group (contemporaries) perform in the same location, are of the same sex, are of similar age, and have been managed alike. For example, a contemporary group for weaning weight in lambs or calves would commonly be defined by breeder (herd or flock), year, season of year, sex of lamb or calf, and management effects such as pasture and feeding regimen. A contemporary group for show, racing, pulling, reining, or cutting performance in horses would typically

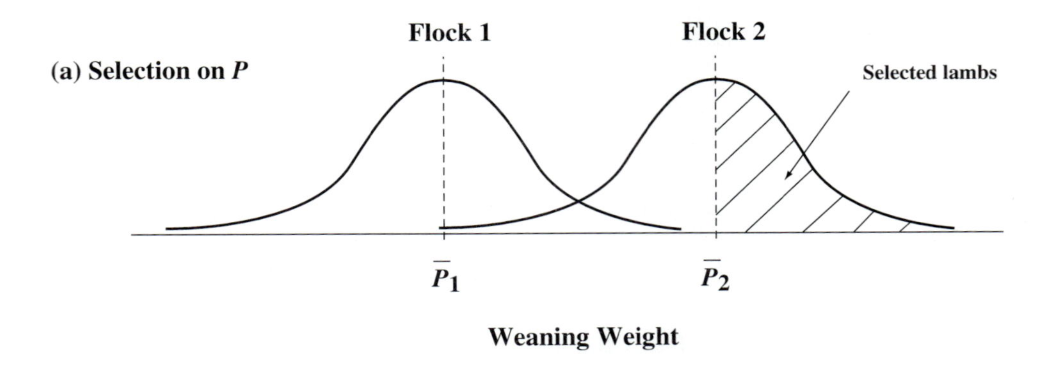

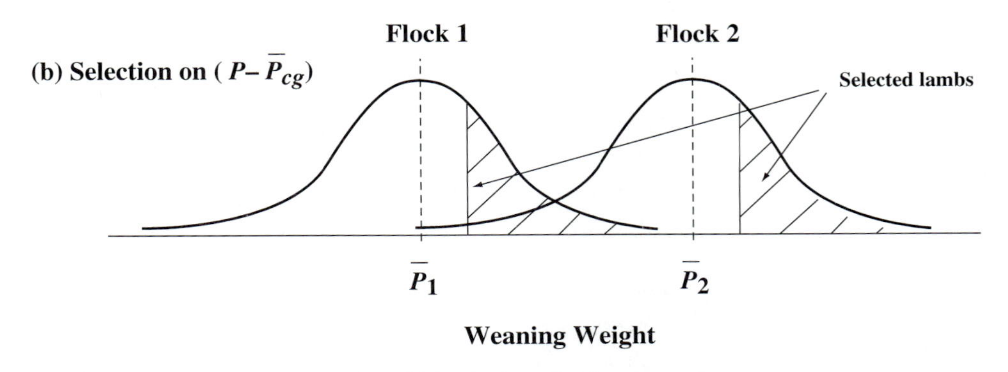

FIGURE 9.6 Graphic representation of (*a*) selection of lambs on weaning weight (*P*), and (*b*) selection of lambs on deviation of weaning weight from a contemporary group mean ($P - \bar{P}_{cg}$). Flocks 1 and 2 are genetically similar, but were raised in different pasture environments. Selection on ($P - \bar{P}_{cg}$) accounts for pasture differences and does a better job of identifying genetically superior lambs.

consist of the animals in the same competition—the same show class, race, pulling contest, or go-round.[11]

Figure 9.6 illustrates the advantage of selecting animals—in this case, ewe lambs—on the basis of their deviation from a contemporary group mean ($P - \bar{P}_{cg}$) as opposed to selection on absolute performance (P). Flocks 1 and 2 belong to a single sheep operation. They are genetically similar, but flock 2 grazed much higher quality pastures than flock 1. Weaning weights are therefore considerably heavier in flock 2. In the upper diagram (a), replacement lambs are chosen on the basis of their own weaning weights (presumably adjusted for age, age of dam, type of birth, and type of rearing). Assuming that 25% of ewe lambs are needed as replacements, selecting the heaviest lambs results in virtually all the selected ewe lambs coming from flock 2. This is patently unfair to the better performing lambs in flock 1. They are likely to have high breeding values for weaning weight, but cannot compete with flock 2 lambs due to poor nutrition.

[11]Contemporary groups for food and fiber species are commonly restricted to animals that have spent their entire lives together. Contemporary groups for competition animals (e.g., horses and dogs) typically include individuals from different origins. The latter type of contemporary group has the disadvantage of being unable to account for precompetition environment.

In the lower diagram (b), ewe lambs are still chosen on phenotype, but this time phenotype is defined as a deviation from a contemporary group mean (contemporary groups, in this example, being synonymous with flocks). With this procedure, equal numbers of lambs are chosen from each flock, which, if the flocks are indeed genetically similar, is as it should be.

Figure 9.6 clearly illustrates the advantage of selecting on the basis of deviations from a contemporary group mean when groups of animals experience different environments. What may not be so clear, however, is how expressing performance as a deviation from a contemporary group mean actually increases the heritability and repeatability of a trait. Think of it this way. When environmental differences exist between groups of animals, the relationship between animals' breeding values and absolute performance is weakened. Some animals with poor breeding values perform relatively well because of a favorable group environment, and some animals with superior breeding values perform poorly because of an unfavorable group environment. But when phenotype is expressed as a deviation from a contemporary group mean, environmental differences between groups are accounted for, and the relationship between animals' breeding values and this new measure of performance is much stronger. Mathematically,

$$r_{BV,(P-\overline{P}_{cg})} > r_{BV,P}$$

Heritability, therefore, is increased by using deviations from a contemporary group mean. For repeated traits, repeatability is increased for the same reason. When phenotype is expressed as a deviation from a contemporary group mean, the relationship between animals' producing abilities and phenotypes is stronger. Mathematically,

$$r_{PA,(P-\overline{P}_{cg})} > r_{PA,P}$$

Contemporary groups are most commonly used to account for environmental differences between groups of animals. They can also be used as an alternative to mathematical adjustments for known environmental effects. Suppose, for example, that the breeder of the sheep in Figure 9.6 felt that the industry standard age-of-dam adjustment factors were not appropriate for her yearling ewes. Perhaps the adjustment factors were too small and put her young ewes at a disadvantage. Rather than accept the industry adjustment factors, she could designate special contemporary groups for the lambs out of her yearling ewes. This way, these lambs would be compared only to lambs out of yearlings, and contemporary group differences caused by faulty adjustment factors would be accounted for. Such an approach is a good idea so long as contemporary groups do not get too small. Very small contemporary groups make meaningful comparisons difficult.

A New Model

contemporary group effect (E_{cg}) An environmental effect common to all members of a contemporary group.

The reason breeders use contemporary groups is to account for **contemporary group effects (E_{cg})**, environmental effects common to all members of a contemporary group. After incorporating contemporary group effects, the genetic model for quantitative traits is:

$$P = \mu + BV + GCV + E_{cg} + E$$

or for repeated traits:

$$P = \mu + BV + GCV + E_p + E_{cg} + E_t$$

Note that the contemporary group effect is another *environmental* effect. It reflects the influence that environmental factors common to all animals in a contemporary group have on an individual's performance. In our sheep example, such environmental factors were primarily associated with forage quality. In racehorses, track conditions could be a large part of contemporary group effects. In poultry, contemporary group effects could arise from differences in temperature, humidity, and ventilation of poultry houses.

The influence of contemporary group effects is illustrated in Figure 9.7. Members of contemporary group 1 (upper diagram) benefit from a better than average contemporary group effect (E_{cg}). Even though these animals are not genetically superior, on average, to the general population, the performance of most of them is higher than the population mean (μ). Animals in contemporary group 2 (lower diagram) are not so lucky. The environment for this group is poor. There is no appreciable genetic difference between the two groups, but almost all members of group 2 have below average performance due to the unfavorable contemporary group effect. As you can see, contemporary group effects tend to obscure performance, making it difficult to compare the performance of individuals in different contemporary groups.

If we could, it would be useful to remove contemporary group effects altogether. Happily, this can be done by expressing performance as a deviation from a contemporary group mean. If we assume that the mean performance of a contemporary group is simply a function of the overall population mean and the contemporary group effect, that is,

$$\overline{P}_{cg} = \mu + E_{cg}$$

then

$$P = \overline{P}_{cg} + BV + GCV + E$$

and our new measure of performance—an animal's phenotypic deviation from its contemporary group mean—can be represented as:

$$P - \overline{P}_{cg} = BV + GCV + E$$

or for repeated traits:

$$P - \overline{P}_{cg} = BV + GCV + E_p + E_t$$

This new measure of performance is no longer obscured by contemporary group effects. They have been removed by subtracting off the contemporary group mean. As with other environmental effects, the removal of contemporary group effects increases heritability and repeatability.

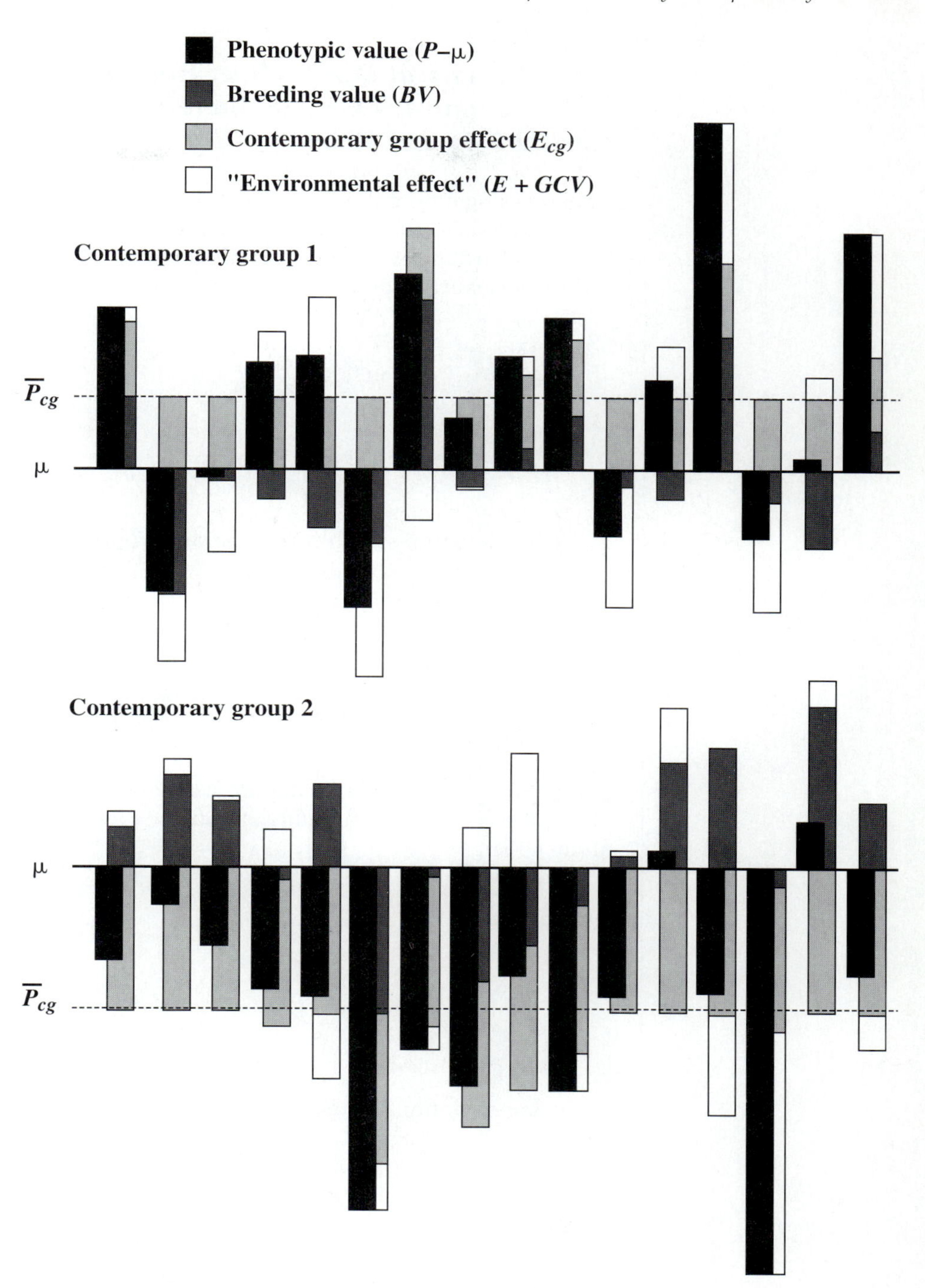

FIGURE 9.7 Schematic representation of performance of animals sampled from two contemporary groups. Contemporary group 1 experiences a favorable environment (E_{cg}), so average performance within the group ($\overline{P}_{cg}$) is above the overall population mean (μ). Contemporary group 2 suffers from a very poor environment, so average performance within the group is well below the population mean.

A Mathematical Perspective: How the Use of Contemporary Groups Increases Heritability and Repeatability

By expressing heritability and repeatability as ratios of variances, you can see how the use of deviations from contemporary group means mathematically increases heritability and repeatability. In the case of heritability, if you use raw performance data (not deviations), then

$$h_1^2 = \frac{\sigma_{BV}^2}{\sigma_P^2}$$

$$= \frac{\sigma_{BV}^2}{\sigma_{BV}^2 + \sigma_{GCV}^2 + \sigma_{E_{cg}}^2 + \sigma_E^2}$$

But using deviations from contemporary group means,

$$h_2^2 = \frac{\sigma_{BV}^2}{\sigma_{P-\bar{P}}^2}$$

$$= \frac{\sigma_{BV}^2}{\sigma_{BV}^2 + \sigma_{GCV}^2 + \sigma_E^2}$$

Because $\sigma_{E_{cg}}^2$ has been eliminated from the denominator, heritability increases.

Likewise, for repeatability, if you use raw performance data (not deviations), then

$$r_1 = \frac{\sigma_{PA}^2}{\sigma_P^2}$$

$$= \frac{\sigma_{BV}^2 + \sigma_{GCV}^2 + \sigma_{E_p}^2}{\sigma_{BV}^2 + \sigma_{GCV}^2 + \sigma_{E_p}^2 + \sigma_{E_{cg}}^2 + \sigma_{E_t}^2}$$

But using deviations from contemporary group means,

$$r_2 = \frac{\sigma_{PA}^2}{\sigma_{P-\bar{P}_{cg}}^2}$$

$$= \frac{\sigma_{BV}^2 + \sigma_{GCV}^2 + \sigma_{E_p}^2}{\sigma_{BV}^2 + \sigma_{GCV}^2 + \sigma_{E_p}^2 + \sigma_{E_t}^2}$$

Eliminating $\sigma_{E_{cg}}^2$ from the denominator increases repeatability.

Expressing performance as a deviation from a contemporary group mean is a useful technique when contemporary group means differ for strictly environmental reasons. But what if contemporary groups are *genetically* different? What if one group is from a genetically outstanding herd or flock and another group is from a genetically poor herd or flock? Is it fair to compare animals from one group with animals from the other on the basis of deviations from contemporary group means? Should an animal with a phenotypic deviation of +20 lb in the first group be considered genetically equivalent to an animal with a +20 lb deviation in the second group?

The answer is clearly no. When there is reason to believe that contemporary groups are genetically different, the use of deviations from contemporary group means for comparing animals from different groups will be misleading. Comparing deviations across contemporary groups works best when contemporary groups are thought to be genetically similar—as, for, example, when they are from the same herd or flock. Comparing deviations across groups does not work well when differences among groups are largely genetic in origin. Understand, however, that for many species and breeds with species, environmental differences among contemporary groups are usually larger than genetic differences among groups. This is especially true for animals such as sheep and beef cattle that depend upon grazed forage and whose environment is largely beyond human control. If you are a breeder of one of the species and are given the choice of comparing animals from different groups on the basis of either absolute performance or deviations from a contemporary group mean, more often than not you will be better off choosing the latter.

There is a way to compare animals across contemporary groups that accounts for both environmental and genetic differences among groups. In Chapter 11, we will discuss Best Linear Unbiased Prediction (BLUP), an advanced statistical methodology for genetic prediction. BLUP does not use deviations from contemporary group means, but does incorporate contemporary group information. It can be used not only to predict genetic values but also to estimate contemporary group effects.

The Importance of Proper Contemporary Grouping

When contemporary groups are correctly formed, they can help to increase heritability and repeatability. When they are incorrectly formed, they can have the opposite effect. More importantly, however, improper contemporary grouping can distort performance records and genetic predictions for individual animals.

The problem usually occurs when some animals receive preferential treatment, yet their performance records are lumped into a contemporary group with those of animals that receive no special treatment. For example, beef cattle being readied for show are typically separated from the herd and given extra feed and care. As a result, they grow faster. If a breeder neglects to form a separate contemporary group for the show string and pools all his performance records in one contemporary group, the show animals will appear to have greater potential for growth rate than they do. Chances are they would have grown faster than average even without preferential treatment (or they would not have been chosen for showing), but their relative performance has been biased upward by improper contemporary grouping.

For a visual example, return to Figure 9.7. Imagine that contemporary group 1 represents animals being readied for show. They have been well fed, so their contemporary group effect (E_{cg}) for weight traits is well above average. Now imagine that the breeder of these animals neglected to create a separate contemporary group for the show string and lumped these animals' performance data in with performance data from contemporary group 2, a group for which E_{cg} is well below average. When the two sets of data are combined, almost all the apparent high performers are show animals, and almost all the apparent low performers are not. Because of the difference in contemporary group effects, the relative performance of all the animals is now badly biased—upward for the show animals, downward for the others. Any genetic predictions involving the performance records of these animals will be biased as well.

Trait Ratios

trait ratio
An expression of relative performance—the ratio of an individual's performance to the average performance of all animals in the individual's contemporary group.

In order to account for contemporary group effects, animal performance is often expressed as a deviation from a contemporary group mean. But in some species, particularly swine, sheep, and beef cattle, an alternative to deviations known as a **trait ratio** is commonly used. Like a deviation from a contemporary group mean, a trait ratio is an expression of relative performance. It is the ratio of an individual's performance to the average performance of all animals in the individual's contemporary group.

To calculate a trait ratio for an animal, that animal's performance record and the performance records of its contemporaries are first adjusted for known environmental effects. Animal i's ratio is then calculated as:

$$\text{Ratio}_i = \left(\frac{P_i}{\overline{P}_{cg}}\right) \times 100$$

For example, if a ewe lamb's 60-day adjusted weaning weight is 56 lb and the average adjusted weaning weight of all lambs in her contemporary group is 50 lb, her weaning weight ratio would be

$$\left(\frac{56}{50}\right) \times 100$$
$$= 1.12 \times 100$$
$$= 112$$

The ratio for a 46-lb ewe lamb from the same contemporary group would be

$$\left(\frac{46}{50}\right) \times 100$$
$$= .92 \times 100$$
$$= 92$$

Trait ratios are simple to understand. The average ratio within a contemporary group is 100, so any ratio above 100 indicates higher than average (not necessarily *better* than average) performance, and any ratio below 100 indicates lower than average performance. Ratios have an additional advantage over deviations from contemporary group means in that their interpretation does not depend on the trait involved and does not require any knowledge of the variability of the trait. In the weaning weight example, the first lamb's 6-lb deviation from her contemporary mean is impressive, but imagine how impressive a 6-lb deviation would be if the trait were birth weight! On the other hand, expressed as a ratio, the first lamb's performance is considered 12% above average and the second lamb's performance is considered 8% below average. Ratios of 112 and 92 would be perceived this same way whether the trait involved were 60-day weaning weight or anything else.

Trait ratios represent a convenient way to portray phenotypic information. But they (and phenotypic measures in general) are limited in their ability to indicate breeding value. Even in species where ratios are commonly used, breeders are paying increasingly less attention to them and increasingly more attention to better indicators of breeding value—EBVs and EPDs.

EXERCISES

Study Questions

9.1 Define in your own words:

heritability (h^2)
heritability in the broad sense (H^2)
repeatability (r)
contemporary group
contemporary group effect (E_{cg})
trait ratio

9.2 **a.** Construct five-record column charts—use Figure 9.1 as a guide—to illustrate:

i. a highly heritable trait.
ii. a lowly heritable trait.

b. Explain how the charts differ.

9.3 Why is heritability in the narrow sense more useful than heritability in the broad sense?

9.4 What general categories of traits tend to be lowly heritable? Highly heritable?

9.5 How can a trait be completely genetically determined and yet not be heritable?

9.6 Students sometimes assume that if a trait is highly heritable, breeding values for the trait are necessarily high. What is wrong with this thinking?

9.7 What is wrong with the following statement: "My dog has high heritability for retrieving instinct"? How should the statement be reworded to reflect its intended meaning?

9.8 Why is it that close relatives tend to have similar performance in highly heritable traits but not in lowly heritable ones?

9.9 Explain why accuracy of selection is generally greater for more heritable traits than for less heritable ones.

9.10 Why is an estimate of heritability needed for the prediction of breeding values, progeny differences, etc.?

9.11 How would information about the heritability of traits help you make decisions about animal management?

9.12 **a.** Construct 5-animal (10-record) column charts—use Figure 9.3 as a guide—to illustrate:
 i. a highly repeatable trait.
 ii. a lowly repeatable trait.
 b. Explain how the charts differ.

9.13 Explain the connection between the two definitions of repeatability: (1) a measure of the strength of the relationship between repeated records for a trait in a population; (2) a measure of the strength of the relationship between single performance records and producing abilities for a trait in a population.

9.14 What is wrong with the following statement: "Ewe 422 has high repeatability for twinning"? How should the statement be reworded to reflect its intended meaning?

9.15 How would information about the repeatability of traits help you make culling decisions?

9.16 Why is an estimate of repeatability needed for the prediction of producing abilities, breeding values, progeny differences, etc. for repeated traits?

9.17 List four methods breeders can use to increase the heritability and repeatability of traits.

9.18 For a species and quantitative trait of your choice:
 a. How would you make the environment more uniform?
 b. How would you increase measurement accuracy and precision?
 c. List environmental effects that can be adjusted for mathematically and explain how you would adjust for them.
 d. What criteria would you use to determine contemporary groups?

9.19 Use the genetic model for quantitative traits (the latest version) to illustrate how a performance record provides a better representation of underlying breeding value if it is expressed as a deviation from a contemporary group mean than if it is not.

9.20 Under what circumstances would expressing performance as a deviation from a contemporary group mean be misleading?

9.21 Describe an example (preferably one you are personally familiar with) of improper contemporary grouping.

9.22 What interpretive advantage does a trait ratio have over a deviation from a contemporary group mean?

Problems

9.1 Siberian racing muskrats are the Russian equivalent of American jumping frogs. The following genetic parameters for time to swim 50 meters (T) and lifetime winnings (W) have been estimated by animal scientists at the Smyatogorsk Polytechnic Institute:

$$\sigma^2_{BV_T} = 4 \text{ sec}^2 \quad \sigma^2_{GCV_T} = 1 \text{ sec}^2 \quad \sigma^2_{E_T} = 11 \text{ sec}^2$$

$$\sigma^2_{BV_W} = 100 \text{ rubles}^2 \quad \sigma^2_{GCV_W} = 0 \text{ rubles}^2 \quad \sigma^2_{E_W} = 2{,}400 \text{ rubles}^2$$

a. Calculate:
  i. phenotypic variance of 50-m time.
  ii. phenotypic variance of lifetime winnings.
  iii. heritability of 50-m time.
  iv. heritability of lifetime winnings.

b. Is a muskrat's single record for 50-m time a good indicator of its breeding value for the trait? Why or why not?

c. Would you expect 50-m time to respond well to selection? Why or why not?

d. Are a muskrat's lifetime winnings a good indicator of its breeding value for the trait? Why or why not?

e. Would you expect lifetime winnings to respond well to selection? Why or why not?

f. A young male muskrat, Pyotr's Oski Doski, swam 50 meters 8 seconds faster than average. Predict his breeding value for 50-m time.

9.2 H. Cushman (Cushy) Pearson IV raises Thoroughbreds. His six two-year-olds posted the following records for lengths behind at the finish in their first two races:

	Lengths Behind	
Horse #	Race 1	Race 2
1	0.0	0.0
2	4.5	3.0
3	9.0	10.5
4	4.0	0.0
5	13.0	9.5
6	5.5	7.0

a. Calculate repeatability for lengths behind from this admittedly small sample.

b. Is first race performance a good indicator of second race performance? How do you know?

c. Should Cushy have sold any of his horses after their first race? Why or why not?

d. If phenotypic variance for lengths behind is approximately 18 lengths^2, what is the variance of producing ability (horse value) for this trait? What is temporary environmental variance for the trait?

e. Are differences in performance in this trait due more to differences in horse value or differences in temporary environmental effects? How do you know?

9.3 Vasily Yevshenko is widely recognized as a master muskrat breeder. A true perfectionist, Vasily has so standardized the management and training of his animals that the variance of environmental effects on 50-m time in his pack is just 6 sec^2. Assuming other genetic parameters as those listed for Problem 9.1, what is the heritability of 50-m time in Yevshenko's pack? What principle is illustrated here?

9.4 Birth weight in a breed of beef cattle average 76.8 lb for heifer calves and 82.2 lb for bull calves. Calculate:

a. the *additive* adjustment factor needed to adjust heifer birth weights to a bull basis.

b. the equivalent *multiplicative* adjustment factor.

9.5 Age-of-dam adjustment factors for weaning weight in a particular breed of beef cattle are:

Cow Age, yr	Adjustment, lb
2	+60
3	+40
4	+20
5+	0

Assume $\sigma_{BV_{WW}} = 30$ lb.

a. Calculate $\sigma_{P_{WW}}$ and h^2_{WW} using unadjusted weights from the following data set.

Calf #	Age of Dam, yr	Unadjusted Weight, lb
1	2	440
2	2	470
3	3	530
4	4	630
5	5	520
6	7	560
7	9	570
8	10	460

b. Do the same using weights that have been adjusted for age of dam.

c. What effects did adjustment for age of dam have?

9.6 Given the following average daily gain (ADG) data on a contemporary group of pigs, calculate on ADG ratio for each pig.

Pig #	ADG, lb/day
1	1.82
2	1.49
3	1.23
4	1.54
5	1.60
6	1.29
7	1.43
8	1.62

CHAPTER 10

Factors Affecting the Rate of Genetic Change

We typically measure the effectiveness of selection by the rate of genetic change that results. In theory, we would like to maximize the rate of genetic change, an achievable goal if we could consistently choose those animals with the best breeding values to be parents. The problem of course, is that we do not know the true breeding values of animals—we must work with *predictions* of breeding values which, in many cases, may not be very informative. The task of selection is therefore not a simple one. Here are some questions that come to mind right away:

- Should I save many female replacements or just a few?
- Should I use many males or just the very best ones?
- Should I use well-proven, older males or promising young ones?
- Should I base selection on individual performance or should I consider information on relatives?
- Should I select strictly within my own herd or flock or should I look to other populations for replacements?

None of these questions is easy to answer. Fortunately, however, we know the general factors that affect the rate of genetic change. An understanding of these factors can help us answer these questions. It can help us develop selection strategies and design breeding programs.

ELEMENTS OF THE KEY EQUATION FOR GENETIC CHANGE

key equation
The equation relating the rate of genetic change resulting from selection to four factors: accuracy of selection, selection intensity, genetic variation, and generation interval.

The factors affecting the rate of genetic change resulting from selection are summarized in what is often called the **key equation** for genetic change. In short, the key equation states that the rate of genetic change is directly proportional to three factors: accuracy of selection, selection intensity, and genetic variation, and is inversely proportional to a fourth factor: generation interval. Let's examine each element of the key equation briefly. (We will return to them later, exploring each one in more detail then.)

Accuracy of Selection

accuracy of selection (accuracy of breeding value prediction) A measure of the strength of the relationship between true breeding values and their predictions for a trait under selection.

Accuracy was defined in Chapter 2 as a measure of the strength of the relationship between true values and their predictions. In the context of selection and genetic change, we are particularly interested in **accuracy of selection** or, more precisely, **accuracy of breeding value prediction:** a measure of the strength of the relationship between true *breeding values* and their predictions for a trait under selection. The reason for this is straightforward: the more accurately we can predict breeding values, the more likely that the animals we choose to be parents will actually *be* the best parents. If accuracy were perfect, we would know every individual's breeding value exactly. Selection would then be a simple matter of selecting the animals with the best breeding values (assuming, of course, that "best" is unambiguous, i.e., that the desired direction of genetic change is known). Accuracy is never perfect, but the higher the accuracy, the better.

Accuracy of selection depends on a number of factors. Heritability, a measure of the strength of the relationship between performance (phenotypic values) and breeding values, is a major one. The higher the heritability of a trait, the better each piece of performance information is as a predictor of underlying breeding value. Any steps that breeders can take to increase heritability—managing animals uniformly, taking careful measurements, adjusting for known environmental effects, or using contemporary groups—will increase accuracy of selection.

Accuracy can also be increased by using more information and more sophisticated genetic prediction technology. In the next chapter we will discuss how performance information on individuals and potentially large numbers of relatives can be combined to provide more accurate predictions of breeding values.

Differences in accuracy of selection can be large. Selection based solely upon each candidate's own phenotypic record, particularly if the trait under selection is lowly heritable, is not very accurate. In contrast, selection of sires on the basis of EPDs derived from large volumes of progeny data is very accurate.

Selection Intensity

selection intensity A measure of how "choosy" breeders are in deciding which individuals are selected.

The rate of genetic change depends also on **selection intensity.** Selection intensity measures how "choosy" breeders are in deciding which individuals are selected. To select very intensely means to choose only the very best individuals, according to whatever criterion (or criteria) selection is based upon. At the opposite extreme, to select with no intensity at all means to choose animals purely at random.

If selection criteria (phenotypic values, predictions of breeding values, etc.) are reasonably accurate, in other words, if they are reasonable indicators of underlying breeding values, then intensely selected parents should be far better than average genetically. Their offspring—the next generation—should be equally superior, and the rate of genetic change should be fast. On the other hand, if parents are chosen with little intensity, they cannot be much better than average genetically. Their offspring will be equally lackluster, and the rate of genetic change will be slow.

Genetic Variation

In the context of the key equation for genetic change, **genetic variation** refers to the variability of breeding values within a population for a trait under selection. You can think of the range of breeding values available for selection as a measure

genetic variation (in the context of the key equation for genetic change): Variability of breeding values within a population for a trait under selection.

of genetic variation (though, technically speaking, it is not the best choice). With lots of genetic variation, this range is large—the best animals are far superior to the worst. With little genetic variation, this range is small—the best individuals are only slightly better than the worst.

If there exists tremendous genetic variation in a population for a particular trait, and if breeders select only the very best individuals based on accurate measures of that trait, then the selected individuals and their offspring will be far better than average, and the rate of genetic change will be fast. However, if there is little genetic variation, then even the best individuals will be only a little better than average, so will their progeny, and the rate of genetic change will be slow.

Generation Interval

generation interval The amount of time required to replace one generation with the next.

The fourth and last factor affecting the rate of genetic change, **generation interval,** is the amount of time required to replace one generation with the next. The shorter the generation interval, the faster the rate of genetic change.

Consider, for example, mice and men. The generation interval in humans is very long. Puberty is late (relative to other species) and, in many societies, customs dictate that young people wait some time before reproducing. Even if humans were selected accurately and intensely (which they are not) and were selected for genetically variable traits, genetic change would be agonizingly slow. The long wait between generations keeps change at a snail's pace. Mice, on the other hand, reproduce within a few months of birth. Their generation interval is so short that they are capable of producing 150 generations in the time it takes humans to produce one. With so many opportunities for selection in such a short period of time, mouse breeders can make extremely fast genetic change. That is one reason why mice are such desirable laboratory animals.

THE KEY EQUATION IN MORE PRECISE TERMS

The factors affecting the rate of genetic change, the elements of the key equation, can be defined precisely enough to allow us to use the key equation for comparing selection strategies. Mathematically, the key equation can be written

$$\Delta_{BV}\Big/_t = \frac{r_{BV,\hat{BV}}\, i\, \sigma_{BV}}{L}$$

where $\Delta_{BV}\Big/_t$ = rate of genetic change per unit of time (t)[1]

$r_{BV,\hat{BV}}$ = accuracy of selection

i = selection intensity

σ_{BV} = genetic variation

L = generation interval

Let's examine each element in more detail.

[1]The Greek letter delta (Δ) is commonly used to signify "change."

The Rate of Genetic Change

rate of genetic change $\left(\Delta_{BV}/_{t}\right)$ **or response to selection**
The rate of change in the mean breeding value of a population caused by selection.

In the key equation, genetic change refers to genetic change *due to selection*, and for this reason the **rate of genetic change** $\left(\Delta_{BV}/_{t}\right)$ is often termed **response to selection.** More specifically, it is the rate of change in the mean *breeding value* of a population caused by selection. Mating systems like inbreeding and crossbreeding cause genetic changes in a population too, but such changes typically involve gene combination value. Because only breeding value is passed from parent to offspring, change due to selection of parents is really change in breeding value.

The unit of time $\left(t \text{ in } \Delta_{BV}/_{t}\right)$ is whatever unit is used to measure generation interval and depends, therefore, on species. In livestock, poultry, and companion animals, the unit of time is usually years. In laboratory species like fruit flies and mice, it might be weeks or even days.

Accuracy of Selection

accuracy of selection $(r_{BV,\hat{BV}})$
The correlation between true breeding values and their predictions for a trait under selection.

selection criterion (*SC*)
An EBV, EPD, phenotypic value, or other piece of information forming the basis for selection decisions.

Accuracy of selection, the strength of the relationship between true breeding values and their predictions for a trait under selection, is represented by the correlation between true breeding values and their predictions or $r_{BV,\hat{BV}}$.

As it is used here, the term "prediction" can be interpreted quite liberally. A prediction could be an estimated breeding value (EBV) calculated from extensive data, it could be a single phenotypic value, or it could be simply a breeder's perception or educated guess. In any case, predictions comprise the information on which selection decisions are based, and from here on I will refer to a prediction used for selection as a **selection criterion** or **SC.**

Accuracy ranges from zero (no information) to almost one (*lots* of information). It is never negative.

Selection Intensity

selection intensity (*i*)
The difference between the mean selection criterion of those individuals selected to be parents and the average selection criterion of all potential parents, expressed in standard deviation units.

selection differential
The difference between the mean selection criterion of those individuals selected to be parents and the average selection criterion of all potential parents, expressed in units of the selection criterion.

Selection intensity (*i*) measures how "choosy" breeders are in deciding which individuals are selected. Mathematically, selection intensity is the difference between the mean selection criterion of those individuals selected to be parents ($\overline{SC}_s$) and the average selection criterion of all potential parents ($\overline{\overline{SC}}$), expressed in standard deviation units.[2] The conversion from units of the selected criterion to standard deviation units is accomplished by simply dividing by the standard deviation of the selection criterion (σ_{SC}). Thus

$$i = \frac{\overline{SC}_s - \overline{\overline{SC}}}{\sigma_{SC}}$$

The numerator of the right-hand expression, the difference between the mean selection criterion of those individuals selected to be parents ($\overline{SC}_s$)and the average selection criterion of all potential parents ($\overline{\overline{SC}}$), expressed in units of the selection criterion (i.e., before conversion to standard deviation units) is called **selection differential.**

[2]The subscript "s" in $\overline{SC}_s$ stands for "selected," and the double bar in $\overline{\overline{SC}}$ signifies an *overall* mean.

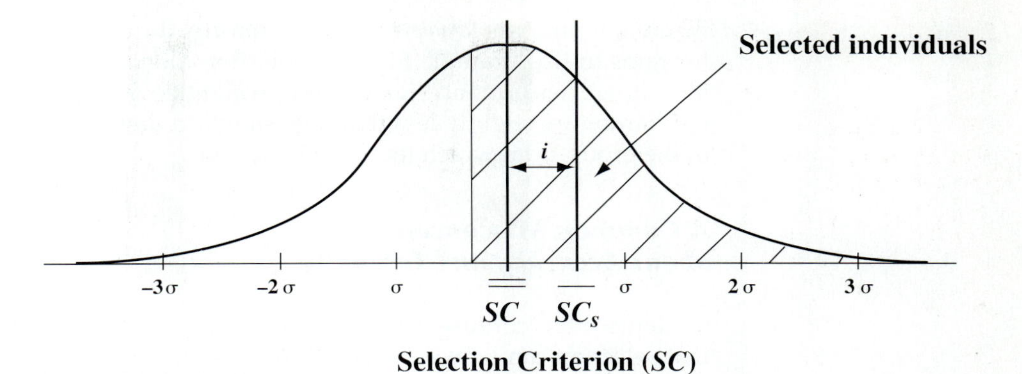

FIGURE 10.1 Graphic depiction of selection intensity. The distribution shown is that of a hypothetical selection criterion (*SC*—phenotypic value, EPD, etc.) expressed in standard deviation units. In this example, the mean selection criterion ($\overline{SC}_s$) of those individuals selected to be parents—the individuals whose selection criterion lies in the hashed area of the distribution—is .6 standard deviations above the mean selection criterion of all potential parents ($\overline{\overline{SC}}$). Thus selection intensity (*i*) is .6.

Selection intensity is depicted graphically in Figure 10.1. The distribution shown is that of a hypothetical selection criterion for all potential parents in a population. Again, the selection criterion could be one of many things—a phenotypic value, an EPD, etc. The units used are standard deviations of the selection criterion. Thus the practical range of the selection criterion is from about −3 standard deviations to about +3 standard deviations. In this example, the mean selection criterion of those individuals selected to be parents ($\overline{SC}_s$) is .6 standard deviations above the mean selection criterion of all potential parents ($\overline{\overline{SC}}$). Thus $i = .6$.

For a more concrete example of selection intensity, suppose you are a sheep breeder and are selecting replacement ewe lambs. Your selection criterion is each ewe lamb's own performance for yearling weight. Of the ewe lambs available from your flock, you plan to keep a fixed number based on this selection criterion. The average yearling weight of all ewe lambs in the flock is 130 lb, and the average yearling weight of the selected ewe lambs is 158 lb. Given the following:

$$\overline{\overline{SC}} = \overline{\overline{P}}_{YW} = 130 \text{ lb}$$

$$\overline{\overline{SC}}_S = \overline{P}_{YWs} = 158 \text{ lb}$$

$$\sigma_{SC} = \sigma_{P_{YW}} = 30 \text{ lb}$$

then

$$i_f = \frac{\overline{SC}_s - \overline{\overline{SC}}}{\sigma_{SC}}$$

$$= \frac{158 - 130}{30}$$

$$= .93$$

where i_f represents *female* selection intensity. (I do not address selection intensity for rams in this example.) Female selection intensity is .93. In other words, using individual performance for yearling weight as a selection criterion, the selected ewe lambs are slightly less than one standard deviation heavier than the average of the group from which they were selected.

A Common Misconception about Accuracy and Intensity

Students often confuse accuracy and intensity. The typical mistake is to assume that selection can only be intense if the selection criterion is a good (i.e., accurate) one. The reasoning seems to be that you can only really be choosy in making selection decisions if you have good information on which to make those decisions.

In fact, accuracy and intensity are independent concepts. You can select intensely (i.e., be very choosy) regardless of the accuracy of the selection criterion. If the selection criterion is not a good indicator of breeding value for the trait you want to improve, you can still choose what *appear* to be the very best animals. You can still select intensely. Unfortunately, intense selection in this case will not be very effective because accuracy is so poor. In other words, when the *product* of accuracy and intensity is a small number, the rate of genetic change will be slow.

Consider the previous example of selection for yearling weight in sheep. The selection criterion is individual performance for yearling weight, and selecting a fixed number of ewe lambs for yearling weight results in female selection intensity of .93 standard deviations. Yearling weight is moderately to highly heritable, and the accuracy of selection ($r_{BV,\hat{BV}}$) using this criterion is approximately .6. When these numbers are combined in the key equation with appropriate measures of genetic variation, generation interval, and male selection intensity, the expected rate of genetic change is about 5 lb per year.[3]

An alternative selection criterion might be expected progeny difference (EPD) for yearling weight. If we assume the following for ewe lambs:

$$\overline{\overline{SC}} = \overline{\overline{\text{EPD}}}_{YW} = +5 \text{ lb}$$

$$\overline{SC}_s = \overline{\text{EPD}}_{YWs} = +19.1 \text{ lb}$$

$$\sigma_{SC} = \sigma_{\text{EPD}_{YW}} = 15.2 \text{ lb}$$

then

$$i_f = \frac{\overline{SC}_s - \overline{\overline{SC}}}{\sigma_{SC}}$$

$$= \frac{19.1 - 5}{15.2}$$

$$= .93$$

Note that selection intensity did not change by using EPDs instead of individual performance figures. The set of ewe lambs chosen on the basis of EPDs is unlikely

[3]A detailed example showing exactly how rates of genetic change are calculated appears later in this chapter.

to be exactly the same as the set chosen on the basis of own performance, but because the same number of animals and only the highest rated animals were selected in either case, they were selected with equal intensity.

EPDs incorporate much more information than just an animal's own performance record, so they are more accurate indicators of true breeding value. The average accuracy of EPDs for yearling weight in lambs could easily be .8. When this level of accuracy is incorporated in the key equation, the expected rate of genetic change is about 6.4 lb per year—a 28% increase over the expected rate using own performance as a selection criterion. Note that this improvement is *not* a result of increased selection intensity. It is a result of better accuracy of selection.

Selection Intensity with Truncation Selection

There are two ways to calculate selection intensity for a population. The most straightforward way is to calculate a deviation from the mean selection criterion (i.e., a selection differential) for each selected individual, average these, then convert to standard deviation units. This method will always work and is the appropriate method when there is not a clear, inviolate level of the selection criterion above which animals are selected and below which they are rejected. For example, if you promised your nephew his pick of your replacements, or if you decided to keep Blossom's daughter, not because she was acceptable in terms of the selection criterion, but because she was . . . well, Blossom's daughter, then there would be no clear level of the selection criterion, no clear *point of truncation,* that determines who is selected and who is not.

When there *is* a clear point of truncation (Figure 10.2), selection is termed **truncation selection,** and a shortcut method exists for calculating selection intensity. In this case, all you need to know is the number of individuals chosen to be parents as a proportion of the number of potential parents—or the **proportion saved (p).** The appropriate selection intensity can then be read from a table such as Table 10.1.

truncation selection
Selection on the basis of a distinct division in the selection criterion (point of truncation) above which individuals are kept and below which they are rejected.

proportion saved (p)
The number of individuals chosen to be parents as a proportion of the number of potential parents.

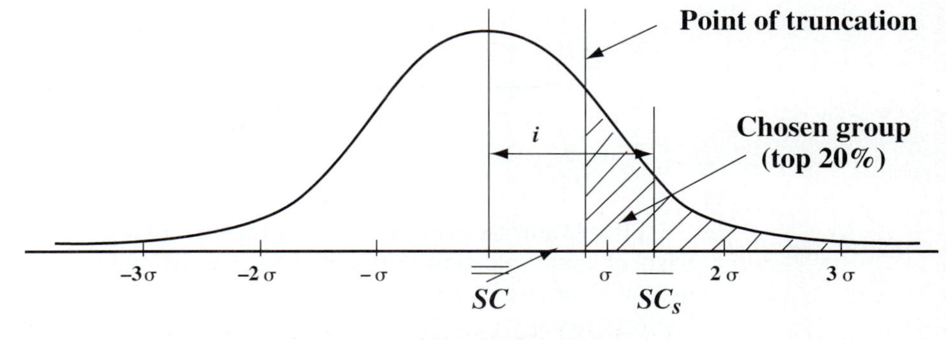

FIGURE 10.2 Selection intensity with truncation selection. Saving the top 20% (p = .2) of potential parents results in a selection intensity (i) of 1.4 standard deviations.

TABLE 10.1 Selection Intensity Expected from Truncation Selection

Proportion Saved (p)	Selection Intensity (i)
.01	2.67
.02	2.42
.03	2.27
.04	2.15
.05	2.06
.06	1.99
.07	1.92
.08	1.86
.09	1.80
.10	1.76
.11	1.71
.12	1.67
.14	1.59
.16	1.52
.18	1.46
.20	1.40
.22	1.35
.24	1.30
.26	1.25
.28	1.20
.30	1.16
.32	1.12
.34	1.08
.36	1.04
.38	1.00
.40	.97
.42	.93
.44	.90
.46	.86
.48	.83
.50	.80
.60	.64
.70	.50
.80	.35
.90	.20
1.00	.00

For example, in Figure 10.2 the top 20% of potential parents are chosen on the basis of some selection criterion. The proportion saved is then .2, and because selection is strictly by truncation, we can use Table 10.1 to find the corresponding selection intensity of 1.4 standard deviations.

Be careful to use this second method *only* with truncation selection. If you save 20% of potential parents, but the chosen animals are not necessarily the best according to your selection criterion, then the proportion saved is indeed .2, but selection intensity would be something less than 1.4 standard deviations.

Selection Intensity with Threshold Traits

Threshold traits were defined in Chapter 5 as polygenic traits that are not continuous in their expression, but rather exhibit categorical phenotypes. Threshold traits present a number of special problems, and perhaps the most immediate of these relates to selection intensity. Selection intensities for threshold traits tend to be small, and the better the performance of a population in a threshold trait, the smaller the selection intensity.

To understand the reason for this, consider the example of fertility (as measured by success or failure to conceive)—the same trait used in the discussion of threshold traits in Chapter 7. Imagine that fertility has a continuous underlying scale of liability as shown in Figure 10.3. Look first at the upper figure, population (a). In this population 20% of females fail to breed. So long as less than 80% (100% − 20%) of females are to remain the breeding population, no nonpregnant females need be selected. The problem, however, is that there is no way to distinguish between those pregnant individuals that are high on the liability scale and those that are low. By selecting from the pregnant group, we are essentially choosing a random sample of animals from that group—not necessarily the most fertile animals. In this case we are selecting a random sample from the best 80%. Selection intensity, the difference between the mean selection criterion of those individuals selected to be parents ($\overline{SC}_s$) and the average selection criterion of all potential parents ($\overline{\overline{SC}}$), is just .35 standard deviations.

Suppose the actual proportion of females saved (p) in this population is 70% or .7. If fertility had been a quantitative trait, one that is continuously expressed, selection intensity (as determined from Table 10.1) would have been .50 standard deviations. But because the 70% saved are actually a random sample chosen from the best 80%, selection intensity is less. The **effective proportion saved (p_e)**, the value that, when substituted for actual proportion saved (p), reflects correct selection intensity, is .8.

effective proportion saved (p_e)
In selection—a value that, when substituted for actual proportion saved (p), reflects correct selection intensity.

Now look at the lower figure, population (b). For whatever reasons (genetic, environmental, or both), fertility is better in this population—only 5% fail to breed. Selection intensity is lower (.11σ), however, because now we are selecting randomly from the top 95% of the population. With threshold traits, the more progress that is made (i.e., the lower the incidence of the undesirable phenotype), the smaller the selection intensity.

Figure 10.3 illustrates the difficulty of selecting for a threshold trait like fertility. Breeders of sheep, swine, and cattle often make a point of culling nonpregnant females. While this may make sense economically, unless fertility rates are very poor, genetic gains, at least in the short term, are likely to be small. There is simply too little selection intensity.

Loss of selection intensity is generally less severe for threshold traits that have more than one threshold. Dystocia, for example, is typically described by several categories (as opposed to just two in the fertility example), with a threshold occurring between each category. The more categories, the closer the threshold trait resembles a continuous trait, and the smaller the threshold effect on selection intensity.

How should we deal with threshold traits? One answer is to select for genetically related continuous traits. Beef cattle breeders, for example, select for smaller birth weight (a continuous trait) in order to reduce dystocia (a threshold trait). Another possibility is to use as the selection criterion a continuously expressed EBV or EPD for a threshold trait. This approach has the advantage of increasing both selection intensity and accuracy of selection.

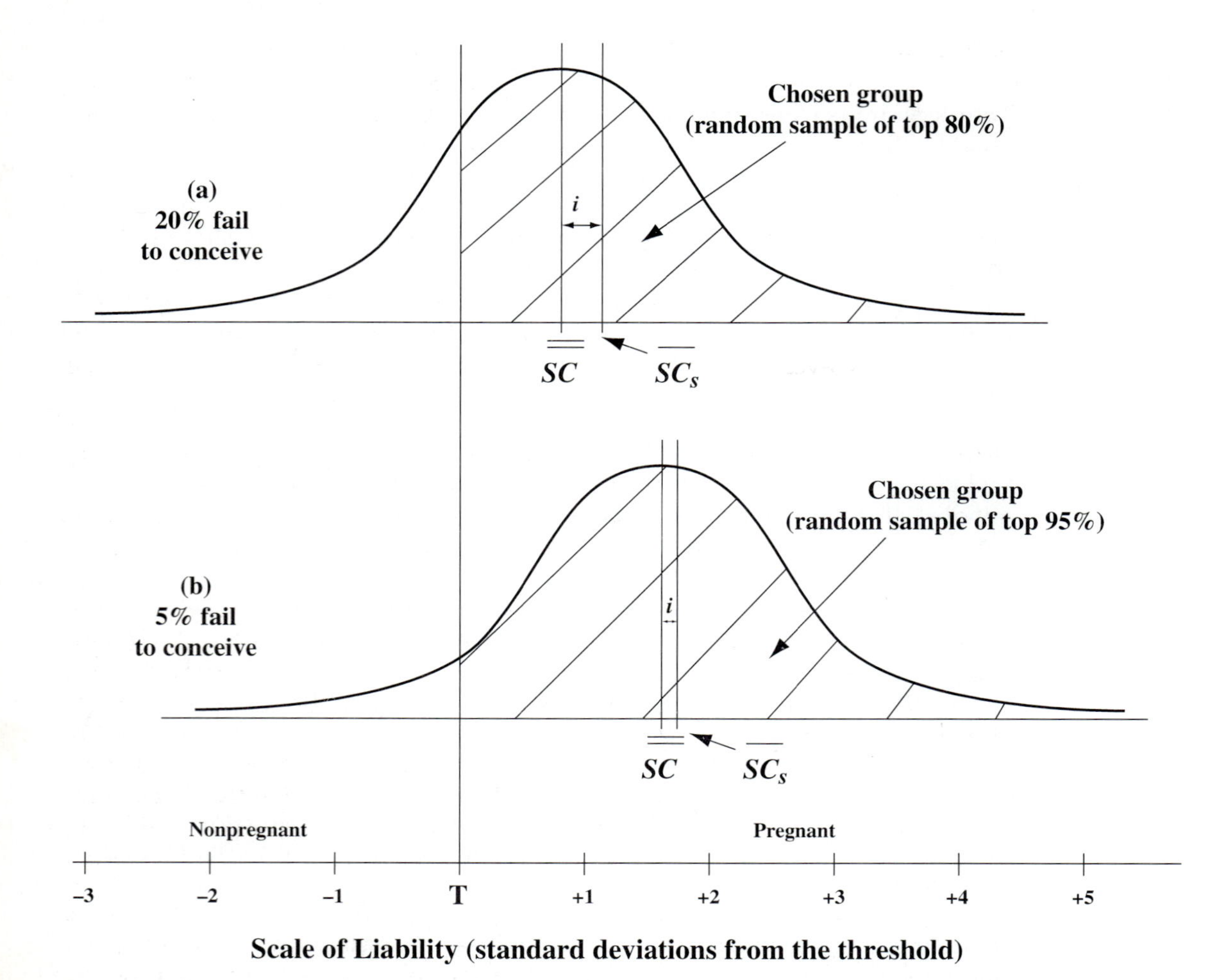

FIGURE 10.3 Schematic representation of selection intensity with a threshold trait—in this case, fertility as measured by breeding success or failure. Animals to the left of the threshold (marked **T** on the liability scale) are nonpregnant. Those to the right are pregnant. Selection intensity is less a function of the actual proportion saved than of the incidence of the undesirable phenotype. In population (*a*), 20% of females fail to breed, and the selection intensity of .35σ reflects random sampling from the top 80% of females. Observed fertility is better in population (*b*)—only 5% fail to breed. Sampling from the top 95% of females, however, results in lower selection intensity (.11σ).

Genetic Variation

Genetic variation is represented in the key equation by the standard deviation of breeding values (σ_{BV}). Unlike accuracy and selection intensity, genetic variation is not something that is easy to manipulate. It tends to be fairly fixed within a population. Outbreeding can increase it somewhat, and inbreeding can have the opposite effect. Selection in one direction over many generations will (in theory) reduce genetic variation, although laboratory experiments suggest that genetic change usually ceases well before genetic variation is exhausted. Furthermore, with the possible exception of broiler chickens, genetic change in domestic animals is slow enough due to either long generation intervals, lack of accuracy, intensity or consistency of selection, or all of the above, that loss of genetic variation is unlikely to be a problem—at least in the near term.

Generation Interval

generation interval (L) (in a closed population): The average age of parents when their selected offspring are born.

closed population A population that is closed to genetic material from the outside.

Generation interval (*L*) is the amount of time required to replace one generation with the next. In **closed populations**—populations that are closed to genetic material from the outside—generation interval can be defined as the average age of parents when their selected offspring are born. For example, beef cattle typically calve first as two-year-olds and calve last at anywhere from two years of age to well into their teens. In any given year, however, the average age of cows calving in a herd is typically between four and six. If we do not restrict in any way the ages of cows whose calves are eligible to become replacements (e.g., if we do not eliminate daughters out of first-calf heifers from consideration or breed older cows to terminal sires and thus rule out their daughters as replacements), then four to six years is the generation interval for females. (Generation intervals for males and females can be quite different.)

The "average age" definition for generation interval works well for herds or flocks that are truly closed (i.e., provide all their own replacements, male and female). It also works well for entire species or subspecies—for example, the entire population of Holstein cattle. It becomes inappropriate, however, when animals are imported from outside a population. For example, if semen from an outside beef bull is introduced into a formerly self-contained herd, the age of that bull should have little bearing on the amount of time required within that herd to replace one generation with another. He might, in fact, be a chronologically old bull, but if he is genetically superior, then with respect to the genetic level of the herd, he is quite "young."

Listed in Table 10.2 are ranges of generation intervals for a number of livestock species. A quick look at the table should make it clear why genetic change is extremely slow in horses (L = 8 to 12 years) and extremely fast in chickens (L = 1 to 1.5 years).

The Key Equation with Phenotypic Selection

In the particular case of phenotypic selection (selection based solely on an individual's own phenotype), the key equation for genetic change reduces to

$$\Delta_{BV}\Big/_{t} = \frac{h^2 i \sigma_P}{L}$$

TABLE 10.2 Common Generation Intervals

Species	Generation Interval (years)
Horses	8 to 12
Dairy cattle	4 to 6
Beef cattle	4 to 6
Sheep	3 to 5
Swine	1.5 to 2
Chickens	1 to 1.5

In this simplified version, accuracy of selection appears to have been replaced by heritability (h^2), and the measure of genetic variation (σ_{BV}) appears to have been replaced by a measure of phenotypic variation (σ_P). (The exchange is not one for one, as the following proof will show.)

Proof of the Formula for Response to Phenotypic Selection

The general form of the key equation is:

$$\Delta_{BV}\Big/t = \frac{r_{BV,\hat{BV}}\, i\sigma_{BV}}{L}$$

With phenotypic selection, accuracy ($r_{BV,\hat{BV}}$) is simply the square root of heritability or h.

Proof

Given:

$$h^2 = r^2_{BV,P}$$

and

$$ACC = r_{BV,\hat{BV}}$$

or, more appropriately for phenotypic selection,

$$ACC = r_{BV,SC}$$

where

$$SC = P$$

then

$$\begin{aligned} ACC &= r_{BV,P} \\ &= \sqrt{r^2_{BV,P}} \\ &= \sqrt{h^2} \\ &= h \end{aligned}$$

The revised key equation is then:

$$\Delta_{BV}\Big/t = \frac{hi\sigma_{BV}}{L}$$

Now,

$$h^2 = \frac{\sigma^2_{BV}}{\sigma^2_P}$$

so

$$h = \sqrt{\frac{\sigma^2_{BV}}{\sigma^2_P}} = \frac{\sigma_{BV}}{\sigma_P}$$

Multiplying the numerator of the key equation by $h\left(\frac{1}{h}\right)$ or $h\left(\frac{\sigma_P}{\sigma_{BV}}\right)$, we have:

$$\Delta_{BV}\Big/t = \frac{h(h)i\sigma_{BV}\left(\frac{\sigma_P}{\sigma_{BV}}\right)}{L} = \frac{h^2 i\sigma_P}{L}$$

phenotypic selection differential (S) The difference between the mean performance of those individuals selected to be parents and the average performance of all potential parents, expressed in units of the trait.

The term $i\sigma_P$ is commonly referred to as **phenotypic selection differential (S).** It is the difference between the mean performance of those individuals selected to be parents and the average performance of all potential parents, expressed in units of the trait. Thus

$$S = i\sigma_P = \overline{P}_s - \overline{\overline{P}}$$

and

$$\Delta_{BV}\Big/t = \frac{h^2 S}{L}$$

For a simple example, if heritability for days to 230 lb in feeder pigs is .25, the animals chosen to be parents are 15 days younger than average when they reach 230 lb, and generation interval is 1.7 years, then

$$\Delta_{BV}\Big/t = \frac{h^2 S}{L} = \frac{.25(-15)}{1.7} = -2.2 \text{ days per year}$$

Partitioning the Key Equation

The key equation is conceptually simple. For actual comparison of selection strategies, however, we need to refine it somewhat (i.e., make it more complicated). In particular, we need to account for the fact that different groups of animals may have different accuracies of selection, selection intensities, and generation intervals. (They may, in fact, have different amounts of genetic variation, though it is commonly assumed that they do not.)

Males versus Females

Accuracy, intensity, and generation interval are often different for males and females. In most species, relatively few sires are needed because each sire can be bred to many females. It is therefore possible to be much "choosier" in sire selection than in dam selection, so selection intensity is considerably higher for sires than dams. Because males typically have many times the progeny that females have and therefore many times the amount of progeny data, accuracy of selection is also generally higher for sires than dams. Generation interval can be longer for sires than dams—it takes time to generate all those progeny data—but this is not necessarily the case and is species and management dependent.

The key equation, when modified to account for different accuracies, intensities, and generation intervals in males and females, becomes

$$\Delta_{BV}\Big/ t = \frac{(r_{BV_m,\hat{BV}_m} i_m + r_{BV_f,\hat{BV}_f} i_f)\sigma_{BV}}{L_m + L_f}$$

where the subscripts m and f refer to males and females, respectively. In the special case of phenotypic selection, it reduces to

$$\Delta_{BV}\Big/ t = \frac{h^2(i_m + i_f)\sigma_P}{L_m + L_f}$$

or

$$\Delta_{BV}\Big/ t = \frac{h^2(S_m + S_f)}{L_m + L_f}$$

This form of the key equation simply averages the genetic superiority of selected males and females, and then divides by the average generation interval of the two sexes. (The 2s used in averaging cancel each other and therefore do not appear in the equation's final form.) The implicit assumption is that, as groups, males and females are equal in their genetic contribution to the next generation—which, in fact, they are.

Use of this version of the key equation is sometimes called the two-path method for predicting the rate of genetic change—one path for males and one for females. A complete example appears later in this chapter.

progeny test
A test used to help predict an individual's breeding values involving multiple matings of that individual and evaluation of its offspring.

The Four-Path Method for Dairy Species

In dairy cattle and goats, many of the most important traits—*all* of the milk traits, obviously—cannot be measured in males. This consideration, combined with the use of **progeny tests** for young males and stringent criteria for selecting the dams of those young males, leads to a four-path form of the key equation. The groups of selected animals defined by each path are (1) sires to produce future sires, (2) sires to produce future dams, (3) dams to produce future sires, and (4) dams to produce future dams. The equation then becomes

$$\Delta_{BV}\Big/ t = \frac{(r_{BV_1,\hat{BV}_1} i_1 + r_{BV_2,\hat{BV}_2} i_2 + r_{BV_3,\hat{BV}_3} i_3 + r_{BV_4,\hat{BV}_4} i_4)\sigma_{BV}}{L_1 + L_2 + L_3 + L_4}$$

where the subscripts 1, 2, 3, and 4 refer to each of the four groups. Like the two-path, male/female form of the key equation, this version simply averages the genetic superiority of each group and divides by the mean generation interval.

Replacements versus Existing Parents

Another way to partition the key equation for genetic change would be to divide the selected animals into two groups: new replacements and existing parents. Accuracy of selection differs between these groups because existing parents already have progeny data, sometimes—as in the case of many males—lots of progeny data. Young replacements have no progeny data. On the other hand, intensity of selection for existing parents, particularly dams, may be low. Once they have entered a breeding herd or flock, they are often more likely to be culled for age or infertility than for the original selection criterion.

The mathematics associated with partitioning the key equation in this way is complicated and will not be presented here. In comparing selection strategies, however, it is sometimes useful to distinguish between the elements of the key equation as they apply to replacements versus existing parents—even if only in a subjective manner.

TRADE-OFFS AMONG ELEMENTS OF THE KEY EQUATION

Ideally, we would like to maximize accuracy of selection, selection intensity, and genetic variation, and minimize generation interval. Doing so would maximize the rate of genetic change. But can we do all these things at once? The answer is clearly no. There are trade-offs among elements of the key equation; a favorable change in one element often dictates an unfavorable change in another. Some of these trade-offs are rather subtle, and what may be a major trade-off in one species may be of little consequence in another. Following are descriptions of some of the better understood, more consistent trade-offs.

Accuracy Versus Generation Interval

A decrease in generation interval usually causes a decrease in accuracy of selection. This is because fewer records, particularly progeny records, are available for use in genetic prediction. For example, we could significantly reduce the generation interval in many species by allowing sires to be used for only one year. Accuracy of selection would be less, however, because there would be no well-evaluated, progeny-tested sires to choose from.

Accuracy Versus Intensity

An increase in accuracy of selection is often accompanied by a decrease in selection intensity and vice versa. To see why, consider the example of young dairy sires. Promising young dairy bulls are typically tested by mating them to an assortment of cows, and then recording the first lactation performance of their daughters. The number of matings is often limited by economics, so there is a choice to be made between testing fewer bulls (and having more records on each bull) or testing more bulls (and having fewer records on each). In the first case, accuracy of selection will be greater because of more data per bull, but selection intensity will be less because there will be fewer tested bulls to pick from. In the second case, the reverse is true. Having fewer records on each bull means lower accuracy, but the larger number of tested bulls allows greater intensity.

Intensity Versus Generation Interval

replacement rate The rate at which newly selected individuals replace existing parents in a population.

To select replacements very intensely is to choose relatively few, only the very best ones. In so doing, the **replacement rate,** the rate at which newly selected individuals replace existing parents, is kept low. If population size is to remain constant, a lower replacement rate dictates that animals remain in the breeding population longer. Thus the generation interval, the average age of parents when their selected offspring are born, increases. The general rule, then, is that an increase in selection intensity is associated with an increase in generation interval and vice versa.[4] The trade-off is usually different for females and males, however.

Females

Selection intensity in replacement females is relatively limited, especially in species like cattle and horses which only occasionally produce twins and almost never produce litters. The need to replace older, nonpregnant, or unsound females requires that a fairly large proportion of young females be kept as replacements. Still, there is often some room to increase selection intensity, and when the intensity of replacement selection increases, generation interval increases—just as you would expect.

The logical question is then: Should I save fewer female replacements, increasing selection intensity but increasing generation interval as well, or should I keep more replacements, sacrificing selection intensity for faster generation turnover? The answer to this question is summarized in the following rule of thumb: If sires are genetically far superior to dams, save many female replacements, and if sires are only marginally better than dams, save fewer replacements and be more selective in your choices.

The rationale behind this rule is illustrated in Figure 10.4. Look first at the upper diagram (a). The dark vertical line represents the breeding value of a single sire

[4]This rule applies if by selection intensity we mean intensity of *replacement* selection. If, on the other hand, we mean intensity of selection of existing parents—call it "culling intensity"—then the opposite is true; the more intensely parents are culled, the higher the replacement rate, and the shorter the generation interval. To avoid confusion, remember that unless explicitly stated otherwise, the term "selection intensity" almost always refers to intensity of replacement selection.

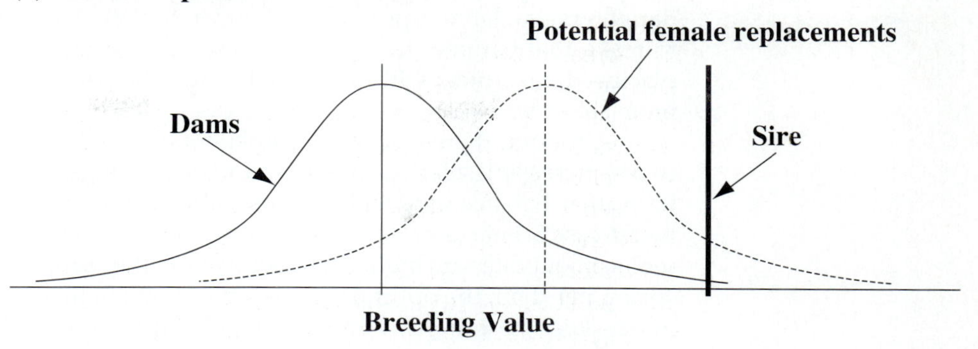

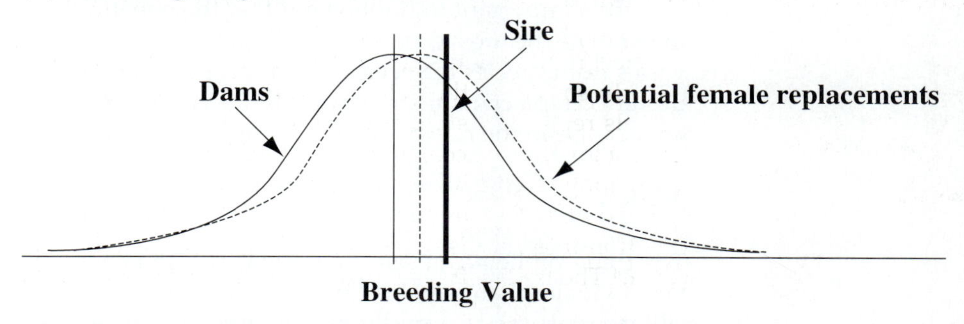

FIGURE 10.4 Schematic representation of the rule of thumb for balancing selection intensity and generation interval in female replacement selection. If sires are genetically far superior to dams (*a*), save many female replacements; if sires are only marginally better than dams (*b*), save fewer replacements and be more selective. (See text for more explanation.)

or the average breeding value of a group of sires. The bell-shaped curves to the left of the sire line represent the distributions of breeding values for dams (solid curve) and their daughters or potential replacements (dotted curve). In this instance, the sire is far superior in terms of breeding value to almost all of the dams to which he is bred. Perhaps the owner of this herd or flock recently assembled the female population by purchasing other breeders' culls, saved money by buying females so cheaply, and invested those savings in the very best semen available. In any case, he would be wise to replace the original females as quickly as possible. Their daughters, having such an outstanding sire, are likely to be superior to their mothers, so it makes sense to keep many of them. In other words, it is better in this situation to turn generations rapidly and not worry about selection intensity.

In the lower diagram (b), the sire is only slightly superior to the average dam. This might be the case for a lowly heritable trait in which accuracy of sire selection is poor. In this instance, daughters are unlikely to be much better than their dams, so there is less of a need to replace the dams, and a breeder can justify being more selective in his choice of replacements. In other words, it is better in this situation to increase selection intensity and be less concerned with generation interval.

The important thing to remember here is that we want to replace dams with daughters that have superior breeding values, and we should do whatever it takes to make that happen. In the situation illustrated in Figure 10.4(a), that meant replacing dams quickly. In the situation depicted in Figure 10.4(b), it meant being more selective.

Although there is a balance to be found in replacement female selection between increasing selection intensity and decreasing generation interval, in the long term, that balance inevitably favors rather fast generation turnover. Even in the worst-case scenario in which sires are genetically no better than dams, optimum replacement rates are high. For species like cattle, which typically have just one offspring per dam, the optimum replacement rate in this scenario is about 30%. That means keeping roughly 70% of available replacements. For litter bearing species such as swine, optimum replacement rates in this situation might be as high as 75%. Because sows produce so many potential replacements, however, this means keeping only 20% to 25% of them.

It is important to understand that the genetically optimum replacement rate may not be the most economically sensible replacement rate. Net replacement cost varies from species to species and from environment to environment. A truly well-designed replacement strategy would account for this cost, balancing the value of an increase in the rate of genetic change against the cost to produce it.

Males

The trade-off between selection intensity and generation interval is much less severe in males than females. So few males are needed as replacements that it is possible to replace sires rapidly and still practice intense replacement selection. In this respect, you can essentially "have your cake and eat it too."

Intensity versus Risk

Males

selection risk
The risk that the true breeding values of replacements will be significantly poorer than expected.

A more important trade-off in sire selection is between selection intensity and **selection risk,** the risk that the true breeding values of replacements will be significantly poorer than expected. Selection intensity can be increased by using only the most promising sires and very few of them. With artificial insemination and frozen semen, you could use just one sire, even if the female population is large. The problem is that this is like putting all your eggs in one basket. What if the one sire's true breeding value is not nearly as good as you had thought? By using more sires, selection intensity is reduced somewhat, but selection risk is reduced as well. In a group of sires, there may be some poor ones, but the *average* of their true breeding values should match expectations.

Females

Selection risk is generally not an issue in female selection. As a rule, breeders keep relatively large numbers of replacement females, so the mean breeding value of replacements is not likely to be much different than anticipated. And because females typically have few offspring relative to males, the consequences of a mistake in female selection are considerably less serious than the consequences of a mistake in male selection.

COMPARING SELECTION STRATEGIES USING THE KEY EQUATION: AN EXAMPLE

To see how the key equation and the various concepts embodied in it can be used to compare selection strategies, consider the following example. Sarah's dad Stan is an old-time breeder of purebred beef cattle and is a real traditionalist. Stan runs a closed herd and selects for essentially yearling weight (though, without a scale on the place, he never actually weighs any animals—just "eyeballs" them). He keeps the top 3% of his young bulls and the top 50% of his heifers for replacements. He mates young bulls and heifers for the first time when they are two years of age so that heifers calve initially as three-year-olds. Because his cattle are registered, Stan keeps extensive pedigree records but, to be perfectly honest, he makes little use of them.

Sarah has just returned from the university and is determined to modernize the breeding program. She tells her father about a number of changes she would like to make. Stan is skeptical, but not wanting to dampen the enthusiasm of his daughter, he agrees to experiment with some new ideas if Sarah can show him what effect they will have. Thus challenged, Sarah lists the following options:

1. Do nothing different.
2. Buy a scale and weigh yearlings, adjusting the weights for known environmental effects and expressing them as deviations from contemporary group means.
3. Use the pedigree information, the newly acquired weights, and Sarah's personal computer to calculate within-herd estimated breeding values (EBVs) for yearling weight.
4. Mate the young bulls and heifers as yearlings so that the heifers calve first at two years of age.
5. Keep more replacement heifers—the top 80%—and cull all but the very best cows after their fifth calf.
6. Do options 2 through 5.

Here is Sarah's reasoning. Her dad is a good cattleman, but his eye is not as good as a scale. And he has never thought to account for factors like age of the calf, age of its dam, or the environmental effect of its contemporary group. She knows that the changes detailed in option 2 will increase the heritability of yearling weight in her father's herd and therefore increase accuracy of selection. Sarah believes that calculating EBVs (option 3) will increase accuracy even further by incorporating information on relatives (in this case, mostly half sibs). Option 4, breeding replacements earlier, should shorten the generation interval, as should option 5, keeping more replacement heifers. Sarah is aware, however, that this last option will reduce selection intensity.

Option 1: Do nothing different. To establish a basis for comparison, Sarah decides to use the two-path version of the key equation, which will allow her to use different values for each of the two sexes, i.e.,

$$\Delta_{BV}\Big/ t = \frac{(r_{BV_m, \hat{BV}_m} i_m + r_{BV_f, \hat{BV}_f} i_f)\sigma_{BV}}{L_m + L_f}$$

She assumes that accuracy of selection is the same for males and females, and because the type of selection used in this herd is simple phenotypic selection, accuracy is just the square root of heritability.[6] Sarah has no reliable information on the current heritability of yearling weight in her father's herd, but guesses it at .2. Thus

$$\begin{aligned} r_{BV_m, \hat{BV}_m} &= r_{BV_f, \hat{BV}_f} \\ &= \sqrt{h^2} \\ &= \sqrt{.2} \\ &\approx .45 \end{aligned}$$

From the table of selection intensities with truncation selection (Table 10.1), Sarah determines that the selection intensities corresponding to the top 3% of bulls and 50% of heifers are 2.27 and .80 standard deviations, respectively. As a measure of genetic variation, she uses a published value for the standard deviation of breeding values for yearling weight ($\sigma_{BV_{YW}}$) of 35 lb and, after some arithmetic, estimates generation intervals to be 6.5 years for females and 5 years for males. Putting all these numbers together, she calculates the current rate of genetic change in her father's herd:

$$\begin{aligned} \Delta_{BV}\Big/ t &= \frac{(r_{BV_m, \hat{BV}_m} i_m + r_{BV_f, \hat{BV}_f} i_f)\sigma_{BV}}{L_m + L_f} \\ &= \frac{[.45(2.27) + .45(.80)](35)}{5 + 6.5} \\ &\approx 4.2 \text{ lb per year} \end{aligned}$$

Option 2: Buy a scale and weigh yearlings, adjusting the weights for known environmental effects and expressing them as deviations from contemporary group means. Sarah predicts that having done these things, heritability of yearling weight ought to be at least the average of published estimates (.4), or double the current heritability of .2. Then

$$\begin{aligned} r_{BV_m, \hat{BV}_m} &= r_{BV_f, \hat{BV}_f} \\ &= \sqrt{h^2} \\ &= \sqrt{.4} \\ &\approx .63 \end{aligned}$$

[6]For a proof, see the previous boxed section.

and

$$\Delta_{BV}\Big/_t = \frac{(r_{BV_m,\hat{BV}_m} i_m + r_{BV_f,\hat{BV}_f} i_f)\sigma_{BV}}{L_m + L_f}$$

$$= \frac{[.63(2.27) + .63(.80)](35)}{5 + 6.5}$$

$$\approx 5.9 \text{ lb per year}$$

Option 2 would, in theory anyway, speed up the rate of genetic change by 40%—from 4.2 to 5.9 lb per year.

Option 3: Use the pedigree information, the newly acquired weights, and Sarah's personal computer to calculate within-herd estimated breeding values (EBVs) for yearling weight. Sarah cannot know for sure, but she guesses that by using EBVs to incorporate sib information (i.e., by basing selection decisions on more than just an animal's own yearling weight), accuracy of selection will increase to about .8. If she is right, then

$$\Delta_{BV}\Big/_t = \frac{(r_{BV_m,\hat{BV}_m} i_m + r_{BV_f,\hat{BV}_f} i_f)\sigma_{BV}}{L_m + L_f}$$

$$= \frac{[.8(2.27) + .8(.80)](35)}{5 + 6.5}$$

$$\approx 7.5 \text{ lb per year}$$

That would be a 78% increase in the rate of genetic change.

Option 4: Mate the young bulls and heifers as yearlings so that the heifers calve first at two years of age. Doing this, and nothing else, should decrease the generation interval of both males and females by a year. Then

$$\Delta_{BV}\Big/_t = \frac{(r_{BV_m,\hat{BV}_m} i_m + r_{BV_f,\hat{BV}_f} i_f)\sigma_{BV}}{L_m + L_f}$$

$$= \frac{[.45(2.27) + .45(.80)](35)}{4 + 5.5}$$

$$\approx 5.1 \text{ lb per year}$$

Option 5: Keep more replacement heifers—the top 80%—and cull all but the very best cows after their fifth calf. From Table 10.1, Sarah determines that retaining so many heifers will decrease female selection intensity from .80 to .35 standard deviations, but hopes that the speedup in generation turnover will more than compensate for the loss in intensity. After some more arithmetic, she estimates that the higher replacement rate will result in a female generation interval of 4.7 years (assuming heifers still calve for the first time at three years of age). Then

$$\Delta_{BV}\Big/_t = \frac{(r_{BV_m,\hat{BV}_m} i_m + r_{BV_f,\hat{BV}_f} i_f)\sigma_{BV}}{L_m + L_f}$$

$$= \frac{[.45(2.27) + .45(.35)](35)}{5 + 4.7}$$

$$\approx 4.3 \text{ lb per year}$$

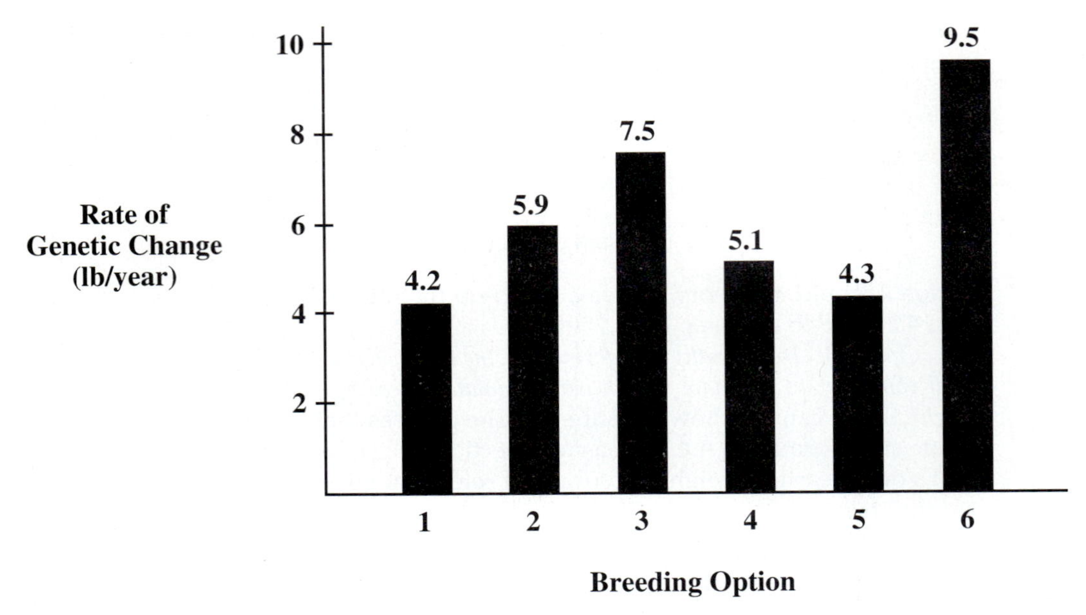

FIGURE 10.5 Rates of genetic change in yearling weight expected in Stan's cattle herd using six breeding options. (See text for details.)

Option 6: Do options 2 through 5. The result of all of Sarah's changes together is

$$\Delta_{BV}\Big/ t = \frac{(r_{BV_m,\hat{BV}_m} i_m + r_{BV_f,\hat{BV}_f} i_f)\sigma_{BV}}{L_m + L_f}$$

$$= \frac{[.8(2.27) + .8(.35)](35)}{4 + 3.7}$$

$$\approx 9.5 \text{ lb per year}$$

or more than double the current rate of genetic change.

Sarah presents her results to her father using a bar graph (Figure 10.5). Stan is impressed. He agrees to try everything Sarah suggests with the exception of saving 80% of the heifers. (Sarah decides not to pursue this particular issue. Her father is no dummy. He knows that keeping more replacements means more calves from two- and three-year-old cows, and because young cows produce less milk than mature cows, their calves will weigh less as weanlings and yearlings. The environmental effect of cow age on *phenotype* for yearling weight did not come into Sarah's calculations of rates of genetic change. She may recommend keeping more heifers again at a later date—after she has convinced her father to use artificial insemination to breed to some *really* good bulls. One thing at a time.)

THE KEY EQUATION IN PERSPECTIVE

The key equation is very instructive. Because it tells us a lot about the factors that affect the rate of genetic change, it can be used to compare different selection strategies. We must be careful, however, not to take the key equation too literally. For example, suppose the sire with the very best breeding value happens to be very old—perhaps long since dead and available only via frozen semen. Should we avoid using him because he will lengthen the generation interval? Not at all. Remember that the fundamental goal of selection is to choose as parents the individuals with the best breeding values. If we are successful in this, then considerations of accuracy, intensity, and generation interval are redundant. In other words, who cares about the sire's age if he is really the best sire?

The example of the old but outstanding sire may seem contrived. If we are doing a good job of selection, the best animals should be the youngest. Still, for various reasons breeders often change selection directions, and old sires once considered passé can enjoy a revival. In situations like this, it is especially important to understand *how* the elements of the key equation work and to know that the key equation is a tool, not a definitive answer.

Manipulating the elements of the key equation to maximize the rate of genetic change may make a great deal of genetic sense but little economic sense. In the example of Stan's herd, implementing all of Sarah's suggestions may not be cost-effective. A scale is probably a good investment, but the additional feed required to develop heifers and young bulls so that they are capable of breeding as yearlings may outweigh the benefit involved. Stan needs to do a careful economic analysis. The optimum rate of genetic change may be something less than the maximum rate.

MALE VERSUS FEMALE SELECTION

In most species, selection of males is much more important than selection of females. The reason for this is not that males as a group contribute any more to the next generation than females as a group do. If you consider the *environmental* effect that females have on their offspring—the effect of milk production, for example—females actually contribute more. The reason is that *individual* males typically contribute much more than do individual females, so selection of those males becomes critical.

In terms of the key equation, the relative importance of male selection can be seen in the advantages that males have over females in accuracy of selection and selection intensity. As noted earlier, few sires are needed, so selection of sires can be very intense. And because sires are capable of having many more progeny than are dams, accuracy of sire selection can be much higher than accuracy of dam selection. For most species and scenarios within species, if you take the key equation and parse out the contributions of males versus females to the rate of genetic change, you will be impressed by the importance of male selection and by the relative unimportance of female selection. Of the 9.5 lb per year improvement in yearling weight expected using all of Sarah's breeding recommendations, over 80% can be attributed to sire selection.

Modern reproductive technology only magnifies the situation. Artificial insemination allows males to have even greater accuracy (more progeny per sire) and selection intensity (a larger population of sires to choose from). Embryo transfer could do the same for females but, for both biological and economic reasons, not to the same degree.[7]

[7]See Chapter 20 for more on artificial insemination and embryo transfer.

Genetic Change when Sires are Purchased

genetic trend Change in the mean breeding value of a population over time.

Almost all commercial producers and many seedstock producers buy their sires from other breeders rather than raise their own. As a result, the rates of genetic change in their herds or flocks are—in the longer term, anyway—completely determined by the genetic change over time or **genetic trend** in the purchased sires. Female selection can influence to some degree the difference or *lag* in breeding values between offspring and sires, but in only a few generations after outside sires have been introduced, the genetic trend in offspring will be the same as the genetic trend in sires.

This concept is illustrated in Figure 10.6. The solid line represents genetic trend in purchased sires. The other lines represent genetic trends in four populations that vary in initial differences in breeding value between sires and dams (the equivalent of three versus the equivalent of six generations of sire trend) and in intensities of female selection (selection so intense that selection differential for females is the same as that for purchased sires versus no female selection differential at all).

In the first few generations, rapid genetic progress is made as each population "catches up" with the sires being used. Soon, however, the genetic trends in all four populations are the same as the sire trend. Breeding values in the two populations with intense selection of females lag one generation behind sire breeding values. Breeding values in the two populations with no selection differential in females lag two generations behind. Thus, while female selection is not without consequence, it is the selection of purchased sires that determines long-term genetic change.

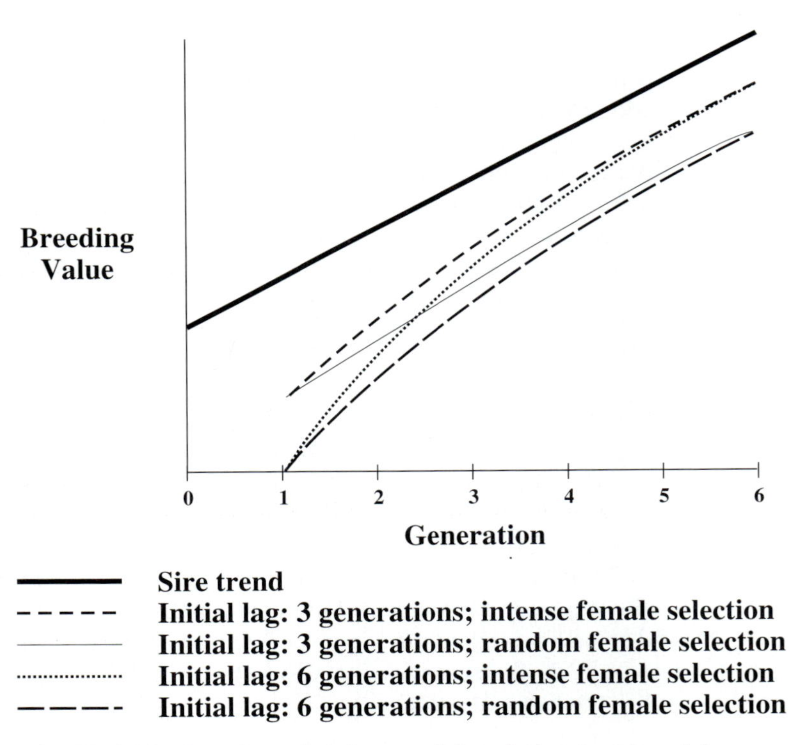

FIGURE 10.6 Genetic trends in four populations that use purchased sires. "Initial lag" refers to the difference in breeding value (expressed in generations of sire trend) between sires and dams in generation zero.

EXERCISES

Study Questions

10.2 Define in your own words:

key equation
accuracy of selection or accuracy of breeding value prediction ($r_{BV,\hat{BV}}$)
selection intensity (i)
genetic variation (σ_{BV})
generation interval (L)
rate of genetic change (Δ_{BV}/t) or response to selection
selection criterion (SC)
selection differential
truncation selection
proportion saved (p)
effective proportion saved (p_e)
closed population
phenotypic selection differential (S)
progeny test
replacement rate
selection risk
genetic trend

10.2 **a.** Explain why an increase in each of the following speeds the rate of genetic change in a population:
  i. accuracy of selection
  ii. selection intensity
  iii. genetic variation

b. Explain why a decrease in generation interval speeds the rate of genetic change.

10.3 For a species and a trait of your choice, describe two breeding scenarios that differ in:

a. accuracy of selection.
b. selection intensity.
c. genetic variation.
d. generation interval.

10.4 Express the key equation mathematically and explain the relationship between the general concept suggested by each term in the equation and the mathematical meaning of each term.

10.5 Can you practice intense selection using inaccurate selection criteria? Why or why not?

10.6 What shortcut can you use to determine selection intensity in the case of truncation selection that you cannot use in the general case?

10.7 Use diagrams of a normal distribution to explain why achieving a high level of selection intensity can be difficult with threshold traits.

10.8 Why is it that with threshold traits, the more progress that is made (i.e., the lower the incidence of the undesirable phenotype), the smaller the selection intensity?

10.9 How can a breeder avoid—to a degree, anyway—the selection intensity dilemma with threshold traits?

10.10 Under what circumstances does the "average age" definition of generation interval become inappropriate?

10.11 Describe two commonly used methods for partitioning the key equation and explain the rationale for each.

10.12 Use an example to explain the following trade-offs among elements of the key equation:

a. accuracy versus generation interval
b. accuracy versus intensity
c. intensity versus generation interval
   i. females
   ii. males
d. intensity versus risk
   i. females
   ii. males

10.13 If, from all available candidates for selection, you are consistently successful in selecting those with the best breeding values, should you be concerned about accuracy of selection, selection intensity, or generation interval? Explain.

10.14 Why is selection of males, as a rule, much more important than selection of females?

10.15 If all sires (but no females) are purchased from sources outside a population, what effect will female selection have on the long-term rate of genetic change in the population? Explain.

Problems

10.1 Calculate the rate of genetic change in feed conversion in a swine population given the following:

Heritability of feed conversion (h^2) = .35
Phenotypic standard deviation (σ_P) = .2 lb
Accuracy of male selection ($r_{BV_m, \hat{BV}_m}$) = .8
Accuracy of female selection ($r_{BV_f, \hat{BV}_f}$) = .5
Intensity of male selection (i_m) = -2.4
Intensity of female selection (i_f) = -1.5
Generation interval for males (L_m) = 1.8 years
Generation interval for females (L_f) = 1.8 years

10.2 In the beef cattle population described in Problem 9.5, adjusting weaning weights for age of dam caused the heritability of weaning weight to increase from .22 to .28. Assuming phenotypic selection for weaning weight alone:

a. What will be the percentage improvement in the rate of genetic change in weaning weight when selection is based on adjusted weights instead of unadjusted weights?
b. What element of the key equation has been changed by adjusting weights?

10.3 A sheep breeder has determined that her ewes are not producing enough lambs and has decided to cull them heavily for twinning, a threshold trait. She keeps the top 35% of her ewes based on number of lambs born. What will be the effective proportion saved and selection intensity (culling intensity) for these ewes if:

a. under current conditions, 36% of the ewes produce twins, 56% produce singles, and 8% fail to breed.
b. management is improved so that 56% of the ewes produce twins, 41% produce singles, and only 3% fail to breed.
c. Why did improving management reduce selection intensity?

10.4 The Dairy Board of Eastern Serbo-Slavonia is reexamining its dairy improvement program. The current program has the following attributes:

Path	Accuracy of Selection ($r_{BV,\hat{BV}}$)	Proportion Saved (p)	Selection Intensity (i)	Generation Interval (L)
Sires to produce future sires	.85	3%	2.27	6 years
Sires to produce future dams	.85	15%	1.55	7 years
Dams to produce future sires	.5	1%	2.67	5 years
Dams to produce future dams	.5	90%	.20	6 years

The phenotypic standard deviation of milk yield in this population is 2,160 lb, and heritability of milk yield is .25.

a. What is the rate of increase in milk yield under the current program?

b. Serbo-Slavonian dairy scientists are considering requiring dams of future sires to have an additional lactation record. They anticipate that this would increase accuracy of selection for these dams from .5 to .6. It would, of course, increase their generation interval by a year. What should be the rate of increase in milk yield under the revised program?

c. Which program should work better? Why?

10.5 A rancher runs a closed herd of breeding cattle. He normally keeps and breeds the top 3% of his bull calves based on individual performance for yearling weight (YW). His sires average three years of age when their offspring are born. He is studying two female replacement strategies:

- saving the top 20% of his heifers based on YW ($L_f = 6.2$ years)
- saving the top 60% based on YW ($L_f = 3.2$ years)

a. If $h^2_{YW} = .5$, and $\sigma_{P_{YW}} = 60$ lb, calculate the expected rate of genetic change in yearling weight for each strategy.

b. What elements of the key equation is the rancher experimenting with?

c. What element appears to be more important?

10.6 Of the 9.5 lb per year improvement in yearling weight expected using all of Sarah's breeding recommendations, over 80% can be attributed to sire selection. Prove it. (See text for details.)

CHAPTER 11

Genetic Prediction

By now it should be clear that the rate of genetic change in a population depends to a large degree on accuracy of selection or, more precisely, accuracy of breeding value prediction. Accuracy can be increased to a certain extent by taking steps to increase the heritability of traits. Managing animals uniformly, taking careful measurements, adjusting for known environmental effects, using contemporary groups—all these things help. To increase accuracy further, however, requires that we use as much information as possible and weight each piece of information appropriately. In short, it requires that we use the technology of genetic prediction.

The purpose of this chapter is to provide an overview of two closely related methodologies commonly used for genetic prediction: the selection index and best linear unbiased prediction. We will not examine these technologies in detail. That would require a background in statistics and matrix algebra that few students have. The emphasis here is not on *how* selection indexes and best linear unbiased prediction work, but rather on when they should be used and what they are capable of. Chapter 12 deals with the presentation and interpretation of genetic predictions produced with these technologies.

COMPARING ANIMALS USING DATA FROM GENETICALLY SIMILAR GROUPS—THE SELECTION INDEX

selection index A linear combination of phenotypic information and weighting factors that is used for genetic prediction when performance data come from genetically similar contemporary groups.

Selection index theory was first developed in the 1930s and 1940s both as a method for genetic prediction and as a means of combining traits in order to select animals in an economically optimal way. In this chapter, we will consider the **selection index** in its role as a prediction methodology. Its role in multiple-trait selection will be discussed in Chapter 14.

A selection index is essentially a linear combination or *index* of various kinds of phenotypic information and appropriate weighting factors. It takes the form:

$$I = b_1x_1 + b_2x_2 + \cdots + b_nx_n$$

where I = an *index value* or genetic prediction
b_i = a weighting factor
x_i = a single item of pheonotypic information—a performance record or the average of a group of performance records
n = the total number of items of phenotypic information

own performance data
Information on an individual's own phenotype.

pedigree data
Information on the genotype or performance of ancestors and(or) collateral relatives of an individual.

progeny data
Information on the genotype or performance of descendants of an individual.

bias
Any factor that causes distortion of genetic predictions.

In a selection index (or, for that matter, in any method of genetic prediction) the information used in calculating genetic predictions for an individual comes from three kinds of sources: (1) the individual's **own performance** record(s), (2) performance records of ancestors and(or) collateral relatives of the individual **(pedigree data),** and (3) performance records of descendants of the individual **(progeny data).** The relative amounts of data from these sources vary. Unborn or very young animals have only pedigree data. As these animals get older, they acquire performance data of their own. (Assuming, of course, that they *can* acquire performance data of their own. Some traits are sex-limited. Dairy sires, for example, do not have own performance records for milk traits.) If the animals are selected to become parents, they will generate progeny data, and if they become popular, they will have large amounts of progeny data.

Records from any of the three types of sources can be records of the particular trait for which predictions are being calculated, or they can be records of other, genetically-related traits. For example, in species for which dystocia is a concern, predictions of genetic susceptibility to dystocia can be determined from a direct measure of dystocia such as dystocia score, from birth weight (a genetically correlated trait), or from both traits.

The data used in a selection index—the x_is—come from many sources within the own performance, pedigree, and progeny data categories. For example, x_1 might be an individual's own performance record for a trait, x_2 might be the average performance of the individual's paternal half sibs for the same trait, x_3 might be the average performance of the individual's progeny, and x_4, x_5, and x_6 might represent the performance of the individual, half sibs, and progeny respectively, for a correlated trait. In any case, each x is a single number.

In a selection index, each item of phenotypic information is normally expressed as a deviation from a contemporary group mean. As we learned in Chapter 9, expressing performance in this way accounts for environmental differences between contemporary groups. The problem with this approach, however, is that it assumes that all contemporary groups are genetically similar. A deviation of +10 units in one contemporary group is assumed to be the genetic equivalent of a deviation of +10 units in any other contemporary group. If the mean breeding values of each group are the same (i.e., if the contemporary groups are indeed genetically similar), this assumption is correct. However, if the mean breeding values of contemporary groups differ, then the use of deviations from contemporary group means creates **bias** in the data. Records from genetically poorer contemporary groups appear better than they should, and records from genetically superior contemporary groups appear worse then they should.

For this reason, *selection indexes should only be used for genetic prediction when performance data come from contemporary groups thought to be genetically similar.* In practice this means that they should probably be used within individual herds or flocks and not across populations. And if a given herd or flock has experienced significant genetic change over time, then data from older contemporary groups—groups with "older" and therefore different mean breeding values—should be excluded from an index.

Selection indexes may be restricted in application, but they are very useful nevertheless. Suppose, for example, that you are breeding sheep and want to compare your rams on the basis of progeny performance within your flock. You could use a selection index to produce progeny-based EBVs or EPDs for each ram. The calculations involved are quite simple, requiring nothing more than a hand calculator. You will need a computer for more complex applications of the selection index, but even then, a typical PC and spreadsheet software will suffice.

Prediction Using Regression: A Review

A selection index is nothing more than a prediction equation. Recall the general form of a prediction equation from Chapter 8:

$$\text{Predicted value} = \text{regression coefficient} \times \text{“evidence”}$$

In a selection index, the index value (I) is a predicted value—normally an EBV, EPD, or MPPA. The x_is in the index are the "evidence," phenotypic information consisting of individual performance records or averages of groups of performance records (expressed as deviations from contemporary group means). The b_is in the selection index are regression coefficients. They are the regressions of true values (*BV*s, *PD*s, or *PA*s) on the evidence. In other words, they measure the expected change in true value per unit change in the evidence. Formulas for the b_is of selection indexes are often quite complicated as you will see, and the most arithmetically challenging part of using a selection index is calculating numerical values for ("solving" for) the b_is.[1]

From the Simple Prediction Equation to the Single-Source Selection Index

The simple prediction equation appears as:

$$\hat{Y}_i = \hat{\mu}_Y + b_{Y \cdot X}(X_i - \hat{\mu}_X)$$

where
$\hat{Y}_i$ = a predicted value for animal i
$\hat{\mu}_Y$ = the expected mean of predictions for animals in the population
$b_{Y \cdot X}$ = the regression of values being predicted (Y) on the evidence (X)
$X_i - \hat{\mu}_X$ = the evidence for animal i, expressed as a deviation from the population mean

In the simplest selection index, an index using a single source of evidence, the predicted value is the index value, or

$$\hat{Y}_i = I$$

[1]Techniques for deriving the formulas for selection index regression coefficients and some example derivations are given in the Appendix.

Breeding values, progeny differences, etc. average zero across a population, and so do their predictions. Thus

$$\hat{\mu}_Y = 0$$

The regression coefficient is

$$b_{Y \cdot X} = b$$

The evidence in a selection index is already expressed as a deviation from a contemporary group mean, so

$$(X_i - \hat{\mu}_X) = x$$

Piecing all this together,

$$I = 0 + bx$$
$$I = bx$$

Prediction Using a Single Source of Information

The simplest selection indexes are those that involve just one source of information—one x. These indexes are of the form:

$$I = bx$$

and are easy to calculate because just one equation is needed to solve for the regression coefficient b. This is the kind of selection index that you determine with just a hand calculator.

Formulas for regression coefficients used in some common single-source selection indexes are listed in Table 11.1. Identified in the far left column of the table is the kind of predicted value—EBV, EPD, or MPPA. The next column lists the kind of true value being predicted. The third column describes the single source of information (x) or the evidence being used, followed by a column of formulas for appropriate regression coefficients (b). The last column lists formulas for accuracy of prediction—the correlation between the true value (*BV, PD,* or *PA*) and its prediction (the index value I).[2]

[2]Genetic predictions for an individual animal are often accompanied by an accuracy value—a correlation between true values and their predictions. Correlations are population measures and, as such, it might seem odd that they can be assigned to an individual. Only values, remember, are supposed to be applied to individuals. Accuracies are in fact population measures, but the "population" is a very theoretical one. It is a population of hypothetical animals having exactly the same kinds and amounts of predictive information as the individual in question. An accuracy value, then, measures the strength of the relationship between true values and predictions in this abstract population and, at the same time, has relevance to an individual animal.

TABLE 11.1 A Sampling of Formulas for (1) Regression Coefficients Used to Calculate Predictions from a Single Source of Information, and (2) Associated Accuracies

Prediction (I)	True Value	Source of Information (x)	Regression Coefficient (b)	Accuracy
EBV	BV	P: a single (nonrepeated) performance record on the individual	h^2	h
EBV	BV	$\overline{P}$: the average of n records on the individual	$\frac{nh^2}{1 + (n - 1)r}$	$\sqrt{\frac{nh^2}{1 + (n - 1)r}}$
MPPA	PA	$\overline{P}$: the average of n records on the individual	$\frac{nr}{1 + (n - 1)r}$	$\sqrt{\frac{nr}{1 + (n - 1)r}}$
EBV	BV	$\overline{P}$: the average of single records on m half sibs	$\frac{mh^2}{4 + (m - 1)h^2}$	$\sqrt{\frac{\frac{1}{4}mh^2}{4 + (m - 1)h^2}}$
EPD	PD	$\overline{P}$: the average of single records on m half sibs	$\frac{\frac{1}{2}mh^2}{4 + (m - 1)h^2}$	$\sqrt{\frac{\frac{1}{4}mh^2}{4 + (m - 1)h^2}}$
EBV	BV	$\overline{P}$: the average of single records on m full sibs	$\frac{mh^2}{2 + (m - 1)(h^2 + 2c^2_{FS})}$	$\sqrt{\frac{\frac{1}{2}mh^2}{2 + (m - 1)(h^2 + 2c^2_{FS})}}$
EBV	BV	$\overline{P}$: the average of single records on p progeny	$\frac{2ph^2}{4 + (p - 1)h^2}$	$\sqrt{\frac{ph^2}{4 + (p - 1)h^2}}$
EPD	PD	$\overline{P}$: the average of single records on p progeny	$\frac{ph^2}{4 + (p - 1)h^2}$	$\sqrt{\frac{ph^2}{4 + (p - 1)h^2}}$
EBV	BV	$\overline{P}$: the average of single progeny records from l litters of size k	$\frac{2lkh^2}{4 + (k - 1)(2h^2 + 4c^2_{FS}) + (l - 1)kh^2}$	$\sqrt{\frac{lkh^2}{4 + (k - 1)(2h^2 + 4c^2_{FS}) + (l - 1)kh^2}}$
EBV	BV	$\overline{P}$: the average of n records apiece on p progeny	$\frac{\frac{1}{2}ph^2}{\frac{1 + (n - 1)r}{n} + (p - 1)\frac{h^2}{4}}$	$\sqrt{\frac{\frac{1}{4}ph^2}{\frac{1 + (n - 1)r}{n} + (p - 1)\frac{h^2}{4}}}$

For practice in using Table 11.1, suppose you wish to calculate a dairy cow's MPPA for milk yield. Let's call this cow Iris. The source of information in this case is Iris's own milking history. An older cow, she has five lactation records averaging 1,072 lb above contemporary group means. The formula for the appropriate regression coefficient, located in the third row of Table 11.1, is

$$\frac{nr}{1 + (n - 1)r}$$

Repeatability (r) of milk yield is typically about .5, and the number of Iris's records (n) is 5, so

$$b = \frac{nr}{1 + (n - 1)r}$$
$$= \frac{5(.5)}{1 + (5 - 1)(.5)}$$
$$= .833$$

For each 1-lb increase in Iris's 5-record average, we increase our expectation of her true producing ability by .833 lb.

To calculate Iris's MPPA, we must now substitute this regression coefficient into the selection index. Thus

$$I = bx$$
$$= .833(+1{,}072)$$
$$= +\ 893 \text{ lb}$$

Iris's MPPA is +893 lb. In other words, we expect her to produce 893 lb more milk than the mean of her contemporaries. If lactation yields in this herd typically average about 14,000 lb, then our prediction of Iris's next record is 14,893 lb of milk.

We can calculate the accuracy of this prediction ($r_{PA,\hat{PA}}$) using the formula in the last column of Table 11.1. Thus

$$r_{PA,\hat{PA}} = \sqrt{\frac{nr}{1 + (n - 1)r}}$$
$$= \sqrt{\frac{5(.5)}{1 + (5 - 1)(.5)}}$$
$$= \sqrt{.833}$$
$$= .91$$

Regression for Amount of Information

regression for amount of information The mathematical process causing genetic predictions to be more or less "conservative" (closer to the mean), depending on the amount of information used in calculating them.

Note that even though Iris averaged 1,072 lb more than the mean of her contemporaries, our prediction of her producing ability is less than 1,072 lb. This is because the prediction has undergone what is sometimes called **regression for amount of information.** This is a mathematical process that causes genetic predictions to be more or less "conservative" (closer to the mean) depending on the amount of information used to calculate them.

To get a better idea of how regression for amount of information works, consider two other cows in Iris's herd: Violet and Rose. Violet has two lactation records averaging 1,204 lb above contemporary group means. In her case,

$$b = \frac{nr}{1 + (n - 1)r}$$
$$= \frac{2(.5)}{1 + (2 - 1)(.5)}$$
$$= .667$$

Violet's MPPA is then

$$I = bx$$
$$= .667(+1{,}204)$$
$$= +803 \text{ lb}$$

with accuracy

$$r_{PA,\hat{PA}} = \sqrt{\frac{nr}{1 + (n - 1)r}}$$
$$= \sqrt{.667}$$
$$= .82$$

Rose has only one record. She produced 918 lb less milk than the average of her contemporaries. In Rose's case,

$$b = \frac{nr}{1 + (n - 1)r}$$
$$= \frac{1(.5)}{1 + (1 - 1)(.5)}$$
$$= .5$$

Rose's MPPA is then

$$I = bx$$
$$= .5(-918)$$
$$= -459 \text{ lb}$$

with accuracy

$$r_{PA,\hat{PA}} = \sqrt{\frac{nr}{1 + (n - 1)r}}$$
$$= \sqrt{.5}$$
$$= .71$$

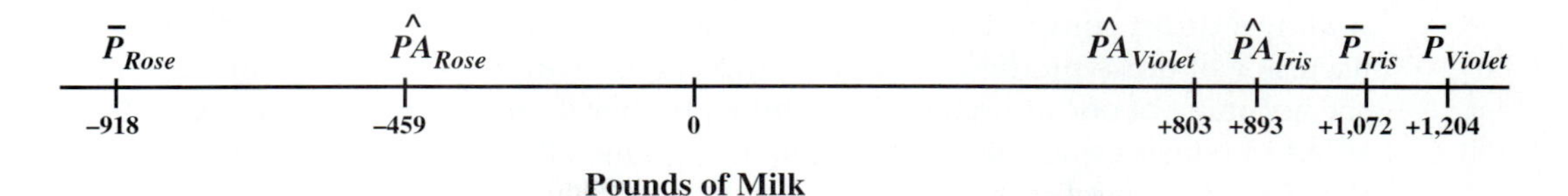

FIGURE 11.1 An example of regression for amount of information: mean lactation yields ($\bar{P}$) and most probable producing abilities ($\hat{PA}$) for three dairy cows, Iris, Violet, and Rose. With five records, Iris's prediction is regressed only slightly—from +1,072 to +893. Despite having the highest average yield, Violet does not have the highest MPPA. With only two records, her prediction has been regressed substantially—from +1,204 to +803. Rose's MPPA has been regressed also, but because her record is negative, her prediction has been regressed *upward*—from −918 to −459.

Lactation averages and MPPAs for Iris, Violet, and Rose are shown graphically in Figure 11.1. Iris's MPPA (+893 lb) is only slightly less than her actual lactation average (+1,072 lb). With five records, Iris has provided us with considerable information (as evidenced by an accuracy of .91), and we can be confident that she does indeed have a high producing ability. Still, five records are not enough records to be totally convincing. It is possible that Iris benefited from a succession of favorable temporary environmental effects (E_t). To account for this possibility, her MPPA has been regressed to a small degree. (The amount of regression, in this case 1,072 - 893 = 179 lb, is actually an estimate of Iris's average temporary environmental effect.)

Violet has the highest lactation average (+1,204 lb), but her MPPA is not as high as Iris's. This is because it has been regressed more. With only two records to go on (accuracy = .82), we cannot be as confident about Violet's ability, so we assign her a more conservative prediction.

Rose has only one record, so her MPPA has been regressed severely—from −918 to −459 lb. Because her first record is negative, Rose's MPPA has been regressed *upward,* this time in Rose's favor. There is a good chance that Rose experienced a poor temporary environmental effect during her first lactation, so we give her the benefit of the doubt.

The dairy cow example shows how MPPAs are regressed for amount of information, but similar examples could be constructed for any of the predictions listed in Table 11.1. In the MPPA example, the degree to which predictions have been regressed is a function of number of records and repeatability of the trait. In examples involving EBVs or EPDs, the amount of regression will usually be a function of number of records, heritability of the trait, repeatability of the trait (in cases involving repeated records), and pedigree relationship between the animals being measured and the animal whose prediction is being calculated. With the exception of pedigree relationship, the formulas for regression coefficients listed in Table 11.1 indicate exactly what factors come into play in each prediction situation.

unbiased
A genetic prediction is considered unbiased if, as more information is used in subsequent predictions for the same animal, those predictions are as likely to change in a positive direction as they are to change in a negative direction.

Regressing genetic predictions for amount of information results in **unbiased** predictions. This simply means that as more information is used in subsequent predictions for the same animal, those predictions are as likely to change in a positive direction as they are to change in a negative direction.

Because genetic predictions are regressed for amount of information, they are, in essence, adjusted for accuracy. In other words, the accuracy of a prediction is accounted for *in the prediction itself.* The benefit of adjustment for accuracy is that it allows direct comparisons of predictions for different animals regardless of the accuracy of those predictions. For example, if two animals have the same EBV, but the accuracy of one animal's EBV is much higher than the other's, you are still justified in considering the animals genetic equals. Of course, there is a greater risk that the true breeding value of the lower accuracy animal is significantly poorer than expected. He represents a greater selection risk. On the other hand, there is also a greater likelihood that this animal's true breeding value is significantly better than expected. In this sense, he represents a greater selection opportunity. The great benefit of accuracy values is that they allow you to select animals in a way that is compatible with your particular attitude toward risk taking.

Accounting for Common Environment

The formulas for regression coefficients and accuracies in Table 11.1 account for similarities in the performance of relatives, similarities that occur because relatives have genes in common. But sometimes relatives perform similarly for other reasons. For example, full sibs have the same dam, so they experience a common environment—in this case, a common maternal environment. Because they have both parents in common, they also share a larger proportion of gene combinations than nonrelatives or even half sibs do. And this source of genetic similarity, unless explicitly accounted for in some other way, is typically considered part of full sibs' "common environment" too. Because of common environment, performance records for full sibs tend to be more alike than you would expect given the proportion of genes they have in common (50%), and additional records on full sibs provide less information than would otherwise be the case.

common environmental effect
An increase in similarity of performance of family members caused by their sharing a common environment. Common environmental effects are particularly important within litters (full sibs).

In theory, **common environmental effects** can occur within families of any kind. They are known to occur, for example, within paternal half-sib families of dairy animals. Dairy farmers sometimes treat daughters of one sire differently than daughters of another sire, and this creates a common environmental effect. The clearest case of common environment, however, is within full-sib families. Common environmental effects are therefore especially important in litter bearing species such as swine, rabbits, cats, dogs, and mice.

When common environmental effects exist but are not accounted for in genetic prediction, regression coefficients (weighting factors) and associated accuracies are biased upward. They are not conservative enough. To properly account for common environmental effects, we incorporate a measure of covariation among relatives caused by common environment, denoted by c^2. Two of the 10 scenarios listed in Table 11.1 (rows six and nine) use information from full sibs. Note the use of c^2 in the denominators of the formulas for the regression coefficients and accuracies. By increasing their denominators, we decrease the size of regression coefficients and accuracies, making them more conservative.

Factors Affecting Accuracy of Prediction

Most of the factors that affect accuracy of prediction can be seen in the accuracy formulas listed in the far right column of Table 11.1. They are the same as the factors that affect the regression coefficients in the table: number of records, heritability, repeatability, and pedigree relationship. To get an idea of the relative importance of these factors to accuracy of prediction, it is useful to examine some different pre-

TABLE 11.2 Accuracy of Breeding Value Prediction from Single Sources of Information

Source of Information	Pedigree Relationship	No. of Records	Heritability .05	.30	.70
Individual	1.00	1	.22	.55	.84
Half sibs	.25	1	.06	.14	.21
		10	.17	.33	.41
		20	.22	.39	.45
		100	.37	.47	.49
		1,000	.48	.49+	.49+
Progeny	.50	1	.11	.27	.42
		10	.34	.67	.82
		20	.45	.79	.90
		100	.75	.94	.98
		1,000	.96	.99	.99+

diction situations. Listed in Table 11.2 are values for accuracy of breeding value prediction from different sources of information. The sources vary in pedigree relationship and number of records. "Individual" refers to the animal whose breeding value is being predicted. In all cases only one record per animal is assumed, and each source is considered independently (i.e., as a single source of information). Accuracies corresponding to three levels of heritability are shown.

The first thing to notice in Table 11.2 is that as heritability increases, so does accuracy of prediction—regardless of the source of information. Pick any row of the table. As you go from left to right (in other words, as heritability increases), accuracy increases as well. This makes sense because heritability measures the strength of the relationship between breeding values and phenotypic values. The stronger this relationship, the better is each animal's performance record as an indicator of that animal's breeding value. When the source of information is the individual's own record, heritability measures the strength of the relationship between the breeding value we are trying to predict and the source of information. Heritability *is* accuracy (or, to be technically correct, the square of accuracy). When the source of information is not the individual itself but some group of relatives (for example, half sibs), then higher heritability causes each sib's performance record to be a better indicator of its own breeding value and therefore a better indicator of the breeding value of its sibling—the individual in question.

Accuracy also increases with pedigree relationship. The numbers in the pedigree relationship column of Table 11.2 measure the proportion of genes held in common by different types of relatives. The individual has 100% of its genes in common with itself, 25% of its genes in common with half sibs, and 50% of its genes in common with progeny.[3] Compare the row for the individual's own record with the rows for one half sib and one progeny. Accuracy is highest for the individual's record, followed by the progeny record, then the record of the half sib. The closer the relationship of the individual to the animals providing the performance records, the better those records are as indicators of the individual's breeding value.

[3]See Chapter 17 for a more detailed explanation of pedigree relationship and its measures.

Accuracy of prediction increases with the number of records. Look at either the rows for half sibs or the rows for progeny in the table. The greater the number of records, the better the accuracy. Clearly, more records provide more information on which to base a prediction.

Table 11.2 offers insight into the value of an individual's own record as opposed to records on relatives. When heritability is high, the individual's own record is especially valuable. At h^2 = .7 (the last column in the table), the individual's own record (accuracy = .84) is more revealing than 10 progeny records (accuracy = .82) and better than a thousand records on half sibs (accuracy = .49). This follows directly from the definition of heritability: the strength of the relationship between performance and breeding values. When heritability is high, an individual's own performance record *should* be a good indicator of its breeding value.

If individual performance is most valuable at high levels of heritability, then it stands to reason that performance records on relatives should be most valuable at low levels of heritability. This is indeed the case. At h^2 = .05 (the first column of accuracies in the table), the individual's own record provides the same amount of information (accuracy = .22) as records on 20 half sibs or between four and five progeny, but cannot come close to providing the equivalent information of higher numbers of relatives.

Progeny records are the ultimate source of information for predicting breeding value. Note the high accuracies of prediction in Table 11.2 when there are large numbers of progeny. With enough progeny, accuracy is high even when heritability is low for the following reason. Progeny records provide a measure of the value of the genes that an individual transmits. A record on any single offspring may not be very revealing for several reasons: (1) environmental effects may have more influence than genetic effects (heritability may be low), (2) the breeding value of the *other* parent is unaccounted for, and (3) the offspring may have received a particularly good or particularly poor sample of genes from the individual—it may have benefited or suffered from Mendelian sampling. However, environmental effects, breeding values of mates, and Mendelian sampling effects tend to even out over a large number of progeny. The *average* performance of many progeny is a good indication of an individual's breeding value.

This is an important point to remember. Breeders often assume that it is virtually impossible to make genetic progress in traits that are lowly heritable because of the difficulty of identifying genetically superior individuals. That is true if we limit our sources of information to individuals' own performance records. However, with progeny records—and large numbers of them—the problem of low heritability can be overcome.[4] We can accurately determine those animals with better breeding values.

Records on siblings can only increase accuracy to a point. Look at the row in Table 11.2 corresponding to 1,000 half-sib records. Even with this huge amount of sib data, accuracy of prediction does not exceed .5. In fact, .5 is the limit of accuracy for predictions derived from half-sib records, regardless of the number of records or the heritability of the trait. This is because half-sib data provide information on only half of an individual's pedigree and do not account for Mendelian sampling.

[4]This statement is an optimistic one and correctly so. A note of caution, however: There is a trade-off between accuracy of prediction and selection intensity. (See Chapter 10.) If each sire produces many progeny, then fewer sires can be evaluated. Given fewer sires to choose from, selection intensity decreases.

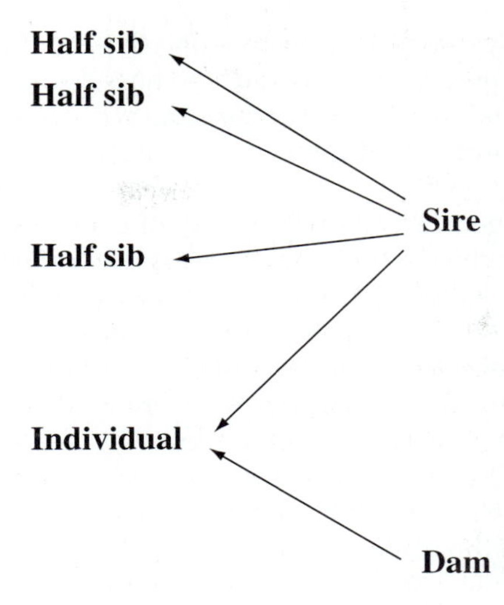

FIGURE 11.2
Arrow diagram showing the relationship between an individual and its half sibs. Paternal half-sib records provide information about the breeding value of the sire of the individual, but provide no information about the breeding value of the dam. Sib records cannot predict how much of either parent's breeding value was inherited by the individual (i.e., they cannot account for Mendelian sampling).

Figure 11.2 illustrates how records on paternal half sibs influence prediction of breeding value for an individual. If there are enough half-sib records, the breeding value of the sire of the individual will be well established. The half-sib records are, after all, *progeny* records for the sire. But establishing the breeding value of the sire is not the same as determining the breeding value of the individual. We cannot tell from the half-sib records whether the individual inherited a better or worse than average sample of genes from its sire—whether it benefited or suffered from Mendelian sampling. In other words, we cannot tell if the individual inherited more or less than half of its sire's breeding value. Furthermore, we have no information about the breeding value of the dam or the sample of genes inherited from her.

In the unlikely event that large numbers of both paternal half-sib and maternal half-sib records are available, it is possible to have prediction accuracy above .5. In this case we have good estimates of the breeding values of both the sire and dam of the individual. But we still have no information about Mendelian sampling, and the supper limit to accuracy turns out to be $\sqrt{.5}$ or .71.

What is true for half-sib data is true for pedigree data in general. Performance records on the parents, ancestors, and collateral relatives of an individual can be helpful, but by themselves can never produce genetic predictions with especially high accuracy.

A Perspective on Progeny Numbers

Animal breeders, especially breeders of the more glamorous species like horses, cattle, and dogs, tend to jump to conclusions. In particular, they tend to form opinions about sires based on insufficient progeny information. If the first few progeny of a sire are outstanding, then that sire is assumed to be outstanding, and if the sire's first few progeny are mediocre, then he is assumed to be mediocre. That breeders should reach premature conclusions about sires is quite understandable; patience is one human virtue often in short supply. Selection would be more effective, however, if the confidence that breeders place in progeny data were in line with progeny numbers.

confidence range
A range of values within which we expect—with a given probability, a given degree of confidence—that a true value of interest lies.

Table 11.3 provides some perspective on progeny numbers using examples from two traits: time required for Standardbred horses to trot one mile and weaned litter size in swine. Listed in each row of the table are values for mean progeny performance, number of progeny, EPD, accuracy (ACC), and 68% **confidence range** for the EPD.

A confidence range is an alternative way of expressing accuracy. It defines a range of values within which we expect—with a given probability, a given degree of confidence—that a true value of interest lies.[5] For example, the EPD in the second row of the table is −.11 seconds. Given the amount of information used to make this prediction, the confidence range for this EPD is −1.22 to +.99 seconds. The chance of the animal's true progeny difference being somewhere between −1.22 and +.99 seconds is slightly better than two out of three or 68%. In other words, we can be 68% confident that the true EPD is within this interval. If 68% seems like a rather arbitrary level of confidence—well, it is. It is the *standard* arbitrary level, however.

Let's look first at the upper half of Table 11.3, the part of the table involved with time to trot one mile. Each row corresponds to different numbers of progeny records for a single horse. For simplicity it is assumed that there is only one record per offspring (i.e., only the first race record of each offspring is used in the calculations) and, regardless of the number of progeny, the progeny mean is the same: −1.0 seconds or one second faster than the average of contemporaries.

You should notice some patterns right away. First, this is a good example of regression for amount of information. Even though the progeny mean does not change, the EPDs do, and the change in EPDs is a reflection of how much they have been regressed for amount of information. The larger the number of progeny, the less the EPDs have been regressed—the farther from zero they become. With just one offspring record of −1.0 seconds, our horse's EPD is −.11 seconds, but with 100 progeny averaging −1.0 seconds, his EPD is −.93 seconds. In the first case, the EPD is regressed severely because little information is available. In the second case, the EPD is regressed very little due to the abundance of information.

You will also notice that as numbers of progeny increase, so do accuracies. And as progeny numbers increase, 68% confidence ranges get *smaller.* With one progeny record, we can expect the horse's true progeny difference to be between −1.22 and +.99 seconds, a total span of 2.21 seconds. With 100 progeny, however, the range is −1.24 to −.61 seconds, for a total span of just .63 seconds. The more progeny, the more closely we are able to pinpoint the horse's true progeny difference, hence the narrower the confidence range.

The most important message you should get from Table 11.3, however, is that *it takes quite a few progeny records before we know much about a sire.* With no progeny records—no information at all—the horse's progeny difference is predicted to be zero or average, and the 68% confidence range for this prediction is −1.17 to +1.17 seconds, a total span of 2.34 seconds. With five progeny records, the confidence range is −1.31 to .53, or 1.84 seconds wide. To be sure, the range is smaller with five progeny records than with no information at all, but not *that* much smaller. Note that the upper limit to the confidence range is still greater than zero (+.53). In other words, with five progeny, we have no assurance that our horse is genetically better than average. Only when the number of progeny is close to 20, when the upper limit to the confidence range sinks below zero, can we be reasonably confident that he will indeed sire faster than average offspring.

[5]The concept of confidence range is closely related to the concept of *possible change,* which is described in detail in the next chapter.

TABLE 11.3 Confidence Ranges for EPDs Derived from Varying Numbers of Progeny Records[a]

Trait (Species)	Progeny Mean	No. of Progeny	EPD	ACC	68% Confidence Range
Time to trot one mile		0	.00	.00	−1.17 to +1.17
(Standardbreds)	−1.0 sec	1	−.11	.34	−1.22 to +.99
h^2=.45	"	2	−.20	.45	−1.25 to +.85
σ_P=3.5 sec	"	5	−.39	.62	−1.31 to +.53
	"	10	−.56	.75	−1.34 to +.22
	"	20	−.72	.85	−1.34 to −.09
	"	50	−.86	.93	−1.30 to −.43
	"	100	−.93	.96	−1.24 to −.61
No. of pigs weaned		0	.00	.00	−.44 to +.44
(Swine)	+.5 pigs	1	+.01	.16	−.42 to +.45
h^2 = .10	"	2	+.02	.22	−.41 to +.46
σ_P=2.8 pigs	"	5	+.06	.34	−.36 to +.47
	"	10	+.10	.45	−.29 to +.50
	"	20	+.17	.58	−.19 to +.53
	"	50	+.28	.75	−.01 to +.57
	"	100	+.36	.85	+.13 to +.59

[a]Assumptions: (Standardbreds) one race per offspring; (swine) one litter per daughter. Units for EPDs and confidence ranges are seconds for racing time, and pigs for number of pigs weaned. ACC stands for accuracy.

The same conclusions can be drawn from the lower half of Table 11.3, the part of the table involving number of pigs weaned. In this example, EPDs are calculated for a boar based upon the weaned litter sizes recorded by his daughters. For simplicity it is assumed that each daughter has produced only one litter.

As in the trotting time example, relatively large numbers of progeny records are necessary before we know much about the sire. But even more records are required for litter size than racing time. That is because number of pigs weaned is less heritable than time to trot a mile (h^2 = .10 versus .45). Each record conveys less information about underlying breeding values, so more records are necessary. Confidence ranges slowly decrease in width from .88 pigs (−.44 to +.44) with no daughters to .83 pigs (−.36 to +.47) with five daughters to .72 pigs (−.19 to +.53) with 20 daughters. Litters from more than 50 daughters are required before the lower limit to the confidence range exceeds zero and we can be at all confident that our boar will sire daughters that are better than average for number of pigs weaned.

Table 11.3 is designed to show the danger of judging sires on the basis of small progeny numbers. You can also conclude from the table that judging *dams* on the basis of progeny data is risky in almost all cases. With the exception of litter bearing species, dams rarely produce 10 offspring, let alone 20, 50, or 100. It is hard to get enough progeny information on dams to be very revealing.[6]

[6]This statement assumes that the trait in question is a *trait of the offspring*—a trait in which each record is attributed to an offspring, not to its dam. Weaning weight in beef cattle, for example, is usually considered a trait of the calf, and it is rare that a cow will have enough calves in a lifetime to accurately predict her breeding value for weaning weight. Some traits can be thought of as *traits of the dam*—traits in which each progeny record is attributed to the dam, not the offspring. Weaning weight can also be considered a trait of the dam (a measure of the dam's ability to produce calf weaning weight), and each calf weight becomes a repeated record on the dam. A dam's progeny data—or, more correctly, her repeated performance data—can be sufficient to provide fairly accurate predictions for traits of this kind.

The Mathematical Relationship between Confidence Ranges and Accuracy

Let T stand for some true value (i.e., BV, PD, PA, etc.). Then

$$68\% \text{ confidence range } = \hat{T} \pm \sqrt{(1 - r^2_{T,\hat{T}})\sigma^2_T}$$

Example

In the scenarios shown in Table 11.3 for time for Standardbreds to trot one mile,

$$T = PD$$
$$\hat{T} = \text{EPD}$$
$$r^2_{T,\hat{T}} = \text{ACC}^2$$

and

$$\sigma^2_T = \sigma^2_{PD}$$

Then

$$\begin{aligned} 68\% \text{ confidence range } &= \hat{T} \pm \sqrt{(1 - r^2_{T,\hat{T}})\sigma^2_T} \\ &= \text{EPD} \pm \sqrt{(1 - \text{ACC}^2)\sigma^2_{PD}} \end{aligned}$$

From Table 11.3,

$$\begin{aligned} \sigma^2_{PD} &= \frac{1}{4}\sigma^2_{BV} \\ &= \frac{1}{4}(h^2\sigma^2_P) \\ &= \frac{1}{4}(.45(3.5)^2) \\ &= 1.38 \text{ sec}^2 \end{aligned}$$

With 10 progeny,

$$\begin{aligned} 68\% \text{ confidence range } &= \text{EPD} \pm \sqrt{(1 - \text{ACC}^2)\sigma^2_{PD}} \\ &= -.56 \pm \sqrt{(1 - (.75)^2)(1.38)} \\ &= -.56 \pm .78 \\ &= -1.34 \text{ to } +.22 \end{aligned}$$

Prediction Using Multiple Sources of Information

We are often interested in predicting a value for an individual based on more than one source of information. For example, we might want to predict an animal's breeding value based on its own performance, the average performance of its paternal half sibs, and the average performance of its progeny. The selection index in this case would appear as:

$$I = b_1x_1 + b_2x_2 + b_3x_3$$

or

$$\text{EBV} = b_1P_{\text{IND}} + b_2\overline{P}_{\text{PHS}} + b_3\overline{P}_{\text{PROG}}$$

To calculate this EBV we need values for the three weighting factors, b_1, b_2, and b_3, and that requires the construction and simultaneous solution of three equations. It is possible to accomplish this by hand, but not without considerable pain. Calculations for selection indexes involving any more than two sources of information are best left to computers.

A Four-Equation Example

Suppose we want to predict an animal's breeding value for a nonrepeated trait based on four sources of information: (1) individual performance, (2) average performance of paternal half sibs, (3) average performance of maternal half sibs, and (4) average performance of progeny. The selection index is then

$$I = b_1x_1 + b_2x_2 + b_3x_3 + b_4x_4$$

or

$$\text{EBV} = b_1P_{\text{IND}} + b_2\overline{P}_{\text{PHS}} + b_3\overline{P}_{\text{MHS}} + b_4\overline{P}_{\text{PROG}}$$

The four equations that must be solved simultaneously to determine values for the regression coefficients (b_1, b_2, b_3, and b_4) are:

$$\frac{1}{h^2}b_1 + R_{\text{IND,PHS}}b_2 + R_{\text{IND,MHS}}b_3 + R_{\text{IND,PROG}}b_4 = R_{\text{IND,IND}}$$

$$R_{\text{PHS,IND}}b_1 + \frac{4 + (n_1 - 1)h^2}{4n_1h^2}b_2 + R_{\text{PHS,MHS}}b_3 + R_{\text{PHS,PROG}}b_4 = R_{\text{IND,PHS}}$$

$$R_{\text{MHS,IND}}b_1 + R_{\text{MHS,PHS}}b_2 + \frac{4 + (n_2 - 1)h^2}{4n_2h^2}b_3 + R_{\text{MHS,PROG}}b_4 = R_{\text{IND,MHS}}$$

$$R_{\text{PROG,IND}}b_1 + R_{\text{PROG,PHS}}b_2 + R_{\text{PROG,MHS}}b_3 + \frac{4 + (n_3 - 1)h^2}{4n_3h^2}b_4 = R_{\text{IND,PROG}}$$

where the Rs refer to pedigree relationships, and n_1, n_2, and n_3 refer to the number of paternal half sibs, maternal half sibs, and progeny, respectively. Specifically,

$$\frac{1}{h^2}b_1 + \frac{1}{4}b_2 + \frac{1}{4}b_3 + \frac{1}{2}b_4 = 1$$

$$\frac{1}{4}b_1 + \frac{4 + (n_1 - 1)h^2}{4n_1h^2}b_2 + 0 + \frac{1}{8}b_4 = \frac{1}{4}$$

$$\frac{1}{4}b_1 + 0 + \frac{4 + (n_2 - 1)h^2}{4n_2h^2}b_3 + \frac{1}{8}b_4 = \frac{1}{4}$$

$$\frac{1}{2}b_1 + \frac{1}{8}b_2 + \frac{1}{8}b_3 + \frac{4 + (n_3 - 1)h^2}{4n_3h^2}b_4 = \frac{1}{2}$$

The regression coefficients—the bs—are functions of heritability, pedigree relationship, and number of records from each group of relatives.

Once the values for the bs have been determined, they are simply substituted in the index equation to calculate the animal's EBV. The accuracy of the index is

$$r_{BV,\hat{BV}} = \sqrt{b_1 + \frac{1}{4}b_2 + \frac{1}{4}b_3 + \frac{1}{2}b_4}$$

For sample applications of such an index, see the data in Table 11.4. They were derived from the index equations listed above.

Weighting Each Source of Information

As you might guess, the values of the bs, the weighting factors needed for a selection index involving multiple sources of information, vary depending on the amount and relevance of data from each source. For example, if an animal has extensive pedigree data and a performance record of its own, but little progeny data, we expect that in calculating that animal's EBV, most of the emphasis will be placed on the pedigree and own performance information. On the other hand, if the same animal acquires vast amounts of progeny data, we expect the emphasis to shift to that source of information. Progeny data, after all, provide the ultimate test of an individual's breeding value. When simultaneous equations are used to solve for selection index weights, each weight automatically reflects the appropriate amount of emphasis that should be placed on its corresponding source of information. You can think of this as the "magic" of simultaneous solution of equations. Table 11.4 has been constructed to illustrate some aspects of this mathematical magic.

Listed in Table 11.4 are decimal proportions representing the relative emphasis being placed on particular sources of information for breeding value prediction. Each row represents a different scenario, with each scenario involving different amounts of data on individual performance (IND), average performance of paternal half sibs (PHS), average performance of maternal half sibs (MHS), and average performance of progeny (PROG). The measured trait is assumed to be nonrepeated (animals may have only one record apiece) with a heritability of .3.

TABLE 11.4 Proportional Emphasis Placed on Different Sources of Records Used for Predicting Breeding Value When Numbers of Records Vary[a]

Number of Records				Proportional Emphasis				
IND	PHS	MHS	PROG	IND	PHS	MHS	PROG	ACC
Pedigree data only:								
0	10	2	0	.00	.76	.24	.00	.38
0	200	2	0	.00	.87	.13	.00	.52
0	400	4	0	.00	.80	.20	.00	.55
Pedigree and own performance data:								
1	10	2	0	.71	.22	.07	.00	.61
1	200	2	0	.54	.40	.06	.00	.67
Pedigree, own performance, and progeny data:								
1	10	2	10	.30	.09	.03	.58	.77
1	200	2	10	.27	.20	.03	.50	.79
1	10	2	200	.03	.01	.00	.96	.97
1	200	2	200	.03	.02	.00	.95	.97

[a]IND = individual; PHS = paternal half sibs; MHS = maternal half sibs; PROG = progeny; ACC = accuracy of prediction; $h^2 = .3$.

The rows (scenarios) in Table 11.4 are grouped in three sections. The first three rows represent situations in which only pedigree information is available on an individual. Perhaps the individual is not yet born or is too young to have a performance record of its own. The next two rows reflect combinations of pedigree and own performance information. The last four rows combine pedigree, own performance, and progeny information—data that would be available only on an older animal. Note that accuracy of breeding value prediction (right-hand column) increases with additional information.

The scenario depicted in the very first row represents a modest amount of pedigree information. The dam of the *individual of interest*—the individual whose breeding value we are predicting—has two previous offspring, and the sire is probably young, having only 10 other progeny. More emphasis is placed on paternal than maternal sib records because there are more paternal sib records, but altogether there is little information to go on, and accuracy is low (.38). In the second row, the sire of the individual of interest, having 200 progeny records, is well evaluated. The emphasis on paternal sib information increases accordingly, and accuracy is higher.

In row three of the table, numbers of records have been doubled. The number of paternal sib records increases from 200 to 400, and the number of maternal sib records increases from two to four. Despite the much larger increase in the number of paternal sibs, the relative emphasis on paternal sibs *decreases.* This is because the paternal side of the pedigree is already well established. With 200 progeny, the breeding value of the sire of the individual of interest is predicted with high accuracy, and adding 200 more progeny will not help a great deal. The dam's side of the pedigree, on the other hand, is the "mystery" side. With only two offspring, we know little about her breeding value. Adding two more offspring may not seem like much of an improvement, but those two records provide information on a part of the pedigree where there was little before. Scarce information carries more weight per record than abundant information.

Accuracy of breeding value prediction based solely on pedigree information is not very high. In the examples in Table 11.4, accuracy peaks at .55 in row three. As explained in the discussion of single-source predictions, pedigree data cannot account for Mendelian sampling. A **pedigree estimate** is useful as a best first guess, but we cannot expect it to be reliable.

pedigree estimate
A genetic prediction based solely on pedigree data.

Rows four and five of the table combine pedigree information with the individual's own performance record. Because the trait is moderately heritable ($h^2 = .3$), the individual's own record carries considerable weight—71% when there is little pedigree data and 54% when there is an abundance of paternal sib data. The individual's record is just one record, but it is a record on the individual's closest relative (itself), and it reflects the Mendelian sampling of genes; it tells us something about the value of the genes that the individual actually received from its parents.

The last four rows of Table 11.4 incorporate all three kinds of information: pedigree, own performance, and progeny data. Note the relative importance of progeny data, even when progeny numbers are small. With large numbers of progeny (last two rows), other sources of information become practically inconsequential, even when they are represented by many records. And with enough progeny data, accuracy of prediction is very high—.97+ in this example.

COMPARING ANIMALS USING DATA FROM GENETICALLY DIVERSE GROUPS—BEST LINEAR UNBIASED PREDICTION

The selection index is a powerful method for genetic prediction. A basic assumption of the selection index, however, is that the performance information used comes from genetically similar contemporary groups. What if we want to make predictions using data from genetically different contemporary groups—groups from different farms and ranches or from different decades? An extension of selection index methodology known as **best linear unbiased prediction** or **BLUP** is designed for just this type of situation.[7]

best linear unbiased prediction (BLUP)
A method of genetic prediction that is particularly appropriate when performance data come from genetically diverse contemporary groups.

The statistical theory behind best linear unbiased prediction is almost as old as the theory behind the selection index. It was not until the early 1980s, however, that computers and mathematical algorithms had advanced to the point that widespread application of BLUP for genetic evaluation became feasible. BLUP requires intensive computation. Like a selection index that uses multiple sources of information, BLUP involves the simultaneous solution of a number of equations. A typical application of BLUP, however, incorporates many times the number of equations that would be used for a corresponding selection index, partly because a single BLUP analysis provides predictions for an entire population of animals—not just for one animal at a time (the case with most applications of the selection index).

Because of its ability to account for genetic differences among contemporary groups, and because it provides genetic predictions for many animals at once,

[7]The distinction made here between the selection index and BLUP reflects differences in how the two technologies have historically been applied. It is theoretically possible to construct selection indexes that could account for genetic differences among contemporary groups and that would be, for practical purposes, every bit as good as the most sophisticated BLUP techniques. Such selection indexes have not been used, however, for the simple reason that they have no particular advantages over analogous BLUP procedures.

large-scale genetic evaluation The genetic evaluation of large populations—typically entire breeds.

BLUP is the preferred method for **large-scale genetic evaluation:** the genetic evaluation of large populations, typically entire breeds.[8] The performance records used in such evaluations usually come from **field data,** data that are regularly reported by individual breeders to breed associations or government agencies.

Types of BLUP Models

field data Data that are regularly reported by individual breeders to breed associations or government agencies.

Best linear unbiased prediction is a technique that can be thought of as a family of **statistical models,** mathematical representations of animal performance that include various genetic and environmental effects and are used for genetic prediction. There are sire models, sire–maternal grandsire models, animal models, repeat measure models, direct-maternal models, multiple-trait models—the list goes on and on. Differences among models have to do primarily with which animals receive genetic predictions (e.g., sires only, all parents, or all animals), the number and kinds of predictions generated, and computational difficulty. In general, the more sophisticated the model, the more equations involved, and the more computer resources required.

Capabilities of Advanced BLUP Models

statistical model A mathematical representation of animal performance that includes various genetic and environmental effects and is used for genetic prediction.

animal model An advanced statistical model for genetic prediction that is used to evaluate all animals (as opposed to just sires) in a population.

Best linear unbiased prediction is a complicated subject, and breeders often have questions about the capabilities of BLUP procedures. In particular, they wonder whether BLUP analyses can overcome the kinds of biases typically found in field data. Following is a discussion of BLUP attributes that is designed to answer many of these questions. It is always dangerous to generalize about BLUP models. Each differs from the next in one way or another, and no two have exactly the same capabilities. The particular model envisioned in constructing this section is the most basic, yet most advanced and currently most popular type of BLUP model—the **animal model.** Animal models have a number of desirable qualities, the most obvious of which is their ability to evaluate *all* animals (as opposed to just sires) in a population.

Genetic Levels of Contemporary Groups

BLUP models account for differences in the mean breeding values of contemporary groups. In other words, they account for the fact that superior performance in a genetically inferior contemporary group is not the equivalent of superior performance in a genetically superior contemporary group. In a BLUP analysis, the performance of the winning horse in the third race at the county fair is not considered as impressive as the performance of the Kentucky Derby winner.

The selection index, you recall, cannot make this distinction. Because it uses deviations from contemporary group means (in the horse racing example, deviations from the mean running time in a given race), the selection index cannot account for the level of competition. A deviation of −5 seconds is a deviation of −5 seconds whether the race is the Kentucky Derby or one of many at the county fair.

[8]For a complete discussion of large-scale genetic evaluation, see Chapter 12.

BLUP models do not use deviations from contemporary group means. Instead they include equations that actually solve for contemporary group effects (E_{cg}), environmental effects common to all members of a contemporary group. They do this by comparing the performance of relatives in different contemporary groups. If the animals in a particular contemporary group do not perform as well as they "should" as determined by the performance of their relatives in other contemporary groups, then the environmental effect common to that contemporary group is estimated to be poorer than average. Likewise, if the animals in the group perform better than they should, E_{cg} is estimated to be better than average. Through simultaneous solution of equations, the estimates of contemporary group effects provide information that is used to produce more reliable genetic predictions for individual animals and vice versa. Thus, BLUP procedures account for the fact that performance information used in prediction comes from contemporary groups that differ for both environmental and genetic reasons. You can directly compare genetic predictions from BLUP models, even if those predictions are derived largely from records produced in very different environments and(or) in contemporary groups that differ widely in average genetic merit.

Genetic Trend

A corollary to the ability of BLUP models to account for the genetic level of contemporary groups is their ability to account for genetic trend. If a population has undergone effective selection for a considerable period of time (i.e., if it has experienced significant genetic trend), then the average breeding value of newer contemporary groups should be better than the average breeding value of older contemporary groups. But because BLUP procedures account for genetic differences among contemporary groups, genetic trend is usually not a problem. (Genetic trend *would* cause biases in a selection index using deviations from contemporary group means.) With BLUP models, we can legitimately use performance records from animals of different eras—from those that are long since dead to newborns.

Use of All Data

Traditional applications of the selection index use records from a relatively small number of sources. The selection index used to generate the values in Table 11.4, for example, used four sources: a record on the individual, records on paternal and maternal sibs, and progeny records. BLUP models can use information from many more sources. In fact, they often use information from *all* animals in a population.

The benefit of this may not be great; records on distant relatives are unlikely to contribute much to a prediction. Still, the ability of BLUP procedures to incorporate information from all kinds of relatives adds to prediction accuracy, even if only slightly. It also clarifies pedigree relationships. What would be considered the record of a half sib in a typical selection index application might be represented more correctly in a BLUP analysis as the record of a relative that is both half sib and first cousin. (See Figure 11.3 and accompanying discussion in the legend.) This record would therefore receive a more appropriate weighting with BLUP than with the selection index.

multiple-trait model
A statistical model used to predict values for more than one trait at a time.

Use of all data can also be interpreted to mean use of data on correlated traits. **Multiple-trait BLUP models,** statistical models used to predict values for more than one trait at a time, allow information on one trait to be used in predicting val-

(a) BLUP animal model

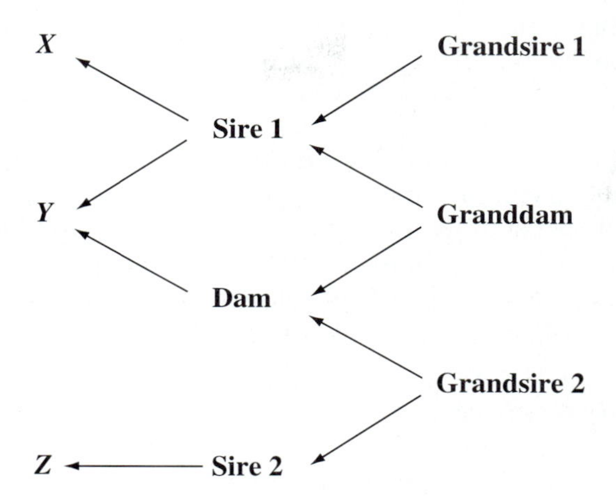

(b) Selection index sire model

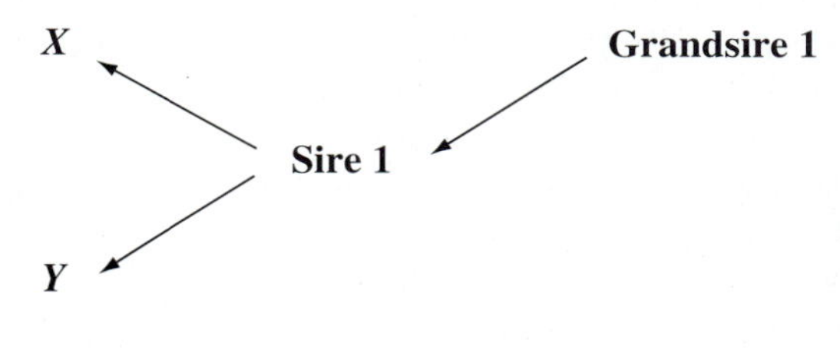

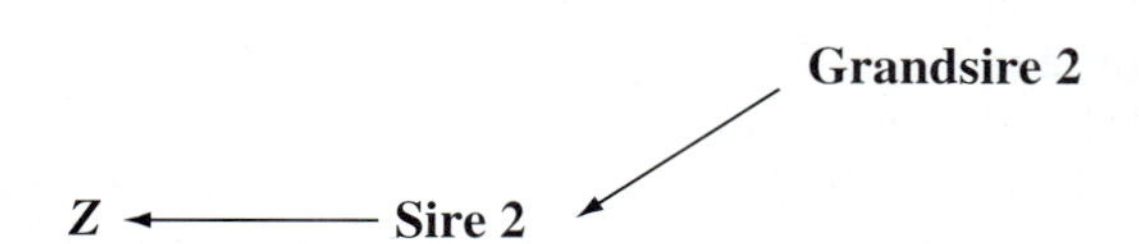

FIGURE 11.3
Arrow diagrams depicting pedigree relationships as they are perceived using two different models: (*a*), a BLUP animal model; and (*b*), a selection index in which the sources of information are sire group averages (means of performance of paternal half sibs). The BLUP animal model uses true pedigree relationships. *X* and *Y* are not only half sibs, they are also cousins because they have a granddam in common. *Y* and *Z* are cousins as well. The selection index that uses sire group averages ignores maternal relationships. *X* and *Y* appear to be half sibs and nothing more, and *Y* and *Z* appear to be unrelated.

ues for another trait. This is especially useful when information on the second trait is scarce. For example, there are typically many more weaning weights than birth weights in beef cattle field data. When both birth weight and weaning weight are incorporated in a multiple-trait BLUP model, weaning weight information helps predict values for birth weight and vice versa. The result is more accurate predictions for both traits.

Nonrandom Mating

Seedstock breeders are reluctant to mate animals randomly (i.e., to assign males to females in a random fashion). Can you blame them? If you were paying a $50,000 stud fee for the service of a famous stallion, would you mate him to just *any* mare?

Of course not. You would maximize the likelihood of producing an outstanding foal by mating the stallion to your very best mare. The problem with this strategy from a genetic prediction standpoint is that it gives some animals an advantage over others—it creates bias. Those sires thought to be the best are mated to the best females and are therefore likely to have better performing progeny. The remaining sires are relegated to lesser females and are therefore likely to have poorer performing progeny.

nonrandom mating
Any mating system in which males are not randomly assigned to females.

BLUP procedures account for **nonrandom mating** of this kind by, in essence, adjusting animals' predictions for the merit of their mates. Genetic predictions for sires and dams are produced simultaneously, and simultaneous solution of equations causes each prediction to account for all other predictions. The happy result (depending on your perspective) is that it is not possible to make an individual look better than he is by assigning him to superior mates.

Culling for Poor Performance

Another kind of bias is caused by culling for poor performance in a repeated trait or in a trait that is recorded prior to recording a genetically related trait. An example of this type of bias as it affects racehorses is illustrated in Figure 11.4. Thoroughbreds typically begin their racing careers as two-year-olds. If they are successful in their first season, they continue to race. Otherwise they are culled for poor performance. Culling of this sort causes bias if the progeny of some sires are culled more severely than the progeny of other sires.

In Figure 11.4, sire *B*'s progeny perform much better than sire *A*'s progeny as two-year-olds. Clearly, sire *B* should have the better breeding value for racing performance. Between racing seasons, some of the progeny of both sires are culled for poor performance, but because sire *A*'s progeny performed poorly in general, a much larger proportion of his progeny are culled than sire *B*'s. Sire *A*'s few remaining progeny race well in later years, so the difference in the average subsequent performance of the two sires' progeny ($\overline{P}_B - \overline{P}_A$) is relativly small. Note that this difference is much smaller than it would have been had no culling occurred. If much of the data used to predict breeding values for these sires comes from progeny that are three years of age or older, sire *A* will have a relative advantage and sire *B* a relative disadvantage.

Multiple-trait BLUP models can account for the bias caused by culling for poor performance. If the trait on which culling is based is included in the analysis, and if the genetic relationship (genetic correlation) between that trait and a subsequent trait is also incorporated, then predictions for the second trait are unaffected by culling. If BLUP is used to evaluate racehorses, predictions of breeding value for racing performance at older ages will not be biased by culling for two-year-old performance.

direct component or **direct effect**
The effect of an individual's genes on its performance.

maternal component or **maternal effect**
The effect of genes in the dam of an individual that influence the performance of the individual through the environment provided by the dam.

Prediction of Direct and Maternal Genetic Components of Traits

All traits have what is called a **direct component,** the effect of an individual's genes on its performance. Some traits have a **maternal component,** the effect of genes in the dam of an individual that influence the performance of the individual through the *environment* provided by the dam.

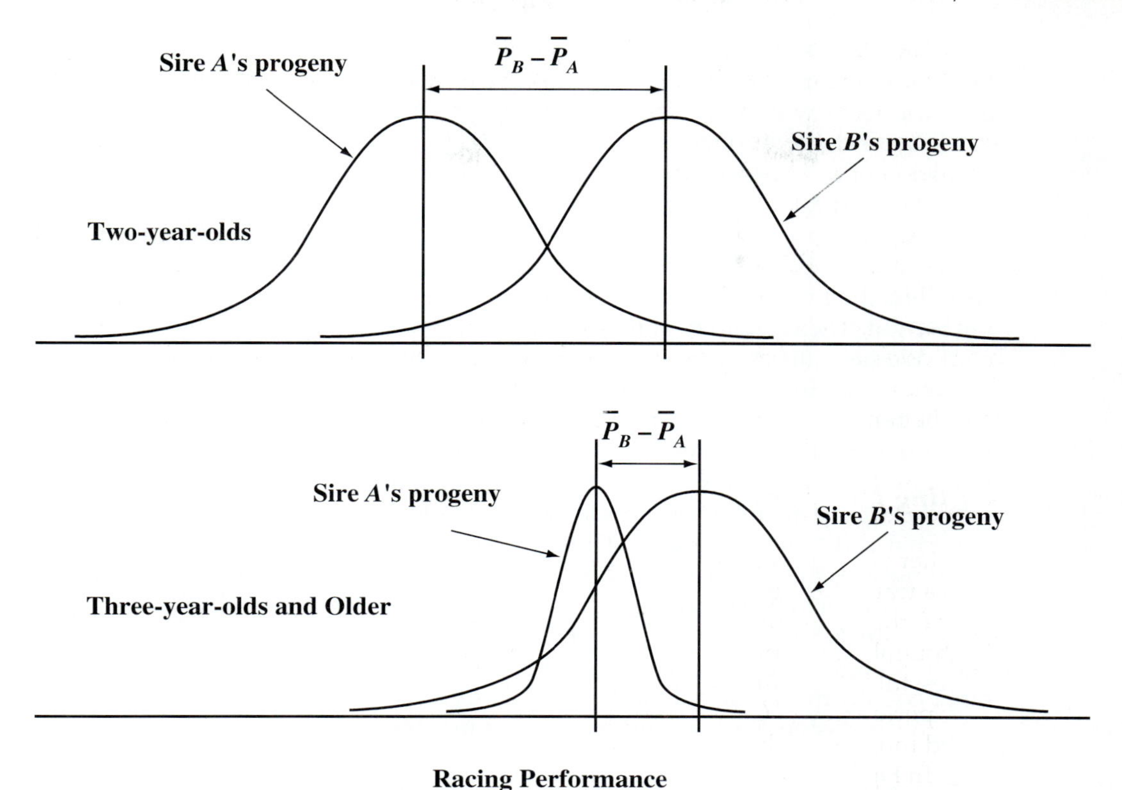

FIGURE 11.4 Schematic representation of bias caused by culling in racehorses. Sire *A*'s progeny were culled heavily for two-year-old racing performance. As a result, his progeny average for subsequent performance is much better than it would have been had the culling not occurred.

The classic example of a trait with both direct and maternal components is weaning weight. An animal's weaning weight is a function of its inherent ability for rate of growth and the milk production and mothering ability of its dam. Inherent growth rate is determined by the animal's genes. It comprises the direct component of weaning weight. Milk production and mothering ability of the dam are determined by her genes (as well as by environment). The dam's genes for these traits do not affect the offspring's growth rate directly, but they do affect the environment experienced by the offspring. Milk production and mothering ability comprise the maternal component of weaning weight.

Other traits having important maternal components include dystocia and survivability. The direct component of dystocia is related to the size and shape of the fetus. The maternal component is associated with the dam's pelvic size and conformation, and other, more subtle physiological and psychological factors. The direct component of survivability is a function of those genes in young animals that affect physical soundness, immune response, and survival instinct. The maternal component relates to the dam's ability to nourish and protect its young.

BLUP procedures are capable of separating the direct and maternal components of a trait, providing genetic predictions for both. In the case of weaning weight, predictions are available for the growth component as well as the milk/mothering ability component.

Breeders often wonder how this is possible, given that the maternal component of weaning weight is not measured directly. No one routinely milks sows, ewes, or beef cows and records milk yield. Instead, the information for prediction of the maternal component comes from weaning weights themselves. The true explanation of this apparent paradox lies in the simultaneous solution of equations, but you can think of it (not incorrectly) as a simple matter of subtraction.

Suppose, for example, that a sire produces a large number of progeny, and from the weaning weights of these progeny we obtain a reliable prediction of the sire's breeding value for preweaning growth rate—the direct component of weaning weight. The daughters of this sire are then bred and produce progeny of their own. We have an expectation of the weaning weights of these progeny—*grand-progeny* of the sire—based on what is already known about their dams' growth potential. If the weaning weights of these grandoffspring exceed expectations, then it is assumed that the difference is due to superior maternal ability (mostly superior milk production) of their dams. On the other hand, if the weaning weights of grandoffspring do not meet expectations, the difference is attributed to the inferior maternal ability of their dams. Thus, predictions for the maternal component of weaning weight are determined by subtracting expectations of weaning performance from actual performance. The whole situation is complicated by the fact that the sire himself may have been bred to better or worse than average females, and his daughters to better or worse than average males, but BLUP takes these and other equally messy considerations into account.

paternal component or **paternal effect** (in rare instances): The effect of genes in the sire of an individual that influence the performance of the individual through the environment provided by the sire. Traits of the dam or offspring that are affected by a male's fertility and physical ability to breed are also said to have a paternal component.

A discussion of direct and maternal components of traits would not be complete without mentioning a third component, the **paternal component.** One definition of a paternal component is analogous to the definition of a maternal component. By this definition, the paternal component of a trait is the effect of genes in the sire of an individual that influence the performance of the individual through the environment provided by the sire. For domestic animals, paternal components as defined this way are rare; sires have little to do with the upbringing of their progeny and therefore have no environmental effect on their offspring's performance. Paternal effects do exist for species of wild birds, and for some—jacanas and emperor penguins, for example—the paternal components of hatchability and survivability of chicks may be far more important than the maternal components.

A second definition of paternal components applies to fertility traits. Fertility measures that are considered traits of the dam or offspring, but are affected by a male's fertility and physical ability to breed, are said to have a paternal component. We speak, for example, of the paternal component of conception rate. This second definition has much more practical application in the breeding of domestic species.

Prediction of a Variety of Values

BLUP procedures typically produce EBVs or EPDs for the direct components of traits and, for traits with important maternal components, EBVs or EPDs for them as well. Predictions from BLUP analyses are not limited to just these, however. For repeated traits, predictions of permanent environmental effects are also possible, and that means that BLUP can be used to predict producing ability. For just as

$$PA = BV + E_p$$

where E_p is assumed to contain both the permanent environmental effect and gene combination value, then

$$\hat{PA} = \hat{BV} + \hat{E}_p$$

We can calculate a dairy female's MPPA for milk production by summing her estimated breeding value and the prediction of her permanent environmental effect. Note how much more sophisticated this MPPA is compared to the MPPAs computed earlier in this chapter using a single-source selection index. In the earlier calculation, the information used was limited to the female's own lactation performance. In contrast, the information used to calculate an MPPA from a BLUP analysis includes lactation records from all sorts of relatives.

total maternal value (BV_{tm}) A combination of breeding values for both the direct and maternal components of a trait. A female's total maternal value represents the heritable part of her ability to produce a quantity measured in her offspring.

Another value commonly predicted by BLUP procedures is called **total maternal value (BV_{tm}).** An individual's total maternal value combines its breeding values for both the direct and maternal components of a trait. For example, a beef cow's total maternal value for weaning weight represents her genetic ability to produce weaning weight. As such, it includes the effects of her genes for milk production and mothering ability (the maternal component of weaning weight) as well as the effects of the genes for preweaning growth rate that she transmits to her calf—the direct component of weaning weight. (I use a beef *female* in this example. We predict total maternal values for males as well. But because a sire's genes for the maternal component of weaning weight are not expressed in the weaning weight of his offspring, an example using a male is harder to understand.)

In equation form,

$$BV_{WW_{tm}} = BV_{WW_m} + \frac{1}{2}BV_{WW_d}$$

where the subscripts *WW*, *m*, and *d* stand for weaning weight, maternal, and direct, respectively. The cow's total maternal value contains all of her breeding value for the maternal component of weaning weight and only half her breeding value for the direct component of weaning weight. This is because all of her genes influencing milk production/maternal ability affect her calf's performance, but because she transmits only half the genes to her calf, only half her genes influencing preweaning growth rate affect calf performance. In terms of genetic predictions,

$$\text{EBV}_{WW_{tm}} = \text{EBV}_{WW_m} + \frac{1}{2}\text{EBV}_{WW_d}$$

and

$$\text{EPD}_{WW_{tm}} = \text{EPD}_{WW_{tm}} + \frac{1}{2}\text{EPD}_{WW_d}$$

Total maternal EBVs and total maternal EPDs are easy to confuse. A female's total maternal EBV is a prediction of her genetic ability to produce something measured in her offspring. It is our expectation of her *own* production (excluding any environmental effects). An individual's total maternal EPD, on the other hand, is our expectation of that individual's *daughters'* production.

A cow's total maternal EBV for weaning weight is a prediction of the *heritable* part of her ability to produce calf weaning weight—her breeding value for weaning weight production. We could go one step further and calculate the cow's

MPPA for weaning weight. This would include any permanent environmental effects on weaning weight production. Mathematically,

$$\begin{aligned} MPPA_{WW} &= EBV_{WW_{tm}} + \hat{E}_{WW_p} \\ &= EBV_{WW_m} + \frac{1}{2}EBV_{WW_d} + \hat{E}_{WW_p} \end{aligned}$$

BLUP analyses also produce estimates of environmental effects common to all members of a contemporary group (E_{cg}).[9] Although these are not routinely used, they could be helpful from a management standpoint. They might indicate situations where animal health programs are ineffective, pastures are being overgrazed, etc. When estimates of contemporary group effects in a population are plotted over time, they reveal **environmental trend**—change in the mean performance of a population over time caused by changes in environment.

environmental trend
Change in the mean performance of a population over time caused by changes in environment.

Predictions for All Animals

BLUP procedures can be used to generate predictions for any animal in a population. Predictions are available for males and females, parents and nonparents, animals not yet conceived and animals long since dead, and animals with performance records and animals without.

EXERCISES

Study Questions

11.1 Define in your own words:

selection index
own performance data
pedigree data
progeny data
bias
regression for amount of information
unbiased
common environmental effect
confidence range
pedigree estimatebest
linear unbiased prediction (BLUP)
large-scale genetic evaluation
field data
statistical model
animal model
multiple-trait model
nonrandom mating
direct component or direct effect
maternal component or maternal effect
paternal component or paternal effect
total maternal value (BV_{tm})
environmental trend

[9]Rarely do BLUP analyses produce pure estimates of E_{cg}. Rather, they produce estimates of $\mu + E_{cg}$. Functionally, there is little difference between the two types of estimates.

11.2 **a.** When is it legitimate to use selection index methodology for genetic prediction as opposed to using best linear unbiased prediction?
b. When should BLUP be used instead of the selection index?

11.3 Describe each element of a selection index equation in statistical terms.

11.4 What is meant by a "single source of information"? Give some examples.

11.5 Why are genetic predictions regressed for amount of information?

11.6 Why must we account for common environmental effects when calculating genetic predictions for litter bearing species?

11.7 List the factors affecting accuracy of prediction for a repeated trait.

11.8 How is it possible to have highly accurate genetic predictions for lowly heritable traits?

11.9 In relative terms, what kind of information is most valuable for genetic prediction when:
a. heritability is high? Why?
b. heritability is low? Why?

11.10 A single progeny record may be misleading, but the average of many progeny records will accurately predict a parent's breeding value. Explain.

11.11 **a.** Why do pedigree data, even large amounts of pedigree data, provide only limited accuracy of prediction?
b. Why is this *not* true of progeny data?

11.12 **a.** Which measure of accuracy do you prefer: a traditional accuracy value or a confidence range?
b. Explain your answer to (*a*).

11.13 What is meant by "multiple sources of information"? Give some examples.

11.14 Why is computing a genetic prediction from multiple sources of information mathematically more difficult than computing one from a single source of information?

11.15 What is meant by the following statement: "Scarce information carries more weight per record than abundant information"? Give an illustrative example.

11.16 If large amounts of progeny data are used in predicting an individual's breeding value, is the individual's own performance an important piece of information? Explain.

11.17 **a.** Explain how each of the following can be a potential source of bias in genetic prediction:
i. differences in genetic levels of contemporary groups
ii. genetic trend
iii. nonrandom mating
iv. culling for poor performance
b. How do advanced BLUP procedures account for (i), (ii), (iii), and (iv) above?

11.18 Describe these additional attributes of advanced BLUP procedures:
a. use of all data
b. prediction of direct and maternal genetic components of traits
c. prediction of a variety of values
d. prediction of values for all animals

Problems

11.1 **a.** Gay Blade, a promising Thoroughbred stallion, was recently retired to stud. His first six foals have just completed their maiden races, averaging two seconds faster than their contemporary group means.

i. Use a single-source selection index to predict Gay Blade's breeding value for racing time. (Assume $h^2 = .35$.)

ii. Calculate accuracy of prediction.

b. Megabuck, the old champion, has sired 120 foals. They have run, on average, 4.2 races apiece, averaging one second faster than contemporary group means.

i. Predict Megabuck's breeding value for racing time. (Assume $h^2 = .35$ and $r = .57$.)

ii. Calculate accuracy of prediction.

c. Why is Megabuck's EBV better than Gay Blade's even though Gay Blade's offspring have run faster than Megabuck's?

d. All else being equal, which sire would you use? Why?

11.2 A boar has sired 20 litters averaging 7.8 weaned pigs each. Postweaning average daily gains of his progeny average .06 lb per day above contemporary group means. Heritability of postweaning gain is .28 and c^2_{FS}, a measure of covariation among postweaning gains of littermates that is caused by common environment (in this case, covariation due to a common dam), is estimated to be about .07.

a. Use a single-source selection index to estimate the boar's breeding value for postweaning average daily gain.

b. Estimate the boar's breeding value assuming there is no environmental covariation among postweaning gains of littermates.

c. Why did his EBV increase when c^2_{FS} was assumed to be zero?

11.3 **a.** Calculate accuracy of breeding value prediction given the following information:

i. a single performance record on the individual; $h^2 = .25$

ii. a single performance record on the individual; $h^2 = .5$

iii. five repeated records on the individual; $h^2 = .25$; $r = .3$

iv. five repeated records on the individual; $h^2 = .25$; $r = .6$

v. single records on five half sibs; $h^2 = .25$

vi. single records on 500 half sibs; $h^2 = .25$

vii. single records on five progeny; $h^2 = .25$

viii. single records on 500 progeny; $h^2 = .25$

b. Why did accuracy change the way it did in the above scenarios?

11.4 **a.** Use your answers to Problem 11.1 to calculate 68% confidence ranges for Gay Blade's and Megabuck's EBVs for racing time. (Assume $\sigma_P = 1.3$ sec.)

b. Which sire represents the greater selection risk? Why?

11.5 A young beef bull is being genetically evaluated for weaning weight. The information used in the analysis includes the bull's own weaning weight, lots of paternal half-sib data, a couple of maternal half-sib records, and limited progeny data. All this information is combined in the following selection index:

$$I = b_1x_1 + b_2x_2 + b_3x_3 + b_4x_4$$

or

$$\text{EBV} = b_1\overline{P}_{\text{IND}} + b_2\overline{P}_{\text{PHS}} + b_3\overline{P}_{\text{MHS}} + b_4\overline{P}_{\text{PROG}}$$

Weighting factors calculated from simultaneous equations are:

$$b_1 = .169$$
$$b_2 = .500$$
$$b_3 = .075$$
$$b_4 = .624$$

Given the following performance data (expressed as deviations from contemporary group means):

$$\overline{P}_{\text{IND}} = +128 \text{ lb}$$
$$\overline{P}_{\text{PHS}} = +22 \text{ lb}$$
$$\overline{P}_{\text{MHS}} = +35 \text{ lb}$$
$$\overline{P}_{\text{PROG}} = +26 \text{ lb}$$

a. Calculate the bull's EBV for weaning weight.
b. Calculate accuracy of prediction.

11.6 A ewe's direct and maternal EPDs for weaning weight (a trait of the lamb) are:

$$\text{EPD}_d = -1 \text{ lb}$$
$$\text{EPD}_m = +2.5 \text{ lb}$$

a. Given just these EPDs, what do we expect her future lambs to weigh (expressed as a deviation from the population mean)?
b. The ewe's permanent environmental effect (E_p) is predicted to be +3 lb. What is her MPPA?
c. Estimates of contemporary group effects ($\mu + \hat{E}_{cg}$) for her flock average 39 lb. What do we expect her future lambs to weigh?

CHAPTER 12

Large-Scale Genetic Evaluation

large-scale genetic evaluation The genetic evaluation of large populations—typically entire breeds.

The last chapter was about genetic prediction. It dealt with the methodologies of genetic prediction, including the most advanced methodology—best linear unbiased prediction (BLUP). This chapter is also about genetic prediction, but it deals with the *application* of BLUP techniques in developed countries. In other words, it deals with **large-scale genetic evaluation.** From this chapter you can learn what a sire summary looks like and where sire summary information comes from, what information is available on nonsires, how data from genetic evaluations should be interpreted, what pitfalls to be aware of, and what alternatives there are to conventional large-scale evaluation programs.

Large-scale genetic evaluation refers to the genetic evaluation of large populations. Typically, these populations are entire breeds within a country or within an even larger geographical area. Because the data used for large-scale genetic evaluation come from many breeders and are processed centrally, large-scale genetic evaluation is a cooperative effort involving breeders, breed associations, and professionals in animal breeding technology.

The purpose of large-scale genetic evaluation is simple—to allow genetic comparison of animals in different herds or flocks. Why is this important? Suppose that you own a sire that you think is outstanding. His own performance and progeny records in your herd or flock are excellent, and you are convinced he is the best you have ever bred. But without some mechanism for comparing him with sires owned by other breeders, you have no objective way of knowing how good he is in the breed as a whole. Large-scale genetic evaluation provides that mechanism.

Large-scale genetic evaluation speeds the rate of genetic change in a population. By allowing direct comparison of animals in different herds or flocks, it effectively enables breeders to select individuals from a larger pool of candidates. Instead of being limited to the animals they themselves own, breeders can select from a much larger population—an entire breed. And just as it is easier to field quality athletic teams at a big school than at a small school because the big school has more athletes to choose from, so it is easier to find truly outstanding breeding animals in a large population than in a small one. In terms of the key equation for genetic change, large-scale genetic evaluation allows increased selection intensity.

Large-scale genetic evaluation also speeds the rate of genetic change by increasing accuracy of prediction. Breed databases contain enormous amounts of information, many times the amount of information available from any one herd or flock. When records from an entire breed are used for prediction, accuracy of prediction increases by virtue of the sheer volume of information available.

A History of Across-Herd and Across-Flock Comparisons

Not counting horse races, which have been used to compare animals from different herds for millennia, the earliest across-herd comparisons were probably those developed for European dairy cattle near the beginning of the twentieth century. These involved progeny tests for dairy sires. Similar progeny tests were initiated in the U.S. dairy industry in the 1930s.

central test
A test designed to compare the performance of animals (usually young males) from different herds or flocks for growth rate and feed conversion by feeding them at a central location.

Some of the first across-herd and across-flock comparisons for swine, sheep, and beef cattle were provided by **central tests.** In a central test young boars, rams, or bulls from different breeding operations are brought to one location where they are fed together for a period of time. They are then compared for daily weight gain, feed conversion (sometimes), and physical measures. Central tests allow breeding animals from different herds or flocks to directly compete against each other, and have historically been used as a marketing tool for seedstock breeders and as a forum for promoting the use of performance information. But the ability of central tests to compare the genetic merit of animals is limited. Comparisons are restricted to the few traits measured at the test station and are based on individuals' own performance data only. And pretest environment can affect performance in the test, giving an advantage to animals from some locations and a disadvantage to animals from other locations. Since the advent of BLUP—with its ability to account for genetic and environmental differences among contemporary groups—and large-scale genetic evaluation using BLUP, central tests have lost much of their genetic justification.

The first true sire evaluations for beef cattle appeared in the 1970s. With the advent of high-speed computers, analyses using BLUP and BLUP-like procedures became the norm for cattle breeds in the 1980s. Change is occurring rapidly, but at this writing (1995), large-scale genetic evaluation is common in industrialized countries for beef and dairy breeds, and programs for breeds of sheep and swine are in the development stages or beyond. Evaluations of performance traits in horses are also available—more technically sophisticated ones in Europe, less sophisticated ones in the United States.

designed test
A carefully monitored progeny test designed to eliminate sources of bias like nonrandom mating and culling for poor performance.

field data
Data that are regularly reported by individual breeders to breed associations or government agencies.

The earliest evaluations were **designed tests,** carefully monitored progeny tests free of sources of bias like nonrandom mating and culling for poor performance. Designed tests were expensive and necessarily limited in size. With the power of modern statistical procedures to account for a number of biases, the general trend in large-scale evaluation has been toward the use of **field data,** data that are regularly reported by individual breeders to breed associations or government agencies. Field data provide huge amounts of information—much more information than designed tests can supply.

Large-scale genetic evaluations using field data require advanced technical expertise, complex software, and powerful computers. They are conducted once a year or more frequently by specialists at breed associations, in government, at universities, or (less commonly) in private companies.

SIRE SUMMARIES

sire summary
A list of genetic predictions, accuracy values, and other useful information about sires in a breed.

The most visible product of large-scale genetic evaluation is the **sire summary.** Sire summaries are lists of genetic predictions, accuracy values, and other useful information about sires in a breed. Summaries vary in format from species to species and from breed to breed. Typically, however, they are comprised of an introductory section followed by a list of sire data. The introductory section of a sire summary is very informative. It may contain an explanation of the data in the list, including a glossary of terms, the qualifications necessary for a sire to be listed in the summary, a table of genetic parameter estimates (heritabilities and correlations used in calculating predictions), distributions of predictions within the breed (in the form of graphs that usually look like normal distributions and(or) tables of percentiles), a table converting accuracies to confidence ranges or possible change values (defined later in this chapter), and graphs of genetic trend.

The sire list itself typically includes three types of data: animal identification; miscellaneous information about the sire including simply-inherited characteristics such as coat color, genetic defects, etc.; and predictions and accuracy measures. A fabricated sample segment of a beef sire list is shown in Table 12.1.

In the example shown in the table, bulls are identified by name, name of sire and maternal grandsire, registration number, and owner(s). Other items of information include the animal's birth date and a code or suffix indicating simply-inherited characteristics. The names of the second, third, and fourth bulls in the list are followed by the suffix "R" connoting red coat color. The trait leader column indicates whether the bull ranks particularly high in the breed for a specific trait. RCN Crescendo 538 has an especially high EPD for yearling weight and is a trait leader for that trait. RCN Prelude 732 has an especially *low* EPD for birth weight

TABLE 12.1 Sample Segment of a Beef Sire List

Name (Suffix)	Reg. #	Birth Date	Owner, State or Province	Trt. Ldr.	Birth EPD ACC	Wean. EPD ACC	Yearl. EPD ACC	Milk EPD ACC	Tot. Mat. EPD
RAB George Washington *Sire:* RAB Sam Adams *MGS:* PBC 737 D2020	129755	3/16/81	N. Maclean, MT		**8.4** .93	**31** .93	**53** .92	**16** .91	**31**
RCN Crescendo 538 (R) *Sire:* RCN Intonation 338 *MGS:* RCN Sonata 008	181650	3/11/85	T. Morrison, NY	Y	**0.9** .87	**39** .86	**71** .85	**−12** .80	**7**
RCN Intonation 338 (R) *Sire:* Copper Kettle *MGS:* Ginny's Chief 105	153082	3/9/83	W. Stegner, VT		**1.3** .86	**24** .85	**43** .84	**11** .81	**23**
RCN Prelude 732 (R) *Sire:* RCN Ensemble 614 *MGS:* Copper Kettle	274698	3/11/87	T. Williams,UT M. Golden, CO	B	**−5.6** .87	**5** .86	**10** .85	**20** .78	**22**
RD Madison Ave 6X *Sire:* PCH Sun Valley 1141 *MGS:* BJR Fireworks 416	329877	3/8/90	J. Salinger, NH E. Hemingway, ID		**1.1** .67	**39** .63	**54** .51	**8** .40	**27**

and is a trait leader in that category. An actual beef sire summary would present considerably more information—including more EPDs—than is shown in this simplified example. Typical dairy summaries present *much* more information.

Predictions

The genetic predictions published in modern, state-of-the-art sire summaries are expected progeny differences or EPDs. Dairy summaries use different terms: predicted differences (PDs) and estimated transmitting abilities (ETAs), but EPD, PD, and ETA all mean the same thing. (I use EPD in this book.) Expected progeny differences (as opposed to estimated breeding values) are listed because they are comparatively easy to interpret. They represent, in relative terms, the expected performance of a sire's progeny.

The sire list shown in Table 12.1 provides examples of EPDs for different components of traits: direct, maternal, and total maternal. The EPDs for birth, weaning, and yearling weight are all measures of direct components. They predict progeny performance that is attributable to genes inherited from the sire. Birth, weaning, and yearling weight EPDs predict the growth potential of the sire's calves at different ages. The milk EPD is a measure of the maternal component of weaning weight. It predicts the milking and mothering ability (measured in pounds of weaning weight) of the sire's daughters. The total maternal EPD measures the combination of direct and maternal components of weaning weight known as total maternal weaning weight. It predicts the overall ability of the sire's daughters to produce calf weaning weight.

Accuracy Measures

repeatability (in dairy publications): Accuracy of prediction.

Accuracy is defined as a measure of the strength of the relationship between true values and their predictions and, in mathematical terms, is the correlation between true values and their predictions. We might call accuracy defined in this way *classical* accuracy. (The term for accuracy that commonly appears in dairy publications is **repeatability—**not to be confused with the concept of repeatability described in Chapter 9.)

In sire summaries, classical accuracy would be the correlation between true progeny differences and EPDs ($r_{PD,\hat{PD}}$). The accuracy values published in sire summaries rarely represent this correlation, however. Instead they are *functions* of classical accuracies. The reason for this is that classical accuracies approach 1.0 rather easily (i.e., with relatively little information) and they do a poor job of differentiating between sires that have a great deal of data and those that have just a moderate amount.

Listed in Table 12.2 are the EPDs and published accuracies of the same sires cataloged in Table 12.1. Also shown are several alternative accuracy measures including classical accuracies. Note how high the classical accuracies are for the first four sires in the list. They are so high, in fact, that it is hard to see much difference in accuracy between the first bull, whose EPDs are derived from a truly large volume of data, and the next three bulls, whose EPDs are derived from moderate amounts of data. The published accuracies are much better in this respect. Like classical accuracies, they range from zero to one, so you can interpret them in much the same way as you would classical accuracies.

TABLE 12.2 Alternative Accuracy Measures for the Sires Listed in Table 12.1

Sire Name		Birth	Weaning	Yearling	Milk
RAB George Washington	**EPD**	**8.4**	**31**	**53**	**16**
	Published accuracy	.93	.93	.92	.91
	Accuracy ($r_{PD,\hat{PD}}$)	.998	.998	.997	.996
	68% confidence range	8.2 to 8.6	30.0 to 32.0	51.3 to 54.7	15.4 to 16.6
	Possible change	±.20	±1.0	±1.7	±.6
RCN Crescendo 538	**EPD**	**.9**	**39**	**71**	**−12**
	Published accuracy	.87	.86	.85	.80
	Accuracy ($r_{PD,\hat{PD}}$)	.992	.990	.989	.980
	68% confidence	.54 to 1.26	37.0 to 41.0	67.8 to 74.2	−13.4 to −10.6
	Possible change	±.36	±2.0	±3.2	±1.4
RCN Intonation 338	**EPD**	**1.3**	**24**	**43**	**11**
	Published accuracy	.86	.85	.84	.81
	Accuracy ($r_{PD,\hat{PD}}$)	.990	.989	.987	.982
	68% confidence range	.91 to 1.69	21.9 to 26.1	39.6 to 46.4	9.6 to 12.4
	Possible change	±.39	±2.1	±3.4	±1.4
RCN Prelude 732	**EPD**	**−5.6**	**5**	**10**	**20**
	Published accuracy	.87	.86	.85	.78
	Accuracy ($r_{PD,\hat{PD}}$)	.992	.990	.989	.975
	68% confidence range	−5.96 to −5.24	3.0 to 7.0	6.8 to 13.2	18.4 to 21.6
	Possible change	±.36	±2.0	±3.2	±1.6
RD Madison Ave 6X	**EPD**	**1.1**	**39**	**54**	**8**
	Published accuracy	.67	.63	.51	.40
	Accuracy ($r_{PD,\hat{PD}}$)	.944	.929	.872	.800
	68% confidence range	−.77 to 2.97	33.8 to 44.2	43.7 to 64.3	3.7 to 12.3
	Possible change	±1.87	±5.2	±10.3	±4.3

confidence range
A range of values within which we expect—with a given probability, a given degree of confidence—that a true value of interest lies.

Another way of expressing accuracy is with **confidence ranges.** As explained in Chapter 10, a confidence range is a range of values within which we expect—with a given probability, a given degree of confidence—that a true value of interest lies. Sixty-eight percent confidence ranges for the yearling weight EPDs of two of the sires from Table 12.2 are illustrated in Figure 12.1. RAB George Washington is well evaluated, having a published accuracy for yearling weight EPD of .92. The 68% confidence range for this EPD is narrow—from 51.3 lb to 54.7 lb. In other words, the chance of his true progeny difference being between 51.3 and 54.7 lb is a little better than two out of three. RAB George Washington's true progeny difference might be outside this range, but even if it is, it is unlikely to be far outside it. RD Madison Ave 6X has relatively little information as evidenced by his published yearling weight accuracy of .51. The 68% confidence range for his yearling weight EPD is much broader—from 43.7 lb to 64.3 lb. Clearly there is more potential error associated with RD Madison Ave 6X's yearling weight EPD than with RAB George Washington's.

Genetic predictions for individual animals change over time as more and more data are included in the calculations used to produce successive evaluations.

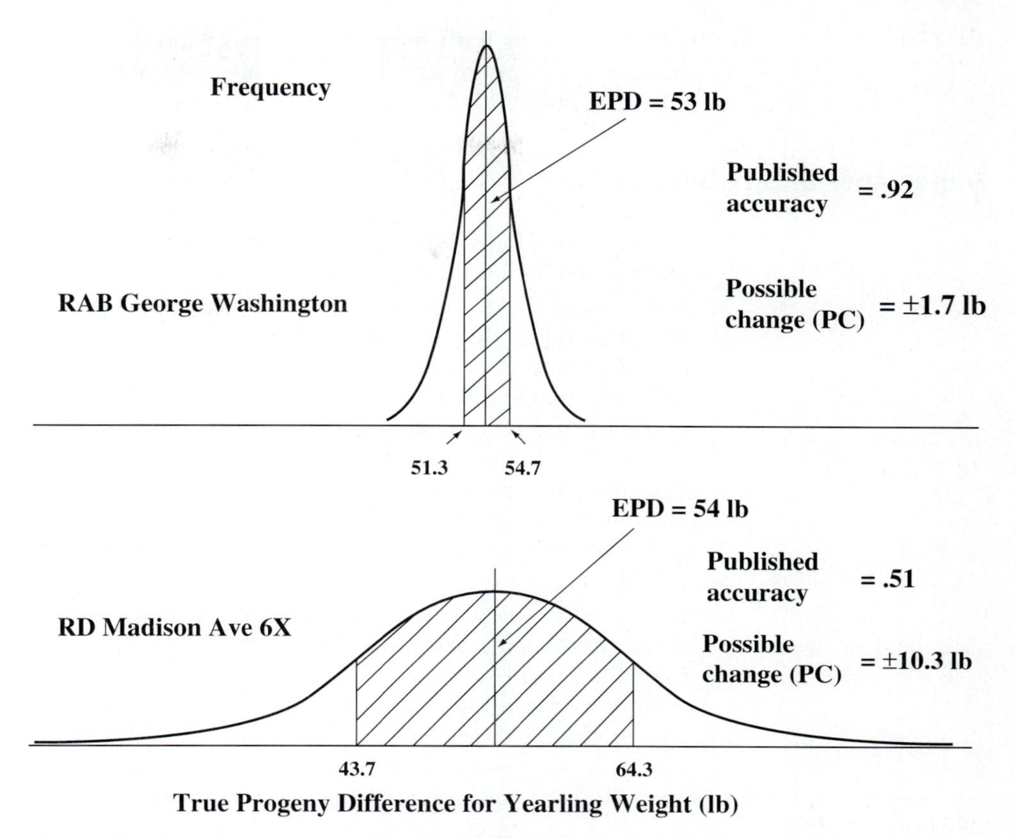

FIGURE 12.1 Distributions of true progeny differences for yearling weight illustrating 68% confidence ranges for EPDs of two of the sires from Table 12.2.

possible change (PC) or **standard error of prediction**
A measure of accuracy indicating the potential amount of future change in a prediction.

A measure of accuracy indicating the potential amount of future change in a prediction is called **possible change (PC).** The statistical term for possible change is **standard error of prediction.** A prediction ± possible change is simply an alternative way of representing a 68% confidence range. For example, RAB George Washington's EPD ± PC for yearling weight is 53 ± 1.7 lb, which corresponds to his 68% confidence range of 51.3 to 54.7 lb (see Figure 12.1). Likewise, RD Madison Ave 6X's EPD ± PC for yearling weight is 54 ± 10.3 lb, corresponding to a 68% confidence range of 43.7 to 64.3 lb. Confidence ranges and possible changes for all five sires listed in Table 12.1 are shown in Table 12.2.

Accuracies, confidence ranges, and possible change values are closely related. The higher (better) the accuracy of a prediction, the smaller the associated confidence range and possible change. Accuracies have some advantage over confidence ranges and possible changes in that you do not need to know anything about the variability of a trait in order to interpret an accuracy. A published accuracy of .95 is a very high accuracy regardless of the trait. The same cannot be said for a confidence range or a possible change. A possible change of ±2.5, for example, means one thing if the trait is mature weight in beef cattle and something quite different if the trait is days to 230 lb in swine. In the first case, it suggests very high accuracy; in the second case, it suggests very low accuracy. On the other hand, if you are familiar with the trait being measured, confidence ranges and possible changes are more graphic and probably more informative indicators of the reliability of a prediction.

Importance of Sire Summaries

Sire summaries are important, first and foremost, for selection. They increase the effectiveness of selection in two ways: (1) by expanding the pool of available sires, thus increasing selection intensity, and (2) by using large amounts of data for genetic prediction, thus increasing accuracy.

Why *sire* summaries and not dam summaries? As explained in Chapter 10, male selection is much more important than female selection in terms of its effect on the rate of genetic change in a population. Sire selection really drives genetic change. Another reason is that sires from someone else's herd or flock are more accessible than dams. Many offspring can be produced for the price of one sire. And if it is biologically and politically feasible within a species or breed to use artificial insemination, access to outstanding sires is even easier.

Sire summaries can also be used to manage selection risk, the risk that the true breeding values of a sire will be significantly poorer than expected. Sires whose published accuracy values are high should breed as advertised in the summary. Sires with low accuracies may not. The accuracy information provided by sire summaries allows breeders to take as much or as little risk as they like.

Sire summaries are also marketing tools. Much like shows and races, they provide a competitive forum for breeders. If a sire has particularly desirable EPDs, those EPDs become advertising points for his owner.

Genetic Evaluation for Nonsires

Large-scale genetic evaluation is not limited to the production of EPDs for sires. Modern BLUP analyses produce predictions for all animals. Of course, there are too many of these to be published in a single document. Typically, breed associations or government agencies send breeders computer printouts containing EPDs and accuracies for the animals they own.

Types of EPDs

parent EPD
An EPD for an animal with progeny data. Parent EPDs typically come with associated accuracy measures.

nonparent EPD
An EPD for an animal without progeny data. Nonparent EPDs typically do not come with associated accuracy measures.

Several different types of EPDs may be reported to breeders. **Parent EPDs** are EPDs for animals with progeny data. They are the immediate product of large-scale analyses, and come with associated accuracy measures. The EPDs published in sire summaries are examples of parent EPDs.

With some BLUP models, EPDs for nonparents are calculated somewhat differently from parent EPDs. **Nonparent EPDs** of this kind typically do not come with published accuracy measures. If a nonparent has no performance information of its own contributing to its EPD, then that EPD is necessarily a **pedigree estimate,** and is simply the average of the EPDs of the individual's sire and dam. Pedigree estimates are always low in accuracy. Other nonparent EPDs are generally low in accuracy, although they may not be if the trait under consideration is highly heritable and(or) is a repeated trait. For example, a gelding (castrated male

pedigree estimate
A genetic prediction based solely on pedigree data.

interim EPD
An updated EPD that is calculated between BLUP analyses and incorporates new information.

horse—clearly a nonparent) with a long racing history could conceivably have a highly accurate EPD for racing ability. (Of course, an expected progeny difference for a gelding is a rather nonsensical idea.)

Because full-blown BLUP analyses are performed months to a year apart, there is often a need for updated EPDs between analyses. For example, many analyses of beef cattle data are performed soon after weaning data are reported so that sire summaries will be available before the next breeding season. In the meanwhile, however, additional information is collected on the postweaning performance of young animals. **Interim EPDs—**updated EPDs calculated between BLUP analyses—incorporate the new information. Interim EPDs for yearling weight combine nonparent EPDs produced by BLUP with more recent postweaning information. In a sense, interim EPDs are "fudges." Only a portion of available data is included in the shortcut procedures used to calculate them. They are usually better, however, than the outdated EPDs they replace.

INTERPRETING GENETIC EVALUATION INFORMATION

The products of large-scale genetic evaluation—EPDs and measures of accuracy—can be confusing and are often misunderstood. The following discussion is designed to address questions about EPDs and accuracies commonly asked by breeders.

Predictions Are for Comparing Animals

EPDs do not predict performance per se. For example, you cannot say that daughters of a boar with an EPD for 21-day litter weight of +5 lb will produce litters averaging 105 lb. Their actual production will depend on other things besides the genes inherited from their sire. Environment is important. If the boar's daughters experience better or poorer than average environments (i.e., if contemporary group effects (E_{cg}) are better or poorer than average), then performance will be affected accordingly. Performance will also be affected by whether the sows and their pigs are inbred or crossbred, i.e., by their gene combination values (GCV), and by the breeding values of the sows' dams and the breeding values of the boars to which they are bred.

The boar's EPD for 21-day litter weight is *the expected difference between the average performance of his daughters and the average performance of daughters of a boar with an EPD of zero—all else being equal.* Of course, unless you are well acquainted with a boar whose EPD for 21-day litter weight is exactly zero, this may not be the most informative comparison. A more meaningful comparison might be between the boar with an EPD of +5 lb and another boar of interest whose EPD is −2 lb. In this case, we expect daughters of the first boar to wean litters 7 lb ($5 - (-2) = 7$) heavier than daughters of the second boar. *EPDs are for comparing animals.*

Knowledgeable breeders have a feeling for the levels of EPDs that best suit their environment, management, mating system, and market. From experience, they know how the offspring of animals with particular progeny differences will typically perform in their breeding operations. For these breeders, an EPD by itself signifies a certain level of performance. However, as soon as we leave the context of a specific farm or ranch, the connection between EPDs and actual performance is lost. Then it is not EPDs themselves that are informative, but rather *differences* in EPDs.

The Meaning of Zero

It would be much easier to interpret EPDs if we knew what an EPD of zero meant. But just as we cannot predict the performance (in an absolute or nonrelative sense) of daughters of the boar with an EPD for 21-day litter weight of +5 lb, neither can we predict the actual performance of offspring of an animal with an EPD of zero. It is a mistake to assume that *any* particular level of EPD will result in a certain level of performance across environments and herds or flocks.

According to the genetic model for quantitative traits discussed in Chapter 7, the performance associated with an EPD of zero ought to be equivalent to μ, the average performance in a population. As handy as this definition might be, it does not apply in the case of large-scale genetic evaluation of populations that contain genetically diverse contemporary groups and have changed over time due to selection.

base
(in large-scale genetic evaluation): The level of genetic merit associated with an EPD of zero.

In reality, definition of the zero point or **base** depends on the statistical model used for genetic prediction and on various characteristics of the data. For practical purposes, the zero point is meaningless. To provide some meaning to the base, those who produce large-scale genetic evaluations sometimes adjust EPDs so that the base represents the average EPD of all animals born in a specified year. For example, in a swine evaluation, EPDs might be adjusted so that an EPD of zero for 21-day litter weight represents breed average for the trait in, say, 1985. EPD adjustments of this kind do not affect comparisons of EPDs in any way. If the difference between two animals' EPDs was 7 lb before adjustment, it will remain 7 lb after adjustment. Tying EPDs to a **base year** simply allows us to have a definition, however inadequate, of zero.

base year
(in large-scale genetic evaluation): The year chosen to represent the base. The average EPD of all animals born in the base year is zero.

Different breeds within a species have different bases. For this reason (and others), EPDs should not be compared across breeds. Even if the EPDs of two breeds are tied to the same base year, their bases will only be the same if both breeds were at the same genetic level in that year. Finding reliable ways to compare EPDs across breeds will be one of the more important challenges for animal breeding specialists in the years to come.

Using Accuracy Measures

Measures of accuracy provide information on selection risk, the risk that the true breeding value or progeny difference of an animal will be significantly poorer than expected. Use of individuals with high accuracy EPDs incurs little selection risk. Use of individuals with low accuracy EPDs is much riskier.

Classical and published accuracies do not always paint a clear picture of the selection risk associated with using a particular animal. Confidence ranges are better in this respect. Suppose, for example, that you are a beef cattle breeder and are interested in purchasing semen from one of two sires listed in Table 12.2: RCN Crescendo 538 and RD Madison Ave 6X. Because of your concern about calving difficulty, you want to be sure that the bull you choose does not have a true progeny difference for birth weight that is greater than +2 lb. The birth weight EPDs of both sires are less than +2. RCN Crescendo's EPD is +.9, and RD Madison Ave 6X's EPD is +1.1. Published accuracies differ, however—.87 and .67, respectively. Despite the fact that their EPDs for birth weight are almost identical, the difference in accuracy tells you that RD Madison Ave 6X represents more of a selection risk. The question is, how much of a risk?

Confidence ranges provide a clue. The 68% confidence range for RCN Crescendo's EPD is +.54 to +1.26. That means that the probability of his true progeny difference for birth weight being within this interval is 68%. The probability of his true progeny difference being greater than +1.26 is relatively small. On the other hand, the 68% confidence range for RD Madison Ave 6X's EPD is −.77 to +2.97. His true progeny difference for birth weight could easily be above +2 lb.

If you are not comfortable with 68% confidence ranges, you could convert them to 95% confidence ranges by simply adding possible change to each end of the 68% ranges. Possible changes for the birth weight EPDs of RCN Crescendo 538 and RD Madison Ave 6X are .36 and 1.87, respectively, so their 95% confidence ranges would be +.18 to +1.62 and −2.64 to +4.84. Clearly, with a 95% confidence range as wide as −2.64 to +4.84, RD Madison Ave 6X is not very well evaluated for birth weight. To use him would be to invite a much higher risk of calving difficulty.

If confidence ranges or possible changes are not published, it is often possible to determine them from published accuracies. Most sire summaries provide conversion tables for this purpose.

Breeders commonly make the mistake of considering small differences in EPDs meaningful. For example, the difference in the birth weight EPDs of RCN Crescendo 538 and RCN Intonation 338 is .4 lb (Table 12.2). The difference between their true progeny differences is probably something other than .4 lb, but even if it were .4 lb, such a small difference would be undetectable in all but the largest herds. In most cases, it is better to relegate EPDs to general categories (e.g., very high, high, medium, low, very low) than to attach particular significance to the numbers themselves.

Confidence ranges provide a means for determining whether the EPDs of two animals are truly different. Confidence ranges for the birth weight EPDs of sires listed in Table 12.2 are shown graphically in Figure 12.2. Confidence ranges for RCN Crescendo 538, RCN Intonation 338, and RD Madison Ave 6X all overlap. This indicates that differences in these EPDs are not significant in a statistical sense. It would be a mistake to assume that these EPDs are really any different from each other. In contrast, confidence ranges for the birth weight EPDs of RAB George Washington and RCN Prelude 732 (+8.2 to +8.6 and −5.96 to −5.24, respectively) are not remotely close to overlapping. You can be sure that the *average* birth weights produced by these two sires will be different from each other and different from the average birth weights produced by the other three sires as well.

Students (and more than a few breeders) sometimes confuse the concept of accuracy of prediction with variability in offspring performance. For example, they might conclude from Table 12.2 that because RAB George Washington's accuracies are high and confidence ranges narrow, he will sire particularly uniform calves. Likewise, they might infer that because RD Madison Ave 6X's accuracies are low and confidence ranges wide, he will sire particularly nonuniform calves. This, of course, is nonsense. We have no reason to believe that calves by the first sire will be any more or less variable in performance than calves by the second sire. The accuracy of an EPD tells us something about the reliability of the EPD as a predictor of the *mean* performance of an individual's offspring. It tells us nothing about *variation* in offspring performance.

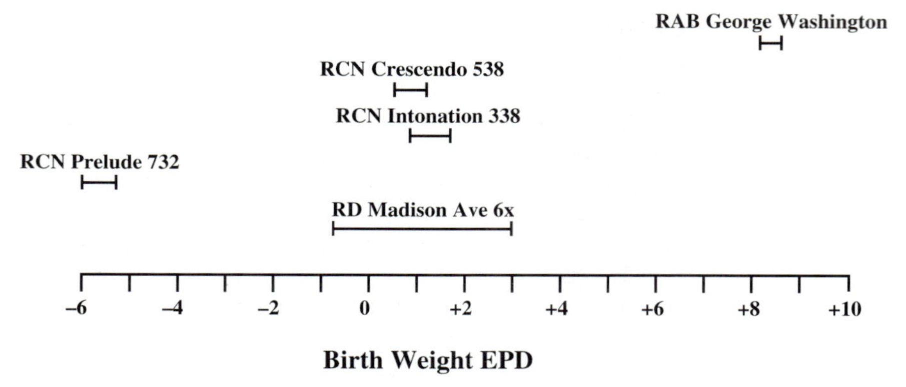

FIGURE 12.2 Graphic representation of confidence ranges for the birth weight EPDs of sires listed in Table 12.2. Confidence ranges for RCN Crescendo 538, RCN Intonation 338, and RD Madison Ave 6X all overlap. Differences in these EPDs are not significant. Confidence ranges for RAB George Washington and RCN Prelude 732 do not overlap. The *average* birth weights produced by these two sires will be different from each other and different from the average birth weights produced by the other three sires as well.

Evaluation or Characterization?

Use of the word "evaluation" in the term "large-scale genetic evaluation" implies that we are somehow assessing the value of animals by calculating genetic predictions for them. In a sense we are, but we must be careful not to take the results of genetic evaluation too literally and confuse EPDs with value. The animal with the highest EPDs is not necessarily the most valuable animal. For many traits there are optimum levels of EPDs that are not the highest or lowest, but somewhere in between. The most valuable animal often represents a balance of EPDs for various traits. EPDs reveal an animal's genetic abilities, but not its worth. Perhaps a better term than "genetic evaluation" would be "genetic characterization."

Pitfalls in Large-Scale Genetic Evaluation

The advent of large-scale genetic evaluation has revolutionized animal breeding for a number of species. It has changed the way breeders operate and accelerated genetic change. That is not to say, however, that large-scale genetic evaluation does not have problems. The following discussion describes some of these pitfalls.

Faulty Data

Genetic evaluation can be no better than the data contributing to it. Field data are reported by many breeders, and those breeders vary greatly in their knowledge of and interest in animal breeding principles. So by its very nature, this type of data is subject to error, intentional or otherwise.

Pedigrees

In modern-day genetic evaluation, genetic predictions are based to a large extent on the performance of relatives. It is important, therefore, that the pedigree relationships used in genetic evaluations are correct. For example, a pedigree estimate for a young animal is useless if that animal's parents have been misidentified.

Parental misidentification is not unusual in many species, and it cause errors in prediction. Fortunately, errors of this sort are generally short-lived. As soon as an individual has enough progeny information, pedigree information is of little consequence. Still, technologies that improve animal identification are sure to benefit genetic prediction.

Performance Records

Sometimes breeders falsify performance records. They succumb to the temptation to make animals look better than they really are. Deliberate falsification is probably rare, however. More common are "guesstimates." Sometimes it is just easier to guess performance than to carefully measure it. Many birth weights in beef cattle, for example, are not measured with a scale. They are simply "eyeballed." Shortcuts like this reduce the reliability of predictions.

incomplete reporting
The reporting of only selected performance records to a breed association or government agency.

A common source of bias in field data is **incomplete reporting.** Incomplete reporting refers to the reporting of only selected performance records to a breed association or government agency. To see the effect of incomplete reporting, imagine that a breeder only reports records on the best 20% of a sire's progeny. By selecting data, the breeder has made the sire appear better than he really is.

Unlike the bias caused by culling for poor performance, the bias caused by incomplete reporting cannot be accounted for with advanced BLUP procedures.[1] While the early records of animals culled for poor performance are accessible, the records of nonreported animals are not. It is as though these animals never existed. *All* records should be reported, both good ones and poor ones.

Data Used to Adjust Performance Records

Performance records can be reported accurately, but if the data used to adjust those records for known environmental effects are in error, the adjusted performance records will be in error too. For example, if the birth date for a set of twin lambs is recorded as being a month later than the true birth date, then adjustment of subsequent weights of the lambs for their age will be faulty. They will appear to have grown faster than they actually did.

Contemporary Groups

Animals are sometimes reported as members of the wrong contemporary groups. An animal raised in contemporary group *A* is rarely reported in group *B*, but breeders sometimes report animals as members of a single contemporary group when, to be correct, they should divide them into two or more contemporary

[1]See Chapter 11 for an explanation of how BLUP technology is used to account for culling for poor performance.

groups. As explained in Chapter 9, misgrouping of animals causes especially severe problems when animals that have received preferential treatment (show animals, for example) are reported in the same contemporary group with animals that have received no preferential treatment. The animals receiving preferential treatment look inordinately good compared to the others.

Lack of Relationship Among Contemporary Groups

BLUP techniques solve for contemporary group effects as well as genetic predictions. It is this ability to separate environmental effects common to all members of a contemporary group from genetic effects that makes BLUP so useful. But in order to get good estimates of contemporary group effects (and, therefore, good genetic predictions), it is necessary that animals in different contemporary groups be related. The best situation would be if all sires in a breed had progeny in all contemporary groups. Then every contemporary group would be connected to every other contemporary group through half-sib families. Of course, in real populations **connectedness** to this degree does not exist. Pedigree relationships between animals in different contemporary groups are often quite distant. Still, the closer these relationships, the better the connectedness among groups, and the better the resulting genetic predictions.

connectedness
The degree to which data from different contemporary groups within a population can be compared as a result of pedigree relationships between animals in different groups.

The structure of a population affects connectedness. Populations in which some sires have progeny in many herds or flocks tend to have good connectedness. Populations in which there is little exchange of germ plasm between herds or flocks tend to have poor connectedness. Artificial insemination is an extremely important tool for increasing the connectedness of a population.

Genotype by Environmental Interaction

genotype by environment ($G \times E$) interaction
A dependent relationship between genotypes and environments in which the difference in performance between two (or more) genotypes changes from environment to environment.

The data for large-scale genetic evaluation often come from very different environments. This is not a problem so long as there is no significant **genotype by environment (G × E) interaction.** Recall that a $G \times E$ interaction occurs when the difference in performance between two (or more) genotypes changes from environment to environment. If a sire's progeny perform well relative to progeny of other sires in one environment, but perform poorly relative to progeny of other sires in another environment, then serious $G \times E$ interaction exists. The sire's published EPDs, which are based on the performance of his progeny in both environments, may not reflect his true progeny difference as it is expressed in either environment.

$G \times E$ interaction causes reranking of individuals (most notably sires) in different geographical areas and management systems. The extent of reranking depends on the species, the trait, and the amount of variation in environment and management. As a rule, lowly heritable traits like fertility and survivability are more susceptible to $G \times E$ interaction than are highly heritable traits.

Alternatives to Large-Scale Genetic Evaluation

Large-scale genetic evaluation is common in industrialized countries where the necessary infrastructure (computers, breed associations, technical expertise, etc.) exists. It is not the only approach capable of accelerating genetic change, however. There are alternatives.

Nucleus Breeding Schemes

nucleus breeding scheme
A cooperative breeding program in which elite animals are concentrated in a nucleus herd or flock and superior germ plasm is then distributed among cooperating herds or flocks.

natural service
Natural mating (as opposed to artificial insemination).

closed nucleus breeding scheme
A nucleus breeding scheme in which germ plasm flows in only one direction—from the nucleus to cooperating herds or flocks.

open nucleus breeding scheme
A nucleus breeding scheme in which the flow of germ plasm is bidirectional—from the nucleus to cooperating herds or flocks and from cooperating herds or flocks to the nucleus.

Nucleus breeding schemes are cooperative breeding programs involving a number of breeders. In the formation of a nucleus breeding scheme, the best males and females owned by the cooperating breeders are combined in a nucleus. Usually these animals are physically gathered together to form a nucleus herd or flock. This is necessarily the case if all matings are by **natural service,** but does not have to be the case if artificial insemination is used. Once the nucleus is formed, all (or *almost* all) sires used in the entire scheme are produced in the nucleus where pedigree and performance records are meticulously kept, genetic prediction techniques are the most sophisticated, and the most intense selection takes place.

A simplified nucleus breeding scheme is diagrammed in Figure 12.3. In a **closed nucleus breeding scheme,** germ plasm flows in only one direction. Animals (particularly sires) produced in the nucleus are used in the cooperating herds or flocks. In an **open nucleus breeding scheme,** the flow is bidirectional. Females that have proven themselves superior in cooperating herds or flocks are transferred to the nucleus.

Much of the advantage of nucleus schemes derives from the initial genetic gain achieved when the elite nucleus population is formed. If nucleus animals are well chosen, the average genetic merit of the nucleus should be much higher than the average merit of contributing herds or flocks. Additional gains result from increased accuracy of selection and (especially in open nucleus schemes) increased

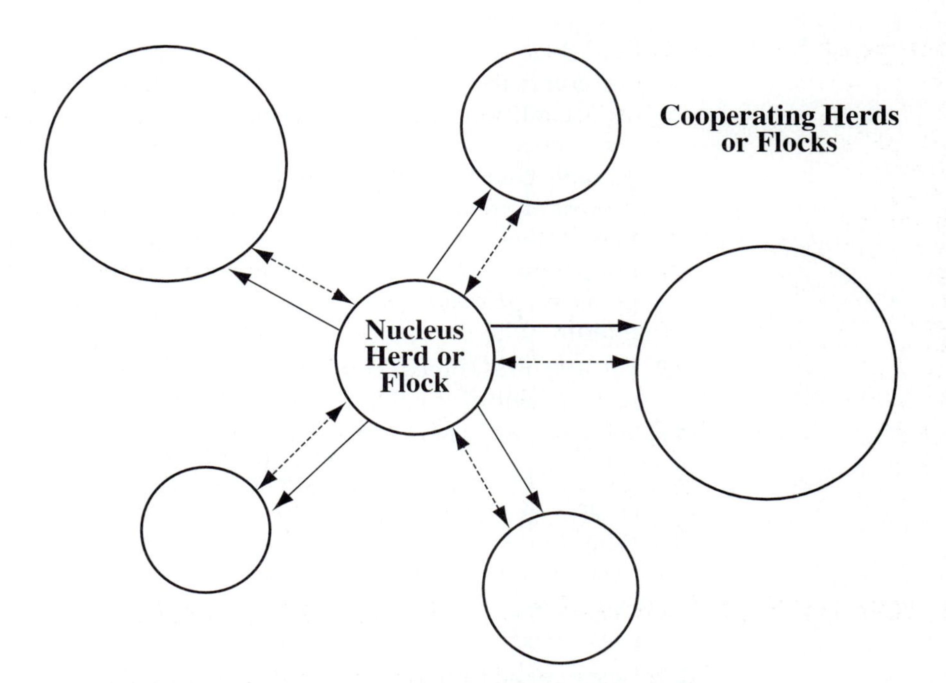

FIGURE 12.3 Diagram of a nucleus breeding scheme. Arrows represent the flow of germ plasm. In closed nucleus schemes, the flow is from the nucleus outward to cooperating herds or flocks (solid arrows). In open nucleus schemes, the flow is bidirectional (dotted arrows).

multiple ovulation and embryo transfer (MOET) Hormonally induced ovulation of multiple eggs followed by transfer of embryos to recipient dams. The term is used in conjunction with breeding strategies designed to increase the rate of genetic change using embryo transfer.

selection intensity. Nucleus breeding schemes combined with **multiple ovulation and embryo transfer (MOET)** have been proposed as a way of further increasing the rate of genetic change. The use of embryo transfer allows elite females to have many progeny, thus increasing both intensity and accuracy of selection among females. If MOET can be successfully used with very young females, then female generation interval can be reduced dramatically. Accuracy may suffer, but the rate of genetic change can be significantly increased. The effectiveness of schemes involving MOET depends to a large degree on the success rate and expense of MOET technology.

Nucleus breeding schemes may not produce individuals with the extremely high accuracies that we see with large-scale genetic evaluation, but they make up for it with initial genetic gain achieved in nucleus formation and in the added advantage (depending on your perspective) of a degree of discipline and consistency imposed on cooperating breeders. And there is no reason that nucleus breeding schemes and large-scale genetic evaluation cannot coexist. Herds or flocks cooperating in a nucleus scheme can be a part of a larger genetic evaluation effort.

EXERCISES

Study Questions

12.1 Define in your own words:

large-scale genetic evaluation
central test
designed test
field data
sire summary
repeatability (in dairy publications)
confidence range
possible change (PC) or standard error of prediction
parent EPD
nonparent EPD
pedigree estimate
interim EPD
base
base year
incomplete reporting
connectedness
genotype by environment ($G \times E$) interaction
nucleus breeding scheme
natural service
closed nucleus breeding scheme
open nucleus breeding scheme
multiple ovulation and embryo transfer (MOET)

12.2 For a species/breed/country of your choice, describe any ongoing large-scale genetic evaluation program.

12.3 What is the purpose of large-scale genetic evaluation?

12.4 What elements of the key equation for genetic change are improved using large-scale genetic evaluation?

12.5 What limits the ability of central tests to compare the genetic merit of animals from different flocks or herds?

12.6 Discuss the relative advantages and disadvantages of designed tests versus use of field data for genetic evaluation.

12.7 Describe the format of a typical sire summary.

12.8 Why are accuracy values published in sire summaries often different from "classical" accuracies (i.e., correlations between true values and their predictions)?

12.9 **a.** How are accuracy, confidence range, and possible change related?
b. Discuss the advantages and disadvantages of each measure.

12.10 Why *sire* summaries and not *dam* summaries?

12.11 Rank the following for typical level of accuracy and explain your ranking:

parent EPD
nonparent EPD
pedigree estimate
interim EPD

12.12 EPDs are for *comparing* animals—not for predicting performance per se. Explain.

12.13 If two breeds of the same species designate the same year as the base year, do they necessarily have the same base—the same level of genetic merit associated with an EPD of zero? Explain.

12.14 What is wrong with the following statement: "This sire's accuracies are high, so his offspring should be especially uniform"?

12.15 Describe the variety of ways in which breeders create faulty data.

12.16 **a.** Why is it important to have good connectedness in a population?
b. What can be done to increase connectedness in a population?

12.17 How do genotype by environment interactions create problems for genetic prediction?

12.18 Describe the structure of a nucleus breeding scheme.

12.19 List potential advantages of a nucleus breeding scheme that incorporates MOET technology.

CHAPTER 13

Correlated Response to Selection

Selection for one trait rarely affects just that one trait. Usually other traits are affected as well. Genetic change in one or more traits resulting from selection for another is termed **correlated response to selection.**

WHAT CAUSES CORRELATED RESPONSE

correlated response to selection
Genetic change in one or more traits resulting from selection for another.

pleiotropy
The phenomenon of a single gene affecting more than one trait.

Correlated response to selection is probably caused by a number of genetic mechanisms. Linkage is one. If major genes affecting two traits are closely linked, they tend to stick together. Selection for one trait increases the frequency of alleles positively influencing that trait and, at the same time, increases the frequency of linked alleles as well. Linked genes do not remain together forever, however. Sooner or later recombination breaks the linkage. For this reason, linkage is only a temporary cause of correlated response to selection.

The major cause of correlated response is **pleiotropy.** A gene is said to have pleiotropic effects if it influences more than one trait. An example of a major gene known to have pleiotropic effects is the halothane gene in swine. The gene causes increased lean yield and feed efficiency, but also decreased litter size, survival rate, and meat quality. The HYPP gene in horses is similarly pleiotropic, causing both increased muscling and decreased survival.

Related polygenic traits are probably influenced by many genes with pleiotropic effects. Illustrated schematically in Figure 13.1 are genes affecting two such polygenic traits, *X* and *Y*. Many genes influence just one trait or the other, but a significant number of genes affect both traits.

Growth traits provide a classic example of pleiotropy. In a number of species we measure the growth rate of animals several times, typically at birth and weaning (for mammals anyway) and often at some later age. There are some genes that affect growth rate only during specific stages of an animal's life. There are also genes that affect growth in general—"growth genes" if you like. These genes have pleiotropic effects on two or more growth traits and, as a result, the traits are genetically correlated. Selection for one causes correlated response in the others.

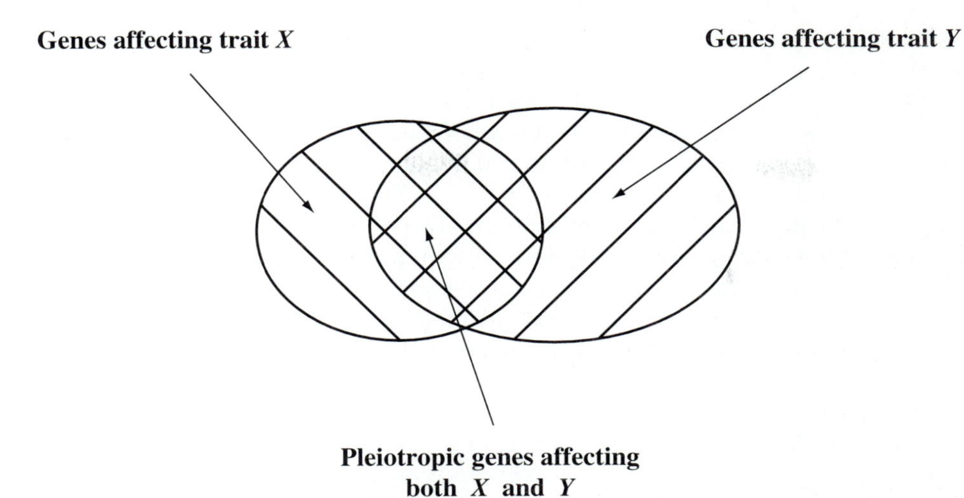

FIGURE 13.1 Schematic representation of pleiotropy. Some genes affect trait *X* only. Others affect only trait *Y*. Pleiotropic genes affect both *X* and *Y* resulting in a genetic correlation between the two traits.

GENETIC, PHENOTYPIC, AND ENVIRONMENTAL CORRELATIONS

genetic correlation (r_{BV_X,BV_Y}) (1) A measure of the strength (consistency, reliability) of the relationship between breeding values for one trait and breeding values for another trait; (2) A measure of pleiotropy.

From a statistical perspective, correlated response to selection results from **genetic correlations** between traits. The genetic correlation (r_{BV_X,BV_Y}) was defined in Chapter 8 as a measure of the strength (consistency, reliability) of the relationship between breeding values for one trait and breeding values for another. This is a good definition, but we can now add some biological meaning to it. A genetic correlation measures the relative importance of pleiotropic effects (and, temporarily anyway, linkage effects) on two traits.

Following are some illustrative examples of genetic correlations. The genetic correlation between keel length (a skeletal measurement) and body weight in turkeys is estimated to be about .5. This fairly high correlation should not be surprising because both traits are measures of body size. Genes that increase keel length are likely to have similar pleiotropic effects on body weight. Selection for increased keel length will increase body weight and vice versa. In contrast, the genetic correlation between body weight and number of eggs in turkeys is much lower (approximately .05). These traits are very different. You would not expect many genes to affect both of them. Selection for body weight should have little effect on number of eggs.

An example of a negative genetic correlation is the correlation between milk yield and percent milk fat in dairy cows. It is estimated at −.3. This too is a sensible correlation in that genes that increase milk yield also increase the water content of milk, and that dictates a decrease in the fat content of milk. Selection for increased yield will decrease percent milk fat.

Some negative genetic correlations are negative only because of the way we measure traits. For example, the genetic correlation between days to 230 lb and feed conversion in swine is thought to be about +.7. (Remember that feed conversion is measured in units of feed required per unit of gain, so the *smaller* the number, the *better* the efficiency.) By selecting for *decreased* number of days required to reach 230 lb we get a *decrease* in the amount of feed required per lb of gain. The correlation between these two traits is positive. However, if we express growth rate in terms of weight per day of age instead of days to 230 lb, the correlation between this alternative measure and feed conversion is negative (−.7). Selection for *increased* weight per day of age causes a *decrease* in the amount of feed required per lb of gain.

phenotypic correlation (r_{P_X,P_Y}) A measure of the strength (consistency, reliability) of the relationship between performance in one trait and performance in another trait.

Genetic correlations are often confused with **phenotypic correlations,** and this confusion leads to a misunderstanding of genetic correlations. The two correlations are not the same. A phenotypic correlation (r_{P_X,P_Y}) is a measure of the strength (consistency, reliability) of the relationship between *performance* in one trait and *performance* in another trait.

The genetic and phenotypic correlations between two traits are often similar, but not always. A typical estimate of the genetic correlation between birth weight and yearling weight in beef cattle is .7. Both traits are growth traits, so a number of the genes that cause faster prenatal growth (heavier birth weights) also cause faster postnatal growth (heavier yearling weights). But a typical estimate of the phenotypic correlations between these traits is just .35. Heavier calves at birth tend to be heavier at a year of age, but the phenotypic relationship between these traits is not as strong as the genetic relationship.

Likewise, the genetic correlation between weaning weight and postweaning weight gain in beef cattle is thought to be about .3. Again, both traits are growth traits, so pleiotropy is expected. Yet the phenotypic correlations is about .1. Heavier weaning calves tend to grow faster postweaning but, as in the case of birth and yearling weights, the phenotypic relationship between these traits is not as strong as the genetic relationship. What causes these differences?

environmental correlation (r_{E_X,E_Y}) A measure of the strength (consistency, reliability) of the relationship between environmental effects on one trait and environmental effects on another trait.

To answer this question, we must understand one more correlation—the **environmental correlation.** An environmental correlation (r_{E_X,E_Y}) is a measure of the strength (consistency, reliability) of the relationship between environmental effects on one trait and environmental effects on another trait. The environmental correlation between birth weight and yearling weight in beef cattle is approximately .1. This suggests that the relationship between prenatal and postnatal environments is positive, but only slightly so. The environment experienced by a calf before it is born has little to do with the environment it will experience from birth to a year of age.

Just as phenotypic values are composed of breeding values and environmental effects, phenotypic relationships between traits are functions of both genetic and environmental relationships. The strong genetic correlation (.7) between birth weight and yearling weight combined with the weak environmental correlation (.1) between the traits results in an overall phenotypic correlation that is positive but moderate (.35).

compensatory gain A relative increase in the growth rate of thin animals after they are placed on adequate feed. They tend to compensate for being underweight.

A similar situation exists for the relationship between weaning weight and postweaning gain in beef cattle. The environmental correlation between these traits is often negative (if only slightly so), typically about −.05. A better than average environment for weaning weight is associated (weakly) with a poorer than average environment for postweaning gain. This is because calves that have experienced a better preweaning environment (i.e., more mother's milk) tend to be fatter at the beginning of the postweaning period. Being fat is actually an environmental handicap with respect to postweaning growth rate. Thinner calves tend to grow faster due to **compensatory gain.** The positive genetic correlation (+.3) and negative environmental correlation (−.05) between weaning weight and postweaning gain result in an overall phenotypic correlation that is positive but weak (+.1).

It is useful to think of the genetic and environmental relationships between two traits as the *functional* relationships between them. Think of the observed relationship between the traits as measured by the phenotypic correlation as simply the net result of underlying genetic and environmental relationships. (Interestingly, the phenotypic correlation is always intermediate to the genetic and environmental correlations. It is rarely the simple average of the two, but always lies somewhere between them.)

An example is illustrated in Figure 13.2. To develop the figure, hypothetical breeding values and "environmental" effects were determined for two traits X and Y, in a sample of 10 animals taken from a larger population. ("Environmental" effects in this case include gene combination values.) Values were generated using a

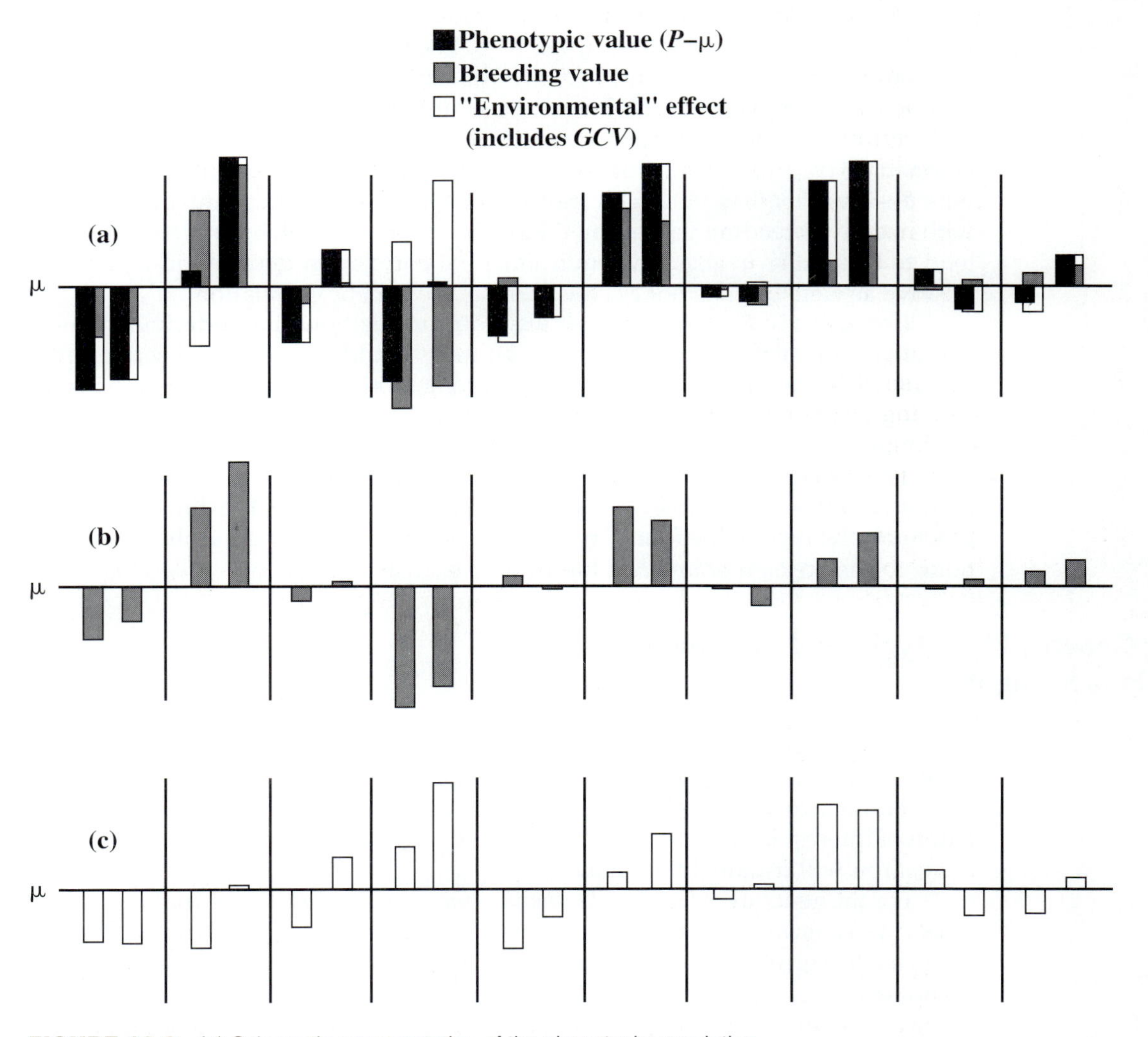

FIGURE 13.2 (*a*) Schematic representation of the phenotypic correlation (r_{P_X,P_Y} = .44) between two traits in a sample of 10 animals taken from a larger population. Each pair of records represents an animal's performance for traits X and Y, respectively. (For clarity, vertical lines separate each pair.) Contributions of breeding values and "environmental" effects ($GCV + E$) are shown in the background. The genetic correlation (r_{BV_X,BV_Y} = .8) and environmental correlation (r_{E_X,E_Y} = .2) between traits X and Y are depicted using the same 10 animals in (*b*) and (*c*).

genetic correlation (r_{BV_X,BV_Y}) of .8 and an environmental correlation (r_{E_X,E_Y}) of .2. The resulting phenotypic correlation (r_{P_X,P_Y} = .44) is shown by the black columns in the upper diagram (a). Each pair of records represents an animal's performance for traits X and Y (for clarity, vertical lines separate each pair), and contributions of breeding values and environmental effects appear in the background.[1]

The phenotypic correlation between traits X and Y in Figure 13.2 is clearly positive. Animals with above average performance in trait X tend to have above average performance in trait Y, and animals with below average performance in one trait tend to have below average performance in the other. The phenotypic relationship is far from perfect, however. There are a number of exceptions to the rule.

The genetic and environmental relationships between traits X and Y are depicted in the next two diagrams, (b) and (c). The breeding values shown in (b) and environmental effects shown in (c) are the same ones that appear in the background in the upper diagram. They have simply been isolated so that the genetic and environmental correlations are easier to see. Note the very high genetic correlation in (b) (r_{BV_X,BV_Y} = .8). Not only are positive breeding values for trait X paired with positive breeding values for trait Y, and negative breeding values for X paired with negative breeding values for Y, but, for a given animal, both breeding values tend to be similar in size. The environmental correlation depicted in (c) appears positive as well, but it is not as consistent as the genetic correlation.

If you study Figure 13.2 carefully, you can see how the underlying genetic and environmental relationships between traits X and Y drive the observed (phenotypic) relationship between them. The strong, positive genetic correlation is offset somewhat by a positive but weak environmental correlation, and the result is a moderate, positive phenotypic correlation.

Typical estimates of genetic, phenotypic, and environmental correlations for a small sample of traits in several species are listed in Table 13.1. Estimates of many other correlations can be found in scientific publications, and a number of correlations, for one reason or another, have never been estimated.

Keeping Genetic, Phenotypic, and Environmental Correlations Straight

Following is a memory device that can help you sort out conceptual differences between genetic, phenotypic, and environmental correlations. Use it with any pair of traits familiar to you. If you can answer yes to the following questions, the correlation of interest is positive. If you answer no, the correlation is either zero or negative. (You will usually know which.)

For the genetic correlation: *Do many of the same genes affect both traits in the same way?* A yes answer to this question clearly indicates positive pleiotropy and a positive genetic correlation between traits. If your answer is no, it could mean that you do not expect any genes to affect both traits, in which case the genetic correlation is zero, or it could mean that you expect that at least some of the genes that cause an *increase* in one trait cause a *decrease* in the other trait. In other words, you expect negative pleiotropy and a negative genetic correlation between the traits.

For the phenotypic correlation: *Is greater than average performance in one trait associated with greater than average performance in the other trait?* A yes answer

[1]If you are having difficulty understanding Figure 13.2, see Figure 7.4 in Chapter 7 and the section on covariation in Chapter 8. Study Figure 8.8 in particular.

TABLE 13.1 Estimates of Genetic (G), Phenotypic (P), and Environmental (E) Correlations for a Sample of Traits in Several Species

Species	Trait		*Correlations*		
			WW	GN	YW (Yearling wt.)
Beef cattle	Birth weight	G	.60	.55	.70
		P	.40	.30	.35
		E	.30	.05	.10
	Weaning weight (WW)	G		.30	.80
		P		.10	.65
		E		−.05	.50
	Postweaning gain (GN)	G			.85
		P			.75
		E			.70
			FY	PY	%F (% fat)
Dairy cattle	Milk yield	G	.45	.80	−.50
		P	.75	.90	−.30
		E	.85	.95	−.20
	Fat yield (FY)	G		.60	.55
		P		.80	.40
		E		.90	.25
	Protein yield (PY)	G			−.15
		P			−.10
		E			−.05
			WD	FC	BF (Backfat thick.)
Swine	Days to 230 lb	G	−1.00	.70	−.25
		P	−1.00	.50	−.30
		E	−1.00	.40	−.40
	Weight per day of age (WD)	G		−.70	.25
		P		−.50	.30
		E		−.40	.40
	Feed conversion (FC)	G			.30
		P			.05
		E			−.15
			SL	KL	BD (Body depth)
Turkeys	Egg number	G	.00	.05	.25
		P	.00	.00	−.10
		E	.00	.00	−.25
	Shank length (SL)	G		.75	.40
		P		.30	.30
		E		.10	.25
	Keel length (KL)	G			.55
		P			.45
		E			.40
			GN	FW	SL (Staple length)
Sheep	Weaning weight	G	.55	.05	−.15
		P	.30	.30	.00
		E	.20	.45	.10
	Postweaning gain (GN)	G		.15	−.20
		P		.20	.00
		E		.25	.15
	Grease fleece weight (FW)	G			.35
		P			.35
		E			.40

suggests a positive phenotypic correlation. A no answer suggests either no phenotypic relationship at all ($r_{P_X,P_Y} = 0$) or a negative phenotypic correlation. In the latter case, *greater* than average performance in one trait would be associated with *less* than average performance in the other trait.

For the environmental correlation: *Is a better than average environment for one trait associated with a better than average environment for the other trait?* Again, a yes answer suggests a positive environmental correlation. A no answer suggests either no environmental relationship at all ($r_{E_X,E_Y} = 0$) or a negative environmental correlation. A negative correlation indicates that some of the environmental effects that *increase* performance in one trait are associated with environmental effects that *decrease* performance in the other trait.

If you know enough about the biological factors affecting two traits, the above questions can often reveal the *sign* (positive or negative) of the correlations between them. Be careful, however, not to be confused by traits for which the signs of correlations seem backward, traits in which smaller is typically better (e.g., feed conversion, days to 230 lb, or age at puberty). To determine the *magnitude* of correlations, you need to ask an additional question in each case: *Is this a strong (consistent, reliable) relationship or a weak one?*

For an example, let's return to birth weight and yearling weight in beef cattle. *Do many of the same genes affect both traits in the same way?* Yes, definitely. Both traits are growth traits. Genes that influence growth rate in general affect both traits in the same way. Many of the genes that promote growth rate before birth promote growth rate after birth as well. Logically, then, the genetic correlation between birth and yearling weights should be positive and probably quite strong.

Is greater than average performance in one trait associated with greater than average performance in the other trait? If you have experience with birth weight and yearling weight in beef cattle, you know the answer to this question—another definite yes. Calves that are heavier than average at birth tend to be heavier than average at a year of age. This relationship is not completely consistent, however. Some calves with light birth weights achieve heavy yearling weights, and some calves with heavy birth weights record light yearling weights. The phenotypic correlation between these traits is positive but moderate.

If you have no prior experience with birth weight and yearling weight in beef cattle, you are probably unprepared to answer the question about the phenotypic relationship between the traits. In that case, the best course of action is to move on to the environmental relationship and return to the phenotypic relationship later.

Is a better than average environment for one trait associated with a better than average environment for the other trait? Environmental relationships between traits are often complicated and sometimes subtle. In this case, you would probably have to answer yes. A good prenatal environment should not hurt postnatal growth. On the other hand, there are many environmental factors influencing a calf's growth in the first year of its life. It seems unlikely that many of these would be associated in any way with prenatal environment. The environmental correlation between birth weight and yearling weight is likely to be positive but weak.

If you do not already have a good feeling for the phenotypic relationship between two traits, knowledge of genetic and environmental relationships should provide some insight. In the birth weight/yearling weight case, the strong, positive genetic correlation and weak, positive environmental correlation dictate a moderate, positive correlation between observed performance in each trait.

More Perspective on Correlations between Traits

Correlations between traits are *population measures.* They reflect relationships between traits in a population and, as such, should not be referred to in the context of a single animal. For example, it is perfectly legitimate to speak of the genetic correlation between 400-yard racing time and conformation index in quarter horses. It is inappropriate, however, to say that a particular horse's breeding value for racing time is highly correlated with its breeding value for conformation index. The horse is not itself a population. What is probably meant by such a statement is that the horse has similar breeding values for both racing time and conformation index.

Correlations between traits are not fixed. They vary depending on the population. In one breed of beef cattle, for example, the phenotypic correlation between milk production and fertility may be quite different from the same correlation in another breed, even if both breeds perform in the same environment. This is because the two breeds are different genetic populations. Genetic potentials for milk production and fertility may not be the same for each breed. In any case, differences in the frequencies of genes affecting traits give rise to different genetic relationships.

Environment can affect correlations too. In a very good nutritional environment, the phenotypic correlation between milk production and fertility in beef cattle is close to zero, perhaps even slightly positive. Heavy-milking cows are as fertile as light-milking cows. In a poor nutritional environment, however, this correlation is decidedly negative. The thin body condition of heavy-milking cows makes it difficult for them to breed regularly.

A correlation between traits can be classified in several ways. First, it can be classified by strength. Is the relationship between a pair of traits a strong, consistent, reliable one, or is it weak? Secondly, a correlation can be classified by sign. Is it positive or negative? This is strictly a mathematical consideration and often depends on how traits are measured (e.g., weight per day of age *versus* days to 230 lb).

A third classification categorizes correlations by whether they are favorable or unfavorable. This is *not* the same as classification by sign. Positive correlations can be either favorable or unfavorable, as can negative correlations. For example, the positive phenotypic correlation between weaning weight and grease fleece weight in sheep is considered favorable. Faster growing lambs not only provide more meat but more fleece as well. In contrast, the positive genetic and phenotypic correlations between birth weight and yearling weight in beef cattle are considered unfavorable. Heavier yearling weights are usually desirable, but heavier birth weights are not because they lead to calving difficulty.

A fourth classification categorizes correlations according to utility. Is a correlation useful or not useful? For example, the genetic correlation between scrotal circumference and age at puberty in cattle is highly negative. Bulls with larger testicles at a year of age reach puberty earlier and so do their female relatives. It is hard to say whether this correlation is favorable or unfavorable. Earlier puberty is valuable in many cases, but unless there is a market for Rocky Mountain oysters (calf testicles), there is little value to large scrotal circumference per se. The genetic correlation between scrotal circumference and age at puberty is an extremely *useful* one, however. It allows us to improve age at puberty, a difficult trait to measure, by selecting for scrotal circumference, an easy trait to measure.

An example of a correlation that is not useful might be the phenotypic correlation between growth rate and ear length in cattle of European origin. Calves with longer ears do appear to grow faster than calves with shorter ears, but because growth rate can be measured so much more accurately by weighing animals than by comparing ear length, the correlation, though interesting to observe, is not very helpful.

FACTORS AFFECTING CORRELATED RESPONSE

direct response to selection
Genetic change in a trait resulting from selection for that trait.

Suppose a breeder selects exclusively for trait *X*. What factors influence correlated response in trait *Y*? The first factors to consider are things we already know about—factors that affect the response of trait *X* to selection for trait *X*. These are sometimes called factors affecting **direct response to selection,** and in the case of selection for trait *X* include accuracy of breeding value prediction for trait *X*, selection intensity, and generation interval. (Genetic variation in trait *X* also affects direct response to selection for trait *X*, but does not directly influence response in trait *Y*.) These factors are elements of the key equation for genetic change discussed in Chapter 10. They affect the rate of genetic change in trait *X*, the trait under selection, and consequently affect the rate of genetic change in any correlated traits.

For example, suppose you are selecting pigs for days to 230 lb, a measure of growth rate. Because of the strong, favorable genetic correlation between this trait and feed conversion (approximately .7), you expect improvement in growth rate to result in improved feed conversion. However, if, for whatever reasons, accuracy of selection for days to 230 lb is poor, intensity of selection is low, or generation interval is long, the population will change very slowly with respect to growth rate. As a result, it will change slowly with respect to feed conversion as well.

Other factors affecting correlated response in trait *Y* are the genetic correlation between traits *X* and *Y* and genetic variation in trait *Y*. Clearly, the strength of the genetic relationship between the traits is important. We expect feed conversion in swine to respond to selection for growth rate because the genetic correlation between the traits is so high. In contrast, we do not expect egg number in turkeys to respond much to selection for keel length. The genetic correlation is too low (approximately .05).

Genetic variation in the correlated trait is important too. If there is little genetic variability in trait *Y*, *Y* cannot change much, even if selection for trait *X* is effective and traits *X* and *Y* are highly genetically correlated.

The General Formula for Correlated Response to Selection

The general formula for correlated response in trait *Y* to selection for trait *X* is:

$$\Delta_{BV_{Y|X}}\Big/_{t} = \frac{r_{BV_X,BV_Y}\, r_{BV_X,\hat{BV}_X}\, i_X \sigma_{BV_Y}}{L}$$

where $\Delta_{BV_{Y|X}}\Big/_{t}$ = the rate of genetic change in trait *Y* per unit of time (*t*) due to selection for trait *X*

r_{BV_X, BV_Y} = the genetic correlation between traits X and Y
$r_{BV_X, \hat{BV}_X}$ = accuracy of selection for trait X
i_X = selection intensity for trait X
σ_{BV_Y} = genetic variation for trait Y
L = generation interval

Proof

You can derive the formula for correlated response using a simple prediction equation approach.

$$\Delta_Y = b_{Y \cdot X} \Delta_X$$

where Δ_Y = the rate of genetic change in trait Y per unit of time (t) due to selection for trait X or $\Delta_{BV_{Y|X}}/t$
$b_{Y \cdot X}$ = the change in trait Y per *unit* change in trait X
Δ_X = direct response to selection for trait X

The regression coefficient ($b_{Y \cdot X}$) is the genetic regression of trait Y on trait X, and we can express this as a function of the genetic correlation. Thus

$$b_{Y \cdot X} = r_{BV_X, BV_Y} \left(\frac{\sigma_{BV_Y}}{\sigma_{BV_X}} \right)$$

and (from Chapter 10) Δ_X is

$$\Delta_{BV_X} \Big/ t = \frac{r_{BV_X, \hat{BV}_X} i_X \sigma_{BV_X}}{L}$$

Altogether,

$$\Delta_{BV_{Y|X}} \Big/ t = r_{BV_X, BV_Y} \left(\frac{\sigma_{BV_Y}}{\sigma_{BV_X}} \right) \left(\frac{r_{BV_X, \hat{BV}_X} i_X \sigma_{BV_X}}{L} \right)$$

$$= \frac{r_{BV_X, BV_Y} r_{BV_X, \hat{BV}_X} i_X \sigma_{BV_Y}}{L}$$

Example

Suppose you are selecting sheep on EPD for grease fleece weight (GFW) and are interested in correlated response in staple length (SL). Assume the following genetic parameters:

$$r_{BV_{GFW}, BV_{SL}} = .35$$

$$r_{PD_{GFW}, \hat{PD}_{GFW}} = r_{BV_{GFW}, \hat{BV}_{GFW}} = .8$$

$$i_{GFW} = 1.9$$
$$\sigma_{BV_{SL}} = .35 \text{ in}$$
$$L = 4 \text{ years}$$

Then

$$\Delta_{BV_{SL|GFW}}/t = \frac{r_{BV_{GFW},BV_{SL}} r_{BV_{GFW},\hat{BV}_{GFW}} i_{GFW} \sigma_{BV_{SL}}}{L}$$
$$= \frac{.35(.8)(1.9)(.35)}{4}$$
$$= 0.47 \text{ in / year}$$

Two- and Four-Path Methods

The general formula works nicely using overall values for accuracy, intensity, and generation interval. If you want to be more precise and use a two- or four-path method for determining rate of genetic change, it is best to use the formula:

$$\Delta_{BV_{Y|X}}/t = r_{BV_X,BV_Y}\left(\frac{\sigma_{BV_Y}}{\sigma_{BV_X}}\right)\Delta_{BV_X}/t$$

Solve for direct response $\left(\Delta_{BV_X}/t\right)$ using whatever method you like, then plug the result into the formula.

Correlated Response to Phenotypic Selection

The factors affecting correlated response to selection in the special case of phenotypic selection are really no different from those in the general case, but they can be expressed in slightly different terms. If we select for trait *X* based strictly on individual performance in trait *X*, we can expect correlated response in trait *Y* to be influenced by factors affecting direct response in trait *X*: the heritability of trait *X*, selection intensity, and generation interval. (Phenotypic variation in trait *X* also affects direct response to selection for trait *X*, but does not directly influence response in trait *Y*.) Other factors affecting correlated response in trait *Y* include the genetic correlation between traits *X* and *Y*, and the heritability of and phenotypic variation in trait *Y*.

The Formula for Correlated Response to Phenotypic Selection

The formula for correlated response in trait Y to phenotypic selection for trait X is:

$$\Delta_{BV_{Y|X}}/t = \frac{r_{BV_X,BV_Y} h_X h_Y i_X \sigma_{P_Y}}{L}$$

where $\Delta_{BV_{Y|X}}/t$ = the rate of genetic change in trait Y per unit of time (t) due to phenotypic selection for trait X

r_{BV_X,BV_Y} = the genetic correlation between traits X and Y

h_X = the square root of heritability for trait X

h_Y = the square root of heritability for trait Y

i_X = selection intensity for trait X

σ_{P_Y} = phenotypic variation for trait Y

L = generation interval

Proof

As with the general formula for correlated response, you can derive this formula using a simple prediction equation approach.

$$\Delta_Y = b_{Y \cdot X} \Delta_X$$

where Δ_Y = the rate of genetic change in trait Y per unit of time (t) due to phenotypic selection for trait X or $\Delta_{BV_{Y|X}}/t$

$b_{Y \cdot X}$ = the change in trait Y per *unit* change in trait X

Δ_X = direct response to selection for trait X

The regression coefficient ($b_{Y \cdot X}$) is no different than before:

$$b_{Y \cdot X} = r_{BV_X,BV_Y}\left(\frac{\sigma_{BV_Y}}{\sigma_{BV_X}}\right)$$

But in the special case of phenotypic selection, Δ_X is

$$\Delta_{BV_X}/t = \frac{h_X^2 i_X \sigma_{P_X}}{L}$$

Altogether,

$$\Delta_{BV_{Y|X}}/t = r_{BV_X,BV_Y}\left(\frac{\sigma_{BV_Y}}{\sigma_{BV_X}}\right)\left(\frac{h_X^2 i_X \sigma_{P_X}}{L}\right)$$

$$= r_{BV_X,BV_Y}\left(\frac{\sigma_{BV_Y}}{\sigma_{BV_X}}\right)\left(\frac{\left(\frac{\sigma^2_{BV_X}}{\sigma^2_{P_X}}\right)i_X\sigma_{P_X}}{L}\right)$$

$$= \frac{r_{BV_X,BV_Y}\left(\frac{\sigma_{BV_X}}{\sigma_{P_X}}\right)i_X\sigma_{BV_Y}}{L}$$

$$= \frac{r_{BV_X,BV_Y}h_X i_X \sigma_{BV_Y}\left(\frac{\sigma_{P_Y}}{\sigma_{P_Y}}\right)}{L}$$

$$= \frac{r_{BV_X,BV_Y}h_X h_Y i_X \sigma_{P_Y}}{L}$$

Example

Let's revisit the grease fleece weight/staple length example—this time with phenotypic selection. Assume the following genetic parameters:

$$r_{BV_{GFW},BV_{SL}} = .35$$

$$h^2_{GFW} = .4$$

$$h^2_{SL} = .5$$

$$i_{GFW} = 1.6$$

$$\sigma_{P_{SL}} = .5 \text{ in}$$

$$L = 4 \text{ years}$$

Then

$$\Delta_{BV_{SL|GFW}}\Big/ t = \frac{r_{BV_{GFW},BV_{SL}}h_{GFW}h_{SL}i_{GFW}\sigma_{P_{SL}}}{L}$$

$$= \frac{.35\sqrt{.4}\sqrt{.5}(1.6)(.5)}{4}$$

$$= .031 \text{ in / year}$$

The Two-Path Method

If you want to use a two-path method for determining rate of genetic change due to phenotypic selection, it is best to use the formula:

$$\Delta_{BV_{Y|X}}\Big/ t = r_{BV_X,BV_Y}\left(\frac{\sigma_{BV_Y}}{\sigma_{BV_X}}\right)\Delta_{BV_X}\Big/ t$$

where direct response $\left(\Delta_{BV_X}\middle/_t\right)$ is

$$\Delta_{BV_X}\Big/ t = \frac{h_X^2(i_{m_X} + i_{f_X})\sigma_{P_X}}{L_m + L_f}$$

SELECTING FOR CORRELATED TRAITS

indirect selection Selection for one trait as a means of improving a genetically correlated trait.

direct selection Selection for a trait as a means of improving that same trait.

There are times when it may be better to select for a correlated trait than to select directly for a trait of interest—when it may be better to use **indirect selection** than **direct selection.** Some traits are too expensive or difficult to measure directly. Feed conversion is a good example. Because feed conversion is a ratio involving feed intake and weight gain, performance records for feed conversion require individual measures of feed intake. That means either keeping animals in individual pens and manually weighing feed or using more sophisticated and expensive technology. The much easier, less expensive alternative is to forget about intake, select for weight gain, and rely on the favorable genetic correlation between gain and feed conversion (in swine, for example, approximately −.7) to improve feed conversion.

Another reason to select for a correlated trait is that accuracy of selection may be greater for the correlated trait than for the trait of interest. In swine, performance records for weight gain are much more numerous than records for feed conversion, so accuracies of genetic predictions are generally higher for weight gain than for efficiency.

Selection intensity can be greater for a correlated trait if the correlated trait is continuous in its expression and the trait of interest is a threshold trait. The classic example in beef cattle is selection for lighter birth weight as a way of reducing calving problems. Calving ease is a threshold trait. Performance records for calving ease typically fall into three categories: no assistance, minor assistance and major assistance. Because the incidence of calving problems is relatively low, especially for calves out of older cows, selection intensity for ease of calving in replacement cattle is necessarily low as well.[2] Birth weight, on the other hand, is a continuous trait that is moderately genetically correlated with calving ease. Breeders can select for lower birth weight without the loss of intensity associated with direct selection for calving ease.

indicator trait A trait that may or may not be important in itself, but is selected for as a way of improving some other genetically correlated trait.

Traits like birth weight in beef cattle are often referred to as **indicator traits,** traits that may or may not be important in themselves but are selected for as a way of improving some other genetically correlated trait such as calving ease. Sometimes an indicator trait serves as a proxy for another, more economically important trait, and selection is restricted to selection for the indicator trait. At other times, a better strategy is to select for both the indicator trait and the trait it indicates.

[2]See *Selection Intensity with Threshold Traits* in Chapter 10 for more explanation.

Ratio of Response

One way to determine whether selecting for an indicator trait is more effective than selecting directly for a trait of interest is to estimate response from both types of selection and express the result as a ratio. You can shortcut the process using the formula:

$$\frac{\Delta_{BV_{Y|X}}}{\Delta_{BV_Y}} = \frac{r_{BV_X,BV_Y} r_{BV_X,\hat{BV}_X} i_X}{r_{BV_Y,\hat{BV}_Y} i_Y}$$

The equivalent formula for the special case of phenotypic selection is

$$\frac{\Delta_{BV_{Y|X}}}{\Delta_{BV_Y}} = \frac{r_{BV_X,BV_Y} h_X i_X}{h_Y i_Y}$$

Proof of the General Formula

Correlated response in trait Y to selection for trait X is

$$\Delta_{BV_{Y|X}} \Big/ t = \frac{r_{BV_X,BV_Y} r_{BV_X,\hat{BV}_X} i_X \sigma_{BV_Y}}{L}$$

and direct response to selection for trait Y is

$$\Delta_{BV_Y} \Big/ t = \frac{r_{BV_Y,\hat{BV}_Y} i_Y \sigma_{BV_Y}}{L}$$

The ratio of response is then

$$\frac{\Delta_{BV_{Y|X}}/t}{\Delta_{BV_Y}/t} = \frac{\dfrac{r_{BV_X,BV_Y} r_{BV_X,\hat{BV}_X} i_X \sigma_{BV_Y}}{L}}{\dfrac{r_{BV_Y,\hat{BV}_Y} i_Y \sigma_{BV_Y}}{L}}$$

and, after canceling common terms,

$$\frac{\Delta_{BV_{Y|X}}}{\Delta_{BV_Y}} = \frac{r_{BV_X,BV_Y} r_{BV_X,\hat{BV}_X} i_X}{r_{BV_Y,\hat{BV}_Y} i_Y}$$

Example

Let's compare indirect selection for staple length (SL) via grease fleece weight (GFW) with direct selection for staple length. Assuming a slightly higher accuracy of selection for staple length than for grease fleece weight—staple length is more heritable—and equal selection intensities, then

$$\begin{aligned}\frac{\Delta_{BV_{SL|GFW}}}{\Delta_{BV_{SL}}} &= \frac{r_{BV_{GFW},BV_{SL}} r_{BV_{GFW},\hat{BV}_{GFW}} i_{GFW}}{r_{BV_{SL},\hat{BV}_{SL}} i_{SL}} \\ &= \frac{.35(.8)(1.9)}{.85(1.9)} \\ &= .33\end{aligned}$$

All else being equal, it makes little sense to select for fleece weight in order to improve staple length. The rate of genetic change in staple length is only about a third what it would be with direct selection for staple length.

In the example, intensity of selection for trait Y (staple length) was no different from intensity of selection for trait X (grease fleece weight), so intensities canceled. Ratio of response was then just a function of the genetic correlation between the traits and relative accuracies of selection. Had Y been a threshold trait for which little selection intensity is possible, i_X and i_Y would have been very different, and the difference in intensities might have had an important effect on the ratio of response.

THE GOOD NEWS/BAD NEWS ABOUT GENETIC CORRELATIONS AND CORRELATED RESPONSE

Genetic correlations between traits and the correlated response to selection brought about by them can be beneficial. There are times when we can use indirect selection to advantage, and, as explained in Chapter 11, multiple-trait statistical models that use genetic correlations reduce bias caused by culling for poor performance and improve accuracy of prediction. That is the good news.

The bad news is that if we are unaware of or choose to ignore unfavorable genetic correlations, selection for one trait can lead to undesirable response in others. In cattle, for example, blind selection for growth rate leads to larger birth weights and more dystocia. If we want faster growth, but cannot tolerate increased dystocia, we must avoid simply selecting for growth or against dystocia. We need a way to select for growth rate and against dystocia at the same time. We need a method of multiple-trait selection—the subject of the next chapter.

EXERCISES

Study Questions

13.1 Define in your own words:

correlated response to selection	compensatory gain
pleiotropy	direct response to selection
genetic correlation (r_{BV_X,BV_Y})	indirect selection
phenotypic correlation (r_{P_X,P_Y})	direct selection
environmental correlation (r_{E_X,E_Y})	indicator trait

13.2 Describe the causes of genetic correlations and correlated response to selection.

13.3 Pick a species and pair of traits of interest to you. Choose what you think should be reasonable values for genetic, environmental, and phenotypic correlations between the traits and explain your reasoning in each case. (Use the line of thinking presented in the section of this chapter entitled *Keeping Genetic, Phenotypic, and Environmental Correlations Straight.*)

13.4 What is wrong with the following statement: "My dog's breeding value for conformation is highly correlated with his breeding value for temperament"? How should the statement be reworded to reflect its intended meaning?

13.5 Use examples to show how correlations between the same two traits may be different for:

a. genetically different populations.

b. genetically similar populations in different environments.

13.6 **a.** List four ways to classify correlations between traits.

b. Give examples of correlations that differ in each classification.

13.7 Are positive correlations between traits always favorable? Explain.

13.8 List the factors affecting correlated response to selection (in general).

13.9 List the factors affecting correlated response to phenotypic selection.

13.10 **a.** For what reasons might a breeder choose to practice indirect selection for a trait rather than direct selection?

b. Describe some situations in which indirect selection might be preferable to direct selection.

13.11 **a.** What are potential benefits of genetic correlations? Give examples.

b. What are potential drawbacks of genetic correlations? Give examples.

Problems

13.1 Recall from Problem 9.1 the following genetic parameters for time to swim 50 meters (T) and lifetime winnings (W) in Siberian racing muskrats:

$$\sigma^2_{BV_T} = 4 \text{ sec}^2 \quad \sigma^2_{GCV_T} = 1 \text{ sec}^2 \quad \sigma^2_{E_T} = 11 \text{ sec}^2$$

$$\sigma^2_{BV_W} = 100 \text{ rubles}^2 \quad \sigma^2_{GCV_W} = 0 \text{ rubles}^2 \quad \sigma^2_{E_W} = 2{,}400 \text{ rubles}^2$$

Animal scientists at the Smyatogorsk Polytechnic Institute have also estimated the following covariances between the two traits:

$$\text{cov}(BV_T, BV_W) = -2.0 \text{ sec}\cdot\text{rubles}$$

$$\text{cov}(E_T E_W) = -118.8 \text{ sec}\cdot\text{rubles}$$

$$\text{cov}(P_T P_W) = -120.8 \text{ sec}\cdot\text{rubles}$$

Calculate:

a. the genetic correlation between 50-meter time and lifetime winnings.

b. the environmental* correlation between the traits. (*For this problem, do *not* include gene combination effects in the environmental category.)

c. the phenotypic correlation between the traits.

13.2 Ace Maverick wants to shorten gestation length (GL), and reduce birth weight (BW) in his herd of registered beef cattle. EPDs for these traits are available, and Ace will use them for both male and female selection. Because many fewer gestation lengths are reported than birth weights, average accuracy of selected animals for gestation length is only .40 compared to .80 for birth weight. Given the following:

$$\sigma_{BV_{GL}} = 2.8 \text{ days} \quad \sigma_{BV_{BW}} = 6.3 \text{ lb} \quad r_{BV_{GL},BV_{BW}} = .25$$

$$i_{GL} = i_{BW} = -1.0 \quad L = 5 \text{ years}$$

a. Calculate:

i. $\Delta_{BV_{GL}}/t$

ii. $\Delta_{BV_{BW}}/t$

iii. $\Delta_{BV_{GL|BW}}/t$

iv. $\Delta_{BV_{BW|GL}}/t$

v. $\dfrac{\Delta_{BV_{GL|BW}}}{\Delta_{BV_{GL}}}$

vi. $\dfrac{\Delta_{BV_{BW|GL}}}{\Delta_{BV_{BW}}}$

b. Interpret your results for (v) and (vi).

c. If you were Ace, which trait would you select for? Why?

13.3 Slim Maverick, Ace's brother, runs commercial cattle just south of Ace's place. Slim wants to improve the probability of conception (PC) in his yearling heifers. But conception is an all-or-none threshold trait for which Slim can achieve little selection intensity. The bulls that Slim buys have EPDs for scrotal circumference (SC), a trait known to be related to heifer conception rate. Slim is contemplating two strategies: (1) practicing phenotypic selection for probability of conception by retaining only those heifers that conceive in their first season, and (2) selecting bulls for scrotal EPD. The following genetic parameters have been estimated for these traits:

$$h^2_{PC} = .10 \quad h^2_{SC} = .5 \quad \sigma_{P_{PC}} = .46 \quad \sigma_{P_{SC}} = 2.0 \text{ cm} \quad r_{BV_{SC},BV_{PC}} = .25$$

Assume:

Slim breeds all his heifers and keeps 40% for replacements.

Typical heifer conception rate is 70%.

The bulls that Slim buys represent the equivalent of the top 5% of his herd for scrotal circumference.

$L_m = L_f = 5$ years

a. Calculate annual selection response in probability of heifer conception using strategy 1.

b. Calculate annual selection response in probability of heifer conception using strategy 2.

c. Which strategy works better? Why?

13.4 Pyotr—remember him, Oski Doski's breeder—has selected for decreased 50-meter time in his muskrats for many generations. By selection alone he has improved the average time of his muskrats by 10 seconds. Use information from Problem 13.1 to answer the following:

a. How much more money should Pyotr's muskrats be winning now than before as a result of genetic improvement?

b. Pyotr's training regimen has resulted in a 10-second improvement in 50-meter time also. How much more money should Pyotr's muskrats be winning now than before as a result of better training?

c. Would Pyotr be better advised to concentrate on his breeding program or his training program? Why?

CHAPTER 14

Multiple-Trait Selection

single-trait selection
Selection for one trait.

multiple-trait selection
Selection for more than one trait.

aggregate breeding value or **net merit**
The breeding value of an individual for a combination of traits.

In this book, the discussion of selection and the examples used to illustrate selection have so far been limited to **single-trait selection,** selection for just one trait. That is because single-trait selection provides a simple framework within which to learn the principles of selection. But in the real world of animal breeding, selection for a single trait is rare. Breeders are typically more interested in improving a number of traits. They practice **multiple-trait selection.** Dairy breeders select for traits related to both milk production and type. Swine breeders select for fecundity (litter size), growth rate, and carcass merit. Breeders of racehorses look for both speed and endurance.

The objective of multiple-trait selection is to improve **aggregate breeding value—**breeding value for a combination of traits—in a population. Another term for aggregate breeding value is **net merit.** To define aggregate breeding value for a particular situation is not only to determine what traits are worthy of selection, but also to assign some relative value to each trait. For example, a breeder of racehorses might identify raw speed and the ability of a horse to be rated (held back) as traits of primary importance. In defining aggregate breeding value, the breeder must also have some concept of how important raw speed is relative to ability to be rated. To define aggregate breeding value, then, is to answer the question posed in the first chapter of this book: What is the "best" animal? Thus, the practice of multiple-trait selection involves more than just genetic theory. It integrates principles of selection with notions of the value of traits (i.e., economics).

METHODS OF MULTIPLE-TRAIT SELECTION

In all but the most industrial types of breeding operations, multiple-trait selection is as much an art as a science. There are no hard-and-fast rules. Typically, breeders conceive of aggregate breeding value in an intuitive way, but do not define it mathematically. They have an idea of what constitutes the best animal and select individuals they think come the closest to that ideal. Despite the fact that multiple-trait selection is, in practice, a highly individualized undertaking, animal breeding theorists like to place methods of multiple-trait selection in three distinct categories: tandem selection, selection using independent culling levels, and selection using economic selection indexes. Understanding the distinctions between these methods and the advantages and disadvantages of each is not only interesting from a theoretical standpoint—it has real practical value.

Tandem Selection

tandem selection
Selection first for one trait, then another.

Tandem selection is simply selection for one trait, then another. (Think of the word "tandem" as it is used in tandem bicycle or tandem horse trailer. In each case, one entity, a bicyclist or horse, precedes another.) For an example—one that will be used repeatedly in this chapter—consider yearling weight and birth weight in beef cattle. Tandem selection for these traits could mean selection for increased yearling weight for a number of years or until a certain level of yearling weight performance is achieved, followed by selection for smaller birth weight.

selection target
A level of breeding value considered optimal in an absolute or practical sense.

Tandem selection is easy. In its pure form, it is just single-trait selection. It often incorporates the idea of a **selection target,** a level of breeding value considered optimal in an absolute or practical sense. Breeders might select for yearling weight until a selection target for that trait is reached, and then switch to selection for birth weight. The difficult question is, What should the target be? How heavy a yearling weight is heavy enough?

The effectiveness of tandem selection depends to a large degree on the genetic correlations between the traits under selection. If two traits are favorably genetically correlated, selection for the first trait will also improve the second trait. Selection for growth rate, for example, should improve feed conversion. If the traits are unfavorably correlated, however, selection for the first trait will cause backsliding in the second trait. Because birth and yearling weights are positively (but unfavorably) correlated, selection for heavier yearling weights will result in larger calves at birth.

Genetic correlations make it impossible to maintain a trait at its optimal level using tandem selection. If the next trait of interest is genetically correlated with the last trait selected, selection for the new trait causes movement in breeding value for the last trait away from the selection target. When a breeder switches from selection for increased yearling weight to selection for lower birth weight, yearling weight suffers.

Despite its deficiencies, tandem selection in one form or another is historically common. There are several reasons why. Sometimes success in selection for one trait causes that trait to assume less importance. For example, in cattle herds having a high level of growth performance already, continued selection for yearling weight may not be very profitable. The improvement in calving ease resulting from selection for decreased birth weight may be more important. Breeders also tend to follow fads and respond to legitimate changes in environments and markets. In so doing, they shift selection emphasis from one trait to another (or from one *set* of traits to another).

Independent Culling Levels

independent culling levels
Minimum standards for traits undergoing multiple-trait selection. Animals failing to meet any one standard are rejected regardless of merit in other traits.

A second method of multiple-trait selection is the use of **independent culling levels.**[1] Independent culling levels are minimum standards for traits undergoing multiple-trait selection. If a breeder selects animals using independent culling lev-

[1]"Independent *culling* levels" is something of a misnomer because culling, strictly speaking, applies only to animals that are parents already. Sometimes we use independent culling levels to cull parents, but more often we use them to reject animals that have yet to become parents. A better term might be "independent rejection levels" or "independent selection standards." "Independent culling levels" is so deeply entrenched in animal breeding vernacular, however, that I make no attempt to reform the language here.

els, those animals that fail to meet any one standard are rejected regardless of merit in other traits. For example, a beef cattle breeder using phenotypic selection might reject all heifer calves with birth weights greater than 95 lb or with yearling weights less than 10 lb below the average of their contemporaries. If EPDs are available, the breeder could use the same strategy with better predictors. In that case, the culling level for birth weight might be an EPD for birth weight of +4 lb and the culling level for yearling weight might be an EPD for yearling weight of +33 lb.

Selection using independent culling levels is illustrated in Figure 14.1. The yearling weight EPDs of a set of beef heifers are plotted against their birth weight EPDs. Culling levels of +33 lb for yearling weight EPD and +4 lb for birth weight EPD are represented by the thin lines in the figure. Those heifers whose yearling weight EPDs exceed the culling level (lower limit) for yearling weight and whose birth weight EPDs are less than the culling level (upper limit) for birth weight are selected as replacements. They appear as black dots on the graph. All other heifers (open circles) are rejected.

Independent culling levels have great intuitive appeal and, as a result, are very popular. They allow you to select simultaneously for more than one trait by applying rather simple rules. They are particularly appropriate when there is a clear distinction between what is acceptable and what is not. For example, unsound animals are clearly unacceptable and should be rejected regardless of performance in other traits. A number of simply-inherited traits and threshold traits have acceptable/unacceptable categories and are good candidates for selection using independent culling levels.

Independent culling levels are also convenient when selection occurs at different stages of an animal's life. In beef cattle, for example, many breeders set

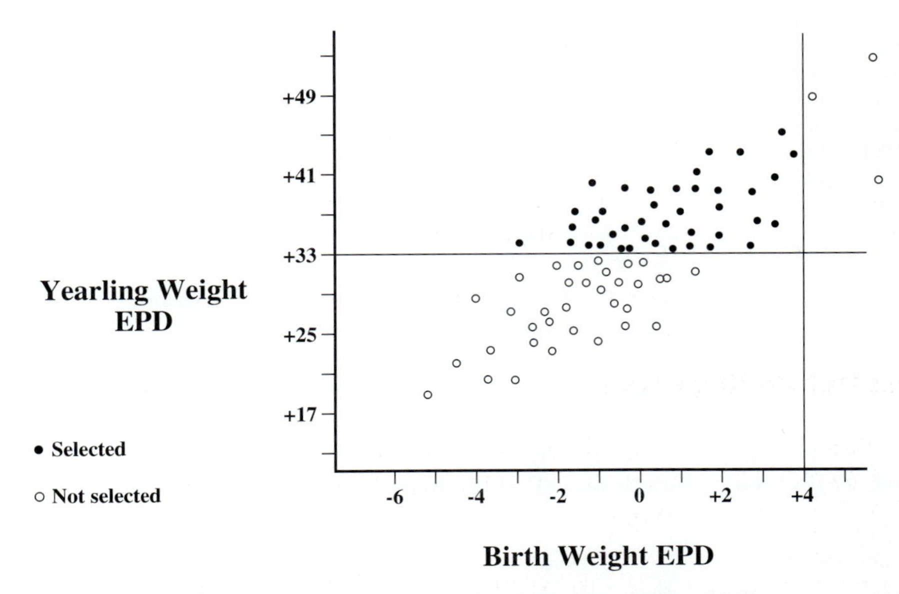

FIGURE 14.1 Illustration of independent culling levels in a set of beef heifers. Yearling weight EPDs are plotted against birth weight EPDs, and culling levels are represented by the thin lines. Selected heifers (black dots) appear in the upper left portion of the plot. Their yearling weight EPDs exceed the culling level (lower limit) for yearling weight, and their birth weight EPDs are less than the culling level (upper limit) for birth weight.

independent culling levels for birth weight, weaning weight, and measures taken on yearlings. Bulls whose birth weights are unacceptably heavy are often castrated as young calves. Calves whose weaning weights are too light are rejected at weaning. A third round of selection occurs at a year of age. Sequential selection of this sort reduces cost by eliminating rejects early on.

The difficulty with independent culling levels comes in determining what the culling levels should be. We are back to the question, What is the best animal? More precisely, we are faced with the question, What is a *good enough* animal from a practical standpoint? If selection standards are too restrictive, we may not be able to find a sufficient number of animals that can meet them.

Most breeders who use independent culling levels take an intuitive/experiential rather than mathematically precise approach when setting those levels. There exist rigorous mathematical and computer-intensive methods for determining culling levels, but they have yet to be applied on a meaningful scale.

If independent culling levels are strictly applied, they may exclude some potentially useful animals. For example, the heifer that appears just to the right of the culling level for birth weight EPD in Figure 14.1 has an extremely high yearling weight EPD. She is disqualified, however, because her EPD for birth weight is too high—though only slightly.

Economic Selection Indexes

economic selection index
An index or combination of weighting factors and genetic information—either phenotypic data or genetic predictions—on more than one trait. Economic selection indexes are used in multiple-trait selection to predict aggregate breeding value.

A third method of multiple-trait selection is the use of an **economic selection index.** Selection index methodology was introduced in Chapter 11 as a means of calculating genetic predictions for a single trait. The same methodology can be used to predict aggregate breeding value in the context of multiple-trait selection.

The index itself looks much as before:

$$I = b_1x_1 + b_2x_2 + \cdots + b_nx_n$$

where I = an *index value* or genetic prediction
b_1 = a weighting factor
x_1 = a single item of information or "evidence"
n = the total number of items of information

However, there are two essential differences between the economic selection index shown here and the selection index described in Chapter 11. First, the index value (I) is no longer a genetic prediction for a single trait. Rather, it is a prediction of aggregate breeding value. It is a single number that predicts the breeding value of an individual for a weighted combination of traits.

breeding objective
A weighted combination of traits defining aggregate breeding value for use in an economic selection index.

The weighted combination of traits defining aggregate breeding value is sometimes called the **breeding objective.**[2] It is represented by another equation that typically appears as:

$$H = v_1BV_1 + v_2BV_2 + \cdots + v_mBV_m$$

[2]Some academic animal breeders define breeding objectives less mathematically. They think of them more as general goals for selection.

where H = aggregate breeding value
v_i = an **economic weight** for a trait in the breeding objective
BV_i = breeding value for a trait in the breeding objective
m = the total number of traits in the breeding objective

economic weight
The change in aggregate breeding value (the change in profit if that is how aggregate breeding value is measured) due to an independent, one-unit increase in performance in a trait.

Aggregate breeding value (H) is measured in dollars or some other monetary unit. You can think of it as a breeding value for a new trait: overall economic merit. This new trait can be measured by profit for an enterprise, profit per animal, or an alternative measure of economic efficiency. The economic weight for each trait in the breeding objective represents the change in aggregate breeding value (the change in *profit* if that is how aggregate breeding value is measured) due to an *independent*, one-unit increase in performance in that trait. "Independent" in this context means independent of changes in breeding values for other traits in the breeding objective.

The traits that appear in the breeding objective should be those that are economically important. The traits that appear in the economic selection index, on the other hand, should be those for which we can collect performance records easily and cheaply and that contribute to or are related to traits in the breeding objective. So the traits that appear in the breeding objective may or may not be the same as those in the index. Economically important but otherwise troublesome traits in the breeding objective are often replaced with indicator traits. For example, a breeding objective for beef cattle might include the following traits: yearling weight, age at puberty, and calf death loss at birth. The corresponding selection index might include yearling weight, scrotal circumference, and birth weight. Scrotal circumference serves as a proxy for age at puberty, and birth weight serves as a proxy for calving losses.

The second difference between the economic selection index and the selection index described in Chapter 11 is in the nature of the xs—the "evidence." In the index that predicts a value for a single trait (the version discussed in Chapter 11), each x represents an individual item of phenotypic information—a performance record or the average of a group of performance records. The xs of an economic selection index may be individual items of phenotypic information as well, but they can also be genetic predictions—EPDs or EBVs.

phenotypic selection index
A form of economic selection index used with phenotypic selection. In the classic form of phenotypic index, the traits in the index are identical to the traits in the breeding objective.

There are several kinds of economic selection indexes. **Phenotypic selection indexes** are used in situations where selection is strictly phenotypic selection (selection on own performance alone—no relatives' information). The xs in a phenotypic index are individual performance records. In the classic form of phenotypic index, the traits in the index are identical to the traits in the breeding objective.

For example, a breeding objective involving yearling weight (YW) and birth weight (BW) in beef cattle might be

$$H = 1.1YW - 6.52BW$$

suggesting that each independent 1-lb increase in yearling weight increases profit by \$1.10, and each independent 1-lb increase in birth weight *decreases* profit by \$6.52. After accounting for genetic and phenotypic relationships between these traits, the resulting selection index is[3]

$$I = YW - 5.8BW$$

[3]The derivation of this equation is explained in the next boxed section.

TABLE 14.1 Yearling Weights, Birth Weights, Index Values, and Index Rankings for a Small Contemporary Group of Bull Calves

Calf ID#	Yearling Weight (*YW*)	Birth Weight (*BW*)	Index Value ($I = YW - 5.8\ BW$)	Rank
7	1,125	80	661	1
4	1,066	77	619	2
10	1,050	79	592	3
5	1,202	108	576	4
3	1,058	84	571	5
8	1,100	94	555	6
1	980	75	545	7
2	922	68	528	8
6	976	87	471	9
9	1,034	102	442	10

Such an index places positive emphasis on yearling weight because of the efficiencies inherent in rapid growth, and negative emphasis on birth weight because of the association of heavy birth weights with calving losses and delayed rebreeding. Listed in Table 14.1 are yearling weights, birth weights, index values, and rankings (according to the index) for a small contemporary group of bull calves. The data have been ordered by ranking in order to better show how each trait affects the index value. Note how the index favors rapid growth yet discriminates against heavy birth weight.

Calculating the Classic Form of Economic Selection Index

In its classic form, the economic selection index is a simple phenotypic index with the same traits in both the index and the breeding objective. Thus,

$$H = v_1 BV_1 + v_2 BV_2 + \cdots + v_n BV_n$$

and

$$I = b_1 P_1 + b_2 P_2 + \cdots + b_n P_n$$

Assuming the economic weights (v_is) are known, we can solve for the index weights (b_is) using a system of n simultaneous equations.

$$\begin{aligned}
\sigma^2_{P_1} b_1 + \mathrm{cov}(P_1,P_2)b_2 + \cdots + \mathrm{cov}(P_1,P_n)b_n &= \sigma^2_{BV_1} v_1 + \mathrm{cov}(BV_1,BV_2)v_2 + \cdots + \mathrm{cov}(BV_1,BV_n)v_n \\
\mathrm{cov}(P_2,P_1)b_1 + \sigma^2_{P_2} b_2 + \cdots + \mathrm{cov}(P_2,P_n)b_n &= \mathrm{cov}(BV_2,BV_1)v_1 + \sigma^2_{BV_2} v_2 + \cdots + \mathrm{cov}(BV_2,BV_n)v_n \\
&\vdots \\
\mathrm{cov}(P_n,P_1)b_1 + \mathrm{cov}(P_n,P_2)b_2 + \cdots + \sigma^2_{P_n} b_n &= \mathrm{cov}(BV_n,BV_1)\, v_1 + \mathrm{cov}(BV_n,BV_2)v_2 + \cdots + \sigma^2_{BV_n} v_n
\end{aligned}$$

Example

Let's calculate the yearling weight/birth weight index for beef cattle where

$$H = 1.1YW - 6.52BW$$

That is, $v_1 = \$1.10/\text{lb}$ and $v_2 = \$-6.52/\text{lb}$. We will need the following parameters.

$$\sigma^2_{P_{YW}} = 3{,}600 \text{ lb}^2 \qquad \sigma^2_{P_{BW}} = 100 \text{ lb}^2 \qquad \text{cov}(P_{YW}, P_{BW}) = 210 \text{ lb}^2$$

$$\sigma^2_{BV_{YW}} = 1{,}440 \text{ lb}^2 \qquad \sigma^2_{BV_{BW}} = 40 \text{ lb}^2 \qquad \text{cov}(BV_{YW}, BV_{BW}) = 168 \text{ lb}^2$$

The equations to solve for the index are then:

$$\sigma^2_{P_{YW}} b_1 + \text{cov}(P_{YW}, P_{BW}) b_2 = \sigma^2_{BV_{YW}} v_1 + \text{cov}(BV_{YW}, BV_{BW}) v_2$$

$$\text{cov}(P_{BW}, P_{YW}) b_1 + \sigma^2_{P_{BW}} b_2 = \text{cov}(BV_{BW}, BV_{YW}) v_1 + \sigma^2_{BV_{BW}} v_2$$

or

$$3{,}600b_1 + 210b_2 = 1{,}440(1.10) + 168(-6.52)$$

$$210b_1 + 100b_2 = 168(1.10) + 40(-6.52)$$

In this very small example, we can solve for the bs by substitution. For larger problems, we would typically use matrix algebra. Either way,

$$b_1 = .2052$$

$$b_2 = -1.191$$

and

$$I = .2052YW - 1.191BW$$

We can simplify the index and still retain the relative emphasis placed on each trait by dividing both weights by .2052. Then

$$I = \frac{.2052}{.2052}YW - \frac{1.191}{.2052}BW$$

$$= YW - 5.8BW$$

The performance records used to calculate the phenotypic index in Table 14.1 were adjusted birth and yearling weights from a single contemporary group. If records come from more than one contemporary group, then deviations from contemporary group means should be used instead. Otherwise, animals from contemporary groups that, for environmental reasons, have light birth weights and(or) heavy yearling weights would have an unfair advantage.

A more general form of economic selection index uses different kinds of phenotypic data (individual performance records and(or) averages of groups of performance records) and is not required to have the same traits in both the breeding objective and the index. This type of index is very similar to the index used to predict genetic values for a single trait as described in Chapter 11. The only real differences are that the economic index is likely to incorporate performance data on more traits and, by including economic weights, predicts overall economic merit rather than breeding value or progeny difference for one trait.

Phenotypic indexes and the more general type of economic index described in the last paragraph are very useful, but they have the same drawback as the selection index of Chapter 11; they are unable to account for genetic differences among contemporary groups. They should therefore be used only when performance data come from contemporary groups thought to be genetically similar. In practice this means that economic indexes of these kinds should probably be used within individual herds or flocks and not across populations. And data used in these types of indexes should be collected over a relatively short period of time so that genetic trend does not cause differences between older contemporary groups and more recent ones.

The most promising kind of economic selection index and, in this age of large-scale genetic evaluation, the kind most likely to achieve broad acceptance, is an index that combines economic weights with genetic predictions calculated with BLUP and BLUP-like procedures. Such an index has all the advantages of BLUP prediction as detailed in Chapter 11. In particular, it can account for genetic differences among contemporary groups. In addition, if genetic predictions are available for all traits in the breeding objective, the equation for the index is the same as the equation describing the breeding objective. No mathematical conversion is necessary.[4] This is because the evidence used for the index is comprised of genetic predictions rather than phenotypic data. These predictions can simply be substituted into the breeding objective to produce a prediction of aggregate breeding value (overall economic merit). Mathematically,

$$H = v_1 BV_1 + v_2 BV_2 + \cdots + v_m BV_m$$

so

$$\hat{H} = v_1 \hat{BV}_1 + v_2 \hat{BV}_2 + \cdots + v_m \hat{BV}_m$$

and therefore

$$I = v_1 \hat{BV}_1 + v_2 \hat{BV}_2 + \cdots + v_m \hat{BV}_m$$

[4]In most cases of this kind, mathematical manipulation is still required for the index weights to be technically correct. But the effort is usually more trouble than it is worth. When genetic predictions are used as the *x*s in an economic selection index, economic weights are typically good approximations of true index weights.

For example, with our birth weight/yearling weight index.

$$H = 1.1YW - 6.52BW$$

so

$$I = 1.1\text{EBV}_{YW} - 6.52\ \text{EBV}_{BW}$$

We can just as easily use EPDs as EBVs in the index. Thus

$$I = 1.1\text{EPD}_{YW} - 6.52\ \text{EPD}_{BW}$$

We can further simplify the index by dividing both economic weights by 1.1. (Division by a constant does not change the *relative* emphasis placed on each trait.) Thus

$$I = \text{EPD}_{YW} - 5.93\ \text{EPD}_{BW}$$

Economic selection indexes in general have very desirable properties. In theory, an economic selection index provides the fastest, most efficient way to improve aggregate breeding value. It uses potentially large amounts of information on several traits to produce a single number—the index value—that predicts the overall economic merit of an individual. Once this number is calculated, it is a simple matter to rank animals for selection.

A particular advantage of an economic selection index over independent culling levels is the ability of the index to balance superiority in one trait against inferiority in another. Figure 14.2 illustrates index selection ($I = \text{EPD}_{YW} - 5.93$

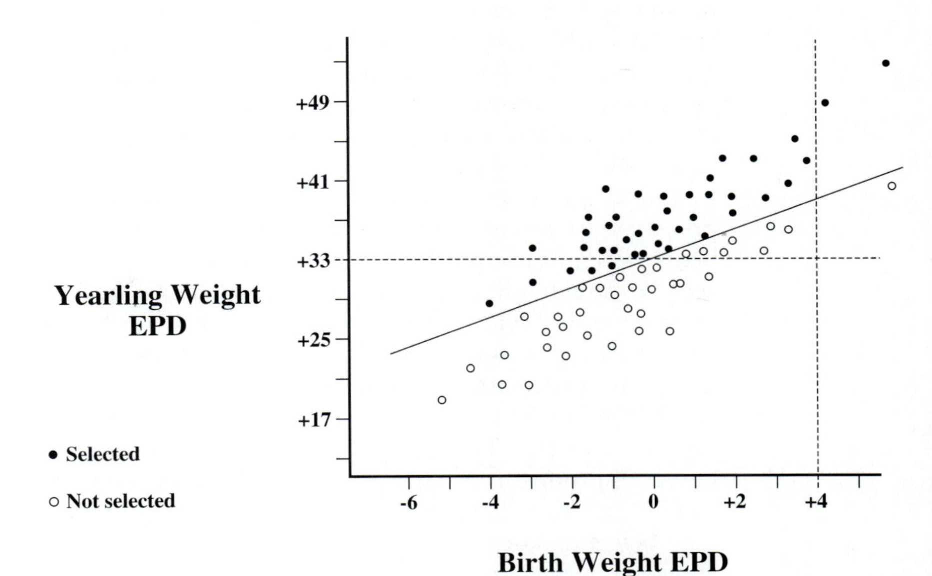

FIGURE 14.2 Illustration of index selection in a set of beef heifers. Yearling weight EPDs are plotted against birth weight EPDs, and the selection index ($I = \text{EPD}_{YW} - 5.93\ \text{EPD}_{BW}$) is represented by the thin diagonal line. Selected heifers (black dots) appear above and to the left of the index line. Culling levels (dotted lines) from Figure 14.1 are shown for comparison of index selection with selection using independent culling levels.

EPD_{BW}) for the same set of beef heifers represented in Figure 14.1. The index appears graphically as the thin diagonal line. Selected heifers are represented by the black dots above and to the left of the index line. Rejects (open circles) are below and to the right of the line. For comparison with selection using independent culling levels, the culling levels from Figure 14.1 appear as dotted lines.

Note that the index, like independent culling levels, favors animals with higher yearling weights and lower birth weights. Unlike independent culling levels, however, the index allows selection of individuals with particularly good EPDs for birth weight or yearling weight even if their performance in the second trait is not up to standard. The heifer that appears just to the right of the culling level for birth weight EPD has a very high yearling weight EPD, but her birth weight EPD is slightly too high to meet the culling level for birth weight. According to the index, however, she is entirely acceptable. Likewise, there are five heifers selected using the index that would have been rejected for low yearling weight EPD using independent culling levels. In Figure 14.2, they appear in the triangular area above the index line and below the culling level for yearling weight EPD. These animals balance less favorable EPDs for yearling weight with very desirable EPDs for birth weight.

Although economic selection indexes are forgiving of inferior ability in one trait if it is offset by superior ability in another, they are tough on individuals that are just marginally acceptable in all traits. In Figure 14.2, the seven heifers that appear within the triangle defined by the index line and the two culling levels have moderate birth and yearling EPDs. With respect to the culling levels, they are acceptable. With respect to the index, they are not.

The chief problem with economic selection indexes is that economic weights are difficult to determine. Economic weights require careful analysis of costs and returns and are likely to be different in different situations. For example, in sheep, the differences in environment, management, and scale of operation that distinguish farm flocks from range flocks are likely to result in different economic weights for some traits.

Economic weights depend on level of production, especially in traits for which there is an optimal level of production that is intermediate—not too low, not too high. Take egg weight in layers, for example. If the average egg weight in a poultry stock is far below optimal, then egg weight should have a substantial economic weight associated with it. If average egg weight is much too heavy, then egg weight should still have a substantial, but this time *negative,* economic weight associated with it. But if average egg weight in a stock is about right, the economic weight for egg weight should be near zero. Stocks that differ in egg weight should therefore use different economic weights for the trait.

Economic weights change over time. Costs and prices change with fluctuations in markets. Management changes with the introduction of new technology. Levels of production change with genetic trend. All these factors affect economic weights, making it necessary to update weights from time to time.

Selecting animals on the basis of a single index value is appealingly simple. However, reducing a many-faceted individual to just one number offends many breeders' sensibilities. That number may indeed be the best predictor of aggregate breeding value, but it does not characterize an animal in any detail. And it is impossible to make corrective matings for specific traits using selection index values alone. These may be some of the reasons why economic selection indexes have found application in large industrial breeding programs for poultry and swine where animals are largely perceived as numbers on paper, but are rarely used in species like horses, dogs, and even cattle, where breeders have a more intimate relationship with their animals.[5]

Another drawback of economic selection indexes is that they provide direction for selection, but give no indication of a selection endpoint for the individual traits in the index. In other words, they do not provide selection targets. Selection targets have intuitive appeal for breeders who want to know not only the direction they are headed but the destination they seek. (Of course, the index itself requires no selection target—the higher the index value, the better.)

Economic selection indexes can occasionally result in selection of individuals that are dangerously extreme in one or more traits. For example, the heifer in the upper right corner of Figure 14.2 is selected largely because of her very high yearling weight EPD. Her birth weight EPD is very high as well, perhaps too high considering that calving difficulty accelerates as birth weight increases. A more sophisticated index, one that places increasing selection pressure against birth weight as birth weight potentials get heavier, could solve this problem.

Combination Methods

There is no rule stating that you must use just one method, either tandem selection, independent culling levels, or an economic index, for multiple-trait selection. You can combine methods as you see fit. For example, by adding an independent culling level for birth weight EPD to a selection index containing yearling weight and birth weight EPDs, you could eliminate the possibility of selecting individuals with an unacceptably high risk of calving difficulty and still retain most of the advantages of selection with an economic index.

For traits with intermediate optimum levels of production, you can breed toward defined selection targets by setting both minimum and maximum culling levels. For example, if a milk EPD of +10 lb is considered optimal in a particular beef production scenario, then a minimum culling level for milk EPD of +5 and a maximum culling level of +15 might be appropriate.

Selection targets and economic selection indexes are an attractive combination. Targets provide breeders with concrete goals, and economic indexes offer the most efficient way to achieve those goals.

Regardless of the method used for multiple-trait selection, it is important that selection be conducted with the end user in mind.[6] In traditional livestock species, this means that the selection targets, culling levels, or economic weights used by seedstock producers should not reflect their own production and economic circumstances, but rather the circumstances of their customers—commercial producers. For example, if you are a sheep breeder with a small farm flock and sell rams to commercial range producers, you should use selection targets, culling levels, or economic weights that best fit range sheep production, even though your own production situation may differ greatly from those of most range flocks.

[5]Use of the TPI or total performance index in dairy cattle is an exception to this rule, and economic selection indexes are being used to varying degrees by sheep and beef cattle breeders in Europe, Australia, and New Zealand.

[6]See the section in Chapter 1 entitled *Breeding Objectives and Industry Structure* for a more complete discussion of the structure of breeding industries in general and end users in particular.

Because of the large number of production and economic factors involved and the great variety of production scenarios possible, determination of appropriate selection targets, culling levels, and economic weights is problematic and is an important research topic in animal breeding. Someday we may have technology in the form of computer simulation models and other mathematical algorithms that will allow these things to be calculated easily, taking much of the guesswork out of multiple-trait selection.

SELECTION INTENSITY AND MULTIPLE-TRAIT SELECTION

An unfortunate but important side effect of multiple-trait selection is that selection intensity for any one trait is reduced from what it would have been had selection been exclusively for that trait. The more traits involved, the lower the selection intensity for each trait. This is because we cannot be as "choosy" with respect to a particular trait if we require selected animals to excel in other traits as well. Loss of selection intensity means less response to selection in each trait. The rate of genetic progress in each trait is slowed.

Loss of selection intensity with multiple-trait selection is illustrated in Figure 14.3. In this population of 77 beef heifers, 40 are required as replacements. If selection is purely for yearling weight EPD, 40 heifers qualify if a culling level for yearling weight EPD (the thin horizontal line on the graph) is set at +34 lb. Likewise, if selection is purely for birth weight EPD, 40 heifers qualify if a culling level for birth weight EPD (the thin vertical line on the graph) is set at −.3 lb. Note, however, that only 12 heifers (black dots) qualify if both culling levels are enforced simultaneously. In order to accommodate selection for both birth weight and yearling weight, yet achieve the required number of replacements, culling levels must be relaxed. Relaxing a culling level is the same as reducing selection intensity.

The independent culling levels from Figure 14.1 appear as dotted lines in Figure 14.3. These culling levels allow 40 heifers to be selected, but they are necessarily less restrictive than the culling levels used in single-trait selection. The distance between culling levels (between thin and dotted lines) suggests the loss of selection intensity in each trait.

The culling level for birth weight EPD is affected the most. It changes from −.3 lb with single-trait selection for birth weight EPD to +4 lb with selection for both birth weight and yearling weight EPDs. The culling level for yearling weight is affected to a smaller degree. It decreases slightly from +34 lb to +33 lb. In this particular example, multiple-trait selection causes little loss in selection intensity for yearling weight, but considerable loss in intensity for birth weight.

Another way to envision the loss of selection intensity in a particular trait is to think in terms of actual proportion of replacements saved and *effective* proportion saved with respect to that trait. In fact, you can often determine just how much selection intensity is lost with this approach.

Recall from Chapter 10 that with truncation selection, the kind of selection implied by the use of culling levels, intensity of replacement selection can be determined from the proportion saved (p) using a conversion table (Table 10.1). In the example illustrated in Figure 14.3, the actual proportion of heifers saved as re-

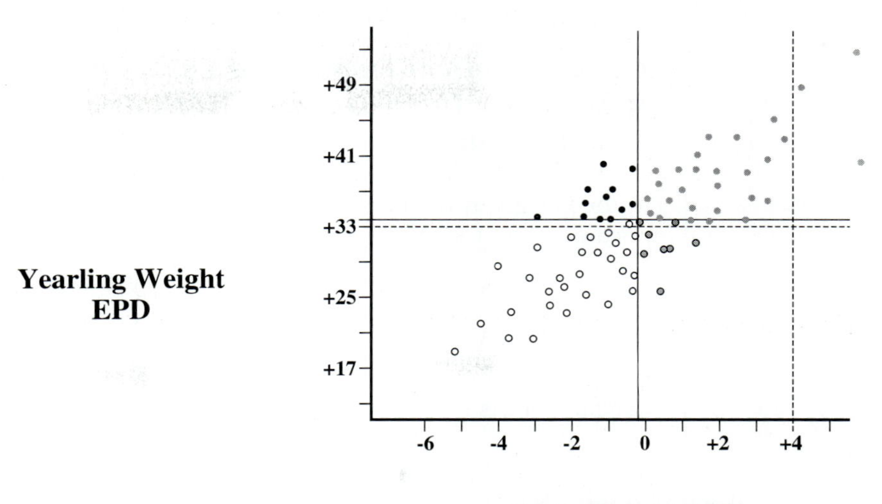

● **Selected with single-trait selection for greater yearling weight EPD**

○ **Selected with single-trait selection for smaller birth weight EPD**

• **Selected with single-trait selection for either greater yearling weight EPD or smaller birth weight EPD**

⊙ **Not selected with single-trait selection for either trait**

FIGURE 14.3 Illustration of loss of selection intensity with multiple-trait selection for birth weight and yearling weight in beef heifers. Thin lines represent culling levels that qualify a sufficient number of replacements if selection is for either birth weight EPD or yearling weight EPD, but not both. Dotted lines represent independent culling levels that qualify the same number of replacements with multiple-trait selection. The distance between culling levels (between thin and dotted lines) suggests the loss of intensity in each trait. See text for a full discussion.

placements is 40 out of 77 or .52. If you interpolate between values listed in Table 10.1, you will find a corresponding selection intensity (i) of .77 standard deviations. That is not very intense selection, but remember that these are females and we are keeping a large proportion of them. This figure for selection intensity is correct if selection is limited to just one trait.

effective proportion saved (p_e)
In selection—a value that, when substituted for actual proportion saved (p), reflects correct selection intensity.

The **effective proportion saved (p_e)** with respect to a particular trait is the value that, when substituted for actual proportion saved, reflects correct selection intensity for the trait. If you select for both birth weight EPD and yearling weight EPD using the independent culling levels (dotted lines) of Figure 14.3, you are *effectively* choosing from the best 74 of 77 heifers with regard to birth weight EPD. That translates to a selection intensity of .08 standard deviations—almost no intensity at all. For yearling weight, the effective proportion saved is 43 out of 77 or .56, resulting in a selection intensity (.71) only slightly smaller than that obtainable from single-trait selection (.77). If you use the independent culling levels depicted in Figure 14.3, you can expect fairly rapid genetic progress in yearling weight, but little, if any, progress in birth weight.

A Mathematical Example of Loss of Selection Intensity with Multiple-Trait Selection

In theory, given n equally important, uncorrelated traits, the effective proportion saved for each trait is

$$p_e = \sqrt[n]{p}$$

If the actual proportion saved is .10, then

Number of Traits	Effective Proportion Saved	Selection Intensity
1	.10	1.76
2	$\sqrt{.10} = .316$	1.12
3	$\sqrt[3]{.10} = .464$	.86
4	$\sqrt[4]{.10} = .562$	.72
5	$\sqrt[5]{.10} = .631$	.60

Loss of Selection Intensity and Correlations between Traits

Under multiple-trait selection, the loss of selection intensity in individual traits (and, therefore, the associated reduction in genetic progress in each trait) depends to a large degree on whether the traits are favorably or unfavorably correlated. More particularly, it depends on whether the selection criteria used for the traits are favorably or unfavorably correlated.

In the beef replacement heifer example, the selection criteria are EPDs for birth weight and yearling weight. The correlation between birth weight and yearling weight EPDs is actually a function of their accuracies, but because the genetic correlation between the traits is strong and positive (approximately .7), the correlation between EPDs is inevitably positive as well. It is, however, *unfavorable* due to our preference for heavier yearling weights but lighter birth weights.

Because of the unfavorable correlation between EPDs, it is difficult to find animals that have both high EPDs for yearling weight and low EPDs for birth weight. In Figure 14.3, only 12 heifers (black dots) have high enough yearling weight EPDs and low enough birth weight EPDs to be chosen with single-trait selection for either yearling weight or birth weight. Therefore, in order to obtain the 40 required replacements, the selection standard for one or both traits must be relaxed considerably.

When selection criteria are *favorably* correlated, loss of selection intensity in individual traits is much less. Figure 14.4 illustrates the same heifer selection example as Figure 14.3, only in this case selection is for yearling weight and *weaning* weight (as opposed to birth weight), and the points on the graph represent EPDs for this new combination of traits. Again, the thin lines represent culling levels for

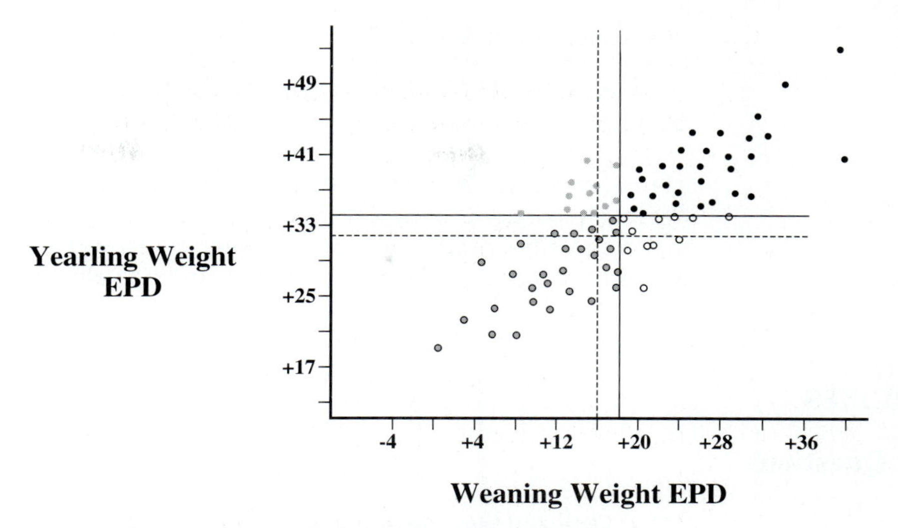

• Selected with single-trait selection for greater yearling weight EPD

○ Selected with single-trait selection for greater weaning weight EPD

• Selected with single-trait selection for either greater yearling weight EPD or greater weaning weight EPD

○ Not selected with single-trait selection for either trait

FIGURE 14.4 Illustration of loss of selection intensity with multiple-trait selection for weaning weight and yearling weight—two *favorably* correlated traits—in beef heifers. Thin lines represent culling levels that qualify a sufficient number of replacements if selection is for either weaning weight EPD or yearling weight EPD, but not both. Dotted lines represent independent culling levels that qualify the same number of replacements with multiple-trait selection. The distance between culling levels (between thin and dotted lines) suggests the loss of selection intensity in each trait. In this case, loss of intensity in each trait is small due to the strong, favorable correlation between weaning and yearling weight EPDs.

single-trait selection, and the dotted lines represent culling levels for multiple-trait selection.

Yearling and weaning weight EPDs are positively correlated. They are also favorably correlated because we prefer higher yearling weights and higher weaning weights. Note that in Figure 14.4, 30 heifers (black dots) have high enough yearling and weaning weight EPDs to be chosen with single-trait selection for either trait—many more than when selection is for yearling weight and birth weight. To achieve the necessary 40 replacement heifers, culling levels need to be relaxed only slightly. If you use the independent culling levels depicted in Figure 14.4, you can expect comparatively rapid genetic progress in both yearling weight and weaning weight.

Loss of Selection Intensity in Perspective

The more traits selected for with multiple-trait selection, the lower the selection intensity and the slower the rate of genetic change in any one trait. This suggests that the number of individual traits included in multiple-trait selection should be kept to a minimum. In practice, it means that frivolous traits should be ignored. Fads and fancies often lead breeders to select for such things as particular shades of coat color. While selection for traits like this may have short-term economic benefits, in the longer term it reduces progress in other, more important traits.

At the same time, no truly meaningful trait should be excluded from multiple-trait selection. Remember that the goal of multiple-trait selection is to improve aggregate breeding value (overall economic merit). So it is progress in aggregate breeding value, not progress in individual traits, that is important. If a trait makes a significant contribution to aggregate breeding value, select for it.

EXERCISES

Study Questions

14.1 Define in your own words:

single-trait selection
multiple-trait selection
aggregate breeding value or net merit
tandem selection
selection target
independent culling levels
economic selection index
breeding objective
economic weight
phenotypic selection index
effective proportion saved (p_e)

14.2 List the pros and cons of tandem selection.

14.3 How do genetic correlations influence the effectiveness of tandem selection?

14.4 List the pros and cons of selection using independent culling levels.

14.5 How does an economic selection index differ from the type of selection index discussed in Chapter 11?

14.6 **a.** In the development of an economic selection index, what criteria are used to determine the traits in the breeding objective?
b. What criteria are used to determine the traits in the index?

14.7 How is a "classic selection index" defined?

14.8 Under what circumstances is it advisable to use an economic selection index that substitutes EBVs or EPDs derived from BLUP and BLUP-like procedures for the *x*s?

14.9 List the pros and cons of selection using an economic selection index.

14.10 For a species and set of traits of your choice, describe the method of multiple-trait selection you think most appropriate. Be sure to consider combination methods. Justify your answer.

14.11 Why is there a loss of selection intensity in individual traits when selection is applied to more than one trait at a time?

14.12 How do correlations between selection criteria for individual traits affect loss of selection intensity in those traits when selection is applied to more than one trait at a time?

14.13 What guidelines should you use to determine how many traits to select for and which ones?

Problems

14.1 A swine breeder is selecting for increased number of pigs weaned (NW) and reduced backfat (BF) in her pigs. She plans to choose three out of the following eight boars based on EBVs for these traits.

Boar #	EBV for Number of Pigs Weaned	EBV for Backfat, in
1	+1.1	−.11
2	+.8	−.25
3	+2.4	−.05
4	+.3	−.36
5	−.5	−.10
6	+3.0	+.20
7	+1.0	+.05
8	−.6	−.40

a. Which boars would she initially select using tandem selection:
  i. when NW is the first trait under selection?
  ii. when BF is the first trait under selection?

b. Which boars would she select using independent culling levels if the levels were set at 0 pigs for NW and −.1 inch for BF?

c. Which boars would she select using an economic selection index if an independent one-pig increase in NW is worth $100 and an independent 1-in *decrease* in BF is worth $1,000?

d. Why was boar 8 selected with the index but not with independent culling levels?

14.2 The following genetic parameters were used in the yearling weight (YW)/birth weight (BW) example in the boxed section entitled *Calculating the Classic Form of Economic Selection Index:*

$\sigma^2_{P_{YW}} = 3{,}600 \text{ lb}^2$ $\sigma^2_{P_{BW}} = 100 \text{ lb}^2$ $cov(P_{YW}, P_{BW}) = 210 \text{ lb}^2$

$\sigma^2_{BV_{YW}} = 1{,}440 \text{ lb}^2$ $\sigma^2_{BV_{BW}} = 40 \text{ lb}^2$ $cov(BV_{YW}, BV_{BW}) = 168 \text{ lb}^2$

Conditions have changed so that a 1-lb increase in yearling weight is now worth $1.22 and a 1-lb increase in birth weight is worth $−4.35. Recalculate the economic selection index accordingly.

14.3 A horse breeder is selecting for a number of equally important, uncorrelated traits. He needs to replace 10% of his mares each year (i.e., keep 20% of his fillies). Calculate effective proportion of females saved (p_{e_f}) and female selection intensity (i_f) for each trait if the number of traits is:

a. 2
b. 3
c. 4
d. 10

Mating Systems

There are two kinds of decisions that animal breeders must make. They must decide which individuals become parents, how many offspring they may produce, and how long they remain in the breeding population. Those, of course, are selection decisions, and the many considerations involved in selection decisions have been the subject of the last nine chapters. Breeders must also decide which males to breed to which females. That is a mating decision, and considerations involved in mating decisions are the subject of the next five chapters.

Chapter 15 deals with mating systems for simply-inherited traits. Chapters 16 though 19 are devoted almost entirely to mating systems for polygenic traits. In these chapters, we will learn how the choice of a breed or line to use in a mating system can, like selection of individuals, change the average breeding value of a population. Just as importantly, we will learn how mating systems can, *unlike* selection, change population averages for gene combination value as well.

CHAPTER 15

Mating Systems for Simply-Inherited Traits

Selection for simply-inherited traits is straightforward. You need only know how many loci are involved, how many alleles at each locus, how those alleles are expressed (whether they are completely dominant, partially dominant, etc.), and the genotypes or probable genotypes of potential parents. Mating systems for simply-inherited traits are equally straightforward. With the same kinds of information, we can determine which mating combinations are capable of producing desirable genotypes in the offspring.

MATING TO PRODUCE PARTICULAR GENE COMBINATIONS

Sometimes breeders are interested in producing homozygotes, heterozygotes, or particular epistatic combinations for a simply-inherited trait. Say, for example, that you are a breeder of Labrador retrievers and want to design matings to produce dogs with a particular coat color. That should be relatively straightforward because we have a good understanding of inheritance of coat color in Labradors. We know that basic coat color is determined by genes at two loci; the B (black) locus and E (extension of pigmentation) locus; that in the Labrador breed only two alleles exist at each of these loci; that the *B* allele is completely dominant to the *b* allele, and *E* is completely dominant to *e*; and that epistasis occurs such that the E locus has no apparent effect except when the individual is homozygous for the recessive allele, in which case its coat will be yellow regardless of what alleles are present at the B locus.[1] Our understanding of Labrador coat color can be summarized as follows:

$$B\_E\_ \Rightarrow \text{black}$$
$$bbE\_ \Rightarrow \text{chocolate}$$
$$\_\,\_ee \Rightarrow \text{yellow}$$

where the dashes in these genotypes indicate that either allele could be substituted without changing the phenotype.

If you want to produce yellow Labs, just mate a yellow to a yellow. As the following Punnett square makes clear, yellow Labs breed true. Regardless of what

[1]See the last section of Chapter 3 for a more detailed explanation of coat color in Labradors.

alleles are present at the B locus, a yellow bred to a yellow produces exclusively yellows.

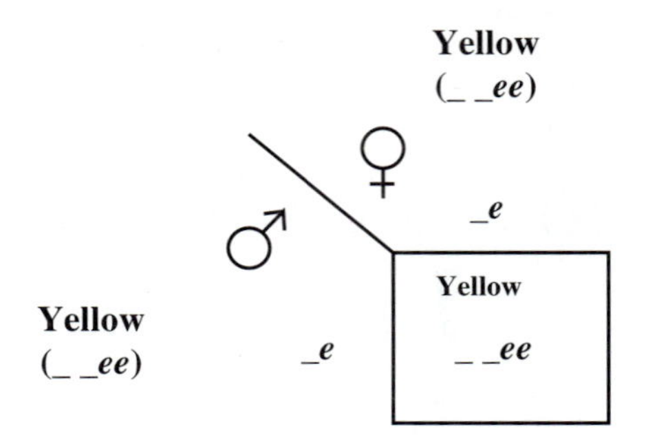

If you want to produce chocolate Labs, mating two chocolates should work most of the time. But as you can see in the next Punnett square, the mating of two chocolates is not guaranteed to produce exclusively chocolate puppies. It could result in some yellow pups.

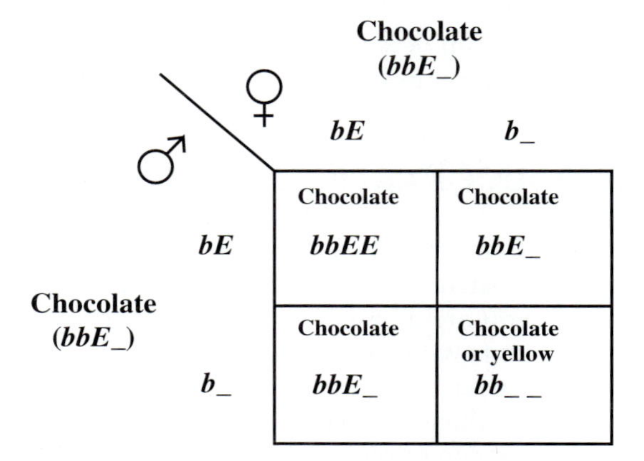

If you want just chocolates, you should use chocolate parents whose pedigrees and past litters indicate that they do not carry the yellow (*e*) allele. In other words, you should use parents whose *probable* genotype is *bbEE.* Consider, for example, the following pedigree of Rachel, a chocolate bitch.

Rachel (*Chocolate*)
- **Ralph** (*Chocolate*)
  - **Suds** (*Chocolate*)
  - **Amanda** (*Chocolate*) (*Yellow progeny*)
- **Meghan** (*Chocolate*)

(I have left out a large part of Rachel's pedigree because it is uninformative. It contains no evidence of any coat color alleles other than the allele for chocolate color.)

In the population of Labradors from which Rachel is descended, the yellow allele is rare, so we can simplify matters by assuming (perhaps not safely) that any dog with no evidence of a yellow allele in its pedigree has the genotype *bbEE*. Because Amanda, Rachel's granddam, produced yellow puppies, we know she carried the yellow allele (*e*). There is therefore a 50% probability that her son, Ralph, was a carrier and a 25% probability (.5(50)) that Ralph's daughter, Rachel, is a carrier.

Now suppose we have a choice of mates for Rachel: Phantom, Murray, and Griz—all chocolate dogs. Phantom's and Murray's pedigrees follow.

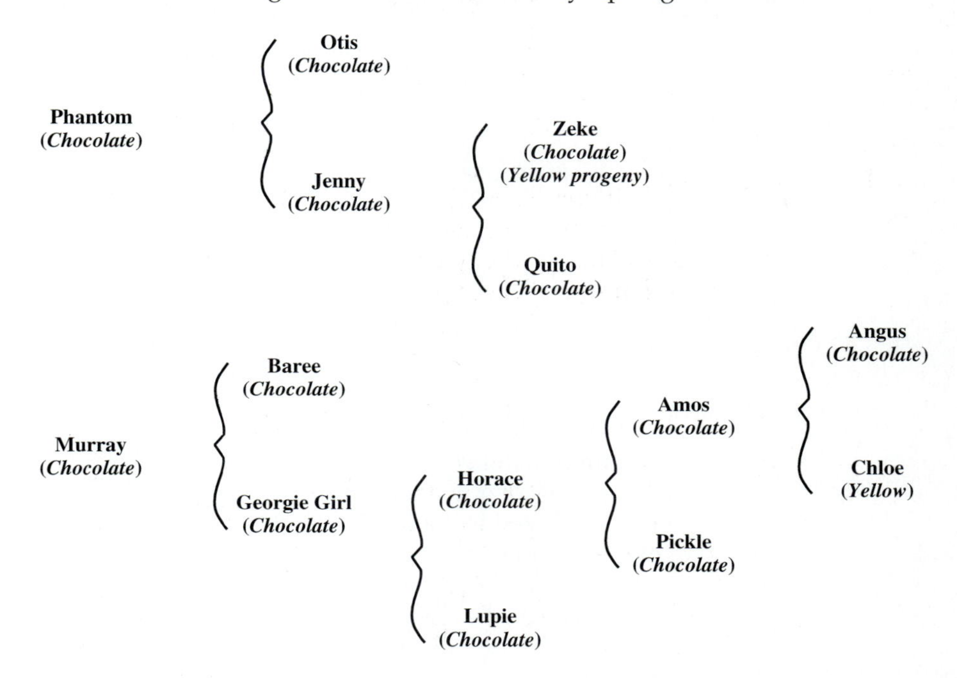

Phantom's pedigree is much like Rachel's. He has a grandparent known to carry the yellow allele, so there is a 25% chance that Phantom is a carrier. Murray's great great granddam, Chloe, was yellow. Her son, Amos, must have been a yellow carrier, and that leaves Murray with a 12.5% probability of carrying the yellow allele. The third dog, Griz, is a known carrier because he sired yellow puppies in a previous litter.

To get yellow puppies from chocolate parents, both parents must be yellow carriers. So a mating of Rachel to Phantom has a .25(.25) = .0625 or 6.25% chance of being the kind of mating that could produce yellow pups. A mating to Murray has only a .25(.125) = .03125 or 3.125% chance of being such a mating. A mating to Griz is the most likely to be such a mating—.25(1) = .25 or 25%. Of the three dogs, Murray's chance of having a *bbEE* genotype is the greatest, making him the best choice if we want exclusively chocolate puppies.

Finally, if you want to produce black Labs, you have several options, the most obvious of which is to mate a black Lab to a black Lab. But you are not guaranteed all black puppies with this mating and, as with the chocolate dogs, you should investigate pedigrees and past litters to try to determine probable genotypes of prospective parents. You will probably want to avoid using animals that are likely to carry either the chocolate (*b*) allele or yellow (*e*) allele. If you study the following Punnett square, you will see that the only way to guarantee exclusively black pups from the mating of two black parents is to determine that (1) at least one parent is homozygous at both loci (*BBEE*) or (2) one parent is homozygous at the black locus (*BBE_*) and the other is homozygous at the E locus (*B_EE*).

♂ \ ♀ Black (B_E_)	BE	B_	_E	_ _
Black (B_E_) BE	Black *BBEE*	Black *BBE_*	Black *B_EE*	Black *B_E_*
B_	Black *BBE_*	Black or yellow *BB_ _*	Black *B_E_*	Black or yellow *B_ _ _*
_E	Black *B_EE*	Black *B_E_*	Black or chocolate *_ _EE*	Black or chocolate *_ _E_*
_ _	Black *B_E_*	Black or yellow *B_ _ _*	Black or chocolate *_ _E_*	Black, chocolate, or yellow *_ _ _ _*

Designing matings to produce particular coat colors in Labradors is easy compared to doing the same in horses. There are probably a dozen or more loci affecting coat color in horses, and the epistatic relationships among them can be complicated. Listed in Table 15.1 are (1) genotypes for four of the more important coat color loci and (2) phenotypes thought to correspond to them.

From Table 15.1 it appears that no colors breed true with the possible exceptions of chestnuts or sorrels (*_ _CCddee*) and cremellos (_ _$c^{cr}c^{cr}$_ _ _ _). (We cannot be absolutely sure of these because there are other loci affecting coat color besides those listed in the table.) Bays, for example, do not breed true. The mating of a bay (*A_CCddE_*) to a bay might produce a bay foal, but it might also produce a brown (a^t_CCddE_), black (*aaC_ddE_*), or chestnut (*_ _CCddee*) foal.

Buckskins and palominos are heterozygous at the C locus, so the most reliable way to produce them is not to mate buckskins to buckskins or palominos to

TABLE 15.1 Four-Locus Genotypes and Corresponding Phenotypes for Coat Color in Horses

Four-Locus Genotype	Phenotype
A_CCD_E_	Yellow dun
A_CCddE_	Bay
A_Cc^{cr}D_E_	Dun buckskin
A_Cc^{cr}ddE_	Buckskin
a^t_CCddE_	Brown
aaC_D_E_	Mouse
aaC_ddE_	Black
_ _CCddee	Chestnut or sorrel
_ _Cc^{cr}D_ee	Dun palomino
_ _Cc^{cr}ddee	Palomino
_ _$c^{cr}c^{cr}$_ _ _ _	Cremello

palominos, but to mate appropriate homozygous types. Buckskins often result from mating bays ($A\_CCddE\_$) to cremellos ($\_\ \_c^{cr}c^{cr}\_\ \_\ \_\ \_$), and palominos are commonly produced by mating chestnuts ($\_\ \_CCddee$) to cremellos. Still, these matings might produce a foal with any one of six coat colors: buckskin ($A\_Cc^{cr}ddE\_$), dun buckskin ($A\_Cc^{cr}D\_E\_$), palomino ($\_\ \_Cc^{cr}ddee$), dun palomino ($\_\ \_Cc^{cr}D\_ee$), mouse ($aaC\_D\_E\_$), or black ($aaC\_ddE\_$). If we have more information about the genotypes of the parents (especially the cremello parent), we can eliminate some of these possibilities. In horses, as in Labradors, the more you know about the genotypes of prospective parents, the easier it is to identify matings that are likely to produce the result you want.

Repeated Backcrossing to Import an Allele

Occasionally there is a need to incorporate a specific allele or alleles existing in one population into another population. For example, when continental European beef breeds, most of which are red, were first imported to North America in the 1960s and 1970s, many North American breeders wanted to develop "purebred" black strains of these breeds. That meant introducing the dominant black (*B*) allele into breeds in which it did not occur naturally. The mating system used to accomplish this is a form of **repeated backcrossing.** Some breeders call it **introgression.** Repeated backcrossing for this purpose has its origins in plant breeding where it has long been used to transfer genes for pest resistance from wild to domestic varieties.

repeated backcrossing or **introgression**
A mating system used to incorporate an allele or alleles existing in one population into another population. An initial cross is followed by successive generations of backcrossing combined with selection for the desired allele(s).

backcrossing
The mating of a hybrid to a purebred of a parent breed or line.

When repeated backcrossing is used to import a specific allele, the population that lacks the allele (population 1) is crossed with another population (population 2) that possesses the allele. Successive generations of offspring are **backcrossed** to purebreds of population 1, and crossbred replacements are chosen from only those individuals carrying the desired allele. After a number of generations, almost all of the genes in the population will trace to population 1, but the desired gene from population 2 (and a few closely linked genes) will have been retained by selection. At this point, further backcrossing is no longer needed, and matings can be made within the new population. If the desired allele is dominant, continued selection is required to maintain the recessive allele at low frequency.

Consider the case of Limousin cattle from France. *Fullblood* (purebred, as the term is defined by European—as opposed to North American—breeders) Limousin are red (*bb*). In order to breed black Limousin, some North Americans crossed fullblood red Limousin bulls on black Angus cows (*BB*).

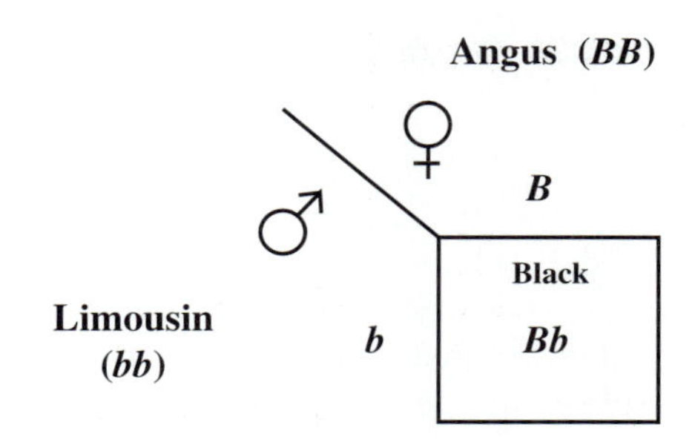

The replacement heifers resulting from this mating were heterozygous *Bb*, and because the black allele (*B*) is dominant, they were all black. In the next generation, these heifers were backcrossed to fullblood Limousin sires to produce, on average, half black and half red offspring that were 75% Limousin.

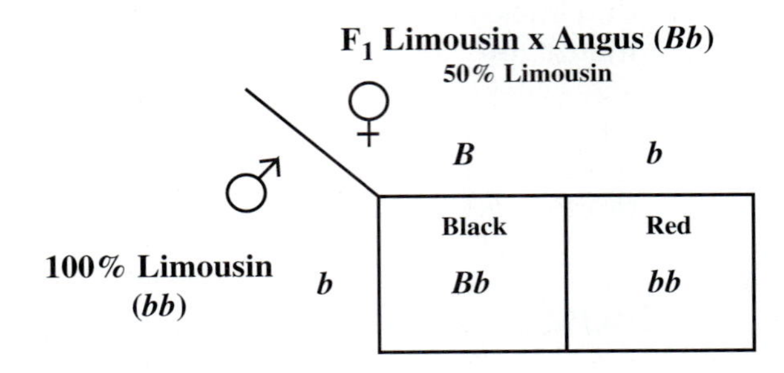

Red heifers were sold, and the remaining black heifers from the first backcross (BC_1) generation were backcrossed again to fullblood red Limousin bulls to produce black and red, 87.5% Limousin offspring.

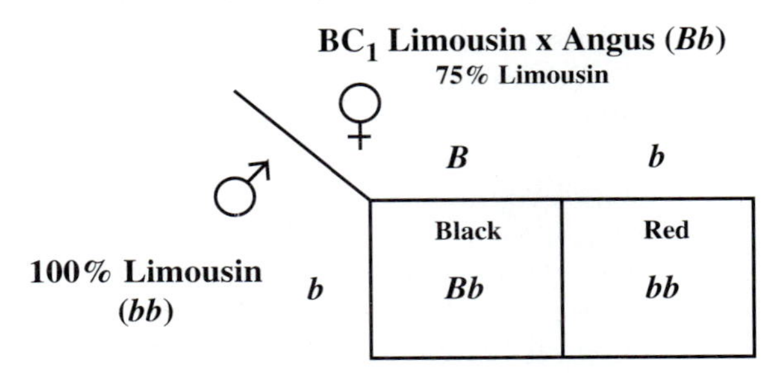

Backcrossing and selection for the black allele continued until the proportion of Limousin in the population was sufficiently high—typically $^{15}/_{16}$ or 94%—that the animals could effectively be considered purebred Limousin. Then phenotypically black young bulls and heifers from the population were mated inter se (among themselves) to produce predominantly black progeny.

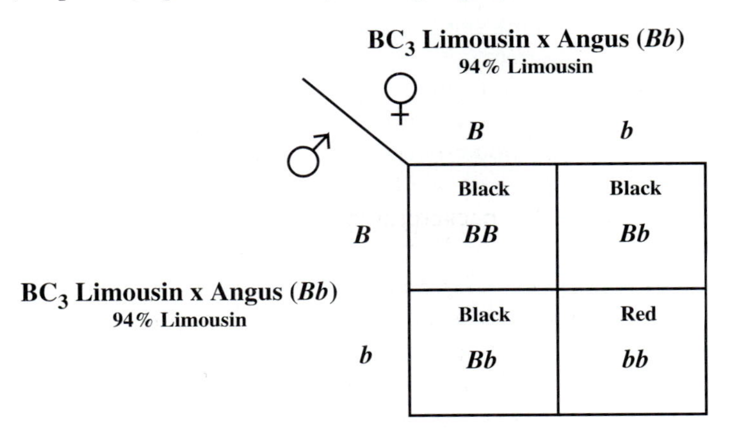

Because the red allele was still very much present in the new Limousin population, breeders of black Limousin cattle continued to cull red animals. Among these breeders, proven homozygous black sires were especially desirable because

TABLE 15.2 Repeated Backcrossing to Incorporate the Black Coat Color Allele (*B*) into Limousin Cattle: A Summary by Generation

	Sires			Dams			Offspring		
Offspring Generation	Genotype	Color	% Lim.	Genotype	Color	% Lim.	Genotype	Color	% Lim.
F_1	*bb*	Red	100	*BB*	Black	0	*Bb*	Black	50
BC_1	*bb*	Red	100	*Bb*	Black	50	*Bb* or *bb*	½ black ½ red	75
BC_2	*bb*	Red	100	*Bb*	Black	75	*Bb* or *bb*	½ black ½ red	88
BC_3	*bb*	Red	100	*Bb*	Black	88	*Bb* or *bb*	½ black ½ red	94
Inter se	*Bb*	Black	94	*Bb*	Black	94	*BB*, *Bb*, or *bb*	¾ black ¼ red	94
Inter se	*BB* or *Bb*	Black	94	*BB* or *Bb*	Black	94	*BB*, *Bb*, or *bb*	¾+ black	94

grading up or **topcrossing**
(1) A mating system designed to create a purebred population by mating successive generations of nonpurebred females to purebred sires; (2) A mating system designed to convert a population from one breed to another by mating successive generations of females descended from the first breed to sires of the second breed.

they produced no red offspring. The example of repeated backcrossing to introduce black coat color into Limousin cattle is summarized in Table 15.2

Students occasionally confuse the kind of repeated backcrossing used to introduce an allele into a population with **grading up.** Grading up, or **topcrossing** as it is sometimes called, involves repeated backcrosses, but does not include any attempt to select for a specific allele. It is simply a mating system designed to convert a population from one breed to another by mating successive generations of females descended from the first breed to sires of the second breed. (Grading up and topcrossing can also refer to a mating system designed to create a purebred population by mating successive generations of nonpurebred females to purebred sires.) In the Limousin example, repeated backcrossing was used to import the allele for black coat color from an Angus population to a Limousin population. At the same time, a population of Angus females was upgraded to "pure" Limousin.

EXERCISES

Study Questions

15.1 Define in your own words:
repeated backcrossing or introgression
backcrossing
grading up or topcrossing

15.2 List four pieces of information needed to design matings that will produce desired gene combinations for simply-inherited traits.

15.3 What is meant by the term *probable genotype?* What is the value of knowing an animal's probable genotype?

15.4 For a species, simply-inherited trait, and genotype of your choice, list the various matings that could produce that genotype.

15.5 Describe, step by step, the process of repeated backcrossing (introgression).

15.6 How does repeated backcrossing to import an allele differ from grading up?

Problems

15.1 In the Labrador example in this chapter, we decided to mate the chocolate bitch, Rachel, to the chocolate dog, Murray, in order to minimize the chances of producing yellow puppies. We have just received new information about Murray's dam, Georgie Girl; at one time she produced a litter containing yellow pups. Should we change mating plans? Support your answer mathematically.

15.2 Cushy Pearson has a thing for bay colored horses. He purchased a single service of a stylish (and expensive) bay stallion in hopes of producing a bay foal. He has four mares available: one brown, one mouse colored, one black, and one chestnut. Assume the following:

The inheritance of coat color in horses is no more complicated than it appears in Table 15.1.

Cushy has no information on the genotypes of his horses other than their phenotypes.

No linkage exists among the four loci shown in Table 15.1

Frequencies of coat color alleles in the Thoroughbred population are estimated to be:

Allele	Frequency
A	.6
a	.3
a^t	.1
C	.7
c^{cr}	.3
D	.2
d	.8
E	.3
e	.7

a. To which mare should Cushy mate his bay stallion in order to maximize the likelihood of producing a bay foal?
b. To which mare should Cushy be sure *not* to mate his stallion?
c. Prove your answers mathematically.[2]

15.3 J. F. Turner owns an exceptional herd of Black Angus (BA) cows that she wants to develop into a herd of red cows, yet still retain as much of her original breeding as possible. She will use purebred Red Angus bulls for one generation to supply the red allele, then backcross repeatedly *via* artificial insemination to black bulls from her foundation herd. Assume the following:

Foundation cows and bulls are homozygous (*BB*) at the black/red color locus.

J. F. breeds 50% of her replacement heifers.

100% conception, no death loss.

Molecular geneticists have located a reliable genetic marker near the black/red locus enabling J.F. to test black animals to see if they are homozygous or heterozygous.[3]

[2]You may want to review the subsection of Chapter 6 entitled *Probabilities of Outcomes of Matings.*
[3]See Chapter 20 for more information on genetic markers.

Show the effects of repeated backcrossing on J.F.'s herd by filling in the following chart. (If there are multiple genotypes or colors within a generation, include the expected proportions.)

	Sires			Dams			Offspring		
Offspring Generation	Genotype	Color	% BA	Genotype(s)	Color(s)	% BA	Genotype(s)	Color(s)	% BA
F_1									
BC_1									
BC_2									
BC_3									
Inter se									
Inter se									
Inter se									

CHAPTER 16

Mating Strategies Based on Animal Performance: Random and Assortative Mating

A *mating system* can be defined as a set of rules for making mating decisions. As such, there is no limit to the number of possible mating systems. There are, however, only a few general *mating strategies.* Some are based on animal performance (or expectation of performance).[1] They include simple random mating, positive assortative mating, and negative assortative mating. These strategies are the subject of this chapter. Two other mating strategies, inbreeding and outbreeding, are based on pedigree relationship. They are the subject of Chapter 17.

We will examine the mating strategies that are based on animal performance in two contexts. The first context is an *individual* context. Here we refer to strategies for mating particular animals. The second context is a *population* context. Here we refer to strategies for crossing breeds or lines.

STRATEGIES FOR MAKING INDIVIDUAL MATINGS

First let's look at strategies for making individual matings—matings of specific sires to specific dams. We will begin with a strategy that is, in a sense, no strategy at all. It is random mating.

Random Mating

random mating
A mating system in which mates are chosen at random.

Random mating is a mating system in which mates are chosen at random. With truly random mating, all conceivable matings are equally likely. To make random matings, a breeder with a statistical bent might assign each female a number from a random number table, then allocate those females with the lowest random numbers to one male, those females with random numbers in the next higher category to another male, and so on. More typical procedures for random mating include

[1]I regularly use the word "performance" in place of "phenotype" or "phenotypic merit" because to many of us, particularly those who breed livestock, performance has immediate meaning, and phenotype is jargon by comparison. I am aware, however, that for some traits—even some quantitative traits—performance is a clumsy term. For example, it is awkward to speak of a horse's performance for cannon bone circumference or a dog's performance for mature size. In your reading, if the trait you have in mind and the word "performance" are not a comfortable match, mentally substitute the word "phenotype."

"gate cutting"—sorting females according to the order they choose to approach a gate—and randomly choosing doses of semen for artificial insemination.

Students sometimes confuse random mating with random selection. They assume that a breeder who randomly mates takes a completely "hands-off" approach to breeding in general, making neither mating decisions nor selection decisions. Random mating has nothing to do with selection, however. A highly select group of individuals can be randomly mated. Random mating can be either a lazy way to breed animals or a deliberate, carefully chosen technique.

Random mating is easy. It requires no performance records or genetic predictions, and little time is involved in making mating decisions. For this reason, random mating is popular in commercial breeding programs where performance information is unavailable or where there are so many animals that other approaches are impractical.

Random mating can be very helpful from the standpoint of genetic evaluation. If a sire is assigned to a sufficiently large number of mates, and those mates are chosen at random, it is unlikely that the sire's evaluation will benefit from having a particularly good set of mates or suffer from having a particularly poor set. As explained in Chapter 11, this is less of a concern if BLUP methodology is used to evaluate sires because the better BLUP models can account for nonrandom mating. It is a concern, however, if BLUP or BLUP-like techniques are not used.

There is no art in random mating. Many breeders feel that to randomly mate is to give up a measure of control, a certain power over nature that is rightfully theirs. Indeed, as we will see in the discussion of assortative mating that follows, there are times when planned matings make good sense. Still, random mating is underrated by many. Given the randomness of inheritance, the ability of a breeder to control the outcome of a specific mating is limited at best. Random mating relinquishes less control than you might think, and the "mistakes" that result from random mating are often balanced by "pleasant surprises."

Assortative Mating

assortative mating
The mating of either similar individuals **(positive assortative mating)** or dissimilar individuals **(negative assortative mating).**

Assortative mating is the mating of either similar individuals **(positive assortative mating)** or dissimilar individuals **(negative assortative mating).** "Similar" in this context traditionally means having similar performance in a trait or set of traits. It can also mean having similar expectation of performance—similar genetic predictions. Any mating strategy that is not random with respect to performance or the expectation of performance is necessarily a form of assortative mating.

Just as students sometimes confuse random mating with an absence of selection, they often make the same mistake with assortative mating. When we speak of breeding the "lowest to the lowest" or the "best to the worst" (forms of positive and negative assortative mating, respectively), it is easy to infer that the lowest and the worst are available for breeding because no selection has taken place. That is not the case, however. The lowest and the worst are simply the animals with the lowest or worst data of those individuals that were selected. They may, in fact, be very select—not particularly low or bad at all.

Assortative mating is more difficult than random mating. It requires performance records, genetic predictions, or some other mating criterion. Animals must be ranked—not always a simple matter, especially when multiple traits are considered.

Assortative mating favors some individuals with respect to progeny performance. A sire mated to only the best females has a distinct advantage over a sire relegated to "bottom end" females. Unless we use prediction technologies that account for nonrandom mating, assortative mating will cause genetic predictions for these sires to be biased.

Positive Assortative Mating

Examples of positive assortative mating include mating the tallest males to the tallest females, or mating males with the highest EPDs to the females with similarly high EPDs. Positive assortative mating means mating the biggest to the biggest, the smallest to the smallest, the fastest to the fastest, and so on.

Positive assortative mating tends to create more genetic and phenotypic variation in the offspring generation than would be found in a comparable randomly mated population. Figure 16.1 shows the distributions of breeding values or phenotypic values in a randomly mated population and in a population undergoing positive assortative mating. Mating the highest to the highest and the lowest to the lowest tends to spread the distribution away from the center and toward the extremes.

Uniformity is usually valuable to breeders, so the increased *phenotypic* variation caused by positive assortative mating is normally considered a drawback of the strategy. As explained in Chapter 10, however, increased *genetic* variation can be beneficial from a selection standpoint. The greater the genetic variation, the faster the rate of genetic change. Positive assortative mating therefore represents a way of speeding genetic change by increasing genetic variation.

Few breeders use this strategy for the express purpose of increasing genetic variation, however. Rather, they mate their best males to their best females in order to increase the probability of producing a truly superior offspring. *They use positive assortative mating to produce extreme individuals.* If those extreme individuals are males, so much the better; they can have a larger impact on the next generation.

For example, a Thoroughbred breeder might mate her very fastest mare to the most highly rated stallion available, paying a fortune for the privilege. She is betting that the mating will produce an extremely fast foal and an extremely fast colt in particular. If the colt is as good as she hopes, his future winnings and stud fees will pay for him many times over.

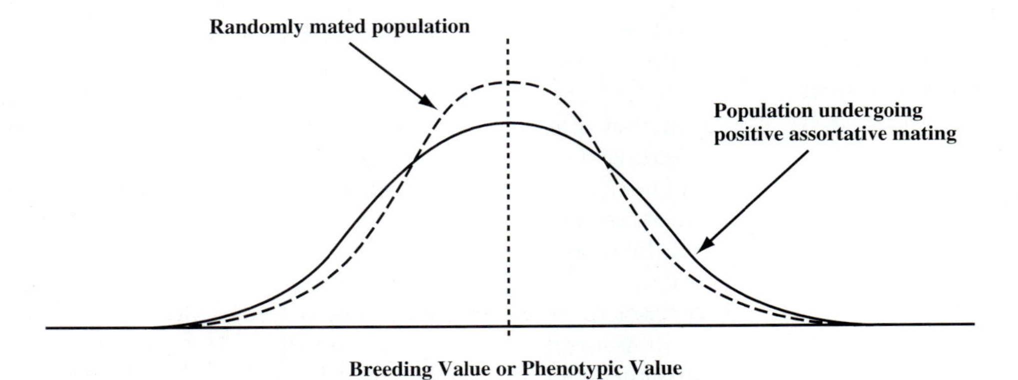

FIGURE 16.1 Graphic representations of increased genetic and phenotypic variation caused by positive assortative mating.

Mating to produce extremes makes sense if the breeding goal is to change the mean performance of a population. For example, if the chief goal of a dairy breeding program is to increase milk yield, then mating the highest producing cows to bulls with the highest predicted differences for milk is a wise approach. On the other hand, if an intermediate level of performance is optimal and uniformity about that optimum is important, positive assortative mating is inappropriate. It makes little sense, for example, to mate hens with extremely high breeding values for egg size to roosters with similarly extreme breeding values if intermediate egg size is optimal.

Negative Assortative Mating

Examples of negative assortative mating include mating the tallest males to the shortest females, or mating males with the highest EPDs to females with the lowest EPDs. Negative assortative mating means mating the biggest to the smallest, the smallest to the biggest, the fastest to the slowest, and so on.

Just as positive assortative mating tends to *increase* genetic and phenotypic variation in the offspring generation, negative assortative mating tends to *decrease* variation. Mating animals that are extreme in one direction to animals that are extreme in the opposite direction tends to produce more intermediate types and reduce the number of extreme offspring. The decreased genetic and phenotypic variation caused by negative assortative mating is illustrated in Figure 16.2.

Negative assortative mating is not a good strategy if you want to speed the rate of directional genetic change. Reduced genetic variation decreases response to selection. However, if your chief goal is to increase phenotypic uniformity about some intermediate optimum, this mating strategy can be beneficial. In the layer example, roosters with high breeding values for egg size bred to hens that produce small eggs (or vice versa) should result in a greater proportion of layers producing moderate-sized eggs. *Negative assortative mating is best used to produce intermediates.*

corrective mating A mating designed to correct in their progeny faults of one or both parents.

Some negative assortative matings can be considered **corrective matings.** These are matings designed to correct in their progeny faults of one or both parents. For example, if you breed horses and your favorite mare is sickle-hocked (too much bend in the hind legs), you might correct the fault in her foals by breeding her to a stallion that is post-legged (too little bend in the hind legs).

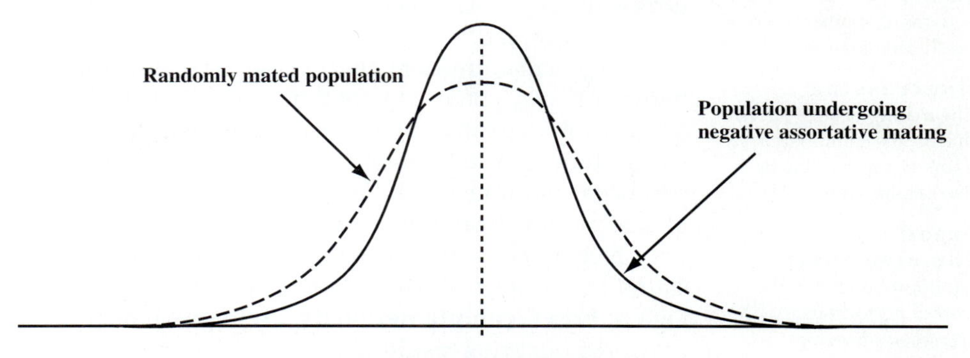

FIGURE 16.2 Graphic representation of decreased genetic and phenotypic variation caused by negative assortative mating.

complementarity
An improvement in the overall performance of offspring resulting from the mating of individuals with different but complementary breeding values.

A corrective mating is an example of breeding for **complementarity,** an improvement in the overall performance of offspring resulting from the mating of individuals with different but *complementary* breeding values. The post-legged stallion complements the sickle-hocked mare. The rooster with the high breeding value for egg size complements the hen whose eggs are too small.

These examples illustrate complementarity for a single trait. Just as important is complementarity for a combination of traits. For example, mating ewes that produce light but very high quality fleeces to rams that produce heavier fleeces of lesser quality can result in progeny that produce fleeces of acceptable quality and weight. The different types of sheep complement each other with respect to both traits.

Complementarity, whether it involves just one trait or a number of traits, results from the prudent combining of breeding values. You can think of it as "mixing and matching" breeding values in such a way that the overall performance of an offspring is superior to the performance of its parents.

Combination Strategies

Breeders commonly use more than one mating strategy at a time. A breeder of registered dairy cattle might use positive assortative mating with his highest producing cows, mating them to expensive A.I. bulls with extremely high predicted differences for milk in hopes of producing especially valuable offspring. At the same time, he might use negative assortative mating to correct in their offspring structural faults of some of his cows. Perhaps he has some cows with loose fore udder attachment and chooses to breed them to a bull whose daughters are particularly strong in that respect. He might randomly mate a portion of his cows to young but promising bulls. Random mating would save him time and provide unbiased data on those bulls when their daughters come into production.

STRATEGIES FOR CROSSING BREEDS OR LINES

crossbreeding
The mating of sires of one breed or breed combination to dams of another breed or breed combination.

linecrossing
The mating of sires of one line or line combination to dams of another line or line combination.

breed complementarity
An improvement in the overall performance of crossbred offspring resulting from crossing breeds of different but complementary biological types.

In theory, the same strategies used in making individual matings—matings of specific sires to specific dams—can be used to design **crossbreeding** or **linecrossing** programs. Breeds or lines within breeds can be randomly crossed or assortatively mated.

In practice, breeders rarely cross populations at random. (After viewing some herds or flocks, you might not think so, however.) There is almost always some assortative strategy involved in crossbreeding programs. Breeders do not commonly use the term "positive assortative mating" in a crossbreeding context either, but the practice is not unusual. Rotational crossbreeding systems work best when breeds of similar biological type are used (i.e., when like is mated to like).[2]

Negative assortative mating of breeds is common in sheep, swine, and beef cattle. The strategy allows breeders, usually commercial producers, to take advantage of **breed complementarity,** an improvement in the overall performance of crossbred offspring resulting from crossing breeds of different but complementary biological types.

[2]See Chapter 19 for details of various crossbreeding systems, including rotational systems.

As with the complementarity associated with individual matings, breed complementarity can involve just one trait or a number of traits. An example of simple, single-trait breed complementarity is the crossing of a beef breed that is low in marbling (a measure of meat quality indicated by flecks of fat in the meat) with a better marbling breed in order to produce market animals with acceptable marbling. More commonly, breed complementarity involves an array of traits. In swine, sheep, and beef cattle, breed complementarity typically comes from crossing **maternal breeds** (breeds that excel in the **maternal traits** of fertility, freedom from dystocia, milk production, maintenance efficiency, and mothering ability) with **paternal breeds** (breeds that are strong in **paternal traits** such as rate and efficiency of gain, meat quality, and carcass yield).

maternal breed A breed that excels in maternal traits.

maternal trait A trait especially important in breeding females. Examples include fertility, freedom from dystocia, milk production, maintenance efficiency, and mothering ability.

paternal breed A breed that excels in paternal traits.

paternal trait A trait especially important in market offspring. Examples include rate and efficiency of gain, meat quality, and carcass yield.

terminal sire crossbreeding system A crossbreeding system in which maternal-breed females are mated to paternal-breed sires to efficiently produce progeny that are especially desirable from a market standpoint. Terminally sired females are not kept as replacements, but are sold as slaughter animals.

terminal sire A paternal-breed sire used in a terminal sire crossbreeding system.

The ultimate in breed complementarity is achieved in **terminal sire crossbreeding systems** in which maternal-breed females are mated to paternal-breed sires to efficiently produce progeny that are especially desirable from a market standpoint. Daughters of **terminal sires** are not kept as replacements, but are sold along with their male counterparts as slaughter animals. In sheep, a terminal sire system might involve crossing Suffolk rams on Columbia ewes. In beef cattle, a similar system might involve crossing Charolais bulls on Hereford × Angus cows.

Like the complementarity associated with individual matings, breed complementarity results from combining breeding values. Biological types are mixed and matched in such a way that the overall performance of crossbred offspring is better than the performance of parent breeds.

Breed complementarity is sometimes an *additive* function of the breeding values of parent breeds. For example, when a light-milking beef breed is crossed with a heavy-milking breed to produce more desirable, moderate-milking females, the mean breeding value for milk production in the crossbred population is simply the average of the mean breeding values for the trait in the parent breeds. The breeding values of the parent breeds combine in an additive fashion. Because moderate milk production is more desirable than either light or heavy milk production, however, the cross is a complementary one. At other times, breed complementarity is a *multiplicative* function of the breeding values of parent breeds. When boars of a breed noted for especially rapid growth are mated to sows of a breed that produces exceptionally large litters, there is a multiplicative effect on weaned litter weight. More pigs weaned and heavier individual weaning weights translate into more litter weight weaned for crossbred pigs than for either of the parent breeds.

Because breed complementarity (and complementarity in general) results from combining breeding values, it is distinctly different from hybrid vigor—the other great benefit of crossbreeding. As we will see in the next chapter, hybrid vigor has nothing to do with breeding value; it is a function of gene combination value.

EXERCISES

Study Questions

16.1 Define in your own words:

random mating
assortative mating
positive assortative mating
negative assortative mating
corrective mating
complementarity

crossbreeding
linecrossing
breed complementarity
maternal breed
maternal trait
paternal breed
paternal trait
terminal sire crossbreeding system
terminal sire

16.2 How does a mating *system* differ from a mating *strategy?*

16.3 If animals are randomly mated, does that mean that selection is random too? Explain.

16.4 Describe the pros and cons of random mating.

16.5 If a breeder practices positive assortative mating, breeding the "best to the best" and the "worst to the worst," does that mean that he has selected his worst animals so that he can mate them to each other? Explain.

16.6 Describe the potential bias in genetic prediction caused by:
a. positive assortative mating.
b. negative assortative mating.

16.7 Describe the effect on genetic and phenotypic variation of:
a. positive assortative mating.
b. negative assortative mating.

16.8 What is the chief purpose of positive assortative mating? Give an example.

16.9 When does positive assortative mating make the most sense: (a) when the breeding goal is to change the mean performance of a population, or (b) when the breeding goal is to increase uniformity about an intermediate optimum level of performance? Explain.

16.10 When does negative assortative mating make the most sense: (a) when the breeding goal is to change the mean performance of a population, or (b) when the breeding goal is to increase uniformity about an intermediate optimum level of performance? Explain.

16.11 How does complementarity for a single trait differ from complementarity for a combination of traits? Give examples.

16.12 Complementarity can be described as "mixing and matching" breeding values. Explain.

16.13 Describe a practical example of a breeding program that uses a combination of mating strategies.

16.14 For a meat producing species of your choice, list the most important:
a. maternal traits.
b. paternal traits.

16.15 For a meat producing species of your choice, list the most important:
a. maternal breeds.
b. paternal breeds.

16.16 Describe an example of breed complementarity when it is:
a. an *additive* function of the breeding values of parent breeds.
b. a *multiplicative* function of the breeding values of parent breeds.

CHAPTER 17

Mating Strategies Based on Pedigree Relationship: Inbreeding and Outbreeding

Assortative mating, the kind of mating described in the last chapter, is a strategy based on animal performance or expectation of performance. Two other mating strategies, inbreeding and outbreeding, are based not on performance, but on pedigree relationship. This chapter is concerned with the effects, both good and bad, of inbreeding and outbreeding. You will learn how these mating strategies cause inbreeding depression and hybrid vigor, respectively. More importantly, you will learn when it is appropriate to use each strategy.

INBREEDING

inbreeding
The mating of relatives.

Inbreeding is the mating of relatives. That is the simplest definition anyway. Because all animals within a population are related to some degree, a more technically correct definition of inbreeding is the mating of individuals more closely related than average for the population.

Effects of Inbreeding

common ancestor
An ancestor common to more than one individual. In the context of inbreeding, the term refers to an ancestor common to the parents of an inbred individual.

Inbreeding has a number of effects, but the chief one and the one from which all others stem is an increase in homozygosity—an increase in the number of homozygous loci in inbred individuals and an increase in the frequency of homozygous genotypes in an inbred population. The connection between the mating of relatives and increased homozygosity was explained in Chapter 4.[1] For a quick review, look at the inbred pedigree and corresponding arrow diagram in Figure 17.1 (a repeat of Figure 4.2). Animal *X* is inbred because its parents (*S* and *D*) are half sibs, having a **common ancestor** in individual *A*. Because *X* could have inherited through its sire and dam identical copies of an allele present in *A*, the inbred mating increases the likelihood that *X* is homozygous for that allele. It also increases the likelihood that *X* is homozygous for every other of *A*'s genes. You can

[1]See the section of Chapter 4 entitled *The Effect of Mating Systems on Gene and Genotypic Frequencies.*

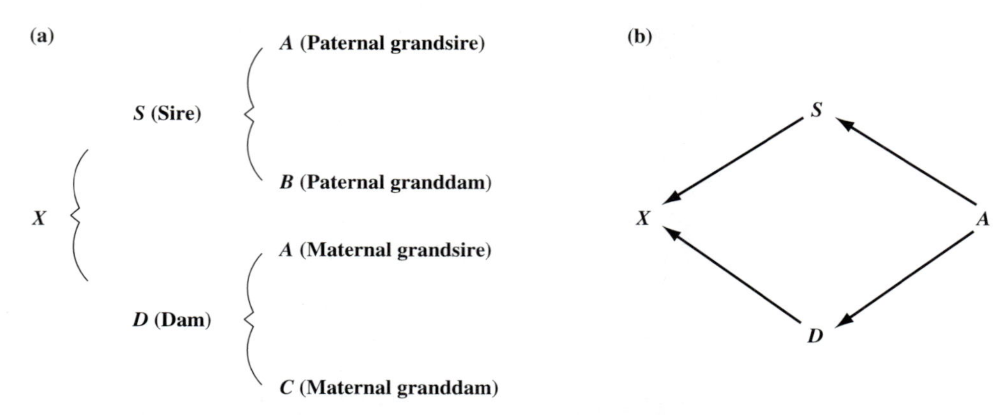

FIGURE 17.1 Pedigree (*a*) and arrow diagram (*b*) showing an inbred mating.

see, then, that inbreeding causes an increase in the proportion of loci at which an individual is homozygous and a corresponding decrease in the proportion of loci at which an individual is heterozygous.

Prepotency

prepotency The ability of an individual to produce progeny whose performance is especially like its own and(or) is especially uniform.

One consequence of the increase in homozygosity caused by inbreeding is greater **prepotency** in inbreds. Individuals are said to be prepotent if the performance of their offspring is especially like their own and(or) is especially uniform. Because inbred individuals have fewer heterozygous loci than noninbreds, they cannot produce as many different kinds of gametes. The result is fewer different kinds of zygotes and therefore less offspring variation.

For a hypothetical example, compare an inbred individual homozygous at three of the four loci affecting a trait versus a noninbred homozygous at only one of the four loci.

Inbred genotype: *AABbCCdd*
Noninbred genotype: *AaBbCCDd*

Possible gametes from the inbred genotype are:

ABCd
AbCd

Possible gametes from the noninbred genotype are:

ABCD
ABCd
AbCD
AbCd
aBCD
aBCd
abCD
abCd

Clearly the inbred produces fewer unique gametes and therefore fewer unique zygotes than the noninbred. The example depicts only four loci, but the same principle holds with the much larger number of loci typical of polygenic traits.

An inbred individual is more likely to be prepotent if its homozygous loci contain chiefly dominant alleles. Its offspring will then have at least one dominant allele at each of these loci. If dominance is complete, the effect of these loci in the offspring will be the same as in the parent, regardless of what genes are contributed by the other parent. The offspring will then more closely resemble the parent and each other.

Breeders tend to overplay the importance of prepotency. True prepotency is likely to be observed only for simply-inherited traits or for highly heritable polygenic traits. When heritability is low, environmental effects influence performance to a much greater degree than genetic effects, overwhelming any consequence of having more uniform gametes.

Expression of Deleterious Recessive Alleles with Major Effects

A second consequence of inbreeding is the expression of deleterious recessive alleles with major effects, and it is this aspect of inbreeding, more than any other, that gives inbreeding a bad reputation. People associate inbreeding with genetic defects such as the spider leg condition in sheep, dwarfism in cattle, and a host of problems in dogs. It is true that defects caused by recessive alleles often surface in inbred populations. But inbreeding does not create deleterious recessive alleles; they must already be present in a population. Inbreeding by itself simply increases homozygosity, and it does so without regard to whether the newly formed homozygous combinations contain dominant or recessive alleles. It therefore increases the likelihood of deleterious recessive alleles becoming homozygous and expressing themselves.

For example, consider the anomaly known as diaphragmatic hernia, a congenital (i.e., occurring during fetal development) defect of the diaphragm in dogs. The recessive allele that causes the problem occurs at low frequency in the general population, so the probability of any noninbred mating producing the condition is extremely low. However, if a dog that carries the recessive allele is mated to a daughter, there is a much higher probability of producing an affected pup.

Three types of matings, two of them noninbred and one inbred, are shown with Punnett squares in Figure 17.2. In each case, the probability that a gamete with a dominant allele will be contributed is designated p, and the probability that a gamete with a recessive allele will be contributed is designated q.[2] The first Punnett square (a) represents noninbred matings of males and females chosen at random from the population. If the frequency of the faulty recessive allele (h) in the population is .02, the frequency of affected (hh) pups produced from matings of this kind is extremely small—.0004 or one in 2,500.

If we look at just those matings involving a carrier (Hh) male, the frequency of affected pups increases. The second Punnett square (b) represents the noninbred matings of a carrier male to unrelated females. The dog, being a heterozygote, has an equal likelihood of contributing either a normal H or faulty h allele, so $p = q = .5$ in his case. But the frequency of affected pups produced from matings of this kind is still small—one in 100.

[2]The use of probabilities and Punnett squares to predict the results of matings is explained in the subsection of Chapter 6 entitled *Probabilities of Outcomes of Matings.*

(a) Matings of randomly chosen males and females

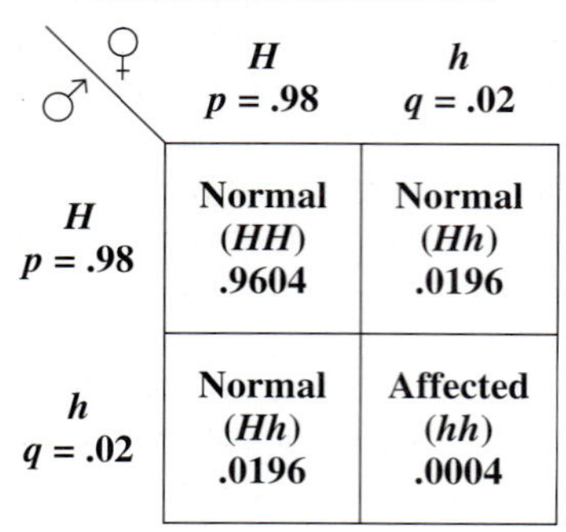

♀
♂
H
p = .98
h
q = .02
H
p = .98
Normal (HH) .9604
Normal (Hh) .0196
h
q = .02
Normal (Hh) .0196
Affected (hh) .0004

(b) Matings of a carrier male to randomly chosen females

♂ \ ♀	*H* $p = .98$	*h* $q = .02$
H $p = .5$	Normal (*HH*) .49	Normal (*Hh*) .01
h $q = .5$	Normal (*Hh*) .49	Affected (*hh*) .01

(c) Matings of a carrier male to his daughters

♂ \ ♀	*H* $p = .75$	*h* $q = .25$
H $p = .5$	Normal (*HH*) .375	Normal (*Hh*) .125
h $q = .5$	Normal (*Hh*) .375	Affected (*hh*) .125

FIGURE 17.2 Comparison of the frequency of pups with diaphragmatic hernia (*hh*) produced from inbred versus noninbred matings. If the frequency of the faulty recessive allele (*h*) in the population is .02, the frequency of affected pups produced from (*a*) the noninbred matings of randomly chosen males and females, (*b*) the noninbred matings of a carrier (*Hh*) dog to unrelated females, and (*c*) the inbred matings of a carrier dog to his daughters (*HH* or *Hh*) is .0004, .01, and .125, respectively.

The third Punnett square (c) represents the inbred matings of a carrier dog to his daughters. Assuming the dams of these daughters did not carry the *h* allele, the daughters are equally likely to be either *HH* or *Hh*. The chances of them contributing an *H* allele are therefore three in four, and the chance of them contributing an *h* allele is just one in four. In this case, $p = .75$ and $q = .25$. The frequency of affected pups produced from inbred matings of this kind is .125 or one in eight—much higher than from the noninbred matings.

The example depicted in Figure 17.2 shows how inbreeding increases the likelihood that deleterious recessive alleles that are present in a population will be expressed. Admittedly, the sire × daughter matings in the example are an extreme case because a sire and his daughters are so closely related. The principle is the same for more distant inbred matings, however. And in populations that have been inbred over a period of time, animals become so related in general that matings of what at first glance appear to be distant relatives can, in reality, be matings of rather close relatives.

Inbreeding increases the incidence of expression of deleterious recessive alleles, and that is a problem. But it is possible to use inbreeding combined with selection to eliminate faulty recessive alleles from a population. The idea here is to inbreed within a small population, continually selecting against undesirable alleles and hoping that, by pure chance, the increased homozygosity brought about by inbreeding will result in fixation of desirable dominant alleles and corresponding elimination of deleterious recessives.

Ridding a population of deleterious recessives by inbreeding is a risky and often wasteful enterprise. Most populations do not survive the process. In practice, breeders that use the procedure work with many small subpopulations (inbred lines), discarding the unsuccessful ones and continuing with the fraction that remain. To be economically viable, inbreeding programs of this kind must involve large numbers of animals that, on an individual basis, are not particularly valuable or expensive to maintain. For this reason, the production of inbred lines is common in plant breeding and was once common in layer flocks, but is rare in populations of all but the smallest mammals.

Inbreeding Depression

Expression of deleterious recessive alleles with major effects, particularly lethal and semilethal genes, is a very visible consequence of inbreeding. It is an example of the effect inbreeding can have on certain *simply-inherited* traits. Less obvious is the expression of unfavorable recessive alleles influencing *polygenic* traits. The individual effects of these genes are small but, taken together, can significantly decrease performance—a phenomenon known as **inbreeding depression.**

inbreeding depression
The reverse of hybrid vigor—a decrease in the performance of inbreds, most noticeably in traits like fertility and survivability.

In the context of the genetic model for quantitative traits, inbreeding depression is the manifestation of poor gene combination value, which is the direct result of increased homozygosity in inbreds. To see how this all works, recall the genetic model for quantitative traits:

$$P = \mu + BV + GCV + E$$

Breeding value (BV) and gene combination value (GCV) make up the genetic portion of the model and together constitute genotypic value (G), i.e.,

$$G = BV + GCV$$

Rearranging, then:

$$GCV = G - BV$$

In other words, an animal's gene combination value is simply the difference between its genotypic value and its breeding value. We will use this concept in the examples that follow.

Several genotypes and associated breeding values, gene combination values, and genotypic values are shown in Table 17.1. The hypothetical trait depicted in the table is influenced by six loci—too few, really, for a polygenic trait, but enough for an example. For simplicity, the independent effect of each dominant allele is assumed to be +4 units, and the independent effect of each recessive allele is assumed to be −2 units. The genotypic value of each homozygous gene pair is assumed to be the same as the sum of the independent effects of the genes at that locus. In this case, the genotypic value of a homozygous dominant pair is 4 + 4 = 8, and the genotypic

TABLE 17.1 Effect of Increased Homozygosity (Decreased Heterozygosity) on Gene Combination Value (GCV) and Genotypic Value (G) for a Hypothetical Trait Influenced by Six Loci Exhibiting Complete Dominance[a,b]

Genotype	BV	G	$GCV\ (G - BV)$
AaBbCcDdEeFf	$6(4) + 6(-2) = 12$	$6(8) = 48$	36
AABbCcddEeFf	$6(4) + 6(-2) = 12$	$8 + 4(8) + (-4) = 36$	24
AAbbCcDDeeFf	$6(4) + 6(-2) = 12$	$2(8) + 2(8) + 2(-4) = 24$	12
AAbbCCddEEff	$6(4) + 6(-2) = 12$	$3(8) + 3(-4) = 12$	0

[a]Simplifying assumptions: Only six loci influence this trait. (Polygenic traits are typically affected by more loci.) The independent effect of each dominant allele equals +4 units. The independent effect of each recessive allele equals −2 units. The genotypic value of each homozygous gene pair is the same as the sum of the independent effects of the genes at that locus. (This is rarely, if ever, the case, but it makes the example much easier to understand.)

[b]For further background, see the section of Chapter 7 entitled *Gene Combination Value.*

value of a homozygous recessive pair is $-2 + (-2) = -4$. (This last assumption is rarely, if ever, true, but makes the example much easier to understand.) Furthermore, complete dominance is assumed at each locus.

Four of many possible six-locus genotypes are listed in Table 17.1, and they vary in the number of heterozygous loci. The first genotype is heterozygous at all six loci and can be considered maximally outbred. The last genotype is homozygous at all six loci and can be considered maximally inbred. The other two genotypes are somewhere in between. Note that each six-locus genotype, regardless of how heterozygous or homozygous, contains six dominant genes and six recessive genes. Because breeding value is just the sum of the independent effects of genes, this means that each genotype has the same breeding value. Numerically, the breeding value of each genotype is $6(4) + 6(-2) = 12$.

The easiest way to determine genotypic value is to proceed locus by locus. The genotypic value of each homozygous dominant gene pair is $2(4) = 8$, and the genotypic value of each homozygous recessive pair is $2(-2) = -4$. With complete dominance, heterozygous loci have exactly the same overall value as homozygous dominant loci, so their genotypic value is 8. The second genotype in Table 17.1 contains one homozygous dominant gene pair, four heterozygous pairs, and one homozygous recessive pair. Its genotypic value is therefore $8 + 4(8) + (-4) = 36$. To find the genotype's gene combination value, simply subtract breeding value from genotypic value. In this case, $GCV = 36 - 12 = 24$.

If you study Table 17.1, it soon becomes apparent that the more homozygous gene combinations an individual has, the less its gene combination value, its genotypic value, and, ultimately, its performance. This reduction in performance is nothing more than inbreeding depression. It is a direct result of the expression of homozygous combinations of unfavorable recessive alleles. In the examples, each homozygous recessive gene pair contributes -4 units to genotypic value, at the same time taking the place of a homozygous dominant or heterozygous pair that would have contributed $+ 8$ units, resulting in an overall loss in genotypic value of 12 units. That is 12 units of inbreeding depression.

hybrid vigor or **heterosis**
An increase in the performance of hybrids over that of purebreds, most noticeably in traits like fertility and survivability.

Inbreeding depression and **hybrid vigor** or **heterosis** are two manifestations of the same phenomenon. Inbreeding depression is simply unfavorable gene combination value. Hybrid vigor is favorable gene combination value. Inbreeding depression comes from the increase in homozygosity brought on by inbreeding and the accompanying *expression of unfavorable recessive alleles occurring in homozygous combinations.* Hybrid vigor derives from the increase in heterozygosity resulting from outbreeding and the attendant *masking of the expression of unfavorable recessive alleles occurring in heterozygous combinations.*

Because inbreeding depression and hybrid vigor are functions of gene combination value and not breeding value, they cannot be inherited. The offspring of a mating between two highly inbred but unrelated individuals that suffer from inbreeding depression is not inbred at all and, in fact, should exhibit a high degree of hybrid vigor. Likewise, the offspring of a mating between two outbred but closely related individuals that enjoy considerable hybrid vigor is inbred and may show signs of inbreeding depression. Inbreeding depression and hybrid vigor are maintained in populations not through inheritance, but through mating systems designed to influence homozygosity and heterozygosity.[3]

[3]The theory explaining how hybrid vigor is lost over time and systems for maintaining hybrid vigor are discussed in Chapters 18 and 19, respectively.

Inbreeding depression and hybrid vigor are affected not only by the relative numbers of homozygous and heterozygous loci influencing a trait, but also by the degree of dominance exhibited at each locus. Table 17.2 depicts the same examples as Table 17.1 except that partial dominance is assumed. Instead of being worth +8 units of genotypic value, each heterozygous gene pair is now worth only +5 units. Note that inbreeding depression and hybrid vigor still occur and are still a function of the relative numbers of homozygous and heterozygous pairs, but their effects are less pronounced. With a lower degree of dominance, the outbred animal shows less hybrid vigor (relative to the inbred animal), and the inbred animal shows less inbreeding depression (relative to the outbred animal).

Two other scenarios are represented in Tables 17.3 and 17.4. In Table 17.3, overdominance causes each heterozygous locus to be worth +11 units of genotypic value. With such a high degree of dominance, differences among genotypes in gene combination value are especially large, resulting in large amounts of hybrid

TABLE 17.2 Effect of Increased Homozygosity (Decreased Heterozygosity) on Gene Combination Value (*GCV*) and Genotypic Value (*G*) for a Hypothetical Trait Influenced by Six Loci Exhibiting Partial Dominance[a]

Genotype	*BV*	*G*	*GCV* (*G* − *BV*)
AaBbCcDdEeFf	6(4) + 6(−2) = 12	6(5) = 30	18
AABbCcddEeFf	6(4) + 6(−2) = 12	8 + 4(5) + (−4) = 24	12
AAbbCcDDeeFf	6(4) + 6(−2) = 12	2(8) + 2(5) + 2(−4) = 18	6
AAbbCCddEEff	6(4) + 6(−2) = 12	3(8) + 3(−4) = 12	0

[a]Same simplifying assumptions as in the example shown in Table 17.1. It is also assumed that partial dominance causes each heterozygous gene pair to be worth + 5 units.

TABLE 17.3 Effect of Increased Homozygosity (Decreased Heterozygosity) on Gene Combination Value (*GCV*)) and Genotypic Value (*G*) for a Hypothetical Trait Influenced by Six Loci Exhibiting Overdominance[a]

Genotype	*BV*	*G*	*GCV* (*G* − *BV*)
AaBbCcDdEeFf	6(4) + 6(−2) = 12	6(11) = 66	54
AABbCcddEeFf	6(4) + 6(−2) = 12	8 + 4(11) + (−4) = 48	36
AAbbCcDDeeFf	6(4) + 6(−2) = 12	2(8) + 2(11) + 2(−4) = 30	18
AAbbCCddEEff	6(4) + 6(−2) = 12	3(8) + 3(−4) = 12	0

[a]Same simplifying assumptions as in the example shown in Table 17.1. It is also assumed that overdominance causes each heterozygous gene pair to be worth +11 units.

TABLE 17.4 Effect of Increased Homozygosity (Decreased Heterozygosity) on Gene Combination Value (*GCV*) and Genotypic value (*G*) for a Hypothetical Trait Influenced by Six Loci Exhibiting No Dominance[a]

Genotype	*BV*	*G*	*GCV* (*G* − *BV*)
AaBbCcDdEeFf	6(4) + 6(−2) = 12	6(2) = 12	0
AABbCcddEeFf	6(4) + 6(−2) = 12	8 + 4(2) + (−4) = 12	0
AAbbCcDDeeFf	6(4) + 6(−2) = 12	2(8) + 2(2) + 2(−4) = 12	0
AAbbCCddEEff	6(4) + 6(−2) = 12	3(8) + 3(−4) = 12	0

[a]Same simplifying assumptions as in the example shown in Table 17.1. It is also assumed that the lack of dominance causes each heterozygous gene pair to be worth 4 + (−2) = 2 units.

vigor and inbreeding depression. In Table 17.4, there is no dominance at any locus. Consequently there are no differences among genotypes in gene combination value and no hybrid vigor or inbreeding depression.

A critical assumption of the theory explaining hybrid vigor and inbreeding depression is that recessive alleles are generally unfavorable or at least less favorable than dominant alleles. It that were not the case, expression of recessive alleles would be beneficial, and performance would actually improve with inbreeding and decline with outbreeding. In reality, most recessive alleles are less favorable than their dominant counterparts, and the reason for this probably has to do with evolutionary forces.

Favorable dominant alleles have a selective advantage over favorable recessive alleles. In the first few generations after the creation (mutation) of new alleles, frequencies of these alleles are necessarily low. At low frequencies, most of them occur in heterozygotes. Dominant alleles are expressed in the heterozygote, and selection should cause the gene frequencies of favorable dominants to increase and the gene frequencies of unfavorable dominants to decrease. Recessive alleles, on the other hand, are not expressed in the heterozygote. There is therefore little selection pressure to cause frequencies of recessives to either increase or decrease. The net result is that favorable dominant alleles are more easily incorporated in a population than equally favorable recessive alleles. The dominance of favorable alleles is a product of evolution.

fitness trait
A trait selected for with natural selection. Fitness traits relate to an animal's ability to survive and reproduce.

This explains, to a degree anyway, why the traits that show the most hybrid vigor and inbreeding depression tend to be **fitness traits,** traits related to an animal's ability to survive and reproduce. Natural selection has operated on these traits for millennia. As a result, many of the favorable alleles that affect fitness traits show a high degree of dominance, and that causes more hybrid vigor and inbreeding depression.

Fitness traits and other traits that show a large amount of hybrid vigor and inbreeding depression tend to be (but are not always) lowly heritable. In contrast, highly heritable traits like those related to skeletal structure tend to show very little hybrid vigor or inbreeding depression. This makes sense if you think in terms of the components of the genetic model:

$$P = \mu + BV + GCV + E$$

Traits that exhibit substantial hybrid vigor and inbreeding depression are heavily influenced by the effects of dominance (i.e., gene combination effects). Animal performance is therefore strongly associated with gene combination value. And if performance is closely tied to gene combination value, it is likely to be much less dependent (at least in relative terms) on breeding value. The relationship between performance and breeding value is therefore weak. In other words, heritability is low.

On the other hand, traits that show little or no hybrid vigor or inbreeding depression are influenced very little by gene combination effects. Performance in these traits is much more likely to be associated with breeding value than gene combination value. That is, heritability of these traits is likely to be higher.

Even though gene combination value, the cause of hybrid vigor and inbreeding depression, has two components—dominance and epistasis—the discussion of gene combination value in this chapter has been limited to dominance alone. In fact, the hypothetical examples in Tables 17.1 through 17.4 ignore epistasis altogether. That is partly because epistasis for polygenic traits is complex and unpredictable; good examples are hard to construct. The role of epistasis has also been played down because much experimental work has shown hybrid vigor and inbreeding depression to be strongly tied to level of heterozygosity. That suggests the overriding importance of dominance. But epistasis can be important too. The hybrid vigor produced in certain crosses of chickens, for example, appears to be almost entirely due to epistasis.

Measuring Inbreeding and Relationship

inbreeding coefficient (F_X)
A measure of the level of inbreeding in an individual; (1) the probability that both genes of a pair in an individual are identical by descent; (2) the probable proportion of an individual's loci containing genes that are identical by descent.

identical by descent
Two genes are identical by descent if they are copies of a single ancestral gene.

alike in state
Genes that are alike in state function the same and have exactly or almost exactly the same chemical structure.

Wright's coefficient of relationship (R_{XY})
A measure of pedigree relationship; (1) the probable proportion of one individual's genes that are identical by descent to genes of a second individual; (2) the correlation between the breeding values of two individuals due to pedigree relationship alone.

The level of inbreeding in an individual is measured by the **inbreeding coefficient (F_X).** In precise terms, the inbreeding coefficient is the probability that both genes of a pair in an individual are **identical by descent.** Two genes in an individual are identical by descent if they are copies of a single gene inherited from an ancestor common to both parents of the individual. Being identical by descent is therefore different from simply being homozygous. Homozygous genes function the same and are likely to have the same chemical structure. They are sometimes referred to as being **alike in state.** But they are not necessarily replicates of a single ancestral gene. Genes that are identical by descent are alike in state, but genes that are alike in state may or may not be identical by descent.

Because the inbreeding coefficient is a probability, it ranges from 0 for noninbred animals to 1 for maximally inbred animals, or 0 to 100 in percentage terms. An individual with an inbreeding coefficient of .25 is said to be 25% inbred. That means that at a given locus in the individual, the probability that the two genes at that locus are identical by descent is .25. And if the probability of having two genes identical by descent at any one locus is .25, then we can expect 25% of the individual's loci to contain pairs of genes that are identical by descent. The inbreeding coefficient can therefore be equivalently defined as the probable proportion of an individual's loci containing genes that are identical by descent.

The probable proportion of an individual's loci containing genes that are *homozygous,* as opposed to identical by descent, is always something greater than the individual's inbreeding coefficient. All animals, even noninbreds, have homozygous pairs of genes. So all animals have a proportion of their loci at which genes are alike in state (homozygous but not identical by descent). Because homozygous loci include loci that are both identical by descent *and* alike in state, the proportion of an individual's loci that are homozygous is necessarily greater than the proportion of loci that are identical by descent.

The level of inbreeding of an offspring is determined by the closeness of the pedigree relationship between its parents. This is one of the reasons we are often interested in knowing how closely related two individuals are. Pedigree relationship is measured by **Wright's coefficient of relationship (R_{XY}),** named for American geneticist Sewall Wright. Wright's coefficient is defined in a similar fashion as the inbreeding coefficient. It is the probable proportion of one individual's genes that are identical by descent to genes of a second individual. As a proportion, Wright's coefficient ranges from 0 for completely unrelated animals to 1 for identical twins.

Wright's coefficient of relationship can also be defined as the correlation between the breeding values of two individuals due to pedigree relationship alone.

We expect that because relatives have genes and therefore the independent effects of those genes in common, they should have breeding values that are, on average, similar—or at least more alike than the breeding values of unrelated individuals. Wright's coefficient measures that similarity as a correlation.

The arrow diagrams in Figure 17.3 illustrate several common matings. Listed beside each mating is the inbreeding coefficient of an offspring from the mating (F_X) and Wright's coefficient of relationship between the parents (R_{SD}). Note that the closer the pedigree relationship between the parents, the higher the inbreeding in their progeny. Note also that in these examples the offspring's inbreeding coefficient is exactly half the parents' relationship coefficient. This is often the case, but not always. If at least one parent is itself inbred, the offspring's inbreeding coefficient may be somewhat greater than half the parents' relationship coefficient.

The mating shown in the bottom right of Figure 17.3 is the mating of inbred but unrelated individuals. Because the parents are unrelated, their offspring is not inbred, even though the parents are. Each parent has pairs of genes that are identical by descent, but there is no way that the offspring can have pairs of identical genes if its parents have no ancestor in common. This example shows how inbreeding can be "undone" by a single mating of unrelated animals.

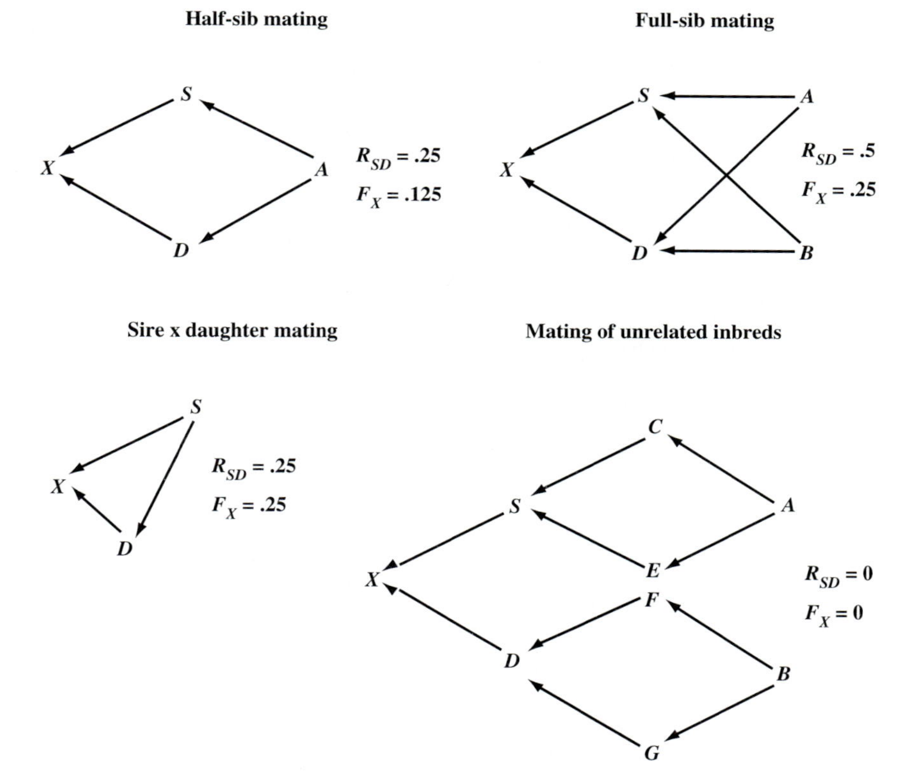

FIGURE 17.3 Arrow diagrams depicting several common matings. Also listed are the inbreeding coefficient of an offspring from each mating and Wright's coefficient of relationship between the parents.

The Inbreeding Formula

The formula for the inbreeding coefficient of animal X is:

$$F_X = \sum_{CA=1}^{k} \left(\frac{1}{2}\right)^{n_1+n_2+1}(1 + F_{CA})$$

where CA = a common ancestor of the sire and dam of X

k = the number of common ancestors in X's pedigree

n_1 = the number of generations separating the common ancestor from the sire of X

n_2 = the number of generations separating the common ancestor from the dam of X

F_{CA} = the inbreeding coefficient of the common ancestor

Proof by Example

Consider the pedigree (a) and corresponding arrow diagram (b) in Figure 17.4. To demonstrate how the inbreeding formula works, let's simulate the flow of genes from the lone common ancestor (A) to the inbred individual (X). Because the inbreeding coefficient is defined as the probability that both genes of a pair in an individual are identical by descent, we need only consider one locus—let's use the J locus—in our simulation.

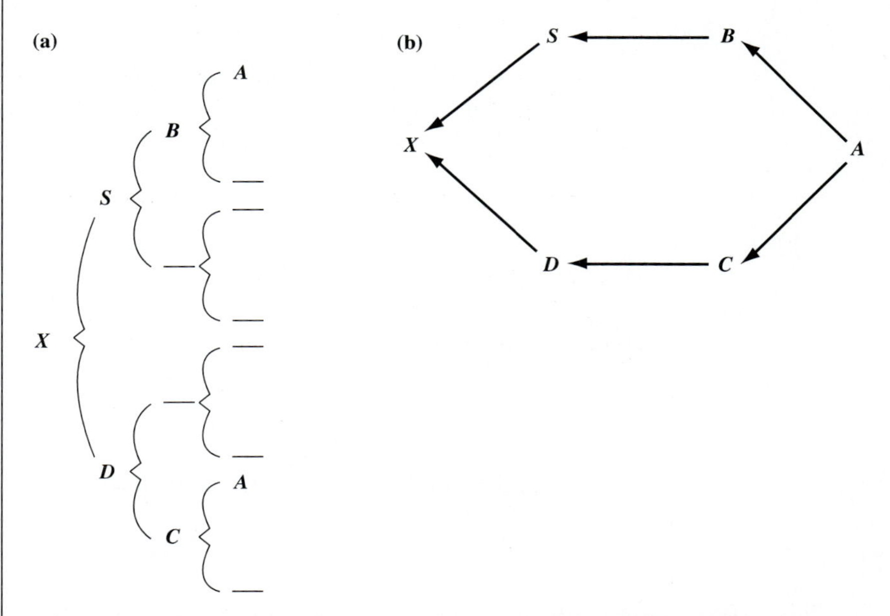

FIGURE 17.4 Inbred pedigree and corresponding arrow diagram. The blanks in the pedigree represent unrelated animals that do not contribute to the inbreeding of *X*.

Common ancestor A has two genes at the J locus: J' and J''. They may be two different alleles, or they may be the same allele. They may even be genes that are identical by descent. We use different notation for these genes simply to identify each one. A transmits a gene to B and a gene to C, so the following events are equally likely:

1. Both B and C receive J'.
2. Both B and C receive J''.
3. B receives J' and C receives J''.
4. B receives J'' and C receives J'.

Chances are ²⁄₄ or ½ that B and C receive the same gene from A. Chances are also ²⁄₄ or ½ that B and C receive different genes from A but, in that case, there is still the probability F_A that J' and J'' are identical. The overall probability that B and C receive identical genes from A is then

$$\frac{1}{2} + \frac{1}{2}F_A = \frac{1}{2}(1 + F_A)$$

The probability that a particular gene is transmitted from B to S is ½, from S to X is ½, and so from B to X is ½ · ½ = $(½)^2$. Likewise, the probability that a particular gene is transmitted from C to X is ½ · ½ = $(½)^2$. The overall probability that X receives identical alleles from its common ancestor is then

$$F_X = \frac{1}{2}(1 + F_A)\left(\frac{1}{2}\right)^2\left(\frac{1}{2}\right)^2$$

$$= \left(\frac{1}{2}\right)^5(1 + F_A)$$

If we define n_1 to be the number of generations separating A from the sire of X, and n_2 to be the number of generations separating A from the dam of X, then $n_1 = n_2 = 2$. By substitution,

$$F_X = \left(\frac{1}{2}\right)^5(1 + F_A)$$

$$= \left(\frac{1}{2}\right)^{n_1+n_2+1}(1 + F_A)$$

Generalizing the formula for k common ancestors,

$$F_X = \sum_{CA=1}^{k}\left(\frac{1}{2}\right)^{n_1+n_2+1}(1 + F_{CA})$$

The Formula for Wright's Coefficient of Relationship

The formula for Wright's coefficient of relationship between individuals X and Y is:

$$R_{XY} = \frac{\sum_{CA=1}^{k}\left(\frac{1}{2}\right)^{n_1+n_2}(1 + F_{CA})}{\sqrt{1 + F_X}\sqrt{1 + F_Y}}$$

In the relationship formula, n_1 and n_2 are defined somewhat differently than in the inbreeding formula. Now n_1 is the number of generations separating the common ancestor from X—not X's parents—and n_2 is the number of generations separating the common ancestor from Y.

If you look closely, the relationship formula looks something like the formula for a correlation: a ratio with a product of square roots in the denominator. That is no coincidence. Recall that Wright's coefficient is defined as the correlation between the breeding values of two individuals due to pedigree relationship alone. We could write the formula in a way that would make the connection between Wright's coefficient and the correlation between breeding values of relatives even more clear:

$$R_{XY} = \frac{\left(\sum_{CA=1}^{k}\left(\frac{1}{2}\right)^{n_1+n_2}(1 + F_{CA})\right)\sigma^2_{BV}}{\sqrt{(1 + F_X)\sigma^2_{BV}}\sqrt{(1 + F_Y)\sigma^2_{BV}}}$$

Expressed this way, the numerator of the formula becomes the covariance of breeding values of two individuals due to pedigree relationship alone. The variances in numerator and denominator cancel each other, of course, and what remains of the numerator $\left(\sum_{CA=1}^{k}\left(\frac{1}{2}\right)^{n_1+n_2}(1 + F_{CA})\right)$ is a computationally handy quantity known as *numerator relationship*.[4]

[4]See the boxed section later in this chapter on the tabular method for calculating inbreeding and relationship coefficients for an application of numerator relationship.

Calculating Inbreeding and Relationship Coefficients

There are several ways to calculate inbreeding and relationship coefficients. The method of choice depends on the number of coefficients to be determined, the size and complexity of the pedigrees, and available computer hardware and software.

Calculating F_X and R_{XY} Using the Path Method

path method
A method for calculating inbreeding and relationship coefficients that simulates the "paths" taken by identical genes as they flow from ancestors to descendants.

If you want to calculate just one or a few inbreeding or relationship coefficients, and if the pedigrees involved are relatively simple (i.e., fairly limited in the number of generations listed and the number of common ancestors), then the **path method** is appropriate. It can be done by hand; no computer is needed. The path method is tricky, requires careful adherence to a strict set of rules, and takes practice. One advantage of the path method, however, is that it follows directly from the definitions of inbreeding and relationship coefficients. It requires you to simulate on paper the "paths" taken by identical genes as they flow from common ancestors to parents of an inbred offspring or to individuals whose relationship is to be determined.

Steps and Rules for Calculating Inbreeding and Relationship Coefficients Using the Path Method

Using the path method to calculate inbreeding and relationship coefficients can be a fun exercise, but you must follow a sequence of steps, and you need to obey a few rules in determining each pair of paths. First the steps:

Step 1. Convert the pedigree to an arrow diagram in which each individual appears only once.

Step 2. Locate common ancestors. For inbreeding calculations, common ancestors are defined to be common to both the sire and dam of the inbred individual (X). In other words, they appear in both the top (paternal) half and bottom (maternal) half of the original pedigree. For relationship calculations, common ancestors are defined to be common to the two individuals of interest (X and Y).

Step 3. Locate inbred common ancestors and calculate the inbreeding coefficient for each. That is, for each inbred common ancestor, go through the entire process outlined here beginning at Step 2. (Step 1 should be done already). If you are calculating R_{XY}, you need to compute the inbreeding coefficients for individuals X and Y as well.

Step 4. Fill in the following table:

Common Ancestor	Paths	$(\frac{1}{2})^{n_1+n_2+1}$ for F_X or $(\frac{1}{2})^{n_1+n_2}$ for R_{XY}	$1 + F_{CA}$	Product of Last Two Columns

Step 5. Sum the last column to compute either F_X or the numerator of R_{XY}.

Step 6. (R_{XY} calculations only) Divide the sum by $\sqrt{1 + F_X}\sqrt{1 + F_Y}$.

The tricky part of calculating F_X or R_{XY} is determining the paths in Step 4. The idea here is to identify every unique pair of paths by which identical genes can be transmitted from the common ancestor to the sire and dam of X (for calculating F_X) or to X and Y (for calculating R_{XY}). Obey the following rules for determining paths:

Rule 1. Each pair of paths must start from the common ancestor and, always following the direction of the arrows in the diagram, proceed to the parents of X (for calculating F_X) or to X and Y (for calculating R_{XY}). There can be no backtracking for the obvious reason that offspring do not transmit genes to their parents.

Rule 2. A pair of paths may not include the same individual (with the exception of the common ancestor). The purpose of this rule is to avoid double counting. If both paths of a pair converge on the same individual, that individual is itself a common ancestor and should be treated as such. This rule is sometimes referred to as the "no bottleneck" rule.

Rule 3. In certain situations an animal is considered to be its own common ancestor (separated from itself by zero generations). You encounter a situation like this when you calculate the inbreeding coefficient of the offspring of a sire × daughter mating. The sire is the common ancestor. Similarly, if you are calculating the relationship between an ancestor and descendant—for example, a grandparent and grandoffspring—the ancestor is its own common ancestor.

Example

Consider the pedigree in Figure 17.5(a). Let's first calculate F_X, the inbreeding coefficient for individual X.

Step 1. Convert the pedigree to an arrow diagram.

The appropriate arrow diagram is shown in Figure 17.5(b).

Step 2. Locate common ancestors of the sire and dam of X.

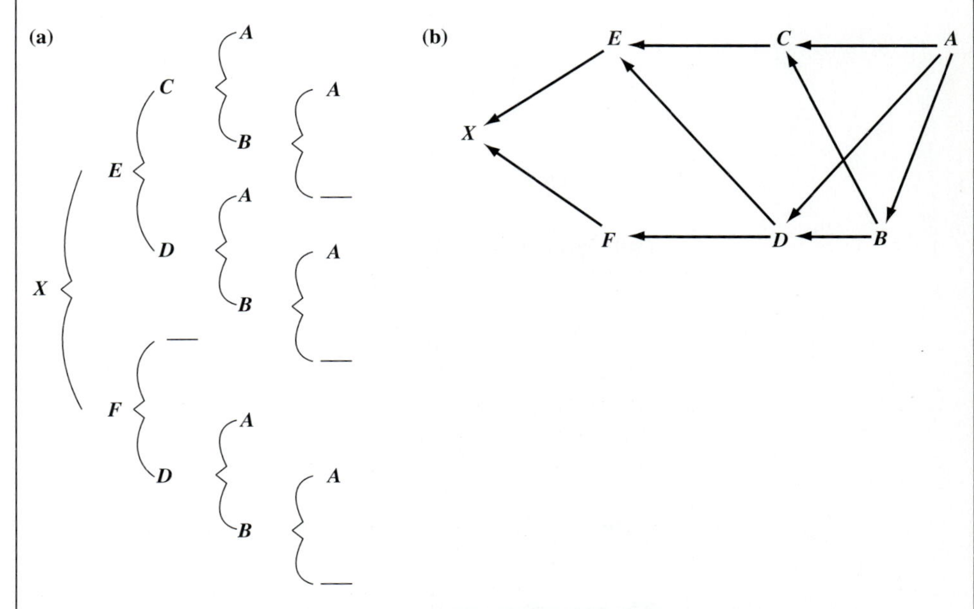

FIGURE 17.5 Inbred pedigree and corresponding arrow diagram for the example.

Common ancestors of E and F are A, B, and D.

Step 3. Locate inbred common ancestors and calculate the inbreeding coefficient for each.

D is the only inbred common ancestor—the product of a sire × daughter mating. C is similarly inbred but is not a common ancestor, so we need not concern ourselves with him. The easy way to calculate D's inbreeding coefficient is simply to recall that the inbreeding coefficient for the offspring of a sire × daughter mating is .25 (assuming no further complications in the offspring's pedigree). To calculate F_D explicitly, we must go back to Step 2 and proceed as if D were the primary individual of interest.

The portion of the arrow diagram representing D's pedigree appears as:

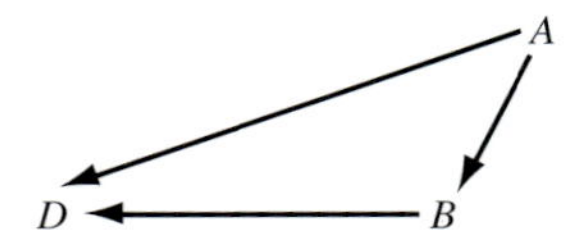

This mating is one for which we can apply Rule 3, the rule stating that an animal can be considered its own common ancestor (separated from itself by zero generations). In this case, A is the sire of the animal whose inbreeding we want to calculate (D). A is also the lone common ancestor of D's parents (A and B). Thus

$$F_D = \left(\frac{1}{2}\right)^{n_1+n_2+1}(1 + F_A)$$
$$= \left(\frac{1}{2}\right)^{0+1+1}(1 + 0)$$
$$= .25$$

Step 4. Fill in the table.

Path Information for Calculating F_X in the Example

Common Ancestor	Paths	$(\frac{1}{2})^{n_1+n_2+1}$	$1+F_{CA}$	Product of Last Two Columns
A	$E \leftarrow C \leftarrow \mathbf{A} \rightarrow D \rightarrow F$	$(\frac{1}{2})^5$	$1 + 0$	$(\frac{1}{2})^5$
	$E \leftarrow C \leftarrow \mathbf{A} \rightarrow B \rightarrow D \rightarrow F$	$(\frac{1}{2})^6$	$1 + 0$	$(\frac{1}{2})^6$
	$E \leftarrow C \leftarrow B \leftarrow \mathbf{A} \rightarrow D \rightarrow F$	$(\frac{1}{2})^6$	$1 + 0$	$(\frac{1}{2})^6$
B	$E \leftarrow C \leftarrow \mathbf{B} \rightarrow D \rightarrow F$	$(\frac{1}{2})^5$	$1 + 0$	$(\frac{1}{2})^5$
D	$E \leftarrow \mathbf{D} \rightarrow F$	$(\frac{1}{2})^3$	$1 + (\frac{1}{2})^{0+1+1} = 1.25$	$1.25(\frac{1}{2})^3$

There are three pairs of paths leading from common ancestor A to the parents of X—three unique ways in which copies of a single gene could be transmitted from A to X. Likewise there is one pair of paths leading from common ancestor B to X's parents and one pair leading from common ancestor D to X's parents.

Note that the following pairs of paths are *not* legitimate:

$$E \leftarrow C \leftarrow A \leftarrow \mathbf{B} \rightarrow D \rightarrow F$$

$$E \leftarrow D \leftarrow \mathbf{A} \rightarrow B \rightarrow D \rightarrow F$$

The first pair of paths violates Rule 1. It neglects the direction of the arrow between A and B, suggesting that B could somehow transmit genes to her sire (A). The second pair violates Rule 2, creating a bottleneck by including D in both paths of the pair.

Each generation in a path is represented by an arrow. So to determine $n_1 + n_2$, just count the arrows in each path. (Of course, if an arrow happens to represent the number of generations separating an individual and itself (zero), do not count it.) The exponent in the third column of the table is just the number of arrows in the paths plus one.

Step 5. Sum the last column.

$$F_X = \left(\frac{1}{2}\right)^5 + \left(\frac{1}{2}\right)^6 + \left(\frac{1}{2}\right)^6 + \left(\frac{1}{2}\right)^5 + 1.25\left(\frac{1}{2}\right)^3$$
$$= .25$$

Note that the largest contribution to the inbreeding of X comes from the last pair of paths—the half-sib mating of E and F involving an inbred common ancestor (D). This illustrates a useful concept in inbreeding: Inbred matings appearing near the front of a pedigree have more effect on an individual's inbreeding coefficient than inbred matings appearing further back in the pedigree.

Now let's calculate R_{CD}, Wright's coefficient of relationship between C and D.

Step 1. Convert the pedigree to an arrow diagram.

Done already.

Step 2. Locate common ancestors of C and D.

Common ancestors of C and D are A and B.

Step 3. Locate inbred common ancestors, calculate the inbreeding coefficient for each, and compute the inbreeding coefficients for C and D as well.

Neither A nor B is inbred. C and D are both the product of a sire × daughter mating. Thus

$$F_C = F_D = .25$$

Step 4. Fill in the table.

Path Information for Calculating R_{CD} in the Example

Common Ancestor	Paths	$(\frac{1}{2})^{n_1+n_2}$	$1 + F_{CA}$	Product of Last Two Columns
A	$C \leftarrow \mathbf{A} \rightarrow D$	$(\frac{1}{2})^2$	$1+0$	$(\frac{1}{2})^2$
	$C \leftarrow \mathbf{A} \rightarrow B \rightarrow D$	$(\frac{1}{2})^3$	$1+0$	$(\frac{1}{2})^3$
	$C \leftarrow B \leftarrow \mathbf{A} \rightarrow D$	$(\frac{1}{2})^3$	$1+0$	$(\frac{1}{2})^3$
B	$C \leftarrow \mathbf{B} \rightarrow D$	$(\frac{1}{2})^2$	$1+0$	$(\frac{1}{2})^2$

Step 5. Sum the last column to compute the numerator of R_{CD}.

$$\text{Numerator of } R_{CD} = \left(\frac{1}{2}\right)^2 + \left(\frac{1}{2}\right)^3 + \left(\frac{1}{2}\right)^3 + \left(\frac{1}{2}\right)^2$$
$$= .75$$

Step 6. Divide the sum by $\sqrt{1 + F_C}\sqrt{1 + F_D}$.

$$R_{CD} = \frac{.75}{\sqrt{1 + F_C}\sqrt{1 + F_D}}$$
$$= \frac{.75}{\sqrt{1 + .25}\sqrt{1 + .25}}$$
$$= .6$$

Calculating F_X and R_{XY} Using the Tabular Method

tabular method
A method for calculating inbreeding and relationship coefficients involving construction and updating of a table relating all members of a population.

base population
The population of animals whose parents are either unknown or ignored for the purposes of inbreeding and relationship calculation—typically the individuals appearing at the back of the pedigrees of the original animals in a herd or flock.

The **tabular method** is a better choice than the path method if you want to calculate inbreeding and relationship coefficients for an entire herd, flock, or kennel. It involves building a table of relationships beginning with members of a **base population,** the population of animals whose parents are either unknown or ignored for the purposes of inbreeding and relationship calculation. These are typically the individuals appearing at the back of the pedigrees of the original animals in a herd or flock. The relationship table is systematically built generation by generation. Once the table is current, it is a relatively simple matter to update it by adding new offspring as they arrive. For all individuals included in the table, pedigree relationships and inbreeding coefficients for offspring of any mating (or offspring of any *prospective* mating) can be determined using table values and a little arithmetic. An additional virtue of the tabular method is that it is fairly easy to computerize.

The concept of a base population is important regardless of the method used to calculate inbreeding and relationship coefficients. That is because these coefficients measure inbreeding and relationship *relative to the level of inbreeding and relationship present in the base population.* For example, if an individual's inbreeding coefficient is determined to be .18, that individual is considered to be 18% inbred relative to members of the base population—the animals at the back of the pedigree. If we could extend pedigrees back further generations (or if older pedigree data were available, and we simply chose to use them), we would probably find that many animals in the previously defined base population were related and even inbred. Accounting for this additional relationship, our 18% inbred individual will turn out to be even more inbred than we thought. However, either by necessity or for simplicity, we assume that base population animals are unrelated. Our individual is then 18% inbred relative to them.

The assumption that members of the base population are unrelated can sometimes make it difficult to compare inbreeding and relationship coefficients across populations. Many breeds, for example, underwent rather severe inbreeding during breed formation. Inbreedings and relationships calculated from modern pedigrees in these breeds significantly underestimate true levels of inbreeding and relationship. In contrast, newly created breeds with complete pedigree data do not have this problem.

Rules and Steps for Calculating Inbreeding and Relationship Coefficients Using the Tabular Method

A relationship table is, in fact, a table of *numerator relationship*—the numerator of Wright's coefficient of relationship. This quantity is especially useful because it is both simple to compute and easy to convert to both Wright's coefficient and the inbreeding coefficient. It is simple to compute because of the following rules:

Rule 1. The numerator relationship between individuals X and Y equals the average of numerator relationships between X and the parents of Y (or between Y and the parents of X). Mathematically, if S and D are the parents of Y, and r_{XY} denotes the numerator relationship between X and Y, then

$$r_{XY} = \frac{1}{2}r_{XS} + \frac{1}{2}r_{XD}$$

Thus, if you know the numerator relationships between animals in earlier generations, you can easily calculate more recent relationships.

Rule 2. The numerator relationship between an individual and itself is 1 plus its inbreeding coefficient. Mathematically,

$$r_{XX} = 1 + F_X$$

To calculate an inbreeding coefficient from numerator relationships, apply Rule 3.

Rule 3. An individual's inbreeding coefficient equals half the numerator relationship between its parents. Mathematically,

$$F_X = \frac{1}{2}r_{SD}$$

where S and D are the parents of X.

Proof

$$F_X = \sum_{CA=1}^{k} \left(\frac{1}{2}\right)^{n_1+n_2+1}(1 + F_{CA})$$

and

$$r_{SD} = \sum_{CA=1}^{k} \left(\frac{1}{2}\right)^{n_1+n_2}(1 + F_{CA})$$

Because S and D are parents of X, the ns in both formulas refer to the same generations. Therefore,

$$\begin{aligned} F_X &= \sum_{CA=1}^{k} \left(\frac{1}{2}\right)^{n_1+n_2+1}(1 + F_{CA}) \\ &= \sum_{CA=1}^{k} \left(\frac{1}{2}\right)\left(\frac{1}{2}\right)^{n_1+n_2}(1 + F_{CA}) \\ &= \frac{1}{2}\sum_{CA=1}^{k} \left(\frac{1}{2}\right)^{n_1+n_2}(1 + F_{CA}) \\ &= \frac{1}{2}r_{SD} \end{aligned}$$

To compute Wright's coefficient of relationship, apply Rule 4.

Rule 4. Wrights coefficient of relationship between individuals X and Y equals the numerator relationship between them divided by $\sqrt{1 + F_X}\sqrt{1 + F_Y}$. Mathematically,

$$R_{XY} = \frac{r_{XY}}{\sqrt{1 + F_X}\sqrt{1 + F_Y}}$$

The proof follows directly from the formula for Wright's coefficient.

Like the path method, the tabular method for calculating inbreeding and relationship coefficients involves a number of steps:

Step 1. Order all animals by birth date, earliest to latest. The order need not be perfect so long as no offspring comes before its parents.

Step 2. Create a square table listing all animals (in order) across the top and again down the left side.

Step 3. Write in the parents of each animal above it at the top of the table. If a parent is unknown, write in a blank (_).

Step 4. Proceeding from left to right, fill in the first row of numerator relationships. The very first element is the diagonal element $(1+F_X)$ for the first animal. This individual is a member of the base population—parents are unknown—so its inbreeding coefficient is presumably zero and its diagonal element is therefore 1. Likewise, numerator relationships among members of the base population are presumably zero. (If you happen to know otherwise, you can fill in known relationships.) Once beyond the base population, use Rule 1 to calculate numerator relationships.

Step 5. Copy the values in the first row of the table to the first column of the table.

Step 6. Fill in the next row of numerator relationships. Use Rule 1 for off-diagonal elements and Rules 2 and 3 for diagonal elements.

Step 7. Copy the values from this row into the corresponding column.

Step 8. Repeat Steps 6 and 7 until the table is complete.

Step 9. To determine an individual's inbreeding coefficient from the completed table, find the individual's diagonal element and subtract 1.

Step 10. To determine Wrights coefficient of relationship between individuals X and Y from the completed table, find the appropriate off-diagonal element and divide by $\sqrt{1 + F_X}\sqrt{1 + F_Y}$.

Example

Let's recalculate F_X and R_{CD} from the previous problem (Figure 17.5), this time using a numerator relationship table.

Step 1. Order all animals by birth date, earliest to latest.

The appropriate order is *A, B, C, D, E, F, X*.

Step 2. Create a table listing all animals across the top and down the left side.

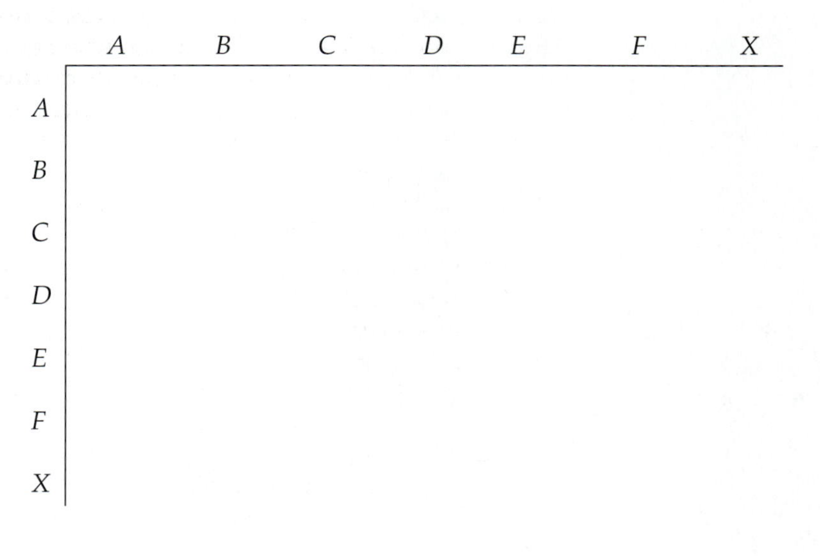

Step 3. Write in the parents of each animal above it at the top of the table.

	_ _	A _	AB	AB	CD	_ D	EF
	A	B	C	D	E	F	X
A							
B							
C							
D							
E							
F							
X							

Step 4. Proceeding from left to right, fill in the first row of numerator relationships.

In this pedigree, individual A is the only true member of the base population—he is the only one with two unknown parents. Without additional information, we must assume he is not inbred. From Rule 2, his diagonal element is

$$\begin{aligned} r_{AA} &= 1 + F_A \\ &= 1 + 0 \\ &= 1 \end{aligned}$$

We can use Rule 1 to compute the remaining elements in the first row. Thus

$$\begin{aligned} r_{AB} &= \frac{1}{2}r_{AA} + \frac{1}{2}r_{A\_} \\ &= \frac{1}{2}(1) + \frac{1}{2}(0) \\ &= .5 \\ r_{AC} &= \frac{1}{2}r_{AA} + \frac{1}{2}r_{AB} \\ &= \frac{1}{2}(1) + \frac{1}{2}(.5) \\ &= .75 \end{aligned}$$

and so on.

	_ _ A	A _ B	AB C	AB D	CD E	_ D F	EF X
A	1	.5	.75	.75	.75	.375	.5625
B							
C							
D							
E							
F							
X							

Step 5. Copy the values in the first row of the table to the first column of the table.

	_ _ A	A _ B	AB C	AB D	CD E	_ D F	EF X
A	1	.5	.75	.75	.75	.375	.5625
B	.5						
C	.75						
D	.75						
E	.75						
F	.375						
X	.5625						

Step 6. Fill in the next row of numerator relationships.

	_ _ A	A _ B	AB C	AB D	CD E	_ D F	EF X
A	1	.5	.75	.75	.75	.375	.5625
B	.5	1	.75	.75	.75	.375	.5625
C	.75						
D	.75						
E	.75						
F	.375						
X	.5625						

Step 7. Copy the values from this row into the corresponding column.

	_ _ A	A _ B	AB C	AB D	CD E	_ D F	EF X
A	1	.5	.75	.75	.75	.375	.5625
B	.5	1	.75	.75	.75	.375	.5625
C	.75	.75					
D	.75	.75					
E	.75	.75					
F	.375	.375					
X	.5625	.5625					

Step 8. Repeat Steps 6 and 7 until the table is complete.

Off-diagonal elements are computed using Rule 1. Diagonal elements require Rules 2 and 3. For example,

$$r_{CC} = 1 + F_C$$
$$= 1 + \frac{1}{2}r_{AB}$$
$$= 1 + \frac{1}{2}(.5)$$
$$= 1.25$$

Altogether,

	__ A	A_ B	AB C	AB D	CD E	_D F	EF X
A	1	.5	.75	.75	.75	.375	.5625
B	.5	1	.75	.75	.75	.375	.5625
C	.75	.75	1.25	.75	1	.375	.6875
D	.75	.75	.75	1.25	1	.5	.75
E	.75	.75	1	1	1.375	.5	.9375
F	.375	.375	.375	.5	.5	1	.75
X	.5625	.5625	.6875	.75	.9375	.75	1.25

Step 9. To determine an individual's inbreeding coefficient, find the individual's diagonal element and subtract 1.

$$F_X = r_{XX} - 1$$
$$= 1.25 - 1$$
$$= .25$$

Step 10. To determine Wright's coefficient of relationship between X and Y, find the appropriate off-diagonal element and divide by $\sqrt{1 + F_X}\sqrt{1 + F_Y}$.

$$R_{CD} = \frac{r_{CD}}{\sqrt{1 + F_C}\sqrt{1 + F_D}}$$
$$= \frac{.75}{\sqrt{1 + .25}\sqrt{1 + .25}}$$
$$= .6$$

Calculating F_X and R_{XY} Using BLUP and Large-Scale Genetic Evaluation

Since the arrival of BLUP methodology and large-scale genetic evaluation, it has become possible to calculate inbreeding and relationship coefficients using the enormous pedigree files available from field data. Inbreeding coefficients are actually a by-product of animal model BLUP analyses. Coefficients or relationship are not, but they can be determined by setting up "phantom matings" of the individuals whose relationship is desired.

The advantage of using BLUP in the context of large-scale genetic evaluation is that it includes relationships between animals in many different herds or flocks. This is especially important when germ plasm is commonly exchanged between herds or flocks. The disadvantage of this approach is that the actual calculations can only be performed by those with access to entire sets of field data—typically university, breed association, or government personnel. Those rare individual breeders that use BLUP software programs for within-herd or within-flock genetic prediction may find it easier to calculate inbreedings and relationships this way than to use a relationship table.

Linebreeding

linebreeding
The mating of individuals within a particular line; a mating system designed to maintain a substantial degree of relationship to a highly regarded ancestor or group of ancestors without causing high levels of inbreeding.

Inbreeding generally has a bad reputation. A mild form of inbreeding called **linebreeding** does not. Linebreeding is the mating of individuals within a particular line. It is a mating system designed to maintain a substantial degree of relationship to a highly regarded ancestor or group of ancestors without causing high levels of inbreeding.

An example of a linebred pedigree is shown in Figure 17.6. Animal X is said to be linebred to ancestor A. Ancestor A (now long since deceased) was presumably an outstanding individual whose genes were worth preserving in future generations. A appears so often in X's pedigree that the relationship between X and A (R_{XA}) is .47, practically the same as the relationship between a parent and its offspring (.5). However, because very close matings have been carefully avoided—no mates shown in the pedigree are more closely related than half sibs—X is only moderately inbred (F_X = .125).

Due to the absence of very closing matings, linebreeding is a slow form of inbreeding. This allows time for selection to offset some of the adverse effects of inbreeding (i.e., inbreeding depression). If the poor performing individuals from a linebred population are systematically weeded out through selection, inbreeding depression may not be as apparent as it would otherwise.

Reasons to Inbreed

There are a number of reasons to inbreed. For one, inbreeding can be used to identify deleterious recessive alleles in a population. A more common reason cited by breeders of registered purebred animals is that they are required to inbreed, at least to a degree, because offspring of matings to nonpurebred stock are often not eligible for registration. The two most important genetic reasons for inbreeding, however, are (1) to increase uniformity and (2) to create an opportunity for hybrid vigor.

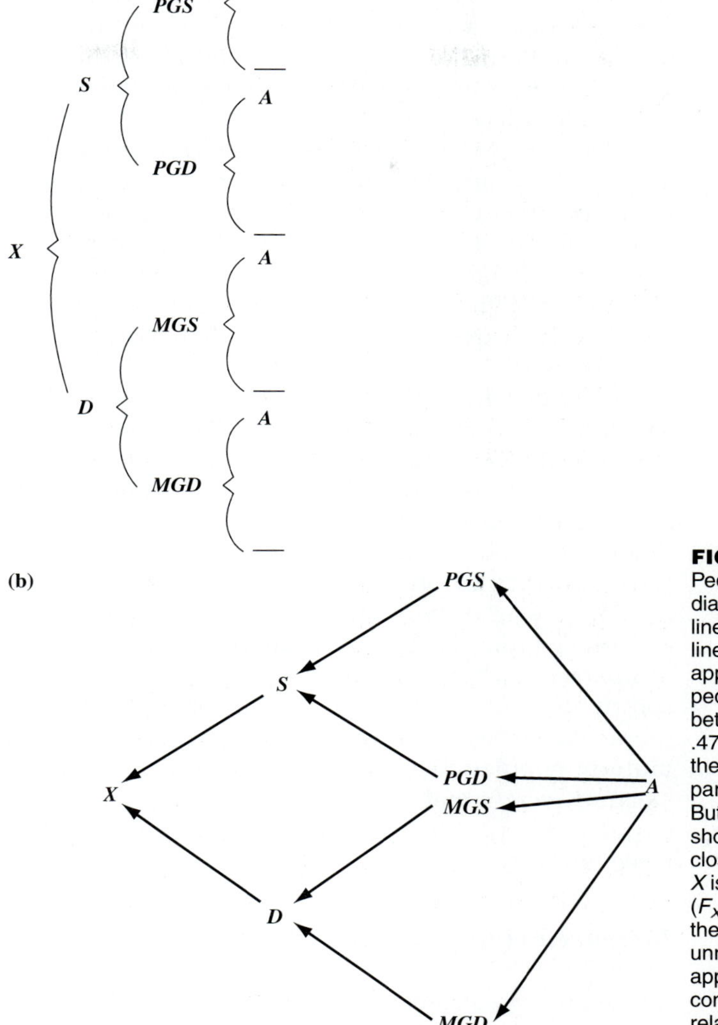

FIGURE 17.6
Pedigree (*a*) and arrow diagram (*b*) illustrating linebreeding. Animal *X* is linebred to ancestor *A*. *A* appears so often in *X*'s pedigree that the relationship between *X* and *A* (R_{XA}) is .47, practically the same as the relationship between a parent and its offspring (.5). But because no mating shown in the pedigree is closer than a half-sib mating, *X* is only moderately inbred (F_X = .125). (The blanks in the pedigree indicate unrelated individuals that appear only once and do not contribute to inbreeding or relationship.)

Increased Uniformity

In most domestic species, inbreeding has historically been used as a way of increasing uniformity within a breed. This has been particularly true for simply-inherited traits like coat and feather color, traits that have a very noticeable effect on an animal's visual appearance. By increasing homozygosity and thus allowing recessive alleles to express themselves, inbreeding enhances the ability of selection to reduce the frequency of or even eliminate from a population unwanted alleles for coat color, feather color, presence of horns, etc.

The effect of inbreeding on uniformity of polygenic traits is less clear. In theory, related animals should be more uniform than unrelated animals because they have many of the same genes in common. They have "similar" breeding values. Breeders sometimes choose to linebreed in order to enhance this similarity by increasing the pedigree relationship among individuals within a population. Such a strategy can, in fact, work. At very high levels of inbreeding, inbred lines are genetically uniform. Members of *maximally* inbred lines (for example, inbred lines of laboratory mice) are the equivalent of identical twins. There is no genetic variation within such lines.

At intermediate levels of inbreeding, however, inbreds may or may not be genetically less variable than noninbreds for polygenic traits. Genetic variability within intermediately inbred populations depends on such factors as the number of loci affecting a trait, the importance of gene combination effects, gene frequencies (particularly of recessive alleles), and whether some loci have major effects. Genetic variability for polygenic traits is generally thought to decrease with inbreeding, but experimental work has shown cases where genetic variability actually increased.

Research also indicates that inbreds are typically more variable than noninbreds in their response to environmental influences. In general, they seem to be more sensitive to stress. For example, the milk production of young, inbred beef cows suffers more in drought years than the milk production of their noninbred counterparts. Such sensitivity tends to increase variation that is environmental in origin. So even if inbreds are genetically more uniform than noninbreds for a polygenic trait, they are often not phenotypically more uniform.

F_1 crosses of two inbred populations are truly uniform, both genetically and phenotypically. That is because their parents, being inbred and therefore comparatively homozygous, are more consistent in the kinds of gametes they contribute, and also because the F_1s, being hybrids, are less sensitive to environmental stresses. For traits with no significant maternal component, the most uniform population is a population of first-cross individuals. For traits such as weaning weight that have an important maternal component, the most uniform population probably consists of F_1 offspring out of F_1 dams (i.e., three-breed or three-line crosses).[5]

Increased Hybrid Vigor

Inbreeding can also be used to increase hybrid vigor—not in the inbreds themselves, but in crosses of inbreds. As you will see in the upcoming section on outbreeding, consistently exceptional hybrid vigor can only come from crossing inbred lines.

Plant breeders (for example, the companies that produce hybrid corn) routinely inbreed a number of lines, then test the crosses of these lines for yield and other performance traits. This technique was once popular among breeders of layers as well. In most domestic animal species, however, the deliberate use of inbreeding to enhance hybrid vigor in crosses in rare. Inbreeding still occurs, though. The inbred lines are simply established breeds, and rather than calling the mating

[5]Historically, the term "F_1" referred to the first cross of two *purebred* populations. More recently, it has taken on a broader meaning, signifying the first cross of two *unrelated* populations whether they are purebred or not. Thus, an F_1 dam can produce an F_1 offspring if she is mated to a sire of a third, unrelated breed, and we can use descriptions of animals like "F_1 four-breed composites." (See Chapter 19 for a description of composite breeds.)

purebreeding or **straightbreeding** The mating of purebreds of the same breed.

of purebreds of the same breed *inbreeding*, we call it **purebreeding** or **straightbreeding.** Most breeds, particularly those that were established long ago, are considerably inbred, and by crossing these breeds we can obtain substantial levels of hybrid vigor.

Inbreeding and Industry Structure

In those livestock species in which crossbreeding is prevalent, inbreeding (purebreeding) is practiced mostly by seedstock producers. Seedstock producers sell breeding animals, but what they really market are breeding value (in the larger sense of the term, which includes breed complementarity) and, in many cases, the ability of their animals to generate hybrid vigor when crossed with animals of other breeds. The latter product can only come from purebreeding. The purebreds themselves may suffer to a degree from inbreeding depression, especially if they are raised in stressful environments. But because seedstock producers do not sell phenotypic value in the form of meat, milk, wool, or eggs per se, and because inbreeding depression disappears with crossbreeding, these breeders are free to inbreed.

OUTBREEDING

outbreeding or **outcrossing** The mating of unrelated individuals.

Outbreeding or **outcrossing** is the opposite of inbreeding. It is the mating of unrelated individuals. Because no animals within a population are completely unrelated, a more technically correct definition of outbreeding is the mating of individuals more distantly related than average for the population. Any mating involving essentially unrelated individuals can be considered an outbred mating but, as a mating strategy, outbreeding more commonly refers to *crossbreeding* (the mating of sires of one breed or breed combination to dams of another breed or breed combination) or *linecrossing* (the mating of sires of one line or line combination to dams of another line or line combination).

Effects of Outbreeding

Just as the primary effect of inbreeding is an increase in homozygosity, the primary effect of outbreeding is an increase in heterozygosity. And just as all the other effects of inbreeding stem from the increase in homozygosity, all (or almost all) other effects of outbreeding result from the increase in heterozygosity.

The increase in heterozygosity and attendant decrease in homozygosity caused by outbreeding is illustrated by the Punnett square in Figure 17.7. The figure shows the potential outcomes at six unlinked loci of mating two inbred but unrelated animals. The parental genotypes are *AABbccDDEeff* and *AabbCCddEEFf*. We can infer that the parents are inbred because of their homozygosity—four out of the six loci shown for each individual are homozygous. Only two loci are heterozygous. We can infer that the animals are unrelated because at loci where one is homozygous for dominant alleles, the other tends to be homozygous for recessive alleles and vice versa. The offspring from this mating are outbred, averaging four heterozygous loci. Some are even heterozygous at all six loci. (The small number in the corner of each cell is the number of heterozygous loci in that particular genotype.)

Parental Genotypes ♂ *AABbccDDEeff*
(each heterozygous at two loci): ♀ *AabbCCddEEFf*

♂ \ ♀	*AbCdEF*	*AbCdEf*	*abCdEF*	*abCdEf*
ABcDEf	4 *AABbCc DdEEFf*	3 *AABbCc DdEEff*	5 *AaBbCc DdEEFf*	4 *AaBbCc DdEEff*
ABcDef	5 *AABbCc DdEeFf*	4 *AABbCc DdEeff*	6 *AaBbCc DdEeFf*	5 *AaBbCc DdEeff*
AbcDEf	3 *AAbbCc DdEEFf*	2 *AAbbCc DdEEff*	4 *AabbCc DdEEFf*	3 *AabbCc DdEEff*
AbcDef	4 *AAbbCc DdEeFf*	3 *AAbbCc DdEeff*	5 *AabbCc DdEeFf*	4 *AabbCc DdEeff*

FIGURE 17.7
Punnett square showing the increase in heterozygosity resulting from outbreeding. Two inbred but unrelated individuals heterozygous at only two of six loci produce offspring that average four heterozygous loci. The small number in the corner of each cell is the number of heterozygous loci in that particular genotype.

Masking of the Expression of Deleterious Recessive Alleles with Major Effects

By increasing homozygosity, inbreeding increases the likelihood of deleterious recessive alleles with major effects becoming homozygous and expressing themselves. The opposite occurs with outbreeding. By increasing heterozygosity, outbreeding tends to keep most deleterious recessives in heterozygous form where they are not expressed. We say their expression is "masked" in the heterozygote. That is why mutts exhibit fewer genetic anomalies than purebred dogs, and why fewer outcross individuals in general tend to suffer from problems of this kind.

It is important to understand that outbreeding does not eliminate deleterious recessive alleles. To the contrary, it perpetuates them by masking their expression, making both natural and artificial selection against them ineffective. If these alleles occur at low frequencies, however, their impact on outbred populations is minimal.

Hybrid Vigor

For polygenic traits influenced to a significant degree by dominance, the result of inbreeding is a loss in gene combination value that we call *inbreeding depression.* For these same traits, the result of outbreeding is a gain in gene combination value that we call *hybrid vigor* or *heterosis.* To see how outbreeding produces hybrid vigor, look at Table 17.5. Listed in the table are the parental genotypes and all 16 possible offspring genotypes from the mating illustrated in Figure 17.5. Also listed are breeding values, gene combination values, and genotypic values for each genotype using the same simplifying assumptions made in earlier examples and assuming partial dominance at all loci such that the genotypic value of each heterozygote is 5 units. The breeding values, gene combination values, and genotypic values of both parents are 12, 6, and 18 units, respectively. Breeding values of offspring range from 0 to 24 units, averaging 12 units—the same as their parents' breeding values.

TABLE 17.5 A Numerical Example Showing How Hybrid Vigor Is Produced by the Mating of Inbred but Unrelated Animals[ab]

Genotype	*BV*	*G*	*GCV* (*G* − *BV*)
Parental genotypes:			
AABbccDDEeff	6(4) + 6(−2) = 12	2(8) + 2(5) + 2(−4) = 18	6
AabbCCddEEFf	6(4) + 6(−2) = 12	2(8) + 2(5) + 2(−4) = 18	6
Average:	12	18	6
Offspring genotypes:			
AABbCcDdEEFf	8(4) + 4(−2) = 24	2(8) + 4(5) = 36	12
AABbCcDdEEff	7(4) + 5(−2) = 18	2(8) + 3(5) + (−4) = 27	9
AaBbCcDdEEFf	7(4) + 5(−2) = 18	8 + 5(5) = 33	15
AaBbCcDdEEff	6(4) + 6(−2) = 12	8 + 4(5) + (−4) = 24	12
AABbCcDdEeFf	7(4) + 5(−2) = 18	8 + 5(5) = 33	15
AABbCcDdEeff	6(4) + 6(−2) = 12	8 + 4(5) + (−4) = 24	12
AaBbCcDdEeFf	6(4) + 6(−2) = 12	6(5) = 30	18
AaBbCcDdEeff	5(4) + 7(−2) = 6	5(5) + (−4) = 21	15
AAbbCcDdEEFf	7(4) + 5(−2) = 18	2(8) + 3(5) + (−4) = 27	9
AAbbCcDdEEff	6(4) + 6(−2) = 12	2(8) + 2(5) + 2(−4) = 18	6
AabbCcDdEEFf	6(4) + 6(−2) = 12	8 + 4(5) + (−4) = 24	12
AabbCcDdEEff	5(4) + 7(−2) = 6	8 + 3(5) + 2(−4) = 15	9
AAbbCcDdEeFf	6(4) + 6(−2) = 12	8 + 4(5) + (−4) = 24	12
AAbbCcDdEeff	5(4) + 7(−2) = 6	8 + 3(5) + 2(−4) = 15	9
AabbCcDdEeFf	5(4) + 7(−2) = 6	5(5) + (−4) = 21	15
AabbCcDdEeff	4(4) + 8(−2) = 0	4(5) + 2(−4) = 12	12
Average:	12	24	12

[a]Listed are the potential genotypes, breeding values (*BV*), gene combination values (*GCV*), and genotypic values (*G*) for offspring of the mating depicted in Figure 17.7. Partial dominance is assumed at each of six unlinked loci influencing a hypothetical polygenic trait.

[b]Simplifying assumptions: Only six loci influence this trait. (Polygenic traits are typically affected by more loci.) The independent effect of each dominant allele equals +4 units. The independent effect of each recessive allele equals −2 units. Partial dominance causes each heterozygous gene pair to be worth +5 units. The genotypic value of each homozygous gene pair is the same as the sum of the independent effects of the genes at that locus. (This is rarely, if ever, the case, but it makes the example much easier to understand.)

Gene combination values of offspring are generally greater than their parents', however, ranging from 6 to 18 units and averaging 12 units. This is a direct result of the increase in heterozygosity produced by crossing unrelated inbreds. The 6-unit increase in gene combination value due to outbreeding causes a similar increase in genotypic value and, presumably, actual performance.

The more unrelated two breeds or lines are, the greater the hybrid vigor expected in crosses between them. Two individuals from closely related populations are likely to be homozygous at many of the same loci. When these individuals are mated, their offspring are necessarily homozygous at those loci. The offspring are not, therefore, particularly heterozygous, and little hybrid vigor is observed. In contrast, two individuals from unrelated populations may be homozygous at some of the same loci, but there is also a good chance that at many loci they are homozygous in opposite ways. For example, where one individual's genotype for the B locus in *BB*, the other individual's genotype is *bb*. When these individuals are mated, their offspring are necessarily heterozygous (*Bb*), and hybrid vigor results.

Represented in Table 17.6 are three six-locus examples showing how hybrid vigor in offspring increases when sire and dam come from increasingly unrelated populations. For simplicity, the parent populations are assumed to be highly inbred so that every parent is homozygous at all six loci and only one offspring

TABLE 17.6 Numerical Examples Showing How Hybrid Vigor in Offspring Increases When Sire and Dam Come From Increasingly Unrelated Populations[ab]

Genotype	*BV*	*G*	*GCV* (*G* − *BV*)
Parents differing at two of six loci:			
AAbbCCddEEff	6(4) + 6(−2) = 12	3(8) + 3(−4) = 12	0
AAbbccddEEFF	6(4) + 6(−2) = 12	3(8) + 3(−4) = 12	0
Offspring:			
AAbbCcddEEFf	6(4) + 6(−2) = 12	2(8) + 2(5) + 2(−4) = 18	6
Parents differing at four of six loci:			
AAbbCCddEEff	6(4) + 6(−2) = 12	3(8) + 3(−4) = 12	0
aabbccDDEEFF	6(4) + 6(−2) = 12	3(8) + 3(−4) = 12	0
Offspring:			
AabbCcDdEEFf	6(4) + 6(−2) = 12	8 + 4(5) + (−4) = 24	12
Parents differing at all six loci:			
AAbbCCddEEff	6(4) + 6(−2) = 12	3(8) + 3(−4) = 12	0
aaBBccDDeeFF	6(4) + 6(−2) = 12	3(8) + 3(−4) = 12	0
Offspring:			
AaBbCcDdEeFf	6(4) + 6(−2) = 12	6(5) = 30	18

[a]Listed are genotypes, breeding values (*BV*), gene combination values (*GCV*), and genotypic values (*G*) for offspring of several matings of highly inbred (highly homozygous) animals. Partial dominance is assumed at each of six loci influencing a hypothetical polygenic trait.
[b]Same simplifying assumptions as in Table 17.5.

genotype is possible from a given mating. The assumptions made in calculating breeding values and genotypic values are the same as those used in Table 17.5.

The first mating shown in Table 17.6 represents a cross between fairly closely-related populations. The parents differ at only the C and F loci. The offspring show some hybrid vigor—6 units of gene combination value more than their parents. The second mating is a cross between more distantly related populations. The parents differ at four loci: A, C, D, and F. The F_1s from this mating exhibit more hybrid vigor—12 units. The last mating represents a "wide" cross, a cross of two very unrelated populations. The parents differ at all six loci, and even more hybrid vigor—18 units—is apparent in the offspring.

As illustrated in Table 17.6, unrelated populations may contain a number of loci where different alleles are fixed. For example, the only genotype possible at the B locus in one population might be *BB,* where as the only possible genotype in a second population might be *bb.* At many loci, however, it is likely that the same genotypes exist in both populations. Population differences at these loci are still present, but they appear as differences in gene and genotypic frequencies. The more unrelated the populations, the greater the differences in these frequencies.

Another way to show the increase in heterozygosity and therefore the increase in hybrid vigor that results from crossing more distantly related populations is to model such crosses using gene frequencies and Punnett squares. Figure 17.8 depicts the outcomes at a single locus of crossing one breed (Breed 1) with four other breeds that are increasingly different from Breed 1 in gene frequencies at that locus. Crossing Breed 1 on Breed 2 (Figure 17.8(a)) produces relatively few heterozygotes at the B locus because the two breeds do not differ in gene frequency. Breeds 3 and 4 are increasingly different from Breed 1 in gene frequencies, so crossing Breed 1 on these breeds ((b) and (c)) produces more heterozygotes. Crossing Breed 1 on Breed 5 (d) produces the most heterozygotes and presumably the most hybrid vigor because the two breeds are so different in gene frequencies.

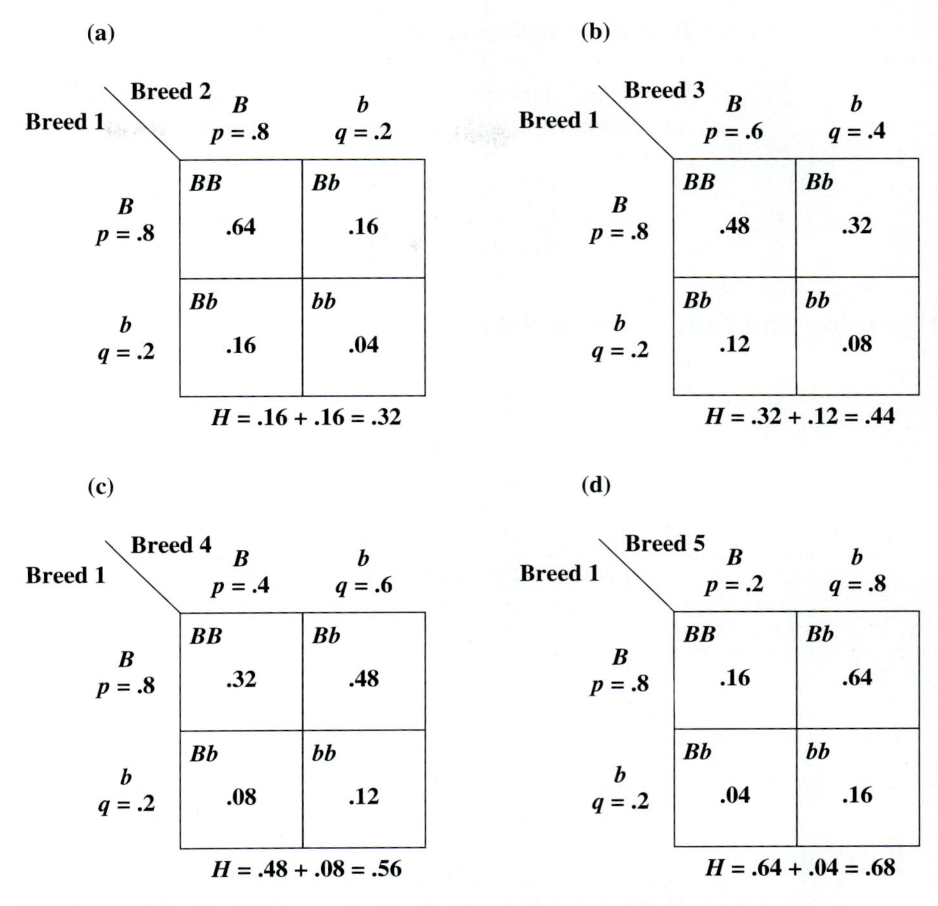

FIGURE 17.8 Single-locus examples showing the increase in heterozygosity (H) of F_1 offspring that occurs when parent breeds are increasingly different in gene frequency.

Reasons to Outbreed

Breeders have a number of reasons for outbreeding. Sometimes they outbreed just to avoid the appearance of inbreeding. At other times they outbreed to cover the existence of some deleterious recessive allele. The two most important genetic reasons for outbreeding, however, are (1) to add hybrid vigor and (2) to take advantage of breed complementarity.

Hybrid Vigor

Hybrid vigor is critically important to production in a number of species. It has major effects on fertility and survivability, so we see it manifested in traits like conception rate, litter size, and weaning rate—traits that are very important economically.[6] For this reason, crossbreeding, the most common form of outbreeding in many species, is typically used to increase hybrid vigor.

[6]For a list of hybrid vigor estimates for several species, see Table 18.2 in the next chapter.

Breed Complementarity

When populations differ in breeding values, outbreeding can also be used to take advantage of breed complementarity, an improvement in the overall performance of crossbred offspring resulting from crossing breeds of different but complementary biological types. Examples of breed complementarity were given in the last chapter. In Chapter 19, you will learn how to design crossbreeding systems that produce both hybrid vigor and complementarity.

Outbreeding and Industry Structure

In livestock species in which crossbreeding is prevalent, outbreeding (crossbreeding) is practiced mostly by commercial producers. Commercial producers sell phenotypic value in the form of meat, milk, wool, or eggs, and therefore want to make use of every tool that can increase production and production efficiency. Because outbreeding boosts production and efficiency by adding hybrid vigor and complementarity, it is an important tool.

Seedstock producers focus on supplying the seedstock inputs to commercial crossbreeding programs and are therefore less concerned about hybrid vigor and complementarity. Hybrid vigor itself is not something they can select for—it is not inherited—and the kind of complementarity that most benefits commercial production is often of the terminal variety. It increases the efficiency of production of animals destined for slaughter, not breeding stock.

This is not to say that all seedstock are purebred. Hybrid seedstock are common in some species and becoming more common in others. For example, seedstock swine breeders often sell F_1 boars and gilts so that commercial pork producers can realize the hybrid vigor present in both parents. Hybrid boars increase conception rates, and hybrid sows wean larger litters and heavier pigs. In beef cattle, there are increasing numbers of *composite* seedstock, a special kind of hybrid.[7]

EXERCISES

Study Questions

17.1 Define in your own words:

inbreeding
common ancestor
prepotency
inbreeding depression
hybrid vigor or heterosis
fitness trait
inbreeding coefficient (F_X)
identical by descent
alike in state
Wright's coefficient of relationship (R_{XY})
path method
tabular method
base population
linebreeding
purebreeding or straightbreeding
outbreeding or outcrossing

17.2 What is the chief effect of inbreeding—the effect from which all other inbreeding effects stem?

[7]For a full discussion of composite breeds, see Chapter 19.

17.3 Explain how inbreeding increases prepotency.

17.4 What kinds of traits are most likely to show prepotency?

17.5 Explain how inbreeding increases the likelihood that deleterious recessive alleles with major effects will be expressed.

17.6 Explain how inbreeding depression and inbreeding's effect on the expression of deleterious recessive alleles with major effects are related.

17.7 Explain inbreeding depression in terms of the components of the genetic model for quantitative traits.

17.8 List gene-level factors that influence the amount of inbreeding depression/hybrid vigor affecting a trait.

17.9 Why must recessive alleles be generally unfavorable or at least less favorable than dominant alleles for the theory explaining hybrid vigor and inbreeding depression to work?

17.10 Explain how the dominance of favorable alleles evolved.

17.11 Why are relatively large amounts of inbreeding depression/hybrid vigor associated with fitness traits?

17.12 Use the genetic model for quantitative traits to explain why animals usually show little hybrid vigor for highly heritable traits and more vigor for lowly heritable traits.

17.13 How do genes that are alike in state differ from genes that are identical by descent?

17.14 The probable proportion of an individual's loci containing genes that are homozygous is always greater than the probable proportion of loci containing genes that are identical by descent. Explain.

17.15 Use pedigrees or arrow diagrams to show:

a. how two noninbred parents can produce an inbred offspring.

b. how two inbred but unrelated parents can produce a noninbred offspring.

17.16 **a.** List the three methods for calculating inbreeding and relationship coefficients.

b. When is it best to use each method?

17.17 Inbreeding and relationship coefficients measure inbreeding and relationship relative to the level of inbreeding and relationship present in the base population. Explain.

17.18 How can breeders avoid substantial inbreeding depression with linebreeding when they cannot do so with faster forms of inbreeding?

17.19 What are the two main reasons for inbreeding?

17.20 Why does inbreeding lead to phenotypic uniformity in simply-inherited traits but not necessarily in polygenic traits?

17.21 For those livestock species in which crossbreeding is prevalent, why is inbreeding (purebreeding) practiced mostly by seedstock producers?

17.22 What is the chief effect of outbreeding—the effect from which all (or almost all) other outbreeding effects stem?

17.23 How does outbreeding cause the expression of deleterious recessive alleles to be "masked" in heterozygotes?

17.24 How does outbreeding perpetuate deleterious recessive alleles in a population?

17.25 Why do crosses of closely related populations show less hybrid vigor than crosses of less related populations?

17.26 What are the two main reasons for outbreeding?

17.27 For those livestock species in which crossbreeding is prevalent, why is it practiced mostly by commercial producers?

Problems

17.1 **a.** Which individual is likely to be more prepotent—one with the genotype:

AaBbCCDdEeFf

or one with the genotype:

AABbccDDeeFF?

b. Prove your answer mathematically. (See the section of Chapter 3 entitled *The Randomness of Inheritance* for hints.)

17.2 A dog is a carrier of the recessive allele (*h*) for diaphragmatic hernia, having inherited it from his grandsire. The frequency of the *h* allele is .05 in this dog's breed. Use Punnett squares to:

a. estimate the frequency of affected (*hh*), noninbred pups from matings of the dog to unrelated females.

b. estimate the frequency of affected, inbred pups from matings of the dog to his first cousins (*HH* or *Hh*—granddaughters of the carrier grandsire). Assume these females, if they carry the *h* allele at all, received it from no other source but their grandsire.

17.3 Two maximally inbred (completely homozygous) mice differ at five loci. Assume the following:

Each dominant allele contributes +3 mg to six-week weight.

Each recessive allele contributes −3 mg to six-week weight.

Partial dominance exists such that each heterozygous locus gains 2 mg in gene combination value (i.e., the genotypic value of each heterozygous locus is 2 mg greater than the breeding value of that locus).

Genetic values for homozygous combinations are the same as breeding values for those combinations.

Environmental effects are as shown.

No epistasis.

a. Fill in the missing values in the following table.

Genotype	*BV*	*GCV*	*G*	*E*	*P*−μ
1. *AAbbCCddEE*				0 mg	
2. *aaBBccDDee*				−3 mg	
3. An offspring of (1) × (2)				2 mg	

b. Which mouse is the heaviest at six weeks?

c. Which mouse is the lightest?

d. Which mouse enjoys the most hybrid vigor?
e. Which mouse is the best bet to produce offspring with heavy six-week weights?
f. Assume complete dominance at all loci and fill in the table again.
g. Do any of your answers for (*b*) through (*e*) change?
h. What does change?

17.4 Given the following pedigree:

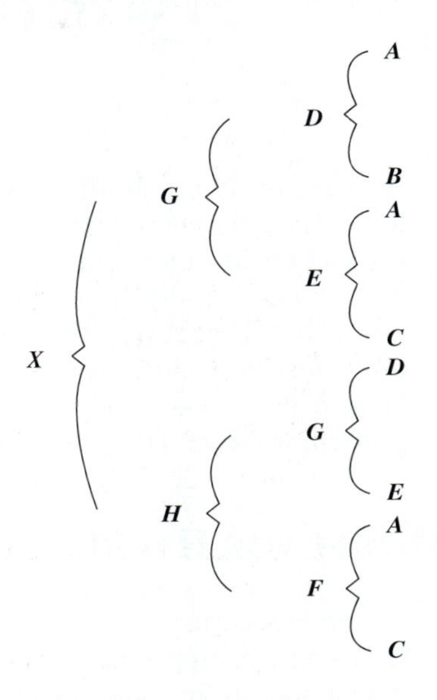

a. Use the path method to calculate F_X.
b. Use the path method to calculate R_{XA}.
c. Use the tabular method to calculate F_X.
d. Use the tabular method to calculate R_{XA}.

17.5 The B locus is a representative locus for a polygenic trait. Gene frequencies at the B locus in three pure breeds are:

Breed	Frequency of B (p)	Frequency of b (q)
X	.4	.6
Y	.9	.1
Z	.8	.2

Assuming that hybrid vigor in crosses of these breeds is proportional to heterozygosity at the B locus, use Punnett squares to determine which of the three crosses of purebreds produces the most hybrid vigor.

CHAPTER 18

Hybrid Vigor

At this point you should have a good understanding of the gene-level origins of hybrid vigor (heterosis). You should know how gene combination value determines hybrid vigor and inbreeding depression, and how crossing unrelated lines or breeds generates hybrid vigor by increasing heterozygosity and gene combination value. There are still a number of questions about hybrid vigor to be answered, however. How is it measured? Does it last beyond the first crossbred generation? How fast is it lost? Can we predict the amount of hybrid vigor generated from a particular cross? These are the questions addressed in this chapter.

A Population Model for Hybrid Vigor

Before answering these questions, let's create a theoretical framework for discussion—a population model for hybrid vigor. The hypothetical or *idealized population* envisioned in this model is illustrated in Figure 18.1. Actually, it consists of a number of populations. First there is an initial, large, randomly bred population—the *base population.* This could be the original population of mice in a laboratory or a wild species before domestic breeds were formed. The base population is not inbred and enjoys a degree of hybrid vigor (or, alternatively, does not suffer from inbreeding depression). In time this population is subdivided either deliberately or by natural circumstances into a number of smaller populations. In the laboratory context, these subpopulations are lines. In the context of a species, they are breeds. Each small population undergoes inbreeding, and during the inbreeding process the lines differentiate in gene and genotypic frequencies and therefore in average breeding value, gene combination value, and overall genotypic value. Eventually the inbred lines are crossed, producing hybrid combinations.

In the context of the idealized population, *the hybrid vigor generated by crossing inbred lines is simply a restoration of the hybrid vigor lost in inbreeding.* It is an "undoing" of accumulated inbreeding depression. To verify this, we need to see how mean breeding values, gene combination values, and genotypic values change as a population progresses from a base population to inbred lines to linecrosses.

A numerical example is given in Table 18.1. For simplicity, the example shows just the B locus, one of many loci influencing a hypothetical polygenic trait. Complete dominance is assumed at this locus. As with the examples in the last chapter, the independent effect of each dominant allele equals +4 units, and the independent effect of each recessive allele equals −2 units. Complete dominance causes each heterozygous gene pair to be worth +8 units. The genotypic value of each homozygous pair is assumed to be the same as the sum of the independent

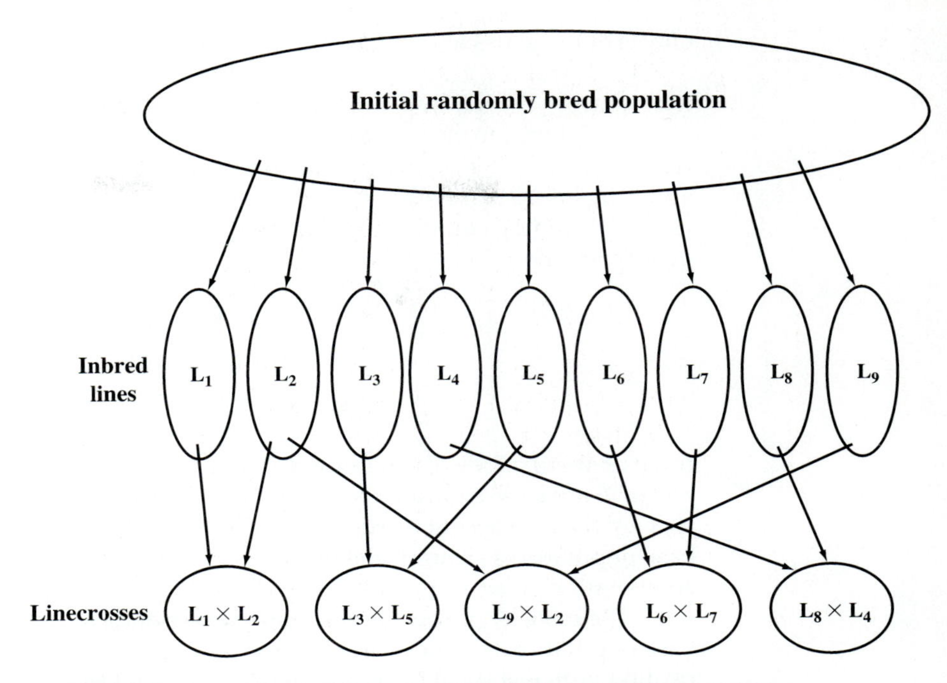

FIGURE 18.1 Schematic representation of an idealized population consisting of an initial, randomly bred base population, inbred lines developed from the base population, and crosses of those lines.

TABLE 18.1 A Numerical Single-Locus Example Showing How Gene Combination Value (Hybrid Vigor) Is Lost in the Formation of Inbred Lines and Restored with Linecrossing[ab]

Genotype	*BV*	*G*	*GCV* (*G* − *BV*)
Initial randomly bred population:			
25% *BB*, 50% *Bb*, and 25% *bb*	.25(8) + .5(2) + .25(−4) = 2	.25(8) + .5(8) + .25(−4) = 5	3
Inbred lines:			
L_1 (*BB*)	8	8	0
L_2 (*bb*)	−4	−4	0
Average	2	2	0
Linecrosses:			
$L_1 \times L_1$ (*BB*)	8	8	0
$L_1 \times L_2$ (*Bb*)	2	8	6
$L_2 \times L_1$ (*Bb*)	2	8	6
$L_2 \times L_2$ (*bb*)	−4	−4	0
Average	2	5	3

[a]Listed are the potential genotypes, breeding values (*BV*), gene combination values (*GCV*), and genotypic values (*G*) associated with the B locus in an initial, randomly bred base population, maximally inbred lines developed from the base population, and all possible linecrosses. Complete dominance is assumed at this locus, one of many loci influencing a hypothetical polygenic trait.

[b]Simplifying assumptions: The base population is in Hardy-Weinberg equilibrium with gene frequencies: $p = .5$ and $q = .5$. The independent effect of each dominant allele equals + 4 units. The independent effect of each recessive allele equals −2 units. Complete dominance causes each heterozygous gene pair to be worth +8 units. The genotypic value of each homozygous gene pair is the same as the sum of the independent effects of the genes at that locus. In this case, the genotypic value of *BB* is therefore 4 + 4 = 8 and the genotypic value of *bb* is −2 + (−2) = −4. (This last assumption is rarely, if ever, true, but it makes the example much easier to understand.)

effects of the genes at that locus. In this case, the genotypic value of *BB* is therefore $4 + 4 = 8$ and the genotypic value of *bb* is $-2 + (-2) = -4$. (Again, this assumption is rarely, if ever, true, but it makes the example much easier to understand.)

To determine the average breeding value, gene combination value, and genotypic value of the base population, we need to know the gene frequencies of the *B* and *b* alleles. Let's say that $p = q = .5$. If we assume that the base population is in Hardy-Weinberg equilibrium, then the genotypic frequencies of the *BB*, *Bb*, and *bb* genotypes are:

$$P = p^2 = .5^2 = .25$$
$$H = 2pq = 2(.5)(.5) = .5$$
$$Q = q^2 = .5^2 = .25$$

In other words, 25% of the individuals in the base population are *BB*, 50% are *Bb*, and 25% are *bb*. Weighting the breeding values and genotypic values of each genotype by these proportions (first row of Table 18.1), the mean breeding value of the base population is 2 units, the mean genotypic value is 5 units, and the mean gene combination value is $5 - 2 = 3$ units.[1]

The next set of rows in the table corresponds to maximally inbred lines. The base population is divided into a large number of inbred lines, but because we are dealing with just the B locus and only two possible alleles, the inbred lines will ultimately be of just two types: *BB* and *bb*. Because the *B* and *b* alleles were equally common in the base population, there should be an equal number of each type of inbred line. (That simplifies the arithmetic to come.) The types differ from each other in both breeding value and genotypic value, with *BB* lines being superior. In fact, the *BB* lines breed and perform better than an average animal from the base population. The *bb* lines do not, however. The average breeding value of *all* the inbred lines is 2—the same as in the base population—but on average the inbred lines are depressed, having a mean genotypic value of 2 and a mean gene combination value of 0. The net effect of inbreeding, then, is a reduction in performance.

Now look at the last set of rows in the table. These correspond to all possible crosses of the inbred lines. The linecrosses vary in breeding value, genotypic value, and gene combination value, but the mean values across all linecrosses are the same as the values in the base population. In other words, the hybrid vigor that was lost in the inbred lines is restored in the linecrosses. By developing inbreds and then crossing them, we have come full circle and re-created the base population.

Well, almost. Instead of one heterogenous population exhibiting considerable genetic variability, we now have four linecross populations that differ from each other genetically, but are each genetically uniform. Three of the four linecrosses perform better than the average animal from the base population, and two of the three, the heterozygous types, have better gene combination value—more hybrid vigor. This suggests an important concept: *By breeding and using just the superior linecrosses, we can consistently achieve more hybrid vigor and performance than was possible with the original base population.* The key to success here is actually selection—not selection of individuals, but selection of inbred lines (breeds). By

[1]If you are uncomfortable using proportions to calculate a mean, imagine there are 100 animals in the base population—25 *BB*, 50 *Bb*, and 25 *bb*. Calculate the average in the standard way. You will get the same answer.

choosing lines that have better breeding values and by crossing them in such a way that the most heterozygous linecrosses result, we can produce consistently high performance.

If you study Table 18.1 carefully, you will notice that the population having the highest breeding value and tied for the highest genotypic value is not a linecross, but the inbred *BB* line. You may then wonder why we bother crossing the *BB* and *bb* inbred lines. Why not just use the *BB* line? In more general terms, why crossbreed if the best population is one that is homozygous for the "right" alleles? Such an approach seems sensible and, in fact, would work well if traits of interest were influenced by just one locus like the B locus in the example. But polygenic traits are affected by many loci, and though inbred lines will be homozygous for the right alleles at some of these loci, it is a rare inbred line that is homozygous for the right alleles at a high proportion of loci. Therefore, given sufficient dominance at enough loci, the best performing populations are likely to be those in which the expression of less desirable recessive alleles is most often masked in heterozygotes. In other words, for traits that respond to hybrid vigor, it is hard to beat good crossbreds.

MEASURING HYBRID VIGOR

In practice, hybrid vigor is measured as the difference between the average performance of crossbreds and the average performance of their purebred parent lines or breeds. Mathematically,

$$HV = \overline{P}_{F_1} - \overline{\overline{P}}_P$$

where HV = hybrid vigor measured in units of a trait

$\overline{P}_{F_1}$ = the average performance of crossbreds

$\overline{\overline{P}}_P$ = the average performance of both parent lines = $\dfrac{\overline{P}_{P_1} + \overline{P}_{P_2}}{2}$

where $\overline{P}_{P_1}$ = the average performance of the first parent line

and $\overline{P}_{P_2}$ = the average performance of the second parent line

Measurement of hybrid vigor is illustrated in Figure 18.2. Note that hybrid vigor is not measured as the superiority of the crosses over the *best* parent line, but rather as the superiority of the crosses over the *average* of the parent lines. The upper diagram in the figure (a) represents a typical situation in which the parent lines perform similarly and hybrid performance is better than that of either parent line. In the lower diagram (b), one parent line is far superior to the other, and even though the crosses exhibit the same amount of hybrid vigor as in (a), their performance is still inferior to the best parent line.

Hybrid vigor is often expressed on a percentage basis—a percentage of the average performance of the parent lines. Mathematically,

$$\%HV = \frac{\overline{P}_{F_1} - \overline{\overline{P}}_P}{\overline{\overline{P}}_P} \times 100$$

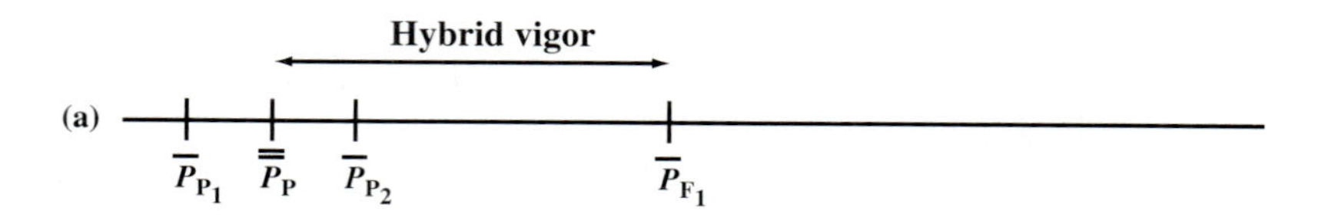

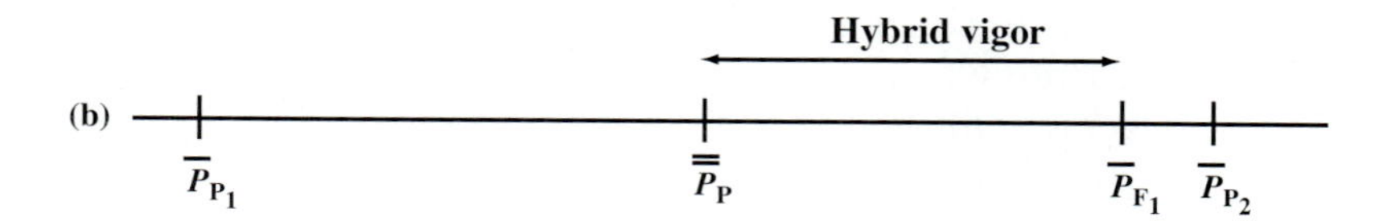

FIGURE 18.2 Schematic representations of hybrid vigor measured as the difference between hybrid performance ($\bar{P}_{F_1}$) and the mean performance of parent lines ($\bar{\bar{P}}_P$). Although the same amount of hybrid vigor is expressed in both the upper and lower diagrams, in (*a*) hybrids are superior to the best parent line; in (*b*) they are not.

For example, if 21-day litter weights average 98 lb for purebred pigs of Breed *A*, 106 lb for purebreds of Breed *B*, and 113 lb for F_1 $A \times B$ cross pigs, then

$$\bar{\bar{P}}_P = \frac{\bar{P}_{P_1} + \bar{P}_{P_2}}{2}$$

$$= \frac{98 + 106}{2}$$

$$= 102 \text{ lb}$$

and

$$\%HV = \frac{\bar{P}_{F_1} - \bar{\bar{P}}_P}{\bar{\bar{P}}_P} \times 100$$

$$= \frac{113 - 102}{102} \times 100$$

$$= 10.8\%$$

Here is another example. If milk production for Breed *A* averages 12,000 lb, for Breed *B* 18,000 lb, and for $A \times B$ crosses 16,000 lb, then

$$\bar{\bar{P}}_P = \frac{\bar{P}_{P_1} + \bar{P}_{P_2}}{2}$$

$$= \frac{12{,}000 + 18{,}000}{2}$$

$$= 15{,}000 \text{ lb}$$

and

$$\%HV = \frac{\bar{P}_{F_1} - \bar{\bar{P}}_P}{\bar{\bar{P}}_P} \times 100$$

$$= \frac{16{,}000 - 15{,}000}{15{,}000} \times 100$$
$$= 6.7\%$$

This last example is similar to the lower diagram in Figure 18.2. The crossbreds exhibit hybrid vigor, but their performance is lower than the performance of the better purebred line. Examples like this point out an important concept: *There is a trade-off to be considered between hybrid vigor (gene combination value) and breeding value.* As much as we might want to maximize hybrid vigor, the effort may not be worthwhile if breeding value is sacrificed in the process. Likewise, it may not make sense to maximize breeding value if an acceptable amount of hybrid vigor cannot be maintained.

In the milk production example, Breed *B* individuals have such high breeding values for milk production relative to Breed *A* individuals that the mean breeding value of $A \times B$ crosses is mediocre by comparison. And even though $A \times B$ crosses display hybrid vigor, they cannot compete with purebreds of Breed *B*. In the context of the U.S. dairy industry, Breed *B* could be Holsteins. Despite the fact that Holstein crosses exhibit considerable hybrid vigor for milk production, the mean breeding value of Holsteins for the trait is so much higher than that of any other breed that the performance of crosses is no match for pure Holsteins. As a result, most dairy cattle in the United States are Holsteins, and crossbred dairy cattle are uncommon.

Individual, Maternal, and Paternal Hybrid Vigor

Direct, maternal, and paternal genetic components of traits were defined in Chapter 11. The direct component of a trait is the effect of an individual's genes on its performance. The maternal component is the effect of genes in the dam of an individual that influence the performance of the individual through the *environment* provided by the dam. The paternal component can be defined much like the maternal component—the effect of genes in the sire of an individual that influence the performance of the individual through the environment provided by the sire—but is more commonly taken to mean the effect of genes in the sire on fertility measures that are considered traits of the dam or offspring.

All traits have a direct component, but not all traits have a maternal component, and relatively few traits have a paternal component. Conception rate makes a good example because it is one trait that has all three components. The direct component of conception rate refers to the effects of genes in the embryo that influence its survival. The maternal component refers to the effects of genes in the dam that influence uterine environment and her ability to conceive. The paternal component refers to genes in the sire affecting his ability to impregnate females.

individual hybrid vigor
Hybrid vigor for the direct component of a trait.

maternal hybrid vigor
Hybrid vigor for the maternal component of a trait.

paternal hybrid vigor
Hybrid vigor for the paternal component of a trait.

Each genetic component of a trait—direct, maternal, and paternal—has potential for hybrid vigor, and we term each kind of hybrid vigor **individual, maternal,** and **paternal hybrid vigor,** respectively. If crossbred offspring perform better than their purebred parents, we attribute the increased performance to individual hybrid vigor. If crossbred dams are better mothers, we infer maternal hybrid vigor. If crossbred sires are more fertile, we say they show paternal hybrid vigor.

Individual hybrid vigor is a function of gene combinations present in the current generation. For example, individual hybrid vigor for conception rate (measured as a trait of the offspring) depends on gene combinations in the embryo, i.e., in the offspring generation. In contrast, maternal and paternal hybrid vigor are functions of gene combinations present in the previous generation. Maternal and paternal hybrid vigor for conception rate depend on gene combinations in dams and sires.

The ultimate mating system for commercial animals would take advantage of all three kinds of hybrid vigor. A good example is provided in swine breeding. Hybrid pigs survive better and grow faster than their purebred counterparts. Hybrid sows produce larger, heavier litters. Hybrid boars increase conception rates. Swine breeding companies typically offer both hybrid gilts and hybrid boars (each sex developed from different breeds) so that producers can benefit from all three kinds of hybrid vigor.

TABLE 18.2 Typical Individual (I), Maternal (M), and Paternal (P) Hybrid Vigor Estimates for a Number of Traits and Species

Species	Trait	$\%HV^I$	$\%HV^M$	$\%HV^P$
Cattle (beef)	Conception rate (trait of cow)	6.0	—	6.0
	Birth weight	3.0	1.5	—
	Weaning weight	5.0	8.0	—
	No. weaned/100 cows exposed	3.0	8.0	5.0
	Weaning weight/cow exposed	7.0	15.0	6.0
	Feed conversion (feed/gain)	−1.0	—	—
	Yearling weight	6.0	2.0	—
	Age at puberty	−5.5	—	—
Cattle (dairy)	Milk yield	6.0	—	—
	Fat yield	7.0	—	—
	Percent fat	—	−1.0	—
	Mature weight	5.0	—	—
	Interval from calving to first service	−1.0	—	—
	Services/conception	−13.0	—	—
	Interval from first service to conception	−17.5	—	—
	Percent calf survival	15.5	—	—
Swine	Conception rate (trait of sow)	3.0	—	7.0
	Number born	2.0	8.0	—
	Number weaned	9.0	11.0	—
	21-day litter weight	12.0	18.0	—
	Days to 220 lb	−7.0	−1.0	—
	Feed conversion (feed/gain)	−2.0	—	—
	Backfat thickness	1.5	4.0	—
	Loin eye area	1.0	1.0	—
Sheep	Conception rate (trait of ewe)	8.0	—	6.0
	Lambing rate (trait of ewe)	3.0	—	8.0
	Number born	3.0	8.0	—
	60-day weaning weight	5.0	9.0	—
	Lambs weaned/ewe exposed	8.0	17.0	6.0
	Grease fleece weight	5.0	—	—
	Staple length	0.0	—	—
	Mature ewe weight	5.0	—	—
Poultry	Age at first egg	−4.0	—	—
	Egg production	12.0	—	—
	Egg weight	2.0	—	—
	Feed conversion (grams of feed per gram of egg)	−5.0	—	—
	Hatchability (trait of chick)	4.0	2.0	—
	Daily gain	5.0	—	—
	Feed conversion (feed/gain)	−11.0	—	—
	Body weight	3.0	—	—

Estimates of individual, maternal, and paternal hybrid vigor (in percentage terms) are listed for a number of traits and species in Table 18.2. Note that for some traits (e.g., age at puberty, feed conversion, or days to 220 lb), hybrid vigor is negative. This does not imply *unfavorable* hybrid vigor. It is simply a function of the way the trait is measured. Hybrid vigor for age at puberty, for example, causes animals to reach puberty at an *earlier* age.

LOSS OF HYBRID VIGOR

F_1 hybrid vigor The amount of hybrid vigor attainable in first-cross individuals—maximum hybrid vigor.

Hybrid vigor is maximized in the F_1 or first cross of unrelated (though not necessarily purebred) populations, and we refer to the amount of vigor gained in this initial cross as **F_1 hybrid vigor.** Can we maintain this amount of hybrid vigor? What happens to hybrid vigor if F_1s are mated to F_1s of the same kind to produce F_2s? What happens if hybrids are mated back to parent lines or breeds?

One way to answer these questions is to simulate these types of matings with Punnett squares and keep track of heterozygosity in the various populations that are produced. This approach requires a couple of simplifying assumptions. First, we will assume that dominance is the overriding cause of hybrid vigor—that epistasis plays only a marginal role. That is, we will assume the *dominance model* for hybrid vigor. For most traits in most populations, the dominance model works well. But there are situations in which interactions among loci affect the amount of hybrid vigor expressed. Second, we will assume that hybrid vigor is *linearly* related to heterozygosity (i.e., for every 1% increase in heterozygosity, there is a fixed increase in hybrid vigor). In practice, this assumption is quite reliable, at least for quantitative polygenic traits like growth rate or milk production. Hybrid vigor for some threshold traits (dystocia score, for example) may be *non*linearly related to heterozygosity. In any case, if we assume that hybrid vigor is proportional to heterozygosity, we will rarely be far off, and determining expected levels of hybrid vigor in different populations becomes a matter of simple arithmetic.

To see how hybrid vigor is affected by various crosses, we need to examine only one hypothetical locus affecting one hypothetical trait. But keep in mind that in reality many loci influence polygenic traits. Let's call our sample locus the J locus. Two purebred parent populations, Breeds *A* and *B*, differ in gene frequencies at the J locus. For Breed *A*, p (the frequency of the *J* allele) is .3, and q (the frequency of the *j* allele) is .7. For Breed *B*, p and q are .7 and .3, respectively. If we assume that both breeds are in Hardy-Weinberg equilibrium, then the proportion of heterozygotes at the J locus for Breed *A* is

$$H_A = 2pq = 2(.3)(.7) = .42$$

and for Breed *B* is

$$H_B = 2pq = 2(.7)(.3) = .42$$

The average heterozygosity in the two parent breeds is then

$$\overline{H} = \frac{H_A + H_B}{2} = \frac{.42 + .42}{2} = .42$$

Remember this number because it represents the baseline for future comparisons.

Now let's cross the two parent breeds to produce F_1 offspring.

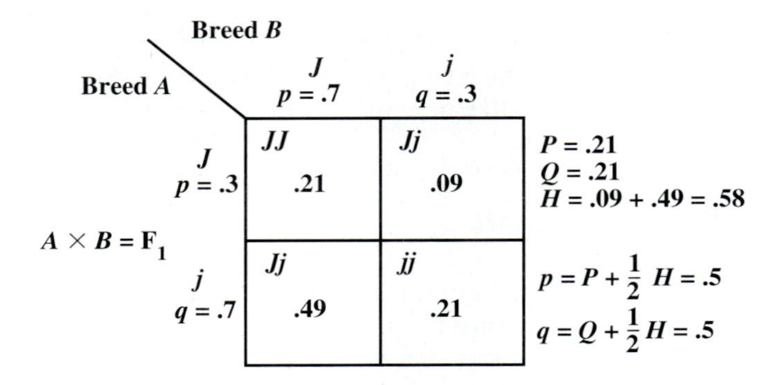

Heterozygosity in the first-cross population is .58, an increase of .16 over the purebred parent populations (.58 − .42 = .16). Assuming that the J locus and other loci influencing our hypothetical trait exhibit a reasonable degree of dominance, this increase in heterozygosity should result in considerable hybrid vigor.

We will need to calculate the gene frequencies in the F_1 population in order to predict the outcomes of future matings. The simplest way to do this is to use the formulas introduced in Chapter 4: $p = P + \frac{1}{2}H$ and $q = Q + \frac{1}{2}H$ (or $q = 1 - p$).[2] In the F_1 population,

$$p = P + \frac{1}{2}H = .21 + \frac{1}{2}(.58) = .5$$

and

$$q = Q + \frac{1}{2}H = .21 + \frac{1}{2}(.58) = .5$$

Now let's mate F_1s to F_1s to produce F_2s. This is like mating black baldies (Black Angus × Hereford cattle) to black baldies.

$F_1 \times F_1 = F_2$

F_1 \ F_1	J ($p = .5$)	j ($q = .5$)
J ($p = .5$)	JJ .25	Jj .25
j ($q = .5$)	Jj .25	jj .25

$P = .25$
$Q = .25$
$H = .25 + .25 = .5$

$p = P + \frac{1}{2}H = .5$

$q = Q + \frac{1}{2}H = .5$

[2]For an explanation of these formulas, see the subsection of Chapter 4 entitled *Outbreeding (Crossbreeding)*.

Heterozygosity in the F_2s is .5, a decrease of .08 from the level of heterozygosity in the F_1s. Note that .5 is exactly halfway between the level of heterozygosity in purebreds (.42) and the level of heterozygosity in F_1s (.58). This is an important result. It indicates that *the hybrid vigor displayed by two-breed F_1 crosses is halved in the corresponding F_2s.* F_2s still exhibit hybrid vigor in comparison to purebreds—just not as much hybrid vigor as F_1s. We call the hybrid vigor that remains in later generations of hybrids—generations subsequent to the first-cross (F_1) generation—**retained hybrid vigor** or **retained heterosis.** Retained hybrid vigor is commonly expressed as a proportion of F_1 (maximum) vigor. In this example, retained hybrid vigor is 50% of maximum.

retained hybrid vigor or **retained heterosis**
Hybrid vigor remaining in later generations of hybrids—generations subsequent to the first-cross (F_1) generation. Retained hybrid vigor is commonly expressed as a proportion of F_1 (maximum) vigor.

What happens when F_2s are mated to F_2s to produce F_3s? Do heterozygosity and hybrid vigor diminish further?

$F_2 \times F_2 = F_3$

F_2 \ F_2	*J* $p = .5$	*j* $q = .5$
J $p = .5$	*JJ* .25	*Jj* .25
j $q = .5$	*Jj* .25	*jj* .25

$P = .25$
$Q = .25$
$H = .25 + .25 = .5$

$p = P + \frac{1}{2}H = .5$

$q = Q + \frac{1}{2}H = .5$

The answer is no. Gene frequencies in F_2s (.5 and .5) are no different than in F_1s. Therefore, mating F_2s to F_2s results in the same gene and genotypic frequencies as mating F_1s to F_1s. Heterozygosity does not change from the F_2 to the F_3 generation. F_2s are actually in Hardy-Weinberg equilibrium, and so will be F_3s, F_4s ($F_3 \times F_3$ crosses), F_5s, and so on. So long as population size in these advanced generations of hybrids is large enough that inbreeding is avoided, no hybrid vigor is lost after the F_2 generation. *Although hybrid vigor is halved between the F_1 and F_2 generations, it remains constant in subsequent generations of two-breed hybrids.* If first-cross black baldies are mated, their offspring should exhibit half of F_1 hybrid vigor, but hybrid vigor should not decline in later generations of black baldies. Retained hybrid vigor remains at 50% of maximum (see Figure 18.3).

The notion of a constant level of retained hybrid vigor in advanced generations of hybrids is useful and generally correct, but it depends on a pair of important assumptions: (1) inbreeding is avoided and (2) the dominance model of hybrid vigor is appropriate. If the hybrid population is large enough, the first assumption is likely to be met. The second assumption appears to be true in a large majority of cases, but not always. Long-term selection in some inbred populations seems to have fixed certain alleles, forming *blocks* of loci containing favorable epistatic combinations. Epistatic blocks of this kind remain intact in an inbred population, but crossbreeding introduces new alleles, causing the blocks to break up. If the loci that are part of an epistatic block are located on different chromosomes or are only distantly linked, the block is broken up in the first cross, and any loss of gene combination value is reflected in the level of F_1 hybrid vigor. But

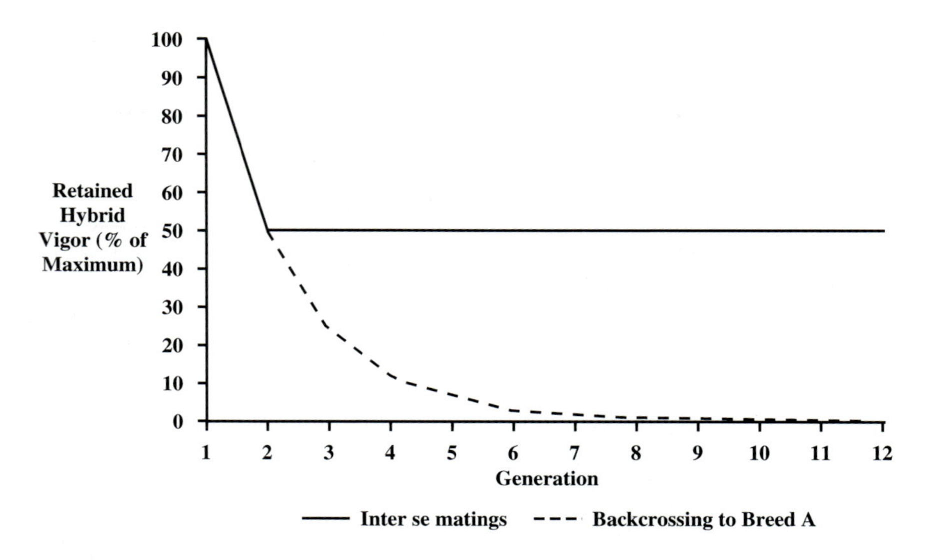

FIGURE 18.3 Retained hybrid vigor (measured as a percentage of maximum vigor) in successive generations of $A \times B$ *inter se* matings (solid line) and backcross matings to Breed A (broken line). The generation marked 1 on the horizontal scale is the F_1 generation. Generation 2 is the F_2 (BC_1) generation, Generation 3 the F_3 (BC_2) generation, and so on.

recombination loss
A loss in gene combination value caused by the gradual breaking up of favorable epistatic blocks of linked loci in advanced generations of certain hybrids.

if some of these loci are closely linked, crossing over causes the block to break up gradually over a number of generations. We call the loss in gene combination value caused by recombination of linked alleles in hybrids **recombination loss.** In those cases in which recombination loss is important, retained hybrid vigor does not remain constant after the F_2 generation, but declines further before eventually reaching equilibrium.

What happens if we backcross—breed hybrids back to a purebred parent breed? Let's breed F_1 $A \times B$ crosses back to Breed A.

Breed A \ F_1	J $p = .5$	j $q = .5$
J $p = .3$	JJ .15	Jj .15
j $q = .7$	Jj .35	jj .35

$A \times F_1 = BC_1$

$P = .15$
$Q = .35$
$H = .15 + .35 = .5$

$p = P + \frac{1}{2}H = .4$

$q = Q + \frac{1}{2}H = .6$

BC_1 (backcross one)
Referring to the first generation of crosses between hybrids and purebreds of a parent breed or line.

Heterozygosity declines from .58 to .5 in the **BC_1** (backcross one) generation. The backcrosses should display less hybrid vigor than the F_1s. Further backcrossing ($A \times BC_1$, $A \times BC_2$, etc.) causes even more decline in heterozygosity and hybrid vigor.

$A \times BC_1 = BC_2$

Breed A \ BC_1	J $p = .4$	j $q = .6$
J $p = .3$	JJ .12	Jj .18
j $q = .7$	Jj .28	jj .42

$P = .12$
$Q = .42$
$H = .18 + .28 = .46$

$p = P + \frac{1}{2}H = .35$

$q = Q + \frac{1}{2}H = .65$

$A \times BC_2 = BC_3$

Breed A \ BC_2	J $p = .35$	j $q = .65$
J $p = .3$	JJ .105	Jj .195
j $q = .7$	Jj .245	jj .455

$P = .105$
$Q = .455$
$H = .195 + .245 = .44$

$p = P + \frac{1}{2}H = .325$

$q = Q + \frac{1}{2}H = .675$

Note how gene and genotypic frequencies in successive generations of backcrosses become more and more like the gene and genotypic frequencies in Breed A. By continually backcrossing to Breed A, we are essentially *grading up* to purebred Breed A.

Retained hybrid vigor (measured as a percentage of maximum vigor) is plotted for successive generations of $A \times B$ *inter se* matings and backcross matings to Breed A in Figure 18.3. The generation marked 1 on the horizontal scale is the F_1 generation. Generation 2 is the F_2 (BC_1) generation, Generation 3 the F_3 (BC_2) generation, and so on. Note that while retained hybrid vigor quickly levels off at 50% of maximum with *inter se* matings, it gradually disappears with backcrossing.

The hybrid vigor implied in Figure 18.3 and in the Punnett squares that precede it is *individual* hybrid vigor, the increase in an individual's performance caused by the individual's own gene combinations. It depends on heterozygosity in the current generation. For example, individual hybrid vigor in an F_2 generation is a function of heterozygosity in F_2 animals. Maternal and paternal hybrid vigor result from gene combinations in the *parents* of an individual and, as such, depend on heterozygosity in the previous generation. F_1 black baldy calves exhibit maximum individual hybrid vigor for the direct (growth rate) component of weaning weight. Because their dams are purebreds, however, they do not benefit from any hybrid vigor for the maternal (milk) component of weaning weight. In contrast, F_2 black baldy calves exhibit only half the maximum level of individual hybrid vigor, but because their dams are F_1s, these calves benefit from maximum maternal hybrid vigor.

PREDICTING HYBRID VIGOR

In animal breeding, hybrid vigor (gene combination value) is not predicted for individuals the way we predict breeding values, progeny differences, or producing abilities. But in order to make crossbreeding decisions, it is often useful to know how much hybrid vigor can be expected from a given cross or mating system. Suppose, for example, that you own hybrid females and want to predict the hybrid vigor likely to result from crossing them with sires of a particular breed or combination of breeds. If you know the breeds involved and the proportions of each breed in sires and dams, if the trait-specific F_1 or maximum hybrid vigor for every two-breed combination of these breeds is the same or at least close enough to being the same that an average value will suffice, if you have a reasonable estimate of F_1 vigor for the trait, and (finally) if you can safely assume the dominance model for hybrid vigor (i.e., recombination loss is not a concern), you can predict retained hybrid vigor for any cross using a fairly simple formula.

The formula combines the proportional breed composition of sires and dams with an estimate of F_1 hybrid vigor to predict retained hybrid vigor in the offspring. In summation notation, the formula is:

$$R\hat{H}V = \left(1 - \sum_{i=1}^{n} p_{s_i} p_{d_i}\right) F_1\hat{H}V$$

where $R\hat{H}V$ = a prediction of retained hybrid vigor (in trait units)
p_{s_i} = the proportion of breed i in sires
p_{d_i} = the proportion of breed i in dams
$F_1\hat{H}V$ = typical F_1 hybrid vigor for the trait
n = the total number of breeds involved

In expanded form, the formula appears as:

$$R\hat{H}V = [1 - (p_{s_1}p_{d_1} + p_{s_2}p_{d_2} + \cdots + p_{s_n}p_{d_n})]F_1\hat{H}V$$

For example, consider a cross of rams that are 50% Breed A and 50% Breed B on ewes that are 25% Breed A, 25% Breed B, and 50% Breed C. We are interested in knowing how much hybrid vigor for 60-day weaning weight is retained with this cross. Let's assume that individual F_1 hybrid vigor for the trait is about 4 lb. Then individual (subscript I) hybrid vigor retained in $(A \times B) \times (C \times (A \times B))$ lambs is

$$\begin{aligned} R\hat{H}V_I &= \left(1 - \sum_{i=1}^{n} p_{s_i} p_{d_i}\right) F_1\hat{H}V_I \\ &= [1 - (.5(.25) + .5(.25) + 0(.5))](4.0) \\ &= .75(4.0) \\ &= 3.0 \text{ lb} \end{aligned}$$

Because weaning weight is affected by both individual and maternal hybrid vigor, and because the dams of these lambs are themselves crossbreds, we need to calculate maternal hybrid vigor also. To do this, we can use the same formula, but we must know the breed composition of the parents of the dams. Assuming that the dams are by purebred Breed C rams and out of 50% Breed A, 50% Breed B ewes, and that F_1 hybrid vigor for the maternal component of 60-day weaning weight is approximately the same as for the direct component (4 lb), maternal hybrid vigor retained in the $C \times (A \times B)$ ewes should be

$$R\hat{H}V_M = \left(1 - \sum_{i=1}^{n} p_{s_i} p_{d_i}\right) F_1\hat{H}V_M$$

$$= [1 - (0(.5) + 0(.5) + 1(0))](4.0)$$

$$= 1(4.0)$$

$$= 4.0 \text{ lb}$$

Combining predictions of individual and maternal hybrid vigor, total hybrid vigor expected in the crossbred lambs is

$$R\hat{H}V = R\hat{H}V_I + R\hat{H}V_M$$

$$= 3.0 + 4.0$$

$$= 7.0 \text{ lb}$$

Sometimes we are less interested in predicting hybrid vigor per se and more interested in comparing the predicted hybrid vigor from a particular cross with a common standard—F_1 vigor. This can be done with a simple change in the formula. If, from the original formula:

$$R\hat{H}V = \left(1 - \sum_{i=1}^{n} p_{s_i} p_{d_i}\right) F_1\hat{H}V$$

we remove the component for F_1 hybrid vigor ($F_1\hat{H}V$), the remainder represents the proportion of F_1 or maximum hybrid vigor attainable in a particular cross. The estimated percentage of retained hybrid vigor is then

$$\%R\hat{H}V = \left(1 - \sum_{i=1}^{n} p_{s_i} p_{d_i}\right) \times 100$$

In the sheep example:

$$\%R\hat{H}V_I = \left(1 - \sum_{i=1}^{n} p_{s_i} p_{d_i}\right) \times 100$$

$$= [1 - (.5(.25) + .5(.25) + 0(.5))] \times 100$$

$$= 75\%$$

In other words, ¾ of maximum individual hybrid vigor is realized in $(A \times B) \times (C \times (A \times B))$ lambs. For their $C \times (A \times B)$ dams:

$$\%R\hat{H}V_M = \left(1 - \sum_{i=1}^{n} p_{s_i} p_{d_i}\right) \times 100$$

$$= [1 - (0(.5) + 0(.5) + 1(0))] \times 100$$

$$= 100\%$$

All of F_1 hybrid vigor is realized in these ewes because they are true F_1s; their sire breed is different from the breeds of their dams.

degree of backcrossing
The proportional amount of backcrossing (in the broader sense) involved in a mating—the proportion of an offspring's loci at which both genes of a pair trace to the same ancestral breed or line.

backcrossing
(1) The mating of a hybrid to a purebred of a parent breed or line; (2) (broader meaning) The mating of an individual (purebred or hybrid) to any other individual (purebred or hybrid) with which it has one or more ancestral breeds or lines in common.

In the formula, $\sum_{i=1}^{n} p_{s_i} p_{d_i}$ represents what is called **degree of backcrossing.** It measures the proportional amount of **backcrossing** involved in a mating. Backcrossing, as it is used here, takes on a broader meaning than was indicated in Chapter 15. It is the mating of an individual (purebred or hybrid) to any other individual (purebred or hybrid) with which it has one or more ancestral breeds or lines in common. We would like to keep the amount of backcrossing to a minimum because minimizing backcrossing means maximizing hybrid vigor. For the $(A \times B) \times (C \times (A \times B))$ lambs, breeds A and B are common to both their sires and dams, and the degree of backcrossing is $.5(.25) + .5(.25) + 0(.5) = .25$ or 25%. For the $C \times (A \times B)$ ewes, no breed is common to both their sires and dams, and the degree of backcrossing is $0(.5) + 0(.5) + 1(0) = 0$—no backcrossing at all.

Another way to understand the proportion of F_1 hybrid vigor retained in a cross—a way that is both conceptually and visually appealing—is to think of it as *the proportion of an individual's loci at which one gene of a pair traces to one parent breed and the other gene traces to a different parent breed.* For F_1s, this proportion is 1 or 100%. 100% of maximum vigor is expressed. For F_2s, F_3s, BC_1s, etc. this proportion is something less than 1 because backcrossing creates an opportunity for both genes at a locus to trace to the same parent breed. When this happens, hybrid vigor is lost (not retained).

You can determine the proportion of maximum hybrid vigor retained or lost in offspring of any mating by constructing a Punnett-like square. The upper square (a) in Figure 18.4 represents the mating in our sheep example: $A \times B$ rams with $C \times (A \times B)$ ewes. We can use it to determine the proportions of individual F_1 hybrid vigor retained and lost in lambs produced from this mating. Each parent breed is assigned a row and column in the square, and the numbers outside the square represent proportional breed composition of the parents. Think of them as the proportion of genes in a gamete that trace to particular parent breeds. The number within a given cell of the square is simply the product of the numbers heading corresponding rows and columns, and indicates the proportion of an offspring's loci expected to contain one paternal gene from the breed represented in that row and one maternal gene from the breed represented in that column. For example, 25% of the loci of lambs resulting from the mating depicted in Figure 18.4(a) should contain one gene of Breed A origin inherited from their sire and one gene of Breed C origin inherited from their dam. (See the upper right cell of the square.)

If the numbers in the off-diagonal cells—the diagonal runs from upper left to lower right—are added together, the sum represents the proportion of an offspring's loci at which one gene of a pair traces to one parent breed and the other gene traces to a different parent breed. In other words, it represents the proportion of F_1 hybrid vigor retained in the cross. In the sheep example, this value is $.125 + .25 + .125 + .25 + 0 + 0 = .75$ or 75%. We expect lambs from this cross to retain 75% of maximum individual hybrid vigor.

The sum of the numbers in the diagonal cells represents the proportion of an offspring's loci at which both genes of a pair trace to the *same* breed. That is, it represents the degree of backcrossing—the proportion of F_1 hybrid vigor lost in the

(a)

$C \times (A \times B)$ ewes

♂ \ ♀		A .25	B .25	C .5
$A \times B$ rams	A .5	.125	.125	.25
	B .5	.125	.125	.25
	C 0	0	0	0

(b)

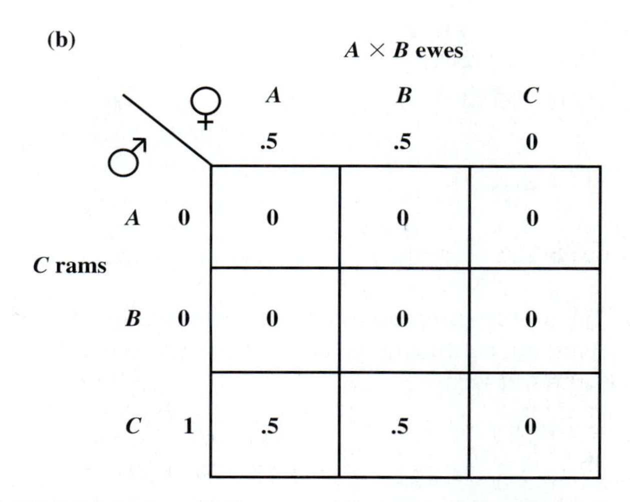

A × B ewes
♀
♂
A
B
C
.5
.5
0
C rams
A 0
B 0
C 1
0
0
0
0
0
0
.5
.5
0

FIGURE 18.4 Punnett-like squares depicting (*a*) the mating of $A \times B$ rams to $C \times (A \times B)$ ewes, and (*b*) the mating that produced these ewes ($C \times (A \times B)$). The sum of off-diagonal values predicts (*a*) the proportion of maximum individual hybrid vigor retained in lambs, and (*b*) the proportion of maximum maternal hybrid vigor retained in ewes. The sum of diagonal values represents degree of backcrossing in each case. (See text for more explanation.)

cross. In the sheep example, this value is $.125 + .125 + 0 = .25$ or 25%. Lambs from this cross lack 25% of maximum hybrid vigor because of backcrossing.

The sum of the numbers in *all* the cells must equal 1. As a practical matter, then, it is often easier to calculate the proportion of F_1 hybrid vigor retained by subtracting the sum of the values on the diagonal from 1 rather than adding up all the off-diagonal values. This is, in fact, exactly the way the calculation is performed using the formula presented earlier.

The lower square (b) in Figure 18.4 represents the mating that produced the ewes in our sheep example, purebred Breed C rams with $A \times B$ ewes. We can use it to determine the proportion of maximum *maternal* hybrid vigor retained by the $C \times (A \times B)$ ewes. Note that the diagonal cells all contain zero values. There is no backcrossing in this mating, so the ewes should exhibit 100% of F_1 maternal hybrid vigor.

A More Rigorous Approach to Predicting Hybrid Vigor

The formula presented here for predicting hybrid vigor takes a shortcut in assuming that F_1 hybrid vigor is roughly the same for all two-breed combinations. What if that assumption is false? What if some of the breeds involved are closely related and therefore show little F_1 hybrid vigor, and others are distantly related and exhibit much more F_1 vigor? In that case we need to use a different formula:

$$R\hat{H}V = \sum_{i=1}^{n}\sum_{j=1}^{n} p_{s_i} p_{d_j}\, F_1\hat{H}V_{ij}$$

where $R\hat{H}V$ = a prediction of retained hybrid vigor (in trait units)
p_{s_i} = the proportion of breed i in sires
p_{d_j} = the proportion of breed j in dams
$F_1\hat{H}V_{ij}$ = typical F_1 hybrid vigor for the trait in crosses of breeds i and j
n = the total number of breeds involved

The double summation in this formula is economical from a notation standpoint, but confusing to many. The formula may be easier to understand in its expanded form:

$$\begin{aligned} R\hat{H}V = {} & (p_{s_1}p_{d_1}\, F_1\hat{H}V_{11} + p_{s_1}p_{d_2}\, F_1\hat{H}V_{12} + \cdots + p_{s_1}p_{d_n}\, F_1\hat{H}V_{1n}) \\ & + (p_{s_2}p_{d_1}\, F_1\hat{H}V_{21} + p_{s_2}p_{d_2}\, F_1\hat{H}V_{22} + \cdots + p_{s_2}p_{d_n}\, F_1\hat{H}V_{2n}) \\ & + \cdots + (p_{s_n}p_{d_1}\, F_1\hat{H}V_{n1} + p_{s_n}p_{d_2}\, F_1\hat{H}V_{n2} + \cdots + p_{s_n}p_{d_n}\, F_1\hat{H}V_{nn}) \end{aligned}$$

In the sheep example, if F_1 individual hybrid vigor for $A \times B$ crosses is 2.7 lb, for $A \times C$ crosses 5.5 lb, and for $B \times C$ crosses 4.0 lb, then individual hybrid vigor retained in $(A \times B) \times (C \times (A \times B))$ lambs is

$$\begin{aligned} R\hat{H}V_I &= \sum_{i=1}^{n}\sum_{j=1}^{n} p_{s_i} p_{d_j}\, F_1\hat{H}V_{ij} \\ &= (.5(.25)(0) + .5(.25)(2.7) + .5(.5)(5.5)) \\ &\quad + (.5(.25)(2.7) + .5(.25)(0) + .5(.5)(4.0)) \\ &\quad + (0(.25)(5.5) + 0(.25)(4.0) + 0(.5)(0)) \\ &= 3.1 \text{ lb} \end{aligned}$$

If the values for two-breed F_1 hybrid vigor for the maternal component of 60-day weaning weight are approximately the same as for the direct component, then maternal hybrid vigor retained in the $C \times (A \times B)$ ewes should be

$$
\begin{aligned}
R\hat{H}V_M &= \sum_{i=1}^{n}\sum_{j=1}^{n} p_{s_i} p_{d_j} F_1\hat{H}V_{M_{ij}} \\
&= (0(.5)(0) + 0(.5)(2.7) + 0(0)(5.5)) \\
&\quad + (0(.5)(2.7) + 0(.5)(0) + 0(0)(4.0)) \\
&\quad + (1(.5)(5.5) + 1(.5)(4.0) + 1(0)(0)) \\
&= 4.8 \text{ lb}
\end{aligned}
$$

Total hybrid vigor retained is then

$$
\begin{aligned}
R\hat{H}V &= R\hat{H}V_I + R\hat{H}V_M \\
&= 3.1 + 4.8 \\
&= 7.9 \text{ lb}
\end{aligned}
$$

EXERCISES

Study Questions

18.1 Define in your own words:

individual hybrid vigor	F_1 hybrid vigor
maternal hybrid vigor	recombination loss
paternal hybrid vigor	BC_1 (backcross one)
retained hybrid vigor or retained heterosis	degree of backcrossing
	backcrossing

18.2 **a.** Describe an *idealized population.*

b. What happens to the mean breeding value, gene combination value, and genotypic value of an idealized population as random mating within the base population is followed by inbreeding within lines, then by linecrossing?

c. What do the changes in mean gene combination value and genotypic value of an idealized population suggest about inbreeding depression and hybrid vigor?

18.3 Why cross selected inbred lines instead of randomly mating within a large, noninbred population?

18.4 In the absence of overdominance, the very best genotype is one that is homozygous for the "right" alleles at all loci. Why then, do we crossbreed to improve performance in polygenic traits instead of using pure lines that are homozygous for the right alleles?

18.5 **a.** How is hybrid vigor measured?
b. If an F_1 cross exhibits significant hybrid vigor, will it necessarily perform better than both the parent lines? Why or why not?

18.6 **a.** Why are most dairy cattle in the United States purebred Holstein and not crossbred?
b. What does this fact suggest about the trade-off between hybrid vigor (gene combination value) and breeding value?

18.7 For a species of your choice:
a. Name two traits that exhibit both individual and maternal hybrid vigor and describe the way in which each kind of hybrid vigor affects the traits.
b. Name two traits that exhibit paternal hybrid vigor and describe the way in which it affects the traits.
c. Outline a mating system that takes advantage of all three kinds of hybrid vigor.

18.8 **a.** How does retained hybrid vigor in advanced generations of *inter se* matings (F_2, F_3, F_4, etc.) differ from hybrid vigor retained in successive generations of backcrosses (BC_1, BC_2, BC_3, etc.)?
b. Why the difference?

18.9 Describe the potential effect of recombination loss on the amount of hybrid vigor retained in F_3, F_4, and subsequent generations of hybrids.

18.10 If a trait exhibits considerable maternal hybrid vigor but little individual hybrid vigor, which is likely to perform better: a two-breed F_1 population or the corresponding two-breed F_2 population? Explain.

18.11 Explain the connection between retained hybrid vigor and degree of backcrossing.

Problems

18.1 Peregrine falcons and gyrfalcons are two species that are prized by breeders but suffer from low reproductive rates. Given that peregrines average 2.3, gyrfalcons 3.2, and F_1 crosses of the two species 3.8 viable offspring per clutch, calculate hybrid vigor and percent hybrid vigor for number of viable chicks per clutch.

18.2 Swine breeds A, B, and C average 1.36, 1.40, and 1.48 lb per day for post-weaning average daily gain (ADG), respectively. Percent hybrid vigor for ADG is typically about 5%.
a. Rank the pure breeds and two-breed crosses for ADG.
b. Did crossbreds always have the best ADG? What principle is illustrated here?

18.3 A sheep breeder has decided to mate his Columbia ewes to Targhee rams. Columbias are ½ Rambouillet and ½ Lincoln. Targhees are ¾ Rambouillet and ¼ Lincoln (approximately).
a. What proportion of F_1 hybrid vigor is retained in this cross?
b. Based upon your answer to (*a*), would you advise a different choice of sire breed? Why or why not?

18.4 A rancher has a herd of ½ Red Angus, ¼ Angus, ¼ Hereford cows (daughters of Red Angus sires and Angus × Hereford dams). She plans to mate them to ½ Charolais, ½ Angus bulls. Typical F_1 individual hybrid vigor for weaning weight is 27 lb. Typical F_1 maternal hybrid vigor for weaning weight is 44 lb.

a. Assuming all four breeds are equally unrelated, calculate the amount of individual, maternal, and total hybrid vigor retained in this cross.

b. Calculate the amount of individual, maternal, and total hybrid vigor retained again—this time accounting for the fact that Angus and Red Angus are so closely related that F_1 individual hybrid vigor for weaning weight in Angus $\times$ Red Angus crosses is only 5 lb, and F_1 maternal hybrid vigor for weaning weight in these crosses is only 8 lb.

CHAPTER 19

Crossbreeding Systems

You shall keep my statutes. You shall not let your cattle breed with a different kind; . . .

Leviticus 19:19[1]

Crossbreeding may have been illegal a few thousand years ago, but today it is commonplace—at least for most species of livestock. Intelligent crossbreeding generates hybrid vigor and breed complementarity, phenomena that are important to production efficiency. Breeders can obtain hybrid vigor and complementarity simply by crossing appropriate breeds. But to sustain acceptable levels of hybrid vigor and breed complementarity in a manageable way over the long term requires a well-thought-out **crossbreeding system.**

crossbreeding system
A mating system that uses crossbreeding to maintain a desirable level of hybrid vigor and(or) breed complementarity.

With few exceptions, crossbreeding systems are the domain of *commercial* animal production. That is largely because they are designed to maintain hybrid vigor (gene combination value), something that is important to food and fiber production but, not being heritable, is not as important to the production of seedstock. In fact, most seedstock breeders are purebred breeders; they do not crossbreed at all. They need a thorough understanding of crossbreeding systems, however, because the animals they produce are destined to be components of those systems.

Not all seedstock are purebred. There are growing numbers of hybrid seedstock, most notably F_1s and *composite* animals—hybrids designed to retain hybrid vigor without crossbreeding. For this reason, I pay considerable attention in this chapter to the role of hybrid breeding animals, and devote a section of the chapter to breeding composite seedstock.

The chapter begins with a list of criteria that can be used to evaluate different crossbreeding systems. These criteria are then applied to several broad categories of systems. The crossbreeding systems discussed are by no means all-inclusive. There are myriad variations and combinations. Nor are these systems

[1] *Holy Bible,* Revised Standard Version.

universally appropriate. Although they could, in theory, be used in any domestic species, in practice they find application primarily in swine, sheep, poultry, and beef cattle.

EVALUATING CROSSBREEDING SYSTEMS

Following is a list of seven criteria useful for evaluating different crossbreeding systems:

1. Merit of component breeds
2. Hybrid vigor
3. Breed complementarity
4. Consistency of performance
5. Replacement considerations
6. Simplicity
7. Accuracy of genetic prediction

The relative importance of each criterion in the list varies depending on the production situation, and in any particular situation there may be other criteria of importance. But if you are trying to choose a crossbreeding system, and you carefully evaluate each candidate system on the basis of these seven items, you will be more likely to make the right choice.

Merit of Component Breeds

For any crossbreeding system to be effective, the breeds in the system must be well chosen. If you are a horse breeder, for example, and wish to create the ultimate crossbred stadium jumper, you would be unlikely to include the Shetland pony as a component breed. Shetlands are simply too small and round to be viable candidates given the needs of stadium jumpers. Each breed included in a crossbreeding system must bring favorable attributes to the mix. Specifically, the mean *breeding values* of each component breed for traits of importance should either be similar to the desired breeding values of crossbred commercial animals, or complement the breeding values of the other breeds in the system.

Merit of component breeds is critically important. It is so important, in fact, that in cases where only one breed is perceived to have acceptable merit, crossbreeding is not recommended. An example (one discussed in some detail in Chapter 18) is the use of Holstein dairy cattle in the United States and other industrialized countries. Another example is the use of White Leghorn chickens in the North American layer industry. Purebred Holsteins and White Leghorns are used commercially because of their unmatched level of production.

Determining the appropriate component breeds for a crossbreeding system is one challenge. Another is locating available animals in those breeds. Import restrictions and small seedstock numbers make availability an issue in many cases. It is not unusual for breeders to design what they think is the optimal combination of breeds, only to find that animals of one or more of these breeds are either unavailable or available at a price they cannot afford.

Although merit of component breeds is extremely important to any crossbreeding system, because no type of crossbreeding system has a particular advantage in this category, breed merit is not the most useful criterion for comparing *kinds* of systems. For example, a rotational crossbreeding system that uses appropriate breeds will inevitably be better than a composite system that uses inappropriate breeds, but that is not to say that, with the right breeds, a composite system would not be the better choice for a particular situation. For this reason, you will not see merit of component breeds mentioned in the system comparisons that follow. Just assume that the breeds used in each crossbreeding system are appropriate choices.

Hybrid Vigor

Generating hybrid vigor is one of the most important (if not the most important) reasons for crossbreeding, so any worthwhile crossbreeding system should provide an adequate amount of hybrid vigor. Generally speaking, the more hybrid vigor the better. But maximum hybrid vigor is only obtainable in F_1s, the first cross of unrelated populations. To sustain F_1 vigor in a herd or flock, a commercial breeder must avoid backcrossing entirely, and that is not always an easy or practical thing to do. Most crossbreeding systems do not achieve 100% of F_1 vigor, but maintain acceptable levels of hybrid vigor by limiting backcrossing in a way that is manageable and economical.

Breed Complementarity

sire × dam complementarity The classic form of complementarity produced by mating sires strong in paternal traits to dams strong in maternal traits. Offspring inherit superior market characteristics from their sires and benefit from the maternal environment provided by their dams.

hybrid seedstock complementarity The form of complementarity produced by crossing genetically diverse breeds to create hybrid breeding animals with a desirable combination of breeding values.

Breed complementarity refers to the production of a more desirable offspring by crossing breeds that are genetically different from each other, but have *complementary* attributes. In beef cattle breeding, we refer to "big bull × small cow" complementarity. The big bull provides growth and leanness to the offspring, the small cow requires less feed to maintain herself, and the result is a desirable market animal economically produced. Breed complementarity is the result of "mixing and matching" the mean breeding values of different breeds.

Big bull × small cow complementarity is breed complementarity in its classic form. It occurs at the commercial level, producing a near-optimal market animal. A more generic name for this kind of complementarity is **sire × dam complementarity.**

Another more subtle but still important form of breed complementarity occurs in the creation of hybrid seedstock. This type of breed complementarity results from crossing genetically diverse breeds to produce a hybrid breeding animal with just the right mix of breeding values. For example, suppose several pure breeds are locally available for use in a crossbreeding program. Suppose also that the mean breeding values for some of these breeds are too high in one important trait and too low in another, and the mean breeding values for the remainder of the breeds are too low in the first trait and too high in the second trait. A hybrid breeding animal derived from crossing these two breed types would have appropriate breeding values for both traits. It would contribute this second form of breed complementarity—**hybrid seedstock complementarity**—to the crossbreeding system.

Consistency of Performance

Ideally, a crossbreeding system should produce a consistent product. It is much easier to market a uniform set of animals than a diverse one. It is also easier to manage a female population that is essentially one biological type than one made up of several types, each with different requirements. Crossbreeding systems vary in their ability to provide this kind of consistency.

Replacement Considerations

In terms of hybrid vigor, the ultimate female is an F_1, and ideally commercial breeders would like to have entire herds or flocks of F_1 females. But how do they produce a continuous supply of F_1s? Either they maintain purebred parent populations (something most commercial producers are reluctant to do) or buy replacements from someone else (something they may be equally reluctant to do). A number of crossbreeding systems manage to overcome the replacement dilemma, allowing commercial breeders to produce replacement females from their own hybrid populations. This convenience comes at a price, however, a price typically paid in hybrid vigor, breed complementarity, or the next item on the list—simplicity.

Simplicity

Crossbreeding systems should be relatively simple. Expensive systems or systems that require an unrealistically high level of management are unlikely to remain in place very long. More complex crossbreeding systems sometimes conflict with important management practices unrelated to breeding. For example, sheep and beef cattle crossbreeding systems that require many breeding pastures make proper grazing management difficult. It is important that crossbreeding systems be in harmony with other aspects of animal production. More often than not, this means that crossbreeding systems should be kept comparatively simple.

Accuracy of Genetic Prediction

The higher the accuracy of genetic prediction, the lower the selection risk and the more predictable the offspring. Because relatively little performance information on commercial animals is recorded and even less is reported for analysis, accuracy of prediction in a commercial context refers to accuracy of prediction for the seedstock inputs to crossbreeding systems—typically sires. In several species, accurate EPDs are available for purebred sires, and crossbreeding systems using purebred sires benefit as a result. At this date, the same cannot be said for many hybrid sires. Even when a hybrid sire would otherwise be the best choice, lack of accurate genetic information makes him less attractive.

In the comparisons of crossbreeding systems that follow, you will notice that each system excels in some criteria, but at the expense of other criteria. Inevitably there are trade-offs to be considered. Some systems sustain very high levels of hybrid vigor but are a management nightmare. Some take full advantage of breed complementarity but cannot produce their own replacements. Some produce replacements but lack consistency of performance. In choosing a crossbreeding system for a particular production situation, the key is to find the system whose upside is especially beneficial in that situation and whose downside is relatively less harmful.

ROTATIONAL SYSTEMS

rotational crossbreeding system A crossbreeding system in which generations of females are "rotated" among sire breeds in such a way that they are mated to sires whose breed composition is most different from their own.

Rotational crossbreeding systems are systems in which generations of females are "rotated" among sire breeds in such a way that they are mated to sires whose breed composition is most different from their own. Such systems produce replacement females internally, yet manage to maintain acceptable levels of hybrid vigor by limiting backcrossing. They come in several flavors. Some use purebred sires; others use crossbred sires. Some use all breeds of sires simultaneously; others use them in sequence.

Spatial Rotations Using Purebred Sires

spatial rotation A rotational crossbreeding system in which all sire breeds are used simultaneously—they are spatially separated. Replacement females leave the location of their birth to be mated to sires with different breed composition.

The classic form of rotational crossbreeding system is a **spatial rotation** using purebred sires. In spatial rotations, all sire breeds are used simultaneously—they are *spatially* separated—and replacement females leave the location of their birth to be mated to sires with different breed composition. The simplest system of this kind, a two-breed rotation, is illustrated schematically in Figure 19.1.

In a two-breed spatial rotation there are two breeding locations (pastures for grazing animals, pens or buildings for animals in confinement), and purebred sires of two breeds are assigned to these locations, one breed per location. Females are allotted to locations according to their breed composition. Those with the least

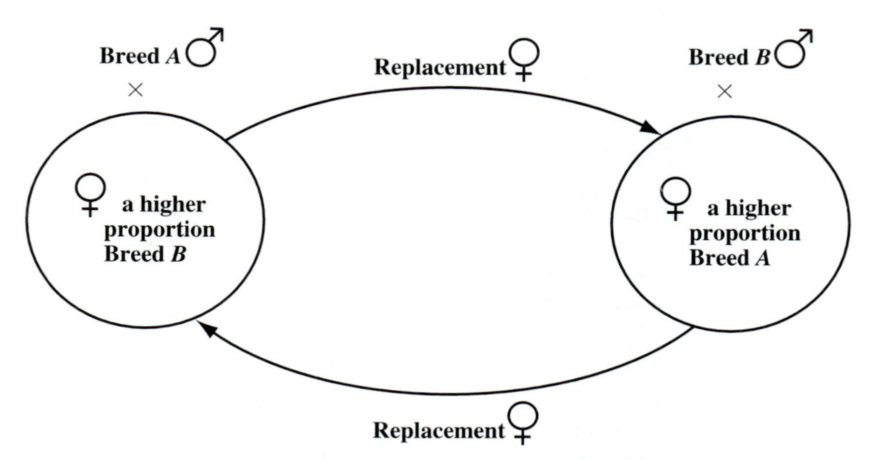

FIGURE 19.1 Schematic representation of a two-breed spatial rotation using purebred sires.

amount of a particular breed in their background are assigned to the location containing sires of that breed. Replacement daughters, having different breed composition than their mothers, are bred to a different sire breed.

As shown in Figure 19.1, the entire scheme appears graphically as a rotation, with replacement females moving from the location of their birth to a second location. *Their* daughters will move to the first location, and so on. In this way, different generations of females are bred in different locations, with no adjacent generations in the same location. In other words, dams should never be in the same breeding location as their daughters. Male offspring and females that are not kept as replacements are sold. They leave the system.

A three-breed spatial rotation using purebred sires is shown in Figure 19.2. It is identical to a two-breed rotation except for the addition of a sire breed and breeding location.

Attributes of Spatial Rotations Using Purebred Sires

Hybrid vigor. Spatial rotations do a good job of maintaining a hybrid vigor in a population. When a rotation is first implemented with purebred females, 100% of F_1 vigor is realized because first-generation animals are true F_1s. If the breed of the foundation females is also a sire breed in the rotation, the second generation of a two-breed rotation (third generation of a three-breed rotation) is a backcross generation, and hybrid vigor declines considerably. Vigor fluctuates in later generations with the change from generation to generation becoming smaller and smaller until breed composition and hybrid vigor reach equilibrium after about seven generations.

Breed composition and percentages of F_1 hybrid vigor retained in successive generations of two- and three-breed spatial rotations using purebred sires are listed in Table 19.1. The foundation generation of females consists of purebreds of a breed

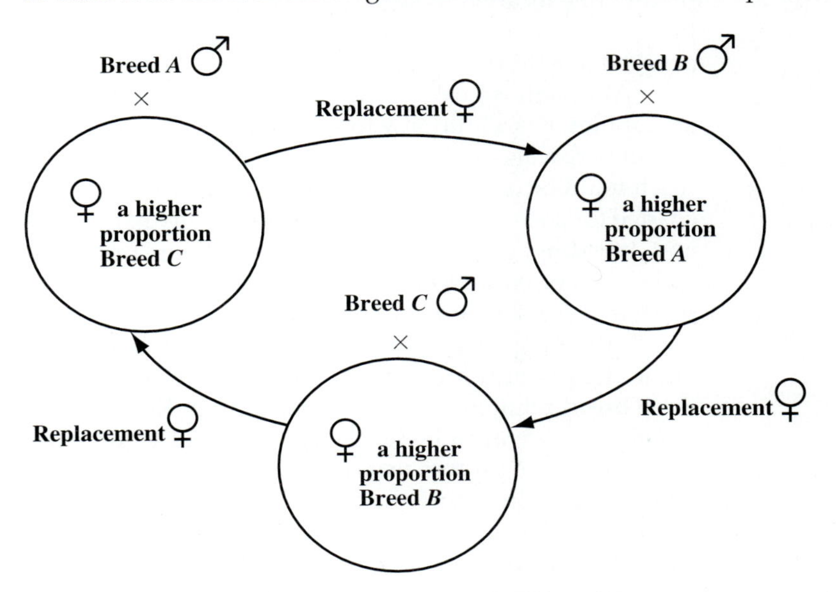

FIGURE 19.2 Schematic representation of a three-breed spatial rotation using purebred sires.

TABLE 19.1 Breed Composition and Percentage of F_1 Hybrid Vigor Retained in Successive Generations of Two- and Three-Breed Spatial Rotations Using Purebred Sires[ab]

	Two-Breed Rotation % Breed Composition				Three-Breed Rotation % Breed Composition				
Generation	Breed of Sire	*A*	*B*	% F_1 HV	Breed of Sire	*A*	*B*	*C*	% F_1 HV
Foundation	*B*	0	100	0	*C*	0	0	100	0
1	*A*	50	50	100	*A*	50	0	50	100
2	*B*	25	75	50	*B*	25	50	25	100
3	*A*	63	37	75	*C*	12	25	63	75
4	*B*	31	69	63	*A*	56	13	31	88
5	*A*	66	34	69	*B*	28	56	16	87
6	*B*	33	67	66	*C*	14	28	58	84
7	*A*	66	34	67	*A*	57	14	29	86
8	*B*	33	67	66	*B*	29	57	14	86
9	*A*	67	33	67	*C*	14	29	57	86
10	*B*	33	67	67	*A*	57	14	29	86

[a]Hybrid vigor is assumed linearly related to heterozygosity.
[b]The breed of the foundation females is also a sire breed in the rotation.

that is also a sire breed in the rotation. Generation 1 is the first generation of crossbred offspring. From this generation are chosen the replacement females that become the dams of Generation 2. Chosen from Generation 2 are the replacement females that become the dams of Generation 3, and so on. For two-breed rotations (see Figure 19.1), females of the foundation generation and of Generations 2, 4, 6, 8, and 10 are bred in one location to sires of Breed *A*. Females of Generations 1, 3, 5, 7, and 9 are bred in a different location to sires of Breed *B*. For three-breed rotations (see Figure 19.2), females of the foundation generation and of Generations 3, 6, and 9 are bred in one location to sires of Breed *A*, females of Generations 1, 4, 7, and 10 are bred in a second location to sires of Breed *B*, and females of Generations 2, 5, and 8 are bred in a third location to sires of Breed *C*.

Although we can conceive of "equilibrium hybrid vigor" from a theoretical standpoint, it is a little silly to speak of equilibrium vigor in species with long generation intervals—cattle, for example. Assuming seven generations are required to reach equilibrium, few rotational systems are likely to remain in place without change for such a long period. The availability of new breeds and shifts in markets and breeding goals prevent most rotations from ever reaching equilibrium. Nevertheless, equilibrium values for hybrid vigor are routinely used in evaluating rotational crossbreeding systems.

Assuming that hybrid vigor is linearly related to heterozygosity,[2] hybrid vigor at equilibrium in a two-breed rotation is 67% of F_1 (maximum) vigor. In a three-breed rotation, it is 86% of maximum. Rotations involving more sire breeds would produce even higher levels of hybrid vigor. Higher-way rotations are unusual, however, for two reasons: (1) they are more difficult to manage, and (2) it is often hard to find more than three compatible breeds with appropriate breeding values. Breeders must consider the trade-off between greater levels of hybrid vigor and better breeding values.

[2]I make this assumption throughout this chapter.

Predicting Equilibrium Hybrid Vigor in Rotational Crossbreeding Systems That Use Purebred Sires

The percentage of F_1 hybrid vigor retained at equilibrium (after about seven generations) in a rotational crossbreeding system that uses purebred sires is predicted by the formula:

$$\% \, R\hat{H}V = \left(\frac{2^n - 2}{2^n - 1}\right) \times 100$$

where n is the number of breeds in the system.

Examples

For a two-breed rotation,

$$\% \, R\hat{H}V = \left(\frac{2^n - 2}{2^n - 1}\right) \times 100$$

$$= \left(\frac{2^2 - 2}{2^2 - 1}\right) \times 100$$

$$= 67\%$$

For a three-breed rotation,

$$\% \, R\hat{H}V = \left(\frac{2^n - 2}{2^n - 1}\right) \times 100$$

$$= \left(\frac{2^3 - 2}{2^3 - 1}\right) \times 100$$

$$= 86\%$$

Breed complementarity and consistency of performance. Rotational systems using purebred sires provide little breed complementarity and may or may not produce consistent performance. Because breed composition varies considerably within the population—in a three-way rotation at equilibrium, females and their offspring can be as much as 57% of a particular breed or as little as 14% of the same breed (Table 19.1)—the only way to be sure of consistent performance is to use breeds that are very similar in biological type. But doing so rules out any possibility of breed complementarity. In beef cattle, for example, you could not use one breed that excels in milk production and another that excels in growth rate (a classic complementary combination) without producing sets of calves that differ a good deal in these traits. Therefore, if complementary breeds are used, consistency suffers, and if breeds are chosen for consistency, breed complementarity is all but eliminated.

Replacement considerations. One of the virtues of a rotational system is that it provides female replacements out of the same dams that produce market animals. There is no need to maintain a special population of dams to produce replacements, nor is there any need to buy replacement females.

Simplicity. Spatial rotations vary in simplicity. A basic two-way system requires only two breeding locations, three or four if young females are bred separately. Three-way rotations need from three to six locations, and higher-way rotations require even more. The larger the number of locations and breeds, the greater the requirements for animal identification, fencing, sorting of animals, monitoring of breeding locations—in short, more investment in facilities, more labor, and higher operating costs. With animals divided among a number of locations, opportunities for high density/short duration grazing and related grazing schemes are limited. Furthermore, spatial rotations are infeasible for very small herds and flocks—those using just one sire.

Accuracy of genetic prediction. For the sires used in crossbreeding systems, accuracy of genetic prediction depends on the status of prediction technology in the species, the willingness of seedstock producers to record and report performance data and use the predictions that are generated, and the size and structure of existing data sets. Accuracy is therefore species and breed dependent. But because state-of-the art EPDs are increasingly common for purebred seedstock, accuracy of prediction should be good in rotational systems that use purebred sires.

Table 19.2 rates most of the crossbreeding systems discussed in this chapter, including two- and three-breed spatial rotations using purebred sires, for the criteria described earlier. Consider these ratings as guides only. They are necessarily subjective and may not be correct in specific production situations.

Spatial Rotations Using Crossbred Sires

Except for differences in the breed composition of sires, spatial rotations using crossbred sires appear identical to spatial rotations using purebred sires (see Figure 19.3). However, the use of crossbred sires has some advantages in terms of hybrid vigor, breed complementarity, and consistency of performance.

TABLE 19.2 Attributes of Various Crossbreeding Systems[ab]

System	HV	Comp	Cons	Reps	Ease	Acc
Two-breed spatial rotation using purebred sires	+	–	varies	+	+	+
Three-breed spatial rotation using purebred sires	+	–	varies	+	–	+
Spatial rotation with crossbred sires	+	+	+	+	varies	?
Rotation in time using purebred sires	+	–	varies	+	++	+
Rotation in time using crossbred sires	+	+	+	+	++	?
Static terminal (buy female replacements)	++	++	+	–	++	+
Static terminal (raise female replacements)	+	+	+	–	–	+
Rotational/terminal	+	+	varies	+	–	+
Pure composite (existing breed)	+	+	+	+	++	?
Pure composite (breed development)	+	+	varies	+	varies	?
Composite/terminal (existing breed)	+	++	+	+	+	?

[a]HV = hybrid vigor; Comp = breed complementarity; Cons = consistency of performance; Reps = replacement considerations; Ease = simplicity; Acc = accuracy of genetic prediction.
[b]– poor; + good; ++ very good.

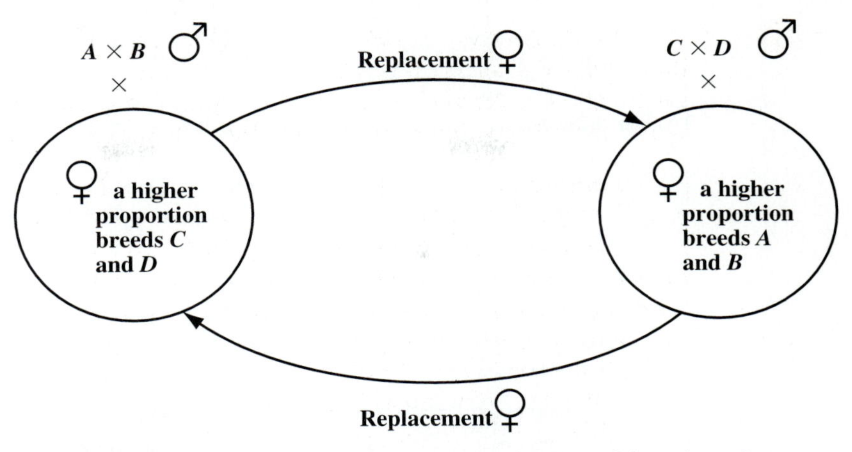

FIGURE 19.3 Schematic representation of a two-breed spatial rotation using crossbred sires.

Attributes of Spatial Rotations Using Crossbred Sires

Hybrid vigor. Rotational crossbreeding systems using crossbred sires usually incorporate more sire breeds than similar systems using purebred sires. As a result, these systems involve less backcrossing and maintain more hybrid vigor. For example, a two-way rotation using $A \times B$ and $C \times D$ sires produces 83% of F_1 hybrid vigor at equilibrium *versus* 67% for a two-way rotation using purebred A and B sires. A three-way rotation using $A \times B$, $C \times D$, and $E \times F$ sires produces 93% of F_1 hybrid vigor *versus* 86% for a three-way rotation using purebred A, B, and C sires. In traits for which paternal hybrid vigor is important—traits such as conception rate—crossbred sires provide an added bonus.

Predicting Equilibrium Hybrid Vigor in Rotational Crossbreeding Systems That Use Crossbred Sires

The percentage of F_1 hybrid vigor retained at equilibrium (after about seven generations) in a rotational crossbreeding system that uses crossbred sires is predicted by the formula:

$$\%\ \mathrm{R\hat{H}V} = \left(\frac{m(2^n - 1) - 1}{m(2^n - 1)}\right) \times 100$$

where n is the number of sire types in the system, and m is the number of breeds present in each sire type. (Assumptions implicit in this formula are that no breed is present in more than one sire type, and that the breeds in each sire type are present in equal proportions.) When $m = 1$, the formula reduces to the simpler formula for rotational systems that use purebred sires.

Examples

For a two-way rotation, two breeds per sire type,

$$\% \ \hat{RHV} = \left(\frac{m(2^n - 1) - 1}{m(2^n - 1)}\right) \times 100$$
$$= \left(\frac{2(2^2 - 1) - 1}{2(2^2 - 1)}\right) \times 100$$
$$= 83\%$$

For a three-way rotation, two breeds per sire type,

$$\% \ \hat{RHV} = \left(\frac{m(2^n - 1) - 1}{m(2^n - 1)}\right) \times 100$$
$$= \left(\frac{2(2^3 - 1) - 1}{2(2^3 - 1)}\right) \times 100$$
$$= 93\%$$

For a two-way rotation, four breeds per sire type,

$$\% \ \hat{RHV} = \left(\frac{m(2^n - 1) - 1}{m(2^n - 1)}\right) \times 100$$
$$= \left(\frac{4(2^2 - 1) - 1}{4(2^2 - 1)}\right) \times 100$$
$$= 92\%$$

Breed complementarity. Because the sires used in any rotational crossbreeding system need to be similar in biological type in order to produce consistent offspring, there is no more opportunity for sire × dam complementarity in rotations using crossbreed sires than in rotations using purebred sires. There is opportunity, however, for hybrid seedstock complementarity—the kind of complementarity that comes into play in the development of crossbred breeding stock. In a beef cattle rotation, for example, sires that are part Hereford (a traditional beef breed) and part Holstein (a dairy breed) might be useful because of their complementary characteristics. Herefords are well adapted to range conditions, but often lack milk and carcass quality. Holsteins are poorly adapted because they milk too heavily, but have excellent carcass quality. Crossbred sires with appropriate proportions of each breed could be just right. Note that this would be an unlikely pair of breeds to use in a rotational system using purebred sires. The change in type of offspring from generation to generation would be too large.

Consistency of performance. Rotations that use crossbred sires have the potential to produce more consistent offspring than rotations using purebred

TABLE 19.3 Minimum and Maximum Percentages of a Single Breed and Percentage of F_1 Hybrid Vigor at Equilibrium for Different Rotational Crossbreeding Systems[a]

System	Min % Breed *A*	Max % Breed *A*	% F_1 Hybrid Vigor
Two-way rotation using purebred *A* and *B* sires	33	67	67
Two-way rotation using $A \times B$ and $C \times D$ sires	17	33	83
Three-way rotation using purebred *A*, *B*, and *C* sires	14	57	86
Three-way rotation using $A \times B$, $C \times D$, and $E \times F$ sires	7	29	93
Two-way rotation using $A \times B$ and $A \times C$ sires	50	50	67
Two-way rotation using $A \times (B \times C)$ and $A \times (D \times E)$ sires	50	50	71
Three-way rotation using $A \times B$, $A \times C$, and $A \times D$ sires	50	50	71

[a]Hybrid vigor is assumed linearly related to heterozygosity.

sires.[3] There are two reasons for this. The first is related to the hybrid seedstock complementarity possible in crossbred sires. Pure breeds may be so diverse that it is difficult to find two or more of them close enough in biological type to provide the necessary consistency of performance in a rotation. The problem can be overcome with proper "mixing and matching" of pure breeds in the development of crossbred sires. Herefords and Holsteins are examples of two rather extreme breeds for which there are few, if any, truly similar counterparts. But there are many breed combinations that are similar to hybrids of Herefords and Holsteins. For every set of diverse and incompatible pure breeds, there are a number of compatible breed combinations.

The second reason why crossbred sires can improve consistency in rotational crossbreeding systems has to do with breed composition. If the breeds included in a rotation vary in biological type, then changes in breed composition between generations will cause differences in offspring performance. Changes in breed composition can be considerable using purebred sires, especially in higher-way rotations. They are much smaller using crossbred sires.

Table 19.3 lists minimum and maximum percentages of a single breed and percentage of F_1 hybrid vigor at equilibrium for different rotational crossbreeding systems. If you look at the upper part of the table, you can see how the use of crossbred sires causes smaller changes in breed composition. The percentage of Breed *A* in a two-way rotation using purebred sires ranges from 33% to 67%—a difference of 34%. With crossbred sires, the difference is $33 - 17 = 16\%$. The percentage of Breed *A* in a three-way rotation using purebred sires ranges from 14% to 57%—a difference of 43%. With crossbred sires, the difference is $29 - 7 = 22\%$.

Sometimes breeders want to maintain the percentage of a particular breed at a constant level. Perhaps that breed has especially desirable characteristics that are lost when the percentage of the breed in the mix is too low. By using crossbred sires in a rotation, breeders can fix the percentage of a given breed and still achieve reasonable levels of hybrid vigor. Examples are shown in the bottom part of Table 19.3. Note that in each case, Breed *A* is represented in each type of crossbred sire, and the percentage of Breed *A* in offspring remains a constant 50%.

[3]Despite the arguments that follow, some breeders remain unconvinced of the potential of crossbred sires for increasing consistency of performance. For more on why breeders feel this way and why they are, in most cases, mistaken, see the discussion of consistency of performance under the subsection entitled *Pure Composite Systems* later in this chapter.

Replacement considerations and simplicity. Spatial rotations that use crossbred sires are no different from spatial rotations that use purebred sires when it comes to replacement considerations and simplicity. Both kinds of systems produce their own replacement females, and in either system management problems are more a function of number of breeding locations than anything else. Rotations using crossbred sires may have an edge in simplicity, however, if you consider that these systems achieve more hybrid vigor per breeding location. For example, a two-way rotation using crossbred sires produces almost as great a percentage of F_1 vigor as a three-way rotation using purebred sires (83% versus 86%, see Table 19.3). By using crossbred sires, a breeder can opt for the simpler system with little sacrifice in hybrid vigor.

Accuracy of genetic prediction. The largest potential drawback of using crossbred sires is lack of genetic information. This need not be the case but, as a rule, EPDs are more prevalent for purebreds than crossbreds. Where crossbred EPDs do exist, accuracies tend to be low. Until accurate genetic predictions for crossbreds are routinely available, rotational systems using crossbred sires will be handicapped.

Rotations in Time

rotation in time
A rotational crossbreeding system in which sire breeds are not used simultaneously, but are introduced in sequence.

Spatial rotations are rotations "in space." Because the population of producing females typically includes individuals from several generations, sire breeds are used simultaneously and kept spatially separated in different breeding locations. **Rotations in time** are rotational crossbreeding systems in which sire breeds are not used simultaneously, but are introduced in sequence. You could say that the breeds are kept *temporally* separated. A three-breed rotation in time is illustrated in Figure 19.4. The chief advantage of rotations in time over spatial rotations is simplicity.

Attributes of Rotations in Time

Hybrid vigor. There is somewhat less hybrid vigor in rotations in time than in spatial rotations due to increased backcrossing. If sire breeds are changed at long intervals, some young females will be bred back to the breed of their sire. If sire breeds are changed at short intervals, the rotation will complete a cycle, and some older females may be bred back to the breed of their sire. However, the difference in the overall levels of hybrid vigor produced by spatial rotations and equivalent rotations in time is not great.

Hybrid vigor can be increased in rotations in time by using crossbred sires and by changing to a higher-way rotation by adding one or more sire breeds to the sequence. Rotations in time are very flexible in this respect.

Breed complementarity, consistency of performance, replacement considerations, and accuracy of genetic prediction. Rotations in time rate essentially the same as spatial rotations for breed complementarity, consistency of performance, replacement considerations, and accuracy of genetic prediction. Replacing purebred sires with crossbred sires has the same effects in a rotation in time that it has in a spatial rotation.

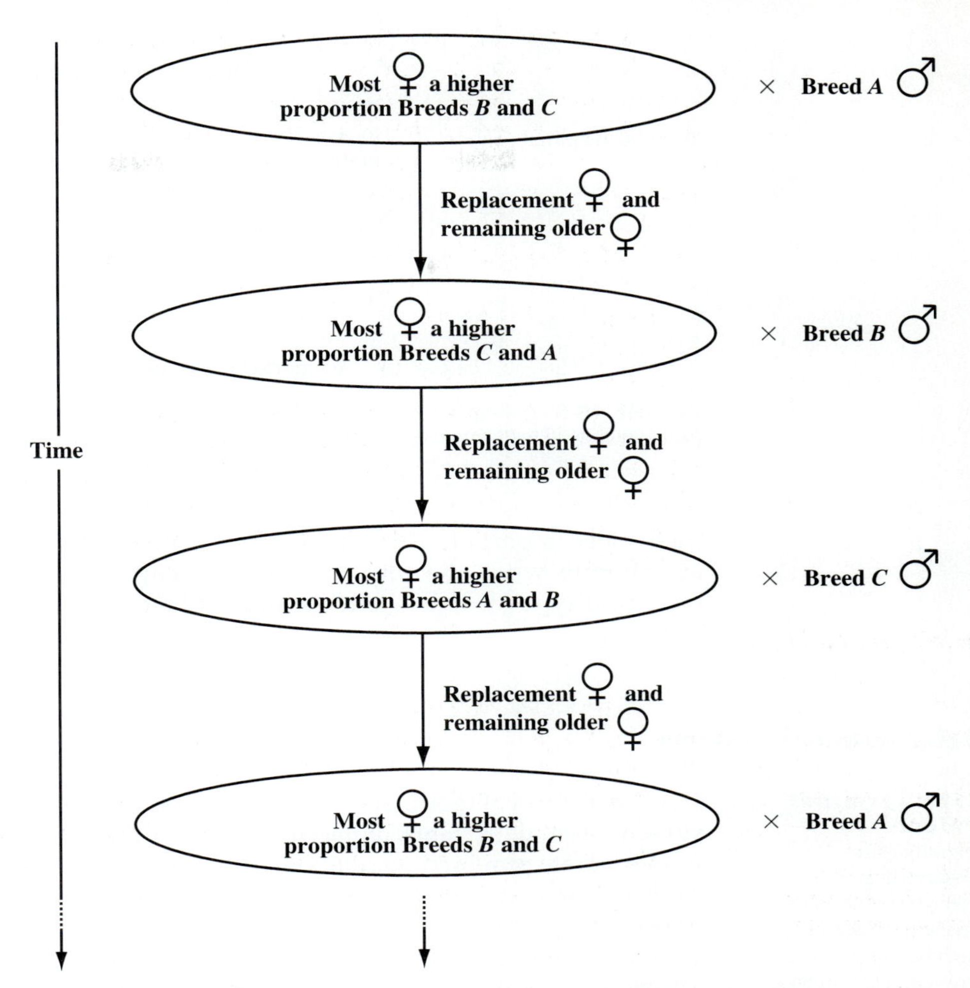

FIGURE 19.4 Schematic representation of a three-breed rotation in time using purebred sires.

terminal sire crossbreeding system
A crossbreeding system in which maternal-breed females are mated to paternal-breed sires to efficiently produce progeny that are especially desirable from a market standpoint. Terminally sired females are not kept as replacements, but are sold as slaughter animals.

Simplicity. Rotations in time are much simpler to manage than spatial rotations. Only a single breeding location is required—two if young females are bred separately. Rotations in time fit nicely with intensive grazing systems. They can be used successfully in the smallest herds or flocks, even those with just one sire.

TERMINAL SIRE SYSTEMS

Terminal sire crossbreeding systems are systems in which maternal-breed females (purebred or crossbred females that excel in maternal traits like conception rate, litter size, milk, and mothering ability) are mated to paternal-breed sires (sires that excel in paternal traits like growth rate and carcass yield) to efficiently produce progeny that are especially desirable from a market standpoint. Terminally sired females are not kept as replacements, but are sold as slaughter animals. These systems produce ample amounts of hybrid vigor, but

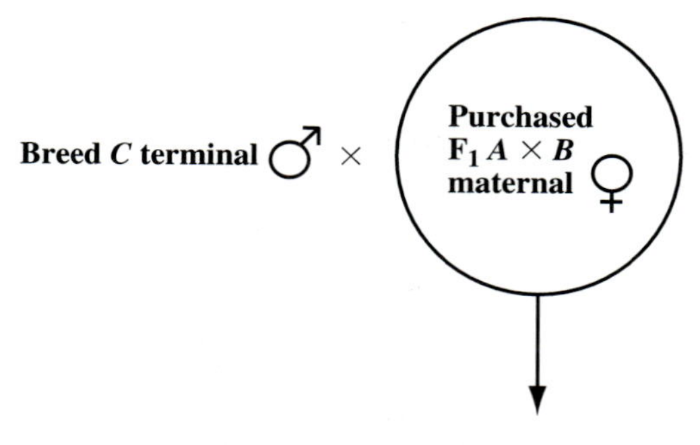

FIGURE 19.5 Schematic example of a static terminal crossbreeding system using purchased replacements.

their most important attribute is breed complementarity. They are especially appropriate when the physical environment, economics, or both favor one biological type for dams, and the market favors a different biological type for offspring.

Static Terminal Systems

static terminal system
A terminal sire crossbreeding system in which replacement females are either purchased or produced from separate purebred populations within the system.

The classic form of terminal sire crossbreeding system is called a **static terminal system.** In static systems, replacement females are either purchased or produced from separate purebred populations within the system. Examples of three-breed static terminal systems are depicted in Figures 19.5 and 19.6. Figure 19.5 illustrates a simple system in which F_1 $A \times B$ replacement females are purchased. They are bred to Breed C terminals sires to produce F_1 $C \times (A \times B)$ market offspring.

Figure 19.6 illustrates a static terminal system in which replacement females are produced in a separate population within the system. This more complex system includes a population of purebred Breed B animals. Excess females from this population are mated to Breed A sires to produce F_1 $A \times B$ replacement females. The F_1 replacements are then bred to Breed C terminal sires to produce F_1 $C \times (A \times B)$ market offspring.

Static terminal systems are common in pig and broiler production. Swine breeding companies often provide their commercial customers with both terminal sires and crossbred females. Most commercial broilers produced today are crosses of Cornish males on females that are themselves crosses of different strains of Plymouth Rocks.

Attributes of Static Terminal Systems

Hybrid vigor. Static terminal systems produce lots of hybrid vigor. In fact, systems that use purchased replacements (like the one in Figure 19.5) produce 100% of both individual F_1 and maternal F_1 hybrid vigor. If F_1 sires are used, these

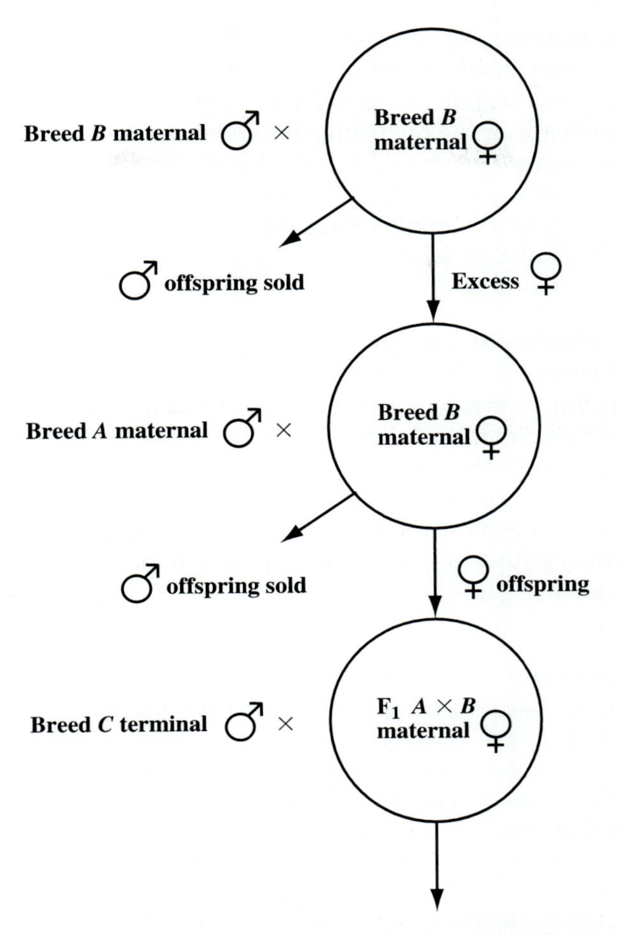

FIGURE 19.6 Schematic example of a static terminal crossbreeding system in which replacement females are produced in a separate population within the system.

systems produce 100% of paternal F_1 vigor as well. Static terminal systems that generate their own replacements (like the one in Figure 19.6) produce 100% of F_1 hybrid vigor in the actual terminal cross. However, if you include the purebred populations involved, the percentage of F_1 vigor in the entire system is considerably less than 100%.

Breed complementarity. The chief purpose of terminal crossbreeding systems is to produce breed complementarity of the sire × dam variety, and these systems do it well. In static systems that use purchased females, every offspring benefits from sire × dam complementarity. In static systems that produce their own replacements, terminal offspring benefit from sire × dam complementarity and F_1 maternal offspring may benefit to an extent, but purebred offspring do not benefit at all.

Consistency of performance. In static terminal systems, the breed composition of terminal offspring is constant. They should, therefore, be consistent. If replacement females are purchased, then all offspring are terminally sired, and the consistency of the offspring should be good. Consistency in the replacement females themselves should be good if they come from a reliable source. If replacement females are raised, several types of offspring are produced: purebreds, maternal F_1s, and terminals. Each type may be quite different from the others, but each type should be uniform.

Replacement considerations. Obtaining replacement females is the most difficult aspect of static terminal systems. Purchasing replacements is the simplest solution, but quality replacements may not be available or, if they are, they may be too expensive. Many breeders do not consider themselves real breeders unless they raise their own replacements. They want their female population to be the result of their own breeding decisions. This attitude is common among cattle breeders, less so among swine breeders. The alternative to buying replacements is to raise them. But there is a price to pay in system-wide hybrid vigor, breed complementarity, and simplicity.

Simplicity. A static terminal system that uses purchased females is very simple from a management standpoint. Only one breeding location is needed. The system works well in large or small populations, even those using just one sire. Static systems that produce their own replacements are necessarily complicated because several different populations are maintained at the same time. These systems are feasible only for large enterprises.

Accuracy of genetic prediction. As in any crossbreeding system, accuracy of genetic prediction in static terminal systems depends on the accuracy of predictions for sires. Because accurate EPDs are more common for purebreds than crossbreds, static terminal systems that use purebred sires are likely to benefit from greater accuracy of prediction than systems using hybrid sires.

Rotational/Terminal Systems

rotational/ terminal system
A crossbreeding system combining a maternal rotation for producing replacement females with terminal sires for producing market offspring.

Rotational/terminal systems are designed to solve the replacement problems inherent in static terminal systems. They combine a maternal rotation for producing replacement females with terminal sires for producing market offspring. The rotational part of the system could be any kind of rotation: a spatial rotation or rotation in time, a rotation using purebred sires or one using crossbred sires. In any case, a portion of the herd or flock is bred to maternal sires to produce replacements, and the remaining females are bred to terminal sires to produce market offspring. If terminal sires are likely to cause dystocia in young females, those females are typically bred to maternal sires, and older females are bred to terminal sires. A two-breed spatial rotational/terminal system (two breeds in the rotation plus a third, terminal sire breed) is shown in Figure 19.7.

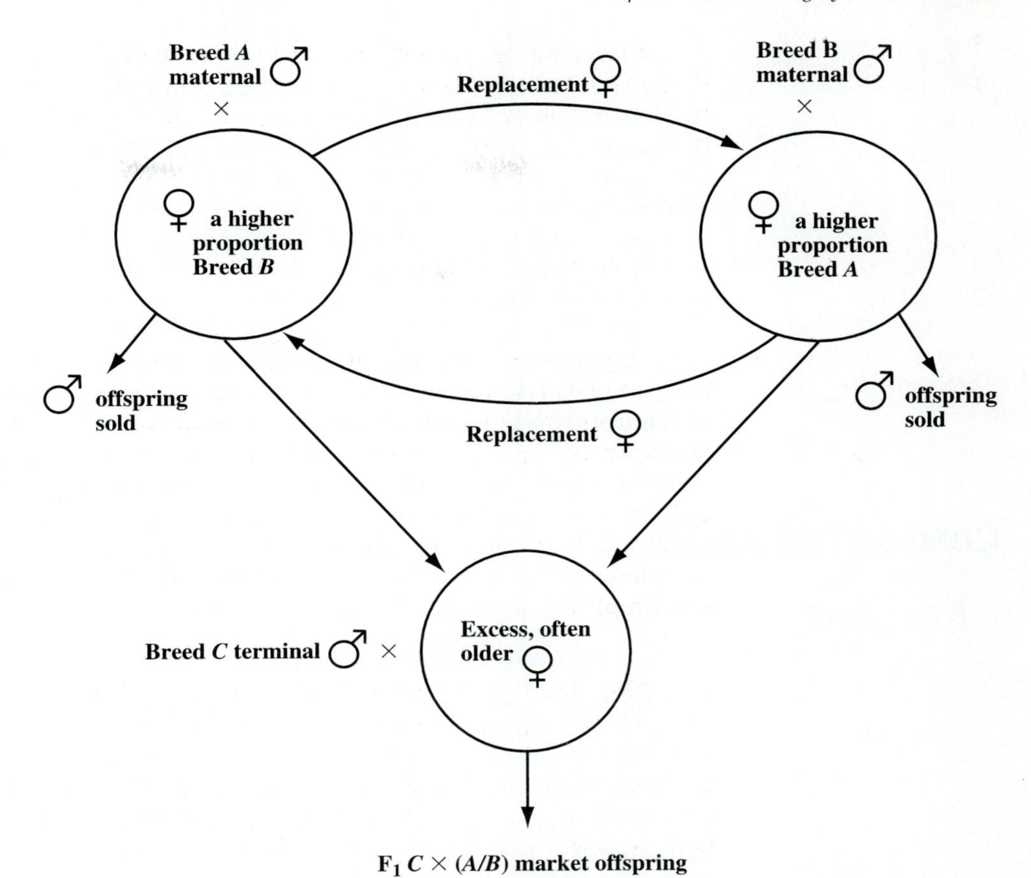

FIGURE 19.7 Schematic representation of a two-breed spatial rotational/terminal system (two breeds in the rotation plus a third, terminal sire breed).

Attributes of Rotational/Terminal Systems

Rotational/terminal systems combine the attributes of their rotational and terminal parts. They provide the breed complementarity missing from pure rotational systems and the crossbred replacements missing from pure terminal systems.

Hybrid vigor, breed complementarity, and replacement considerations. Rotational/terminal systems provide more hybrid vigor and breed complementarity than comparable rotational systems, but less than comparable static terminal systems. They are also intermediate in the replacement category. Even though these systems produce their own replacement females, the task is made harder by the fact that only a portion of the population is bred to replacement producing sires. Compared to pure rotational systems, a greater proportion of young maternal females must be kept as replacements, causing selection differential in replacement females to be smaller.

Consistency of performance and accuracy of genetic prediction. Rotational/terminal systems produce two kinds of animals, maternal and terminal, that are likely to be quite different from each other. Terminal offspring should be uniform. Uniformity of maternal offspring depends on how well the breeds in the rotation are matched. As usual, accuracy of genetic prediction depends on the accuracy of the predictions for the sires used.

Simplicity. Whenever two crossbreeding systems are combined, you can expect the combination to be more complex than its parts. Rotational/terminal systems are no exception. Beyond the requirements of the rotation, an additional location is needed to accommodate terminal matings. The entire system can be made less complicated by using a simpler rotation—a rotation in time, for example—or (sometimes) by using artificial insemination and reducing the number of breeding locations.

COMPOSITE ANIMALS

On the lecture circuit, Mark Twain would tell a story involving a hybrid dog:

> . . . and a mighty good dog too; he wa'n't no common dog, he wa'n't no mongrel; he was a composite. A composite dog is a dog that's made up of all the valuable qualities that's in the dog breed—kind of a syndicate; and a mongrel is made up of the riffraff that's left over.[4]

Twain probably did not understand the word "composite" as it is used in modern animal breeding, but he understood enough to know that a composite is a legitimate animal, an animal with considerably more status than a mongrel. And he was right. Today we use composites to gain many of the benefits of crossbreeding—without crossbreeding. Bred and managed much like purebreds, composites retain enough hybrid vigor to be viable commercial animals.

composite (synthetic animal) A hybrid with at least two and typically more breeds in its background. Composites are expected to be bred to their own kind, retaining a level of hybrid vigor normally associated with traditional crossbreeding systems.

Like Mark Twain's dog, **composite animals,** sometimes called **synthetics,** are hybrids. They have at least two breeds in their background and often more. What distinguishes them from typical crossbreds is not their genetic makeup per se, but rather the way in which they are used. Composites are expected to be bred to their own kind, retaining a level of hybrid vigor we normally associate with traditional crossbreeding systems, but without crossbreeding.

For example, consider the standard black baldy cow. She is a hybrid, typically the result of mating a purebred Angus bull to a purebred Hereford cow or vice versa. In all likelihood she will be bred back to a purebred bull of one of the parent breeds or perhaps of a third breed. Because she is to be used as part of a conventional crossbreeding system (e.g., a rotation of some kind), we would not consider her a composite animal. However, if her owner decided to breed her to black baldy sires, saving daughters and perhaps even sons as replacements, we would have to label her a composite. She became a composite, as opposed to simply a crossbred, because the breeder chose to mate her to her own hybrid

[4]Clemens, Samual Langhorne, *The Autobiography of Mark Twain,* edited by Charles Neider. Copyright © 1959 by Charles Neider and The Mark Twain Company. HarperCollins Publishers Inc., New York.

kind with the expectation of retaining a degree of hybrid vigor without further crossbreeding.

Admittedly, this definition leaves a little to be desired. What if I have a population of composite animals and one day I decide to breed them to terminal sires or make them part of a conventional rotational crossbreeding system? Are they still composites? Whether you answer yes or no depends on how strict you want to be in your definition of a composite. I would say yes because these animals were bred to be part of a composite breeding system and still have that potential.

Most of our experience with composites comes from plants. Plant breeders developed composites as a practical way for farmers in third world countries to take advantage of hybrid vigor. Composite plants may not yield as well as F_1 hybrids, but they do much better than pure varieties. And because further crossing of composites is not necessary, farmers can save their own composite seed for next year's planting. They are not dependent on seed companies to supply them with F_1 seed every year. Composite plant populations are termed *synthetic varieties.* The analogous term in animal populations is **composite breeds.** A composite breed is a breed made up of two or more component breeds and designed to benefit from hybrid vigor without crossing with other breeds.

composite (synthetic) breed A breed made up of two or more component breeds and designed to benefit from hybrid vigor without crossing with other breeds.

Pure Composite Systems

pure composite system A mating system limited to matings within a single composite breed.

The simplest way to use composite animals in commercial breeding is with a **pure composite system.** Such a system is "pure" in the sense that it involves just one composite breed, and all matings are within-breed matings. There is no crossbreeding.

Attributes of Pure Composite Systems

Hybrid vigor. Pure composite systems can produce considerable hybrid vigor. This comes as something of a surprise to many of us because we have been taught that the only way to get hybrid vigor is by crossbreeding. Recall from Chapter 18, however, that when two-breed F_1s are mated to produce F_2s, half of F_1 hybrid vigor is lost but, barring significant inbreeding, half remains in the F_2, F_3, and subsequent generations. This remaining hybrid vigor is retained in what is now a two-breed composite.

If composites were limited to half of F_1 hybrid vigor, they would not be the most attractive alternative to many crossbreeding systems. But we can do much better than half of F_1 vigor. The amount of vigor retained depends on the number and proportions of component breeds in the composite. You can see this mathematically using the formulas for predicting retained hybrid vigor from Chapter 18:

$$R\hat{H}V = \left(1 - \sum_{i=1}^{n} p_{s_i} p_{d_i}\right) F_1\hat{H}V$$

and

$$\%R\hat{H}V = \left(1 - \sum_{i=1}^{n} p_{s_i} p_{d_i}\right) \times 100$$

Because sires and dams of the same composite breed have identical breed composition, the proportion of a given breed in sires (p_{s_i}) equals the proportion of that breed in dames (p_{d_i}). Thus, for composites,

$$R\hat{H}V = \left(1 - \sum_{i=1}^{n} p_i^2\right) F_1\hat{H}V$$

and

$$\%R\hat{H}V = \left(1 - \sum_{i=1}^{n} p_i^2\right) \times 100$$

where p_i is the proportion of the ith breed in a composite made up of n component breeds.

For example, consider a two-breed composite that is 50% Breed *A* and 50% Breed *B*.

$$\begin{aligned}\%R\hat{H}V &= \left(1 - \sum_{i=1}^{n} p_i^2\right) \times 100\\ &= [1 - ((.5)^2 + (.5)^2)] \times 100\\ &= [1 - (.25 + .25)] \times 100\\ &= 50\%\end{aligned}$$

As expected, the two-breed composite retains half of F_1 hybrid vigor.

Now consider a four-breed composite with equal fractions of each component breed.

$$\begin{aligned}\%R\hat{H}V &= \left(1 - \sum_{i=1}^{n} p_i^2\right) \times 100\\ &= [1 - ((.25)^2 + (.25)^2 + (.25)^2 + (.25)^2)] \times 100\\ &= [1 - (.0625 + .0625 + .0625 + .0625)] \times 100\\ &= 75\%\end{aligned}$$

The four-breed composite is expected to retain 75% of F_1 hybrid vigor. A similar eight-breed composite should retain 88% (see Figure 19.8). These are respectable amounts of vigor. We could get even more hybrid vigor with more breeds in the composite, but finding that many appropriate breeds could be difficult. Again, we must ask whether the hybrid vigor gained by adding another breed to the mix is worth the potential loss in breeding value caused by adding that breed.

Composite breeds achieve expected levels of retained hybrid vigor in most cases, but not always. Exceptions occur for some traits in some breed combinations. As explained in Chapter 18, the equations for retained hybrid vigor are valid to the extent that the dominance model for hybrid vigor is valid. But if there is significant recombination loss, hybrid vigor in advanced generations of composites will continue to decline before eventually reaching equilibrium.

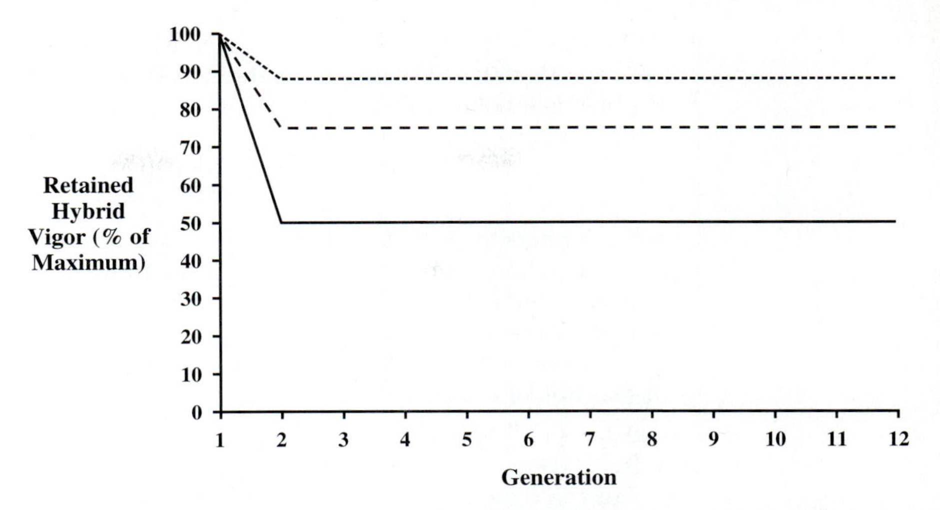

FIGURE 19.8 Retained hybrid vigor (measured as a percentage of maximum vigor) in successive generations of two-breed (solid line), four-breed (broken line), and eight-breed (dotted line) composites. Each composite population is assumed to contain equal fractions of component breeds. The generation marked 1 on the horizontal scale is the F_1 generation—the first generation of the composite in which all component breeds are represented. Generation 2 is the F_2 generation, Generation 3 the F_3 generation, and so on.

A More Rigorous Approach to Predicting Crossbred Performance

You can use the formula developed below to predict more precisely the performance of a composite population or, for that matter, the performance of any crossbred population. Begin with a version of the genetic model for quantitative traits:

$$P = \mu + BV + GCV + E_{cg} + E$$

If we modify the model to represent *average* performance for a population, individual environmental effects cancel, and we are left with:

$$\overline{P} = \mu + \overline{BV} + \overline{GCV} + E_{cg}$$

From experience we can guess the mean performance of a contemporary group ($\mu + E_{cg}$). Then, substituting a prediction of retained hybrid vigor for GCV, we have:

$$\hat{\overline{P}} = (\mu + \hat{E}_{cg}) + \hat{\overline{BV}} + \mathrm{R\hat{H}V}$$

We can calculate $\hat{\overline{BV}}$ using a weighted average of mean breeding values of the breeds involved. Thus

$$\hat{\overline{BV}} = \sum_{i=1}^{n} p_i \overline{BV_i^*}$$

where i refers to the ith breed of n breeds. The asterisk (*) signifies that the n mean breeding values have somehow been transformed to a common scale. In other words, these breeding values can be used to compare the genetic potentials of the various breeds.

We can predict retained hybrid vigor using the formula shown in the last section of Chapter 18:

$$\mathrm{R\hat{H}V} = \sum_{i=1}^{n} \sum_{j=1}^{n} p_i p_j \mathrm{F_1 \hat{H}V}_{ij}$$

Putting all this together,

$$\begin{aligned} \hat{\overline{P}} &= (\mu + \hat{E}_{cg}) + \hat{\overline{BV}} + \mathrm{R\hat{H}V} \\ &= (\mu + \hat{E}_{cg}) + \sum_{i=1}^{n} p_i \overline{BV_i^*} + \sum_{i=1}^{n} \sum_{j=1}^{n} p_i p_j \mathrm{F_1 \hat{H}V}_{ij} \end{aligned}$$

Example

Let's return to the sheep example from Chapter 18. Suppose you decided to develop a composite breed from $(A \times B) \times (C \times (A \times B))$ lambs and want to predict the performance of the composite for 60-day weaning weight. We know from experience that the mean 60-day weaning weight of purebred, average performing (meaning $\overline{BV^*}$ for the breed is near zero) sheep in your environment is about 42 lb. Thus

$$(\mu + \hat{E}_{cg}) = 42 \text{ lb}$$

Let's assume that for the direct component of weaning weight,

$$\overline{BV^*_{d_A}} = -1 \text{ lb}'$$

$$\overline{BV^*_{d_B}} = +4 \text{ lb}$$

$$\overline{BV^*_{d_C}} = +6 \text{ lb}$$

and for the maternal component of weaning weight,

$$\overline{BV^*_{m_A}} = +4 \text{ lb}$$

$$\overline{BV^*_{m_B}} = -2 \text{ lb}$$

$$\overline{BV^*_{m_C}} = +1 \text{ lb}$$

The composite sheep are ⅜ Breed A, ⅜ Breed B, and ¼ Breed C, so

$$\hat{\overline{BV}}_d = \sum_{i=1}^{n} p_i \overline{BV^*_{d_i}}$$

$$= \frac{3}{8}(-1) + \frac{3}{8}(4) + \frac{1}{4}(6)$$
$$= +2.6 \text{ lb}$$

and

$$\overline{\hat{BV}_m} = \sum_{i=1}^{n} p_i \overline{BV^*_{m_i}}$$
$$= \frac{3}{8}(4) + \frac{3}{8}(-2) + \frac{1}{4}(1)$$
$$= +1 \text{ lb}$$

Then

$$\overline{\hat{BV}} = \overline{\hat{BV}_d} + \overline{\hat{BV}_m}$$
$$= 2.6 + 1$$
$$= +3.6 \text{ lb}$$

Given F_1 individual hybrid vigor for $A \times B$ crosses: 2.7 lb: for $A \times C$ crosses: 5.5 lb: and for $B \times C$ crosses: 4.0 lb, individual hybrid vigor retained in the composite breed should be

$$R\hat{H}V_1 = \sum_{i=1}^{n}\sum_{j=1}^{n} p_i p_j F_1 \hat{H}V_{I_{ij}}$$
$$= \left(\frac{3}{8}\left(\frac{3}{8}\right)(0) + \frac{3}{8}\left(\frac{3}{8}\right)(2.7) + \frac{3}{8}\left(\frac{1}{4}\right)(5.5)\right)$$
$$+ \left(\frac{3}{8}\left(\frac{3}{8}\right)(2.7) + \frac{3}{8}\left(\frac{3}{8}\right)(0) + \frac{3}{8}\left(\frac{1}{4}\right)(4.0)\right)$$
$$+ \left(\frac{1}{4}\left(\frac{3}{8}\right)(5.5) + \frac{1}{4}\left(\frac{3}{8}\right)(4.0) + \frac{1}{4}\left(\frac{1}{4}\right)(0)\right)$$
$$= 2.54 \text{ lb}$$

Assuming similar values for maternal hybrid vigor (i.e., $R\hat{H}V_M = 2.54$ lb as well), then

$$R\hat{H}V = R\hat{H}V_I + R\hat{H}V_M$$
$$= 2.54 + 2.54$$
$$= 5.1 \text{ lb}$$

Altogether,

$$\hat{P} = (\mu + \hat{E}_{cg}) + \overline{\hat{BV}} + R\hat{H}V$$
$$= 42 + 3.6 + 5.1$$
$$= 50.7 \text{ lb}$$

Breeders often ask: After a while, won't a composite breed become just another breed? In other words, will a composite population lose its ability to retain hybrid vigor over time? The answer is no *if* inbreeding is kept to a minimum. On the other hand, if the composite breed is allowed to become significantly inbred—as purebreds are—it will indeed become just another pure breed.

Breed complementarity. Because the animals within a pure composite system are all of the same basic biological type, there is little opportunity for breed complementarity of the sire × dam variety. But there is opportunity for hybrid seedstock complementarity—the kind of complementarity realized in the *formation* of composite breeds. Just as we can use Hereford × Holstein sires (crosses of two very different but complementary breeds) in a rotational crossbreeding system, we can use these breeds in a single composite breed. While purebreds from breeds like Holstein or Hereford may be too extreme in one trait or another, a composite population containing fractions of each breed might be just right.

Consistency of performance. The performance of composite animals is about as consistent as the performance of purebreds. This comes as a surprise to many, perhaps because classical genetics texts are full of examples showing increased variation in the progeny of hybrids. The books are not wrong, but the examples are misleading because they inevitably involve simply-inherited traits with categorical phenotypes.

One such example involves coat color and horns in crosses of Black Angus and Horned Hereford cattle. Purebred Black Angus are, with some exceptions, homozygous black and homozygous polled (*BBPP*). Horned Herefords are red and horned (*bbpp*). Because the black and polled alleles are completely dominant, F_1 Angus × Herefords or black baldies are black and polled. They are consistent with respect to coat color and absence of horns.

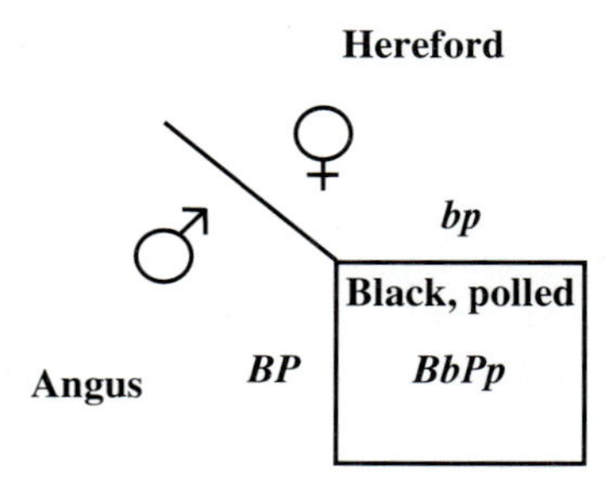

But when F_1 black baldies are mated to F_1 black baldies, segregation of genes is clearly visible in their F_2 offspring. Most calves will be black and polled, but some will be black and horned, some red and polled, and a few red and horned. Because of the increase in the number of distinct phenotypes involving coat color and presence or absence of horns, there is now little consistency with respect to these traits.

Angus x Hereford (♀ across the top; ♂ Angus x Hereford down the side)

♂ \ ♀	*BP*	*Bp*	*bP*	*bp*
BP	**Black, polled** *BBPP*	**Black, polled** *BBPp*	**Black, polled** *BbPP*	**Black, polled** *BbPp*
Bp	**Black, polled** *BBPp*	**Black, horned** *BBpp*	**Black, polled** *BbPp*	**Black, horned** *Bbpp*
bP	**Black, polled** *BbPP*	**Black, polled** *BbPp*	**Red, polled** *bbPP*	**Red, polled** *bbPp*
bp	**Black, polled** *BbPp*	**Black, horned** *Bbpp*	**Red, polled** *bbPp*	**Red, horned** *bbpp*

Without selection for coat color and the presence or absence of horns, a herd of two-breed Angus/Hereford composites can be expected to show all four of the phenotypes apparent in F_2s. The herd will be inconsistent with respect to these simply-inherited traits.

The same is *not* true for polygenic traits, the traits that tend to be the most important economically. Because phenotypes for most of these traits exhibit continuous expression, F_2 and later generations of composites show no increase in the *number* of distinct phenotypes. We expect statistical measures of variability to be somewhat greater for composites than purebreds or F_1s, but the more loci affecting a trait, the smaller the increase in variability. Experimental data suggest that composites are as consistent for most polygenic traits as purebreds. For traits influenced heavily by hybrid vigor, they may be somewhat more consistent. When compared to hybrids from a rotational crossbreeding system, composites are inevitably more consistent because they do not vary in breed composition.

Replacement considerations. Like straightbreds, composites produce their own female replacements, so composites score well in this category. Composites have the potential to produce their own replacement *males* as well, though for most commercial producers the extra level of management and record keeping required to do a good job of home-raised sire selection is probably impractical. Most buy composite sires from composite seedstock producers.

Simplicity. From a management standpoint, breeding an existing population of composites is like breeding straightbreds; only one breeding location is needed (two if young females are bred separately). All the problems associated with having multiple breeds are eliminated, and for this reason, the greatest virtue of a composite breeding program may well be simplicity. Composites can be used successfully in small herds or flocks—even those with only one sire—and with composites there should be no conflict between the breeding program and forage management.

Simplicity is often *not* a virtue of composite breed development. Assembling a composite breed can be very complex, involving many breeds at once. On the other hand, it can be quite simple. For example, one easy way to develop a four-breed composite is to mate an existing population of two-breed F_1 females to purchased sires that are a cross of two other breeds. In this way, you can create a four-breed composite population in just one generation.

Accuracy of genetic prediction. Just as crossbred sires are handicapped by a lack of genetic information, so are composites. Some composite populations have EPDs, but EPDs are more common for purebreds. That is partly for technical reasons—it is often more difficult to compute EPDs for hybrids than for purebreds—and partly because composite breeders are typically not allowed access to data on the purebred ancestors of their animals.

Additional considerations. Composite animals can be designed to fit a specific environment. By carefully selecting breeds and individuals within breeds, we can create composites specifically adapted to the desert, to the tropics and subtropics, or to cold country. Wherever the environment poses special challenges, there is an opportunity for an appropriately designed composite breed.

Composites are unique in that they can play both commercial and seedstock roles. They retain enough hybrid vigor to be viable commercial animals, but are also needed as seedstock to be used in commercial composite herds or flocks. The dual roles of composite animals allow their breeders a degree of flexibility. Commercial breeders can become seedstock breeders and vice versa. Or breeders can be both at the same time.

Composite/Terminal Systems

composite/terminal system A crossbreeding system combining a maternal composite breed for producing replacement females with terminal sires for producing market offspring.

Simply breeding composites to composites as though they were purebreds is not the only way to use composites commercially. A modified scheme is the **composite/terminal system.** In this system, some of the composite females, typically the younger ones, are bred to composite sires, and the rest are bred to terminal sires (Figure 19.9). Replacement females come from the composite × composite matings, and all terminally sired offspring are marketed.

Attributes of Composite/Terminal Systems

A composite/terminal system is more complex than a pure composite system. Use of terminal sires means an additional breeding location. However, this modest loss in simplicity comes with an additional measure of complementarity (of the sire × dam variety) and hybrid vigor.

Breeding Composite Seedstock

Breeding composite seedstock is different from breeding purebred seedstock. There are two reasons for this. The first is that in composite breeding there are two distinct breeding stages: (1) forming the composite and (2) breeding the composite once it is formed. The second reason is that in composite breeding we are interested

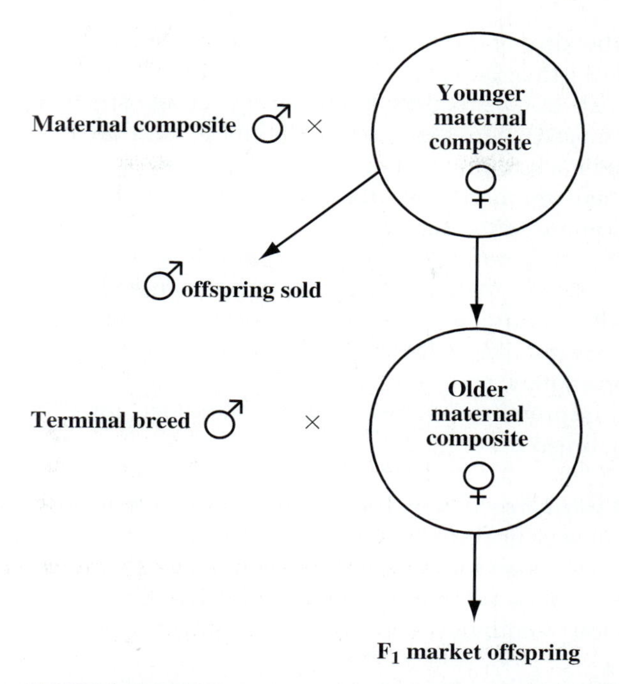

FIGURE 19.9 Schematic representation of a composite/terminal system.

not only in improving breeding value through selection—our sole genetic goal in purebred breeding—but also in maintaining a high level of hybrid vigor. These basic, theoretical differences between the two types of breeding spawn a number of practical differences.

Forming the Composite

Between-breed selection. Selection of the breeds and the proportions of those breeds going into a composite is *the* critical step in composite breed formation and may well determine whether a composite breed succeeds or fails. If composite breeders do a conscientious and carefully researched job of between-breed selection, the newly formed composite may not need much genetic change—change that can now only be achieved through slow-paced within-breed selection. In other words, if the composite is put together in such a way that it exhibits close to optimum performance in the economically important traits when it is first formed, then any genetic change following breed formation can be considered fine-tuning.

In choosing the breeds to go into a composite, breeders should know how the composite will be used by commercial producers. They should define the geographical areas and environmental/management scenarios appropriate for the new breed. They should also determine the mating systems in which the composite will be used. Many composite breeds are likely to be general-purpose breeds because that is the kind of animal required for commercial pure composite systems. Other composites may be specialized maternal breeds designed to be bred to terminal sites. In all cases, breeders should design composite breeds for optimum levels of performance (where they can be identified), and use breed differences to gain hybrid seedstock complementarity.

There are many important traits for which EPDs and other performance information are rare. This is usually because performance in these traits is hard to measure. Temperament is a typical example. It is an important trait, but a difficult one to quantify. Without objective performance data, it is hard to change these traits with traditional within-breed selection. Therefore, it is precisely these traits that we would like to "get right" in the breed formation stage of composite breeding. Fortunately, breed differences in many of these traits are large. Considering hard-to-improve traits during composite breed formation is another way of exploiting breed differences.

Within-breed selection. The choice of what breeds to include may be the single most critical step in composite breed formation, but the breeder's responsibility does not end there. Composite breeders should be selective in their choices of individual foundation animals. A good job of composite breed formation means a good job of both between- and within-breed selection.

Maintaining hybrid vigor. In purebred breeding, we pay little attention to hybrid vigor. Because purebreds are likely to be crossed with animals of other breeds at the commercial level, hybrid vigor within purebred populations is not important. Composites, on the other hand, are designed to be used commercially *without* crossbreeding. Any loss of hybrid vigor among seedstock composites can therefore mean a loss of hybrid vigor at the commercial level.

The way to maintain hybrid vigor within composite seedstock populations is to prevent inbreeding, and the key to preventing inbreeding during breed formation is to establish as broad a genetic base as possible. From a practical standpoint this means including in the foundation population a number of unrelated sires or daughters of unrelated sires from each component breed. To see why, consider the extreme case of a four-breed composite in which one component breed's contribution came in the form of semen from a single sire. Every member of the first generation of composite animals will be a grandson or granddaughter of that animal. He will appear repeatedly in the pedigrees of future generations, and the result is an increase in inbreeding.

The need to sample broadly in composite formation can make it difficult to intensely select foundation animals. Few really good candidates may be available, or those that are available may be related. The problem is especially acute for breeds with small population size. To accommodate the trade-off between sampling broadly and selecting intensely, you must be willing to relax selection standards, accept narrower sampling of a breed, or do a little of both.

Breeding the Formed Composite

Once a composite breed has been formed, breeding composite animals is much like breeding purebreds. The principal objective is to improve breeding value through within-breed selection. The tools are the same: pedigree and performance records,

EPDs, etc. Selection of composites differs from purebred selection only with regard to certain practices needed to avoid inbreeding within the composite population.

Maintain a large population. The rate of inbreeding is much faster in small populations than in large ones. If you have created your own composite herd or flock and keep it closed (no animals from outside allowed in), the population needs to be large enough that inbreeding accumulates very slowly. How large is large enough? That depends upon the rate of inbreeding you are willing to accept. It also depends on the average number of offspring per sire—the fewer progeny per sire (and, therefore, the more sires), the better from an inbreeding standpoint. Given an average of 75 offspring over a sire's lifetime, a minimum population size of 500 is reasonable for beef cattle. Larger numbers are needed for sheep and swine because their shorter generation intervals increase the annual rate of inbreeding.

Cooperate with other breeders. Your herd or flock may be small, but if you work with other composite seedstock producers by exchanging semen, sires, or females, the **effective population size** of the composite—the size of the population as reflected by its rate of inbreeding—can be kept large. Cooperative arrangements of this sort essentially replicate the structure of pure breeds.

effective population size The size of a population as reflected by its rate of inbreeding.

Avoid linebreeding. Linebreeding, the mating of individuals within a particular line, is a time-honored practice in purebred breeding. Purebred breeders do not hesitate to make a half-brother × half-sister matings or to build pedigrees laced with sons and daughters of a particular sire. However, if certain lines become prominent within a composite population and within-line matings are the rule, the effective population size is reduced and the rate of inbreeding increases.

Avoiding linebreeding basically means not overusing any one sire or line of sires. This is a departure from purebred breeding where whole breeding programs have been built around one outstanding animal. Compared to purebred breeding, composite breeding places less emphasis on selecting superior *individuals* and more emphasis on selecting better *groups* of animals.

It is possible for a number of lines to be developed within a composite breed and then for commercial breeders to avoid inbreeding within their commercial composites by judiciously choosing sires from the various lines. The problem with this idea is that it misses the main point of the commercial composite breeding system—simplicity. If commercial producers must keep sires from one composite line separate from sires from another line, then, from a management perspective, we are back to rotational crossbreeding and the headaches associated with it. The other advantage of composites (e.g., hybrid seedstock complementarity and consistency of performance) are still present, however.

Reconstitute the composite from time to time. One fortunate aspect of inbreeding is that it can be "undone." As soon as inbreds are mated to unrelated animals, the offspring are no longer inbred. In a composite context, inbreeding can be undone by adding to the composite population new first-generation composites, particularly animals whose purebred parents or grandparents are relatively unrelated to the purebreds that formed the foundation for the original composite population. "Reconstituting" the composite in this way is not easy, but it is the best solution for a composite population that is nearing the point of too much inbreeding.

EXERCISES

Study Questions

19.1 Define in your own words:

crossbreeding system
sire × dam complementarity
hybrid seedstock complementarity
rotational crossbreeding system
spatial rotation
rotation in time
terminal sire crossbreeding system
static terminal system
rotational/terminal system
composite (synthetic) animal
composite (synthetic) breed
pure composite system
composite/terminal system
effective population size

19.2 Why are crossbreeding *systems* (as distinguished from crossbreeding itself) necessary?

19.3 Why are crossbreeding systems found primarily in commercial animal production?

19.4 List and explain the importance of the seven criteria used in this book to evaluate crossbreeding systems.

19.5 How does sire × dam complementarity differ from hybrid seedstock complementarity?

19.6 In two- and three-way rotational crossbreeding systems, why does hybrid vigor vary over time before eventually reaching equilibrium after about seven generations?

19.7 If retained hybrid vigor increases with more breeds in a rotation, why not include many breeds?

19.8 Why is it often difficult to achieve both breed complementarity and consistency of performance in a crossbreeding system?

19.9 In the context of rotational crossbreeding compare the use of crossbred sires relative to the use of purebred sires with respect to:

a. hybrid vigor
b. breed complementarity
c. consistency of performance
d. simplicity
e. accuracy of genetic prediction

19.10 What is the chief advantage of rotations in time over spatial rotations?

19.11 What is the single most important attribute of terminal sire crossbreeding systems?

19.12 Describe the psychological barrier that prevents some breeders from using static terminal sire systems.

19.13 Why is female selection intensity less in rotational/terminal systems than in pure rotations?

19.14 Why do breeders often mistakenly believe that composite sires (and hybrid sires in general) produce offspring that are noticeably more variable for polygenic traits than offspring of purebred sires?

19.15 What is unique about composites that allows them to play both commercial and seedstock roles?

19.16 List two reasons why breeding composite seedstock is different from breeding purebred seedstock?

19.17 If you were to breed composite seedstock, describe the approach you would take with respect to:

a. choice of biological type for the formed composite.
b. choice of traits to emphasize in determining component breeds and their proportions in the composite.
c. breeding objectives for within-breed selection of foundation animals.
d. prevention of inbreeding during composite formation.
e. breeding objectives for selection within the formed composite.
f. prevention of inbreeding in the formed composite.

19.18 Summarize much of this chapter by rating the following crossbreeding systems for hybrid vigor, breed complementarity, consistency of performance, replacement considerations, simplicity, and accuracy of genetic prediction:

a. two-breed spatial rotation using purebred sires
b. three-breed spatial rotation using purebred sires
c. spatial rotation with crossbred sires
d. rotation in time using purebred sires
e. rotation in time using crossbred sires
f. static terminal (buy female replacements)
g. static terminal (raise female replacements)
h. rotational/terminal
i. pure composite (existing breed)
j. pure composite (breed development)
k. composite/terminal (existing breed)

Problems

19.1 Rank the following rotational crossbreeding systems for proportion of F_1 hybrid vigor retained at equilibrium. (For systems using hybrid sires, assume that the breeds in each sire type are present in equal proportions.)

a. Four-way rotation using purebred *A*, *B*, *C*, and *D* sires.
b. Three-way rotation using hybrid $A \times B \times C \times D$, $E \times F \times G \times H$, and $I \times J \times K \times L$ sires.
c. Two-way rotation using hybrid $A \times B \times C \times D \times E \times F \times G \times H$ and $I \times J \times K \times L \times M \times N \times O \times P$ sires.

19.2 Assuming negligible inbreeding and recombination loss, calculate the proportion of F_1 hybrid vigor retained in advanced generations of each of the following composite beef breeds:

a. Brangus ($\frac{5}{8}$ Angus, $\frac{3}{8}$ Brahman)
b. Charbray ($\frac{13}{16}$ Charolais, $\frac{3}{16}$ Brahman)
c. RX_3 ($\frac{1}{2}$ Red Angus, $\frac{1}{4}$ Hereford, $\frac{1}{4}$ Red Holstein)
d. Barzona (approximately $\frac{1}{4}$ Hereford, $\frac{5}{64}$ Angus, $\frac{1}{4}$ Afrikaner, $\frac{15}{64}$ Shorthorn, $\frac{3}{16}$ Brahman)

19.3 Use the genetic and environmental sheep data and assumptions listed in the last boxed section of this chapter to rank the following pure and composite breeds for 60-day weaning weight performance:

a. *A*
b. *B*
c. *C*
d. $\frac{1}{2}A$, $\frac{1}{2}C$
e. $\frac{1}{4}A$, $\frac{1}{4}B$, $\frac{1}{2}C$

PART V

New Techniques, Old Strategies

This last section contains just two chapters. The first chapter describes new biological techniques that have the potential to transform the way we breed animals. The second chapter is an essay about old strategies that will always work, no matter how animal breeding evolves.

As animal breeders, we need to stay abreast of new techniques because the practice of animal breeding changes with advances in technology. In the first 100 years after Darwin's *The Origin of Species*, we altered the way we measure and evaluate animals as we came to understand Mendelian mechanisms and as we developed statistical procedures for dealing with quantitative traits. In the second half of the twentieth century, advances in information technology and computing took us a step further. We gained the ability to select animals using information synthesized from enormous numbers of performance records. And we are far from seeing the last effect of the computer revolution on animal breeding.

Now we are entering a new age—the age of biotechnology. It is an age of great promise, yet there is little consensus on how biotechnology will influence animal breeding in the future. The most shameless optimists predict we will soon be creating designer animals, building them gene by gene. The most hardened skeptics doubt that biotechnology will have much effect on animal breeding at all. The truth, as it usually does, lies somewhere between these extremes. Chapter 20 is an attempt to put the impact of biotechnology on animal breeding in perspective.

The more animal breeding changes, however, the more it, well, stays the same. Breeders will always select what they hope are the best animals. They will always design matings. And the most successful, longest-lived breeding programs will always be those that use time-tested, commonsense animal breeding strategies. Those strategies are the topic of Chapter 21.

CHAPTER 20

Biotechnology and Animal Breeding

biotechnology The application of biological knowledge to practical needs. The term usually refers to (1) technologies for altering reproduction and (2) technologies for locating, identifying, comparing, or otherwise manipulating genes.

Biotechnology can be broadly defined as the application of biological knowledge to practical needs. From an animal breeding perspective, biotechnologies fall into two categories (with some potential overlap between them). The first category is comprised of reproductive technologies such as artificial insemination, embryo transfer, and sex control. The second category consists of molecular technologies, which can be used to locate, identify, compare, or otherwise manipulate genes. These include such techniques as DNA fingerprinting, marker assisted selection, and gene transfer.

In this chapter I discuss those biotechnologies that are likely to affect the practice of animal breeding. Some are being used now; some are not yet feasible. I make no attempt to address the technical details of each technology. Others have written thick books for that purpose. My intent is to provide a general (and, I hope, not *too* simplistic) overview of each technology so that you can appreciate its possibilities and pitfalls. Biotechnology is changing rapidly. New discoveries and technical breakthroughs occur every month. Keep in mind that some of what I have written in this chapter may be outdated by the time you read it.

Advances in biotechnology will undoubtedly affect the way we go about breeding animals. But the extent of their impact will depend on their effectiveness, practicality, and cost—things that breeders need to remember if they are to maintain perspective. The most glamorous technologies may be technically feasible but prohibitively expensive. Comparatively pedestrian but cheap technologies may be very helpful. Before passing judgment on a new technology, find out how well it works, how easy it is to use, how much it costs now, and how much it is likely to cost in the future.

REPRODUCTIVE TECHNOLOGIES

People often confuse animal breeding with reproduction. In reality, animal breeding and reproduction are two distinct disciplines. Animal breeding is a branch of genetics, and reproduction is an aspect of physiology. Still, the two fields are inextricably tied. Although the underlying genetic principles of animal breeding remain largely independent of the physiology of reproduction, the practice of animal breeding does not. Selection and mating decisions are made in the context of available reproductive technologies. For an example, consider the effect on animal breeding of one of the oldest, most widely used, and economically important reproductive technologies—artificial insemination.

Artificial Insemination

artificial insemination (A.I.)
A reproductive technology in which semen is collected from males, then used in fresh or frozen form to breed females.

Artificial insemination (A.I.) is a reproductive technology in which semen is collected from males and then used in fresh or frozen form to breed females. The popularity of A.I. varies from species to species and country to country. In general, it is used widely in poultry, swine, and cattle, and to a lesser extent in sheep and horses. The benefits of A.I. have been huge, especially for those species in which semen can be frozen and stored indefinitely. Frozen semen makes it possible for one animal to sire literally thousands of offspring. Imagine the accuracy of selection made possible by such large progeny numbers. A.I. also increases selection intensity by providing access to the very best sires from the very best herds and flocks. In some species, A.I.'s ability to increase accuracy of selection and selection intensity, combined with the technology's relatively low cost, have resulted in a marked increase in the rate of genetic change.

multiple-sire pasture
A breeding pasture (or pen) containing more than one sire at a time.

connectedness
The degree to which data from different contemporary groups within a population can be compared as a result of pedigree relationships between animals in different groups.

Artificial insemination has other genetic benefits. It makes it easier and less expensive to import new breeds. A.I. allows the sire of individual offspring to be identified—something that is difficult when the alternative to artificial insemination is a **multiple-sire pasture,** a breeding pasture (or pen) containing more than one sire at a time. Freezing semen is a way to preserve germ plasm, saving genes that may be useful in the future. In those species in which artificial insemination is used extensively, designed matings to test a sire for deleterious recessive alleles are rare. More commonly, sires that carry one or more deleterious alleles are identified after they have been mated artificially to many females from the general population. Tests of this kind are not the most efficient (they sometimes spread undesirable genes widely before carrier sires are detected) but they work. Finally, A.I. profoundly increases **connectedness,** the degree to which data from different contemporary groups within a population can be compared as a result of pedigree relationships between animals in different groups. When the same sire is used artificially in different herds or flocks, those populations become genetically connected, and large-scale genetic evaluation using BLUP and BLUP-like procedures is made possible.

estrus synchronization
The administration of hormones to a group of mammalian females causing them to come into heat (estrus) at or near the same time.

Despite its popularity, artificial insemination is not easy, nor is it always practical. Mammalian females must be detected in heat, moved to a breeding area, and properly inseminated at the right time. The technology known as **estrus synchronization** makes A.I. easier. Hormones are administered to a group of females causing them to come into heat (estrus) at or near the same time. This reduces the number of days needed to inseminate the group and often simplifies heat detection, sometimes making it unnecessary.

Embryo Transfer and Related Technologies

What artificial insemination is to sires, **embryo transfer (E.T.)** is to dams—at least to a degree. The analogy is not perfect because, while A.I. allows a sire to have potentially thousands of progeny, E.T. by itself allows a dam to have tens. Embryo

embryo transfer (E.T.)
A reproductive technology in which embryos from donor females are collected and transferred in fresh or frozen form to recipient females.

superovulation
The administration of a hormone causing a female to develop and release more eggs than normal.

transfer involves the collection of embryos from donor females and the transfer of those embryos to recipient females. Typically the donor is **superovulated—**given a hormone injection causing her to develop and release more eggs than normal. She is then inseminated, and after an interval of time (the length of which depends on species), embryos are collected and either transferred immediately to recipients or frozen for transfer at a later date.

As with artificial insemination, embryo transfer allows an animal to have many more offspring than normal and provides breeders access to select individuals, thus increasing accuracy of selection and selection intensity—only for females instead of males. But because potential numbers of offspring per parent are far fewer with E.T. than with A.I., accuracy and intensity remain correspondingly smaller. And because females normally have relatively few offspring and therefore little progeny data, identifying superior candidates for embryo transfer can be difficult.

Embryo transfer creates certain problems for genetic evaluation, especially for traits with important maternal effects—traits like weaning weight, for example. Because the embryo donor does not raise her own offspring, progeny performance cannot contribute directly to a prediction of her maternal ability. Likewise, if little information exists on the genetic merit of recipient females—as would usually be the case if recipients are commercial (nonregistered) animals—their maternal ability can cloud prediction of direct (growth) effects for the progeny and the donor. These problems usually work themselves out in time as the offspring acquire progeny data themselves.

Embryo transfer is probably the safest way to import or export germ plasm because embryos are less likely to harbor disease organisms than either frozen semen or live animals. Freezing embryos is also an excellent way to preserve germ plasm for the future. Unlike a sperm cell, an embryo is a whole individual, so freezing and storing an embryo not only preserves individual genes but gene combinations as well. To reconstitute an otherwise extinct population from semen alone requires generations of grading up. But a population can be re-created from embryos in just one generation.

Embryo transfer is considerably more difficult and costly than artificial insemination. Donor and recipients must be in good breeding condition and, if fresh (nonfrozen) embryos are transferred, ovulation of recipients must be synchronized with that of the donor. Success rates are highly variable due primarily to differences in response to superovulation. Some attempts produce many pregnancies. Others produce none at all.

in vitro fertilization
A technique by which eggs are collected from donor females, then matured and fertilized in the laboratory.

A developing technology that may make embryo transfer more flexible and cost-effective (at least for cattle and horses) is ***in vitro* fertilization** or fertilization in a test tube. Eggs are collected from donor females, then matured and fertilized in the laboratory. The resulting embryos can either be transferred immediately to recipients or frozen.

The chief advantage of in vitro fertilization over traditional methods of embryo recovery is an increase in the number of pregnancies possible. With *in vitro* fertilization, you can collect eggs repeatedly from the same donor at relatively short intervals. She need not be at a particular stage of her estrous cycle, and can be in the first third or so of pregnancy. It is even possible to retrieve eggs from the ovaries of animals that have just died on the farm or at the slaughterhouse.

A second advantage of *in vitro* fertilization is its potential for decreasing generation interval. Eggs can be collected from young, prepubertal females and then matured, fertilized, and transferred, producing first offspring in considerably less time than would otherwise be possible. Incorporated into nucleus breeding schemes using MOET (multiple ovulation and embryo transfer), this technology could, in theory, decrease generation interval so much that the rate of genetic change would be increased substantially.[1]

In vitro fertilization is expensive, success rates to date are not very good (the average for cattle is two to three pregnancies per month), and there is some concern about abnormalities in the offspring. The technology is young, however, and improvements may come quickly.

Sex Control

It is now possible to determine the sex of an embryo by physically removing a few cells and examining the chromosomes. It is also possible to sort sperm carrying a male sex chromosome from sperm carrying a female sex chromosome, though sorting rates are currently too slow to make sexed sperm commercially viable.[2,3] Someday it may even be possible to develop males that are capable of siring offspring of just one sex, or of whatever sex the breeder desires at the moment. Scientists have predicted the imminent use of sex control for decades, yet practical sex control continues to elude us. Still, it may not be far away.

Why sex control? The main reason for sex control is that one sex is often more valuable than the other. For example, males of meat animal species are typically worth more commercially than females because males grow faster and produce more meat. Sex control would allow breeders to produce more animals of the more valuable sex and fewer of the less valuable sex.

The chief reason for sex control is an economic one, but the ability to know the sex of offspring ahead of time will inevitably affect selection and mating decisions. If we know, for example, that a particular sire is strong in maternal traits and weaker in paternal traits, we will be more inclined to use him if we can be assured of only female offspring. And knowing that he will produce only daughters will affect our choice of mates. We will likely mate him to dams with the best potential for producing good replacement females.

Sex control gives an advantage to certain crossbreeding systems. Terminal sire systems become more attractive if market offspring can be limited to the more valuable sex. Combination systems—rotational/terminal systems, composite/terminal systems, and similar variants—benefit from sex control too. If the maternal-breed sires in these systems produce only daughters and the paternal-breed sires produce only sons, all offspring are of the "right" sex. There are no by-products in the form of sons of maternal sires or daughters of paternal sires. And fewer dams are needed for the maternal portion of these systems because all their offspring, not just half of them, are potential replacement females. This leaves more dams for the terminal part of these systems, the part that should be the most profitable.

[1]See Chapter 12 for a description of nucleus breeding schemes.

[2]The idea of sexing sperm applies to mammals but not to birds. In birds, the gamete produced by the *female,* not the male, determines the sex of the offspring.

[3]*In vitro* fertilization requires much smaller numbers of sperm than standard artificial insemination. While current sorting rates are too slow to produce the sperm numbers required for A.I., they are adequate for *in vitro* fertilization.

single-sex system
A proposed production system for beef cattle that increases overall feed efficiency by combining sex control (all females) with slaughter at young ages.

In beef cattle, **single-sex systems** have been proposed to dramatically increase the efficiency of beef production. In such a system, each young female is bred with female-producing semen (or receives a female embryo) soon after puberty, producing a daughter which is weaned after just a few months. The dam is then fed for slaughter, and her daughter replaces her in the herd. The efficiency of the system derives from the fact that no (or very few) older animals are maintained. Every animal is young and growing. Compared to a conventional system, a much larger proportion of the feed consumed is used for growth (i.e., meat production) instead of being used to maintain mature cows. Feed is used more efficiently. If single-sex systems become common in beef cattle production, the demand for *all-around* animals (as opposed to terminal or maternal types) should increase, and breeders will probably place greater selection emphasis on early puberty and calving ease.

In addition to producing more offspring of the more valuable sex, sex control can be used to influence the elements of the key equation and speed the rate of genetic change. Consider the example of young dairy sires. Today promising young bulls are bred to a number of cows and evaluated on the basis of first-lactation records of their daughters. The number of test matings of this kind is limited for economic reasons. With sex control, twice as many daughters can be produced with the same number of matings. Therefore, a bull could generate twice the progeny data (increasing accuracy of selection), his test matings could be completed in a shorter period (decreasing generation interval), or twice as many bulls could be evaluated (increasing selection intensity). By whatever route, genetic change occurs faster.

Cloning

cloning
A reproductive technology for producing genetically identical individuals.

clone
(1) An individual member of a clonal line; (2) a clonal line.

clonal line
A population of genetically identical individuals (also referred to as a clone).

embryo splitting (bisection)
A reproductive technology in which embryos are mechanically cut in half to produce twin embryos.

nuclear transplantation
Surgical removal of an egg's nucleus followed by insertion of an individual cell extracted from an embryo.

No reproductive technology has greater potential to change the way we breed animals than **cloning,** the production of genetically identical individuals **(clones).** If cloning becomes feasible, the commercial herds and flocks of today could be replaced by **clonal lines,** populations of individuals (presumably highly select ones) that are the genetic equivalent of identical twins.[4]

Small-scale cloning is possible now. **Embryo splitting (bisection)** is a relatively simple mechanical technique for cutting an embryo in half to produce twin embryos which can then be transferred to recipient females. Embryo splitting produces only two (or possibly a few) identical offspring. It is therefore more a way of increasing the efficiency of embryo transfer than a method for producing clones.

Cloning by **nuclear transplantation** is entirely different from embryo splitting. Eggs are matured in vitro and their nuclei are removed surgically. Individual cells from a multicell embryo are then inserted in each egg, producing a number of identical embryos. The embryos can be transferred or frozen, or can serve as cell donors for repeated cycles of cloning by nuclear transplantation. In theory, the number of possible clones produced by this method is unlimited.

Hypothetical scenarios for large-scale cloning vary from species to species. In dairy species, clones would presumably be females. In meat animals, we might want maternal-type clones for dams and a different type of clone for market offspring. Where hybrid vigor is important, clones should be crossbred, probably F_1s.

[4]The word "clone" can mean both a clonal line and an individual member of a clonal line. To avoid confusion, I use "clone" to refer to the individual and "clonal line" to refer to the group.

A typical cloning scenario would begin with matings of highly select, accurately evaluated sires and dams followed by limited cloning of the resulting embryos. Most of the embryos from each clonal line are transferred to recipients, but a few are preserved by freezing. After birth, cloned individuals are measured for performance and genetically evaluated using conventional statistical procedures (allowing for the fact that they were raised by recipient dams and incorporating the extremely high pedigree relationship among fellow clones—they are, after all, copies of a single individual). Once certain clones have proved superior, their frozen embryo twins are thawed and cloned on a large scale, resulting in many genetically identical copies for sale. The process could take years to complete, depending primarily on the time required to adequately measure performance.

A single cycle of cloning as I have just described would cause an immediate, sizable increase in the average genetic merit of many herds and flocks. But it would not be a recipe for continued genetic improvement. For that to occur, the best clonal lines should be crossed, creating new genetic combinations, and the process begun again. The cycle of limited cloning, testing and selection, large-scale cloning, and crossing of clones is repeated over and over.

The potential benefits of cloning are many. First there is the initial jump in genetic merit as current herds and flocks convert to using clones. In just one generation, these populations change from genetically average or below to genetically elite. Second, uniformity of performance increases dramatically for many traits. If a herd or flock is comprised of a single clonal line, genetic variation within the herd or flock is nonexistent. Observed variation is necessarily environmental in origin and depends on heritability. The higher the heritability of a trait, the more uniform the population for that trait. (Actually, observed variability within a single clonal line is a function of **broad-sense heritability (H^2)** because genetically identical animals have the same gene combination value as well as the same breeding value. In other words, they have the same *genotypic* value.) Third, large-scale cloning can increase accuracy of selection by virtue of the potentially large amounts of performance data available for each clonal line, and can increase selection intensity by providing virtually unlimited access to the very best clonal lines.

heritability in the broad sense (H^2)
A measure of the strength of the relationship between performance (phenotypic values) and genotypic values for a trait in a population.

Cloning has its problems. If a population is reduced to a small number of clonal lines, it may lose the genetic variation necessary for future improvement, and excessive inbreeding may result. And if a single clonal line comprises a significant proportion of a population, there is always the risk that the line could be susceptible to a particular pathogen or environmental stress, resulting in an epidemic or widespread loss of production. Because each clonal line should be genetically evaluated before it is offered commercially, cloning cycles are time consuming. Cloning is not uniformly successful. Cloning by nuclear transplantation tends to produce abnormal newborns, many of which die. Those that survive usually normalize within a few days. Cloning is also expensive. The technique is costly in itself (even without including the expense of embryo transfer), but what makes a system of cloning really costly is the amount of waste involved. To preserve genetic variation and provide enough selection intensity, many clonal lines should be created, but only a relative few will be selected and allowed to propagate. The rest become very expensive, failed experiments.

Much of the time and expense of cloning could be eliminated if we could go beyond the cloning of embryos to the cloning of mature animals—if we could, for example, clone an individual starting from a hair sample. We could then avoid the initial limited cloning process altogether, making copies of only those animals that had already proven superior. Unfortunately, such a cloning technology, if it ever becomes feasible, is still years away.

Same-Sex Mating

same-sex mating A reproductive technology for mating individuals of the same sex.

nuclear fusion Artificial fertilization—a laboratory technique for combining the nuclei of two gametes.

Another twist in reproductive technology with implications for animal breeding is **same-sex mating,** the mating of individuals of the same sex. Using procedures for nuclear transplantation and embryo transfer, and a laboratory technique known as **nuclear fusion** (a kind of artificial fertilization that combines the nuclei of two gametes), it is theoretically possible to create individuals whose parents are both male or both female. Same-sex mating is not currently feasible in mammals, and some scientists think it will never work. They believe that chromosomes from both male and female origins are necessary for normal fetal development.

The genetic rationale for same-sex mating is much the same as for other reproductive technologies: to increase the rate of genetic change by increasing accuracy of selection and selection intensity. Because the most accurately evaluated and intensely selected individuals today are sires, breeders could benefit by mating the very best proven sires to the very best proven sires. (With clonal lines of females, female × female matings might be just as useful.) In mammals, male × male matings should result in one female to every two male offspring. A third kind of offspring containing two male sex chromosomes would not be viable. Female × female matings would produce all females.[5] We could even mate individuals to themselves **(selfing),** quickly creating inbred lines that could be used to make especially productive hybrids in much the same way hybrid plant varieties are produced today.

selfing The mating of an individual to itself.

Molecular Technologies

Reproductive technologies operate at the level of the sperm cell, egg cell, or embryo. Molecular technologies go a step further; they operate at the level of individual genes, at the level of DNA. The rest of this chapter is devoted to three molecular technologies that are likely to influence animal breeding: DNA fingerprinting, marker assisted selection, and gene transfer.

DNA Fingerprinting for Animal Identification

DNA fingerprinting A laboratory method for graphically characterizing an individual's DNA, creating a unique genetic "fingerprint."

DNA fingerprinting is a laboratory method for graphically characterizing an individual's DNA, creating a genetic "fingerprint" unique to each individual. The process begins with a small sample of blood or other tissue. Specific regions

[5]The reverse is true in poultry because the *female* carries both kinds of sex chromosomes.

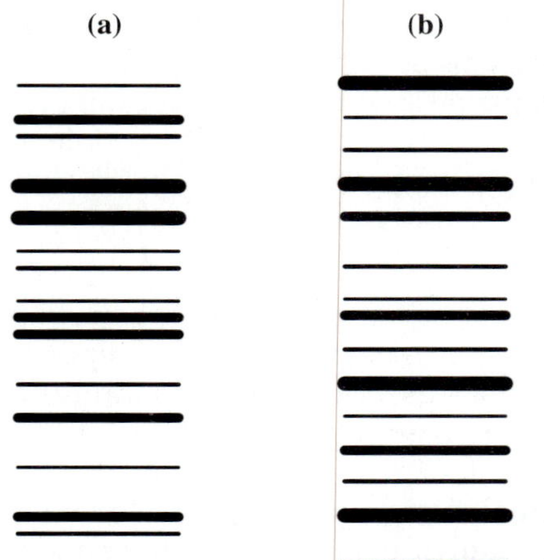

FIGURE 20.1
Schematic representation of banding patterns revealing the DNA fingerprints of two distinct individuals: (*a*) and (*b*).

polymorphic (with respect to genes and DNA fragments) Having at least two alternative forms or alleles in a population.

of extracted DNA that are **polymorphic** (i.e., that have at least two alternative forms or alleles occurring in the population) are chemically copied and placed on a gel where they are sorted using an electric current. The resulting pattern resembles a product bar code (Figure 20.1). The pattern of bands represents the presence of specific alleles at a number of loci.[6] Although we often know little about the chromosomal location or function of the DNA fragments indicated by the bands, their combinations are unique for every individual (except identical twins). The banding pattern is then the DNA equivalent of a conventional fingerprint, accurately identifying individual animals.

The more closely related two animals are, the more alleles they have in common, and the more their DNA fingerprints look alike. Because of this, DNA fingerprinting can be used to establish parentage. Comprehensive fingerprints could even be used to determine an animal's breed composition.

DNA fingerprinting has implications for the management of pedigreed populations. In order to know the sire of each offspring, breeders of pedigreed animals usually put just one sire in each breeding pasture (pen) or use artificial insemination. Neither method is easy to manage. With cost-effective DNA fingerprinting, breeders could use multiple-sire pastures, determining paternity from blood or tissue samples once the offspring are born. They could also use the technique to measure differences between sires in their ability to compete and impregnate females in multiple-sire situations.

From a genetic perspective, the chief benefit of DNA fingerprinting would be an increase in accuracy of selection. Sires (and, occasionally, dams) are often misidentified, and the faulty pedigrees that result cause genetic predictions to be less reliable, especially for animals with little progeny data. DNA fingerprinting would prevent errors of this kind. And if breeders who must use multiple-sire pastures could identify sires through fingerprinting, the amount of genetically useful information would increase significantly. Breed data banks could conceivably be

[6]To call the different forms of DNA fragments detectable by DNA fingerprinting techniques "alleles" is, strictly speaking, incorrect. Many fragments do not appear to be functional genes at all and therefore do not have *alleles*, only different *versions*. Because the distinction is not important enough to warrant the awkwardness of continually making it, I use "alleles" in all cases.

expanded to include information from commercial herds and flocks. Commercial data would be especially useful for traits not normally measured in seedstock—carcass traits, for example.

Other Applications of DNA Fingerprinting

DNA fingerprinting can be used for more than just animal identification. In theory, it can be used to decrease the time required for a population to achieve "purebred" status in grading-up programs or in programs of repeated backcrossing to import an allele. In these programs the *average* breed composition of a crossbred population is easy to predict. For example, two-breed BC_1 populations necessarily average 75% of one breed and 25% of another. But segregation of alleles causes the breed composition of individual animals to vary from the average. As little as 60% of the genes of one BC_1 individual may derive from a particular parent breed, and as many as 90% of the genes from another BC_1 individual may derive from the same breed. It may be possible to compare the DNA fingerprints of crossbred individuals to a typical DNA profile of purebreds of the desired breed to determine those individuals carrying more genes from that breed. Selecting these animals could then reduce the number of backcross generations required.

DNA fingerprinting can (in theory, anyway) also be used to predict hybrid vigor for specific crosses. Hybrid vigor is greater when the parent breeds or lines are more distantly related. By comparing the representative fingerprints or DNA profiles of particular populations, we may be able to tell which populations are least related. That information would indicate which crosses should produce the most hybrid vigor. We could accomplish the same thing for individual matings by comparing the fingerprints of individual sires and dams.

Marker Assisted Selection for Simply-Inherited Traits

The key to selecting for simply-inherited traits is knowing the genotypes of prospective parents. Sometimes that is easy. We know, for example, that a chocolate Labrador has the genotype *bb* at the locus affecting black/chocolate coat color. But at other times we can only guess an animal's genotype. A black Labrador can have either a *BB* or *Bb* genotype, and unless one of its parents is known to be chocolate, we cannot know which genotype it has until it produces offspring. The best we can do is assign it a *probable* genotype based on what is known about the coat color genotypes of its ancestors. The less certain we are about a prospective parent's probable genotype, the more difficult selection becomes. Selection for or against a specific allele, particularly when dominance is complete, would be much easier if we had a method for ferreting out the genotypes of individuals before mating them. **Marker assisted selection** is one such method.

marker assisted selection
Selection for specific alleles using genetic markers.

genetic marker
A detectable gene or DNA fragment used to identify alleles at a linked locus.

Marker assisted selection is selection for specific alleles using **genetic markers,** detectable genes or DNA fragments that are linked closely enough to loci of interest that they can be used to identify alleles at those loci. "Detectable" in this context means that we are in some way able to identify different versions (alleles) of the marker. Some markers are detectable because their alleles produce easily distinguishable phenotypes. Alleles for other markers can be differentiated by their protein products. A rapidly increasing number of genetic markers are now detected using DNA fingerprinting techniques.

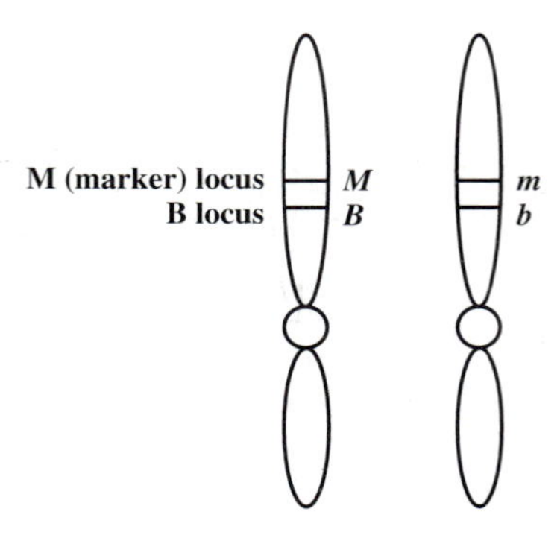

FIGURE 20.2
Schematic drawing of a chromosome pair in which the M (marker) and B loci are linked so that detectable genes at the M locus serve as markers for genes of interest at the B locus. Detection of the *M* allele indicates the presence of the *B* allele, and detection of *m* indicates the presence of *b*.

The concept of genetic markers is illustrated schematically in Figure 20.2. The figure shows a homologous pair of chromosomes with linked M and B loci. The B locus affects some simply-inherited trait of interest—say, black/chocolate coat color in Labs. The M locus is of little interest in itself, but alleles at this locus are detectable with DNA fingerprinting and serve as markers for genes at the B locus. Detection of the *M* allele indicates the presence of the *B* allele, and detection of *m* indicates the presence of *b*. By obtaining a black Labrador's DNA fingerprint from a blood sample and determining its alleles at the marker locus, we can predict whether it is homozygous black (*BB*) or a heterozygote (*Bb*) carrying the gene for chocolate coat color.

A practical example of the use of marker assisted selection involves the halothane gene in swine.[7] Boars can now be tested for the presence of the halothane allele using genetic markers and can be certified heterozygous or halothane-free. The test makes the old, slow method of progeny testing unnecessary.

linkage analysis
A mathematical procedure that uses information from pedigreed populations to determine whether two loci are linked and, if so, how closely.

A detectable gene becomes a genetic marker once it has been proven to be linked to another gene of interest. This is usually done by **linkage analysis,** a mathematical procedure that uses information from pedigreed and often specially bred populations to determine whether two loci are linked and, if so, how closely. Geneticists estimate linkage distance by examining the rate at which genes segregating at two loci recombine.

The value of a marker, and therefore the effectiveness of marker assisted selection using that marker, depends on how closely linked the marker is to the gene of interest. The more distant the linkage, the more likely that during meiosis a crossover will occur between the marker and the gene it is identifying, causing the genes to recombine. Recombination of this sort is depicted in Figure 20.3. Before a crossover event, detection of the *M* allele at the marker locus indicates the presence of the *B* allele at the black/chocolate locus, and detection of *m* indicates the presence of *b*. Recombination causes the linkage relationship to change. Detection of *M* now indicates the presence of *b*, and detection of *m* indicates the presence of *B*. The practical consequence of recombination of this kind is that for a period of time a particular marker may indicate the presence of one allele, and later the same marker may indicate the presence of a different allele. If a marker is used to indicate a favorable allele, if the marker and gene of interest recombine in the formation of a gamete, and if chromosomes from the gamete are incorporated in an offspring, the marker will subsequently indicate precisely the *wrong* allele in the offspring and many of its de-

[7]For more on the halothane gene, see the section of Chapter 13 entitled *What Causes Correlated Response?*

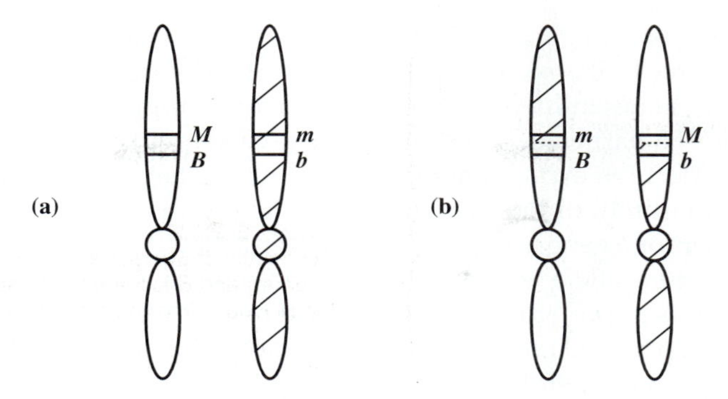

FIGURE 20.3 Arrangement of alleles at the M (marker) and B loci (*a*) before any crossing over and (*b*) after a crossover occurs between the two loci. A crossover of this kind changes the linkage relationship. Before the crossover, detection of the *M* allele indicates the presence of the *B* allele, and detection of *m* indicates the presence of *b*. After the crossover, detection of *M* indicates the presence of *b*, and detection of *m* indicates the presence of *B*.

scendants. A corollary to this is that the marker will likely mean one thing for one group of closely related individuals within a population and something quite different for another group.

Clearly, marker assisted selection is most trustworthy when recombination of markers and genes of interest is minimal (i.e., when markers and the genes they indicate are *very* closely linked). The best markers (if they are detectable) are in fact alleles of the genes of interest themselves or fragments of DNA *within* them. For these markers, recombination is not a worry. To make it easier to find closely linked markers, molecular geneticists are rapidly developing **gene maps** (alternatively called **linkage maps** or **chromosome maps**) for livestock species. These are diagrams showing the chromosomal locations of specific markers and genes of interest. The more markers on a map, the more likely that a gene of interest will be closely linked to a known marker or, better yet, closely linked to known markers on either side of it on a chromosome. Once gene maps become "saturated" with markers, finding reliable markers for a gene of interest should be much easier.

gene map (linkage map, chromosome map)
A diagram showing the chromosomal locations of specific genetic markers and genes of interest.

Marker assisted selection can sometimes be used to increase accuracy of selection for an imported allele in a repeated backcrossing program. If the allele is completely recessive and a closely linked marker is available, marker assisted selection will identify heterozygous replacements. Suppose, for example, that you want to introduce the gene for red coat color into an all black cattle population. Because the allele for black color is completely dominant and all F_1 and backcross individuals are black, it is not possible to tell which potential replacements carry the red allele by looking at them. But they could be identified with a genetic marker.

Marker Assisted Selection for Polygenic Traits

Marker assisted selection is possible (at least in theory) for polygenic traits, including economically important quantitative traits such as conformation, growth rate, litter size, racing ability, milk production, etc. But it is likely to be more difficult than marker assisted selection for simply-inherited traits. Recombination is a problem in

family
A group of related individuals within a population. We often speak of half-sib and full-sib families, but the term can refer to less related groups—even including all descendants of a particular ancestor.

either case, causing markers to mean different things in different **families.** For a polygenic trait, a second problem arises if there are important epistatic relationships among genes influencing the trait. Different groups of relatives may share the same allele at a particular locus, but because each group has different alleles at other loci that affect the expression of the gene, it may have a different degree of expression or an entirely different effect in one family than in another. Therefore, even if a marker for a gene is reliable—in the sense that it consistently indicates the presence of the same allele—the marker may ultimately be unreliable because the allele it indicates is important to performance in one family, but not in another.

A practical consequence of recombination and epistasis is that studies of major genes and their linkage relationships must be conducted within families. Unless epistatic effects are small and linkage so close that recombination can be ruled out, results from one family cannot be extended to other families, even within the same breed.

A third problem for marker assisted selection for polygenic traits is that differences in performance caused by any one gene may be too small to detect. A polygenic trait is, by definition, affected by many genes, and the usual assumption is that no one gene has an overriding effect on the trait. In that case, performance differences caused by a single gene should be so small, and the "noise" caused by environmental effects proportionately so great, that identifying the effect of the gene will be difficult even with a saturated gene map and a large experimental population for performance and linkage analysis. Still, geneticists are betting that some quantitative traits are not as "polygenic" as we think they are, and that one or more important **quantitative trait loci (QTLs)** can be identified that will allow marker assisted selection to be effective.

quantitative trait locus (QTL)
A locus that affects a quantitative trait.

A common argument questioning the usefulness of marker assisted selection for many polygenic traits is as follows. Natural and(or) artificial selection has operated on these traits for a long time. Therefore, any alleles with major, favorable effects (i.e., favorable major genes) have already received strong selection pressure. Their frequency should be, if not 1.0, very close to it. At such high frequencies, these genes contribute little to variation in a trait. In other words, traditional selection has so promoted favorable major genes that there is little left for these genes to contribute, and further marker assisted selection will be ineffective.

For some traits, the argument is probably sound. But there are many potentially important polygenic traits on which breeders have not placed much selection pressure (disease resistance is one example) or for which selection has been inconsistent, even contradictory. Gene frequencies of major genes affecting these traits may be intermediate, and marker assisted selection could be useful. There is also the possibility that some major genes show overdominance, heterozygotes being favored over either homozygote. In this case, selection causes gene frequencies to remain intermediate.[8] Marker assisted *mating* might then be used to increase the number of heterozygotes by mating individuals homozygous for one allele to individuals homozygous for another allele.

Realistically, marker assisted selection for polygenic traits, though it may enjoy some success, should not be thought of as a replacement for selection using traditional genetic prediction. But it could supplement traditional techniques. For example, marker information might be used to screen sires before progeny testing or it might be incorporated in the statistical models used for genetic prediction. Its main benefit will be to increase accuracy of selection.

[8]See the section of Chapter 6 entitled *Factors Influencing the Effectiveness of Selection* for an explanation.

When combined with technologies like *in vitro* fertilization and embryo transfer, marker assisted selection could also increase selection intensity and decrease generation interval. The following is one scenario. A sire known to be heterozygous for a rare but economically important major gene is bred to superovulated females. Cells are removed from the resulting embryos, sexed, and DNA fingerprinted to identify those embryos that received the desired allele from their sire. Heterozygous female embryos are then transferred to recipients and carried to term, producing a number of daughters. While these daughters are still very young (prepubertal), eggs are collected from them, matured, and inseminated *in vitro* with sperm from a second carrier sire. The resulting embryos are screened, and those homozygous for the desired allele are transferred.

In this scenario, selection intensity increases because superovulation and *in vitro* fertilization produce more offspring to choose from. Generation interval decreases because females too young to reproduce naturally do so *in vitro.* In other words, reproductive technologies cause improvements in these elements of the key equation for genetic change. In a larger sense, however, marker assisted selection is also responsible because it makes accurate selection of embryos possible. Without it, there would be little reason to trouble with the reproductive techniques.

Gene Transfer

gene transfer
Transplantation of specific genes from one individual to another using laboratory techniques.

Gene transfer is transplantation of specific genes from one individual to another using laboratory techniques. Much of the gene transfer work done with animals involves the transfer of genes coding for economically important proteins from domestic animals to bacteria. The genetically altered bacteria then produce the proteins very cheaply. It is also possible to transfer genes within and across species of domestic animals. The most common method of doing this entails the physical insertion of foreign DNA into the nucleus of a newly fertilized egg.[9]

Gene transfer is an alternative to repeated backcrossing for incorporating a specific allele existing in one population into another population. Repeated backcrossing works, but it is slow, requiring generations of backcrossing and selection for the desired allele. It also creates an opportunity for undesirable alleles (for example, lethal recessives) to migrate from one population to another. Gene transfer provides a way to obtain the allele of interest without importing a host of other genes from the donor population.

genome
The entire set of an individual's genes.

Unfortunately, gene transfer is a slow process—perhaps as slow as repeated backcrossing—because it takes generations to develop homozygotes and spread the gene throughout a population. Only one copy of a gene is transferred at a time. The transferred gene could be physically incorporated anywhere in the **genome,** so obtaining a homozygote is not as simple as transferring the gene to two individuals and then mating them. Homozygotes are more likely to result from matings of relatives whose common ancestor received the gene by gene transfer. But if you want to transplant a gene from one species to a reproductively incompatible species, gene transfer is the *only* workable method.

Gene transfer could become an important tool for increasing genetic variation within populations. Imagine the savings in the cost of vaccines and disease treatment if we could incorporate disease resistance genes from wild bird species

[9]Strictly speaking, DNA is inserted into one of two *pronuclei* existing within the egg. At this stage, the pronuclei—one derived from the egg itself, the other from the sperm—are still distinct.

transgenic
(1) Referring to gene transfer; (2) An individual that has received genetic material by gene transfer.

into domestic poultry. Or imagine the savings in winter feed costs if we could put the genes responsible for pseudohibernation in bears into beef cattle. **Transgenic** techniques could also be used to delete genes or repair faulty ones. Experiments using these techniques may provide information that can be applied with or without transgenic methods.

But glamorous as it may be, gene transfer is far from commercial application. Success rates are poor, and gene transfer procedures are very expensive. Most animal transgenics produced to date are deficient in some way. The reason may be that genes work in complex and subtle ways. The *structural* parts of genes (the parts that specify amino acids) depend on *regulatory* parts to function correctly. Sometimes they depend on entirely different genes for proper regulation. Successful gene transfer therefore requires more than transplantation of structural parts of genes; the regulatory parts and sometimes other genes must be transferred. Identifying, isolating, and transferring these genes and parts of genes are difficult tasks. Gene transfer (and molecular techniques in general) have been more successful in plants than in animals. That may be because many important traits in plants (disease and herbicide resistance, for example) are simply-inherited and simply-regulated. Their mechanisms are less complex.

The beauty of traditional selection, either natural or artificial, is that it allows only fully functional animals, animals with workable genes and gene combinations, to survive and reproduce. Most natural "experiments," mutant animals or simply extreme types, are weeded out. Of course, natural selection operates on transgenics too. But that is precisely what makes success rates for gene transfer so low. Few transgenics seem to survive natural selection. And low success rates and high costs for the technology are a discouraging combination.

Still, we would be unwise to dismiss gene transfer or, for that matter, any biotechnology. As molecular biologists and reproductive physiologists learn more, success rates will improve. Costs will go down. Revolutionary change in the way we breed animals is not out of the question.

EXERCISES

Study Questions

20.1 Define in your own words:

biotechnology
artificial insemination (A.I.)
multiple-sire pasture
connectedness
estrus synchronization
embryo transfer (E.T.)
superovulation
in vitro fertilization
single-sex system
cloning
clone
clonal line
embryo splitting (bisection)
nuclear transplantation
heritability in the broad sense (H^2)
same-sex mating
nuclear fusion
selfing
DNA fingerprinting
polymorphic
marker assisted selection
genetic marker
linkage analysis
gene map (linkage map, chromosome map)
family
quantitative trait locus (QTL)
gene transfer
genome
transgenic

20.2 From an animal breeding perspective, what are the two major categories of biotechnologies?

20.3 List three criteria that determine the impact of a biotechnology.

20.4 **a.** How do animal breeding and reproduction differ?
b. How are the two disciplines tied?

20.5 Discuss the effects of artificial insemination on:
a. accuracy of selection
b. selection intensity
c. sire identification
d. connectedness

20.6 Why is embryo transfer unlikely to have as far-reaching effects as artificial insemination?

20.7 What problems does embryo transfer create for genetic evaluation?

20.8 What advantage does embryo transfer have over artificial insemination for preserving germ plasm?

20.9 How can *in vitro* fertilization complement embryo transfer?

20.10 Discuss the potential effects of sex control on:
a. economics of commercial production
b. individual mating decisions
c. mating systems
d. production efficiency (via single-sex systems)
e. accuracy of prediction
f. generation interval
g. selection intensity

20.11 Describe a hypothetical scenario for large-scale cloning.

20.12 Describe the effects of cloning on:
a. short-term genetic change
b. uniformity of performance
c. accuracy of selection
d. selection intensity

20.13 What are some potential dangers of cloning?

20.14 What would be the advantage of cloning mature animals, as opposed to embryos?

20.15 Explain the genetic rationale for same-sex mating.

20.16 **a.** Describe the implications of DNA fingerprinting (as a means of animal identification) for managing pedigreed populations.
b. Describe its effects on accuracy of selection.

20.17 How might DNA fingerprinting be used to:
a. decrease the amount of time required for a population to achieve "purebred" status in grading-up programs or in programs of repeated backcrossing to import an allele?
b. predict hybrid vigor for specific crosses or individual matings?

20.18 Explain how marker assisted selection works.

20.19 Describe a scenario showing how marker assisted selection can be used to help select for a simply-inherited trait.

20.20 Why is it important that a genetic marker be as closely linked to the gene of interest as possible?

20.21 Why are molecular geneticists developing increasingly detailed gene maps?

20.22 Describe how marker assisted selection can make importing an allele by repeated backcrossing more efficient.

20.23 Discuss the effect of each of the following on the potential success of marker assisted selection for a polygenic trait:

a. recombination
b. epistasis
c. size of gene effects
d. previous selection

20.24 Under what conditions is marker assisted selection and(or) marker assisted mating for polygenic traits likely to be the most effective?

20.25 Discuss the potential benefits of gene transfer.

20.26 Discuss the drawbacks of gene transfer.

CHAPTER 21

Commonsense Animal Breeding

The technical aspects of animal breeding can be overwhelming. If you have just waded through this book, you know that. But animal breeding is much more than the sum of its technical details. It is important for breeders to understand genetic theory and animal breeding technology, but knowledge alone does not guarantee success. Success has more to do with common sense in the application of theory and technology.

The breeders who run the longest-lived breeding programs—whether those programs be small or large, seedstock or commercial—tend to display the following attributes:

Understanding
Good information
Deliberation
Consistency
Simplicity
Patience

There is nothing particularly profound about these qualities. They are dictated by common sense. In this chapter I refer to them in the context of making animal breeding decisions, but you could apply them to almost any endeavor. You could, for example, apply them to the *business* of animal breeding, something not covered in this chapter.

BE KNOWLEDGEABLE

Being knowledgeable about animal breeding means understanding the more important concepts outlined in this book, for example:

- Genotype by environment interactions—"environment" is used here in its broadest sense—and their importance to breeding objectives
- The randomness of inheritance and how it causes both limitations and opportunities

- The necessary differences in approach to breeding for simply-inherited versus polygenic traits
- Components of the genetic model for quantitative traits
- Elements of the key equation for genetic change and trade-offs among them
- The importance of sire selection
- Breed roles and the power of between-breed selection
- The need to balance hybrid vigor and breeding value in designing crossbreeding systems.

These are *big* subjects that require a good deal of background to understand. But they are important—much more important than being able to calculate an inbreeding coefficient or divine the precise number of mates needed to prove a sire free of some recessive allele. Those things you can look up.

Being knowledgeable also means understanding many things that are not in this book. Each species has its own idiosyncrasies, its own lore, and its own market factors. You can learn some of these through coursework, some from books and articles, some from talking with breeders, and many more through personal experience in breeding. It is important to know as much as possible about the genetic capabilities of your own animals and the limits to production imposed by the local environment. If you produce seedstock, you should also know your customers' animals and environments.

To be knowledgeable is to keep up-to-date. Genetic theory, at least the part of genetic theory most applicable to animal breeding, is not evolving rapidly. Occasionally we amend it somewhat. But animal breeding technology is changing fast, and research continually increases our knowledge of $G \times E$ interactions, genetic relationships, breed attributes, and so on. To stay current, you must read the latest publications and interact with breeders who understand a breeding industry well and stay up-to-date themselves.

USE GOOD INFORMATION

Good decisions are based on good information. In animal breeding, that means good information about breed performance and individual animals. Use the best genetic predictions you can find. For some species and traits, the best predictions come in the form of EPDs, EBVs, ETAs, and MPPAs derived from vast amounts of data and sophisticated statistical analyses. Other species and traits are not so blessed. In any case, look for the most meaningful information available, and if you are a commercial producer, patronize those seedstock breeders that provide good information on their animals.

Having good information means measuring animal performance and keeping good records. Within-herd or within-flock performance information is not cheap—it takes time and labor to collect, store, retrieve, and synthesize into usable form. It must be good quality information. Procedural details (proper contemporary grouping, for example) need close attention. Many commercial producers may find the benefit of performance records outweighed by the effort required to maintain them. They should probably limit their record keeping to the most basic items. Seedstock breeders, on the other hand, are obligated to do as much as they can to document the performance of their animals because of the value of seedstock derives not only from the animals themselves, but also from the information that goes with them. Seedstock breeders also have the social responsibility of contributing their records to breed or government databases. By doing so, they improve the quality of genetic information available to themselves and everyone else.

TAKE TIME TO THINK

Clear technical understanding and quality genetic information will be of little value if you do not make good use of them. Take the time to think carefully about your breeding program. How is the "best" animal characterized? What is the fastest, easiest, most economical way to obtain animals with those characteristics? All too often breeders display a pack mentality, doing what other breeders do with little appreciation for what makes their own set of circumstances different from the circumstances of others. Try to see the big picture. Be analytical.

BE CONSISTENT

Lack of consistency is the bane of far too many breeding programs. Breeders are often influenced by fads that may seem important at the time but have little long-term justification. They alter their selection programs to meet the new need, only to find later that they have been wasting time or even backtracking. In some breeding industries, consensus as to what constitutes the ideal animal shifts like a pendulum, and inevitably the pendulum swings too far before enough breeders revolt and its direction is reversed.

To be consistent, you must be able to see into the future and determine what kind of animal will be most useful over the longer term. This means studying and thinking, setting goals and sticking to them. That is not to say that once set, breeding objectives cannot be amended. There are legitimate reasons to alter the direction of a breeding program. Markets change and new information may cause you to reconsider long-held notions. Still, the more consistent your goals, the more definable and therefore the more marketable your product.

KEEP IT SIMPLE

Successful breeding programs are usually simple. Simple does not mean primitive; these programs do not avoid using advanced breeding technology. They are simple in concept. They have well-established goals and remain unencumbered by elaborate rules for individual matings and other grand schemes for beating the Mendelian odds.

Breeders can simplify their programs by clearly defining the roles their animals best fit. Not all breeders should provide a specialty product, but neither should they try to make their animals all things to all people. They should limit the traits under selection to those that are truly important and (in food and fiber species, anyway) ignore those traits whose value is purely aesthetic.[1]

[1]Some breeders may object to this last statement because part of the satisfaction of being an animal breeder comes from enjoying the aesthetic aspects of animals. We all appreciate a handsome individual. But there need not be a conflict (or at least not *much* of a conflict) between function and looks. We can and do alter our notions of what constitutes a pretty animal, and in time those notions tend to describe very functional types. Perhaps we just need to be faster to adjust our aesthetic sense to the realities of animal production.

Complex breeding programs are difficult to maintain. They require more time and energy than their perceived advantages justify. Simple programs, however, tend to remain in place. Simplicity breeds consistency.

Be Patient

The pace of genetic change, at least when compared to the pace of so many things in contemporary life, is slow. Much slower, in fact, than most of us would like. We become restless, overly eager for the next great sire to appear. We are quick to jump to conclusions about animals we really know little about. To be successful, we need patience.

Because gene segregation and gamete selection are largely random processes, inheritance is necessarily unpredictable. Knowing this, the wisest (though not necessarily the most flamboyant) breeders patiently play the averages, assured that if they follow the rules, the next generation of animals will be better than the last, and the generation after next even better. The random nature of inheritance also creates opportunities. Occasionally, gene segregation and gamete selection conspire to produce a truly outstanding animal, a gift from nature to be sought and exploited. And the most successful breeders do just that. Rejecting formula breeding, they are patient opportunists.

Exercises

Study Questions

21.1 List six attributes that successful (in the long-term) animal breeders tend to have in common.

21.2 What genetic principles do you think are most important for a breeder to understand?

21.3 How can a breeder best keep up-to-date?

21.4 For a species and seedstock or commercial operation of your choice, list the animal measurements you would make and the performance records you would keep.

21.5 Why do seedstock breeders have an obligation to track as much information on their animals as possible?

21.6 Why is it important to take time to think about your breeding program?

21.7 What causes inconsistency in breeding programs?

21.8 What can be done to foster consistency?

21.9 Why is simplicity important for a breeding program?

21.10 What can be done to foster simplicity?

21.11 Why is lack of patience a danger for animal breeders?

21.12 What does it mean to be a "patient opportunist"?

Glossary

accuracy: A measure of the strength of the relationship between true values and their predictions.

accuracy of selection or **accuracy of breeding value prediction ($r_{BV,\hat{BV}}$):** A measure of the strength of the relationship between true breeding values and their predictions for a trait under selection.

additive gene effect: *See* **independent gene effect.**

additive genetic value or **additive value:** Breeding value.

aggregate breeding value or **net merit:** The breeding value of an individual for a combination of traits.

A.I.: *See* **artificial insemination.**

alike in state: Genes that are alike in state function the same and have exactly or almost exactly the same chemical structure.

allele: An alternative form of a gene.

allelic frequency: *See* **gene frequency.**

animal model: An advanced statistical model for genetic prediction that is used to evaluate all animals (as opposed to just sires) in a population.

arrow diagram: A form of pedigree depicting schematically the flow of genes from ancestors to descendants.

artificial insemination (A.I.): A reproductive technology in which semen is collected from males, then used in fresh or frozen form to breed females.

artificial selection: Selection that is under human control.

assortative mating: The mating of either similar individuals (positive assortative mating) or dissimilar individuals (negative assortative mating).

backcrossing: (1) The mating of a hybrid to a purebred of a parent breed or line. (2) (broader meaning) The mating of an individual (purebred or hybrid) to any other individual (purebred or hybrid) with which it has one or more ancestral breeds or lines in common.

base: In large-scale genetic evaluation, the level of genetic merit associated with an EPD of zero.

base population: The population of animals whose parents are either unknown or ignored for the purposes of inbreeding and relationship calculation—typically the individuals appearing at the back of the pedigrees of the original animals in a herd or flock.

base year: In large-scale genetic evaluation, the year chosen to represent the base. The average EPD of all animals born in the base year is zero.

BC_1: Backcross one. Referring to the first generation of crosses between hybrids and purebreds of a parent breed or line.

best linear unbiased prediction (BLUP): A method of genetic prediction that is particularly appropriate when performance data come from genetically diverse contemporary groups.

between-breed selection: The process that determines the breed(s) from which parents are selected.

bias: Any factor that causes distortion of genetic predictions.

biological type: A classification for animals with similar genotypes for traits of interest. Examples include heavy draft types (horses), prolific wool types (sheep), large dual-purpose types (cattle), and tropically adapted types (many species).

biotechnology: The application of biological knowledge to practical needs. The term usually refers to (1) technologies for altering reproduction and (2) technologies for locating, identifying, comparing, or otherwise manipulating genes.

bisection of embryos: *See* **embryo splitting.**

BLUP: *See* **best linear unbiased prediction.**

breed: A race of animals within a species. Animals of the same breed usually have a common origin and similar identifying characteristics.

breed complementarity: An improvement in the overall performance of crossbred offspring resulting from crossing breeds of different but complementary biological types.

breeding objective: (1) A weighted combination of traits defining aggregate breeding value for use in an economic selection index. (2) A general goal for a breeding program—a notion of what constitutes the best animal.

breeding value (BV): (1) The value of an individual as a (genetic) parent. (2) The part of an individual's genotypic value that is due to independent and therefore transmittable gene effects.

breed true: A phenotype for a simply-inherited trait is said to breed true if two parents with that phenotype produce offspring of that same phenotype exclusively.

categorical or **qualitative trait:** A trait in which phenotypes are expressed in categories.

central test: A test designed to compare the performance of animals (usually young males) from different herds or flocks for growth rate and feed conversion by feeding them at a central location.

chromosome: One of a number of long strands of DNA and associated proteins present in the nucleus of every cell.

chromosome map: *See* **gene map.**

clonal line: A population of genetically identical individuals (also referred to as a clone).

clone: (1) An individual member of a clonal line. (2) A clonal line.

cloning: A reproductive technology for producing genetically identical individuals.

closed nucleus breeding scheme: A nucleus breeding scheme in which germ plasm flows in only one direction—from the nucleus to cooperating herds or flocks.

closed population: A population that is closed to genetic material from the outside.

coefficient of relationship: *See* **Wright's coefficient of relationship.**

collateral relatives: Relatives that are neither direct ancestors nor direct descendants of an individual (e.g., siblings, aunts, uncles, nieces, and nephews).

commercial producer: An animal breeder whose primary product is a commodity for public consumption.

common ancestor: An ancestor common to more than one individual. In the context of inbreeding, the term refers to an ancestor common to the parents of an inbred individual.

common environmental effect: An increase in similarity of performance of family members caused by their sharing a common environment. Common environmental effects are particularly important within litters (full sibs).

compensatory gain: A relative increase in the growth rate of thin animals after they are placed on adequate feed. They tend to compensate for being underweight.

complementarity: An improvement in the overall performance of offspring resulting from mating individuals with different but complementary breeding values.

complete dominance: A form of dominance in which the expression of the heterozygote is identical to the expression of the homozygous dominant genotype.

composite (synthetic) animal: A hybrid with at least two and typically more breeds in its background. Composites are expected to be bred to their own kind, retaining a level of hybrid vigor normally associated with traditional crossbreeding systems.

composite (synthetic) breed: A breed made up of two or more component breeds and designed to benefit from hybrid vigor without crossing with other breeds.

composite/terminal system: A crossbreeding system combining a maternal composite breed for producing replacement females with terminal sires for producing market offspring.

confidence range: A range of values within which we expect—with a given probability, a given degree of confidence—that a true value of interest lies.

connectedness: The degree to which data from different contemporary groups within a population can be compared as a result of pedigree relationships between animals in different groups.

contemporary group: A group of animals that have experienced a similar environment with respect to the expression of a trait. Contemporaries typically perform in the same location, are of the same sex, are of similar age, and have been managed alike.

contemporary group effect (E_{cg}): An environmental effect common to all members of a contemporary group.

corrective mating: A mating designed to correct in their progeny faults of one or both parents.

correlated response to selection: Genetic change in one or more traits resulting from selection for another.

correlation or **correlation coefficient ($r_{X,Y}$):** A measure of the strength (consistency, reliability) of the relationship between two variables.

covariance ($\mathrm{cov}(X,Y)$): The basic measure of covariation.

covariation: How two traits or values vary together in a population.

crossbred: Having parents of different breeds or breed combinations.

crossbreeding: The mating of sires of one breed or breed combination to dams of another breed or breed combination.

crossbreeding system: A mating system that uses crossbreeding to maintain a desirable level of hybrid vigor and(or) breed complementarity.

crossing over: A reciprocal exchange of chromosome segments between homologs. Crossing over occurs during meiosis prior to the time homologous chromosomes are separated to form gametes.

culling: The process that determines which parents will no longer remain parents.

dam: A female parent.

degree of backcrossing: The proportional amount of backcrossing (in the broader sense) involved in a mating—the proportion of an offspring's loci at which both genes of a pair trace to the same ancestral breed or line.

designed test: A carefully monitored progeny test designed to eliminate sources of bias like nonrandom mating and culling for poor performance.

direct component or **direct effect:** The effect of an individual's genes on its performance.

direct response to selection: Genetic change in a trait resulting from selection for that trait.

direct selection: Selection for a trait as a means of improving that same trait.

DNA: Deoxyribonucleic acid, the molecule that forms the genetic code.

DNA fingerprinting: A laboratory method for graphically characterizing an individual's DNA, creating a unique genetic "fingerprint."

dominance: An interaction between genes at a single locus such that in heterozygotes one allele has more effect than the other. The allele with the greater effect is **dominant** over its **recessive** counterpart.

dystocia: Difficulty in giving birth or being born.

economic selection index: An index or combination of weighting factors and genetic information—either phenotypic data or genetic predictions—on more than one trait. Economic selection indexes are used in multiple-trait selection to predict aggregate breeding value.

economic weight: The change in aggregate breeding value (the change in profit if that is how aggregate breeding value is measured) due to an independent, one-unit increase in performance in a trait.

effective population size: The size of a population as reflected by its rate of inbreeding.

effective proportion saved (p_e): In selection—a value that, when substituted for actual proportion saved (p), reflects correct selection intensity.

embryo: An organism in the early stages of development in the shell (bird) or uterus (mammal).

embryo splitting (bisection): A reproductive technology in which embryos are mechanically cut in half to produce twin embryos.

embryo transfer (E.T.): A reproductive technology in which embryos from donor females are collected and transferred in fresh or frozen form to recipient females.

environmental correlation (r_{E_X,E_Y}): A measure of the strength (consistency, reliability) of the relationship between environmental effects on one trait and environmental effects on another trait.

environmental effect (*E*): The effect that external (nongenetic) factors have on animal performance.

environmental trend: Change in the mean performance of a population over time caused by changes in environment.

epistasis: An interaction among genes at different loci such that the expression of genes at one locus depends on the alleles present at one or more other loci.

estimated breeding value (EBV): A prediction of a breeding value. *See* **breeding value**.

estimated transmitting ability (ETA): *See* **expected progeny difference.**

estrus synchronization: The administration of hormones to a group of mammalian females causing them to come into heat (estrus) at or near the same time.

expected progeny difference (EPD), predicted difference (PD), or **estimated transmitting ability (ETA):** A prediction of a progeny difference. *See* **progeny difference.**

F_1: Referring to the first generation of crosses between two unrelated (though not necessarily purebred) populations.

F_1 hybrid vigor: The amount of hybrid vigor attainable in first-cross individuals—maximum hybrid vigor.

F_2: Referring to the generation of crosses produced by mating F_1 (first-cross) individuals among themselves.

family: A group of related individuals within a population. We often speak of half-sib and full-sib families, but the term can refer to less related groups—even including all descendants of a particular ancestor.

fertility: The ability (of a female) to conceive or (of a male) to impregnate.

field data: Data that are regularly reported by individual breeders to breed associations or government agencies.

fitness: The ability of an individual and its corresponding phenotype and genotype to contribute offspring to the next generation. The term refers to the number of offspring an individual produces—not just its ability to be selected.

fitness trait: A trait selected for with natural selection. Fitness traits relate to an animal's ability to survive and reproduce.

fixation: The point at which a particular allele becomes the *only* allele at its locus in a population—the frequency of the allele becomes one.

gamete or **germ cell:** A sex cell, a sperm or egg.

gamete selection: The process that determines which egg matures and which sperm succeeds in fertilizing the egg.

gene: The basic physical unit of heredity consisting of a DNA sequence at a specific location on a chromosome.

gene combination effect: The effect of a combination of genes (i.e., a dominance or epistatic effect).

gene combination value (GCV): The part of an individual's genotypic value that is due to the effects of gene combinations (dominance and epistasis) and cannot, therefore, be transmitted from parent to offspring.

gene frequency or **allelic frequency:** The relative frequency of a particular allele in a population.

gene map (linkage map, chromosome map): A diagram showing the chromosomal locations of specific genetic markers and genes of interest.

generation interval (L): (1) The amount of time required to replace one generation with the next. (2) In a closed population, the average age of parents when their selected offspring are born.

genetic correlation (r_{BV_X,BV_Y}): (1) A measure of the strength (consistency, reliability) of the relationship between breeding values for one trait and breeding values for another trait. (2) A measure of pleiotropy.

genetic marker: A detectable gene or DNA fragment used to identify alleles at a linked locus.

genetic prediction: The area of academic animal breeding concerned with measurement of data, statistical procedures, and computational techniques for predicting breeding values and related values.

genetic trend: Change in the mean breeding value of a population over time.

genetic variation (σ_{BV}): In the context of the key equation for genetic change, variability of breeding values within a population for a trait under selection.

gene transfer: Transplantation of specific genes from one individual to another using laboratory techniques.

genome: The entire set of an individual's genes.

genotype: (1) The genetic makeup of an individual. (2) The combination of genes at a single locus or at a number of loci. We speak of one-locus genotypes, two-locus genotypes, and so on.

genotype by environment ($G \times E$) interaction: A dependent relationship between genotypes and environments in which the difference in performance between two (or more) genotypes changes from environment to environment.

genotypic frequency: The relative frequency of a particular one-locus genotype in a population.

genotypic value (G): The effect of an individual's genes (singly and in combination) on its performance for a trait.

germ cell or **gamete:** A sex cell—a sperm or egg.

germ plasm: Genetic material in the form of live animals, semen, or embryos.

grading up or **topcrossing:** (1) A mating system designed to create a purebred population by mating successive generations of nonpurebred females to purebred sires. (2) A mating system designed to convert a population from one breed to another by mating successive generations of females descended from the first breed to sires of the second breed.

half sibs: Half brothers and sisters.

Hardy-Weinberg equilibrium: A state of constant gene and genotypic frequencies occurring in a population in the absence of forces that change those frequencies.

heritability (h^2): A measure of the strength of the relationship between performance (phenotypic values) and breeding values for a trait in a population.

heritability in the broad sense (H^2): A measure of the strength of the relationship between performance (phenotypic values) and genotypic values for a trait in a population.

heterosis or **hybrid vigor:** An increase in the performance of hybrids over that of purebreds, most noticeably in traits like fertility and survivability.

heterozygote (heterozygous genotype): A one-locus genotype containing functionally different alleles.

homolog: One of a pair of chromosomes having corresponding loci.

homozygote (homozygous genotype): A one-locus genotype containing functionally identical genes.

hybrid: An individual that is a combination of species, breeds within species, or lines within breeds.

hybrid seedstock complementarity: The form of complementarity produced by crossing genetically diverse breeds to create hybrid breeding animals with a desirable combination of breeding values. *See also* **complementarity.**

hybrid vigor or **heterosis:** An increase in the performance of hybrids over that of purebreds, most noticeably in traits like fertility and survivability.

identical by descent: Two genes are identical by descent if they are copies of a single ancestral gene.

inbreeding: The mating of relatives.

inbreeding coefficient (F_X): A measure of the level of inbreeding in an individual: (1) The probability that both genes of a pair in an individual are identical by descent. (2) The probable proportion of an individual's loci containing genes that are identical by descent.

inbreeding depression: The reverse of hybrid vigor—a decrease in the performance of inbreds, most noticeably in traits like fertility and survivability.

incomplete reporting: The reporting of only selected performance records to a breed association or government agency.

independent assortment: The independent segregation of genes at different loci during gamete formation.

independent culling levels: Minimum standards for traits undergoing multiple-trait selection. Animals failing to meet any one standard are rejected regardless of merit in other traits.

independent gene effect: The effect of a gene independent of the effect of the other gene at the same locus (dominance) and the effects of genes at other loci (epistasis).

indicator trait: A trait that may or may not be important in itself, but is selected for as a way of improving some other genetically correlated trait.

indirect selection: Selection for one trait as a means of improving a genetically correlated trait.

individual hybrid vigor: Hybrid vigor for the direct component of a trait.

interaction: A dependent relationship among components of a system in which the effect of any one component depends on other components present in the system.

interim EPD: An updated EPD that is calculated between BLUP analyses and incorporates new information.

intermediate optimum: An intermediate level of performance that is optimal in terms of profitability and(or) function.

introgression: *See* **repeated backcrossing.**

***in vitro* fertilization:** A technique by which eggs are collected from donor females, then matured and fertilized in the laboratory.

key equation: The equation relating the rate of genetic change resulting from selection to four factors: accuracy of selection, selection intensity, genetic variation, and generation interval.

large-scale genetic evaluation: The genetic evaluation of large populations—typically entire breeds.

line: A group of related animals within a breed.

linebreeding: The mating of individuals within a particular line; a mating system designed to maintain a substantial degree of relationship to a highly regarded ancestor or group of ancestors without causing high levels of inbreeding.

linecrossing: The mating of sires of one line or line combination to dams of another line or line combination.

linkage: The occurrence of two or more loci of interest on the same chromosome.

linkage analysis: A mathematical procedure that uses information from pedigreed populations to determine whether two loci are linked and, if so, how closely.

linkage map: *See* **gene map.**

locus: The specific location of a gene on a chromosome.

major gene: A gene that has a readily discernible effect on a trait.

marker: *See* **genetic marker.**

marker assisted selection: Selection for specific alleles using genetic markers.

maternal breed: A breed that excels in maternal traits.

maternal component or **maternal effect:** The effect of genes in the dam of an individual that influence the performance of the individual through the environment provided by the dam.

maternal hybrid vigor: Hybrid vigor for the maternal component of a trait.

maternal trait: A trait especially important in breeding females. Examples include fertility, freedom from dystocia, milk production, maintenance efficiency, and mothering ability.

mating: The process that determines which (selected) males are bred to which (selected) females.

mating system: A set of rules for mating.

mean: An arithmetic average.

meiosis: The process of germ cell formation.

Mendelian sampling: The random sampling of parental genes caused by segregation and independent assortment of genes during germ cell formation, and by random selection of gametes in the formation of the embryo.

migration: The movement of individuals into or out of a population.

most probable producing ability (MPPA): A prediction of producing ability. *See* **producing ability.**

multiple alleles: More than two possible alleles at a locus.

multiple ovulation and embryo transfer (MOET): Hormonally induced ovulation of multiple eggs followed by transfer of embryos to recipient dams. The term is used in conjunction with breeding strategies designed to increase the rate of genetic change using embryo transfer.

multiple-sire pasture: A breeding pasture (or pen) containing more than one sire at a time.

multiple-trait model: A statistical model used to predict values for more than one trait at a time.

multiple-trait selection: Selection for more than one trait.

mutation (specifically point mutation): The process that alters DNA to create new alleles.

natural selection: Selection that occurs in nature independent of deliberate human control.

natural service: Natural mating (as opposed to artificial insemination).

negative assortative mating: The mating of dissimilar individuals.

net merit: *See* **aggregate breeding value.**

no dominance: A form of dominance in which the expression of the heterozygote is exactly midway between the expressions of the homozygous genotypes.

nonadditive gene effects: Gene combination effects.

nonadditive genetic value or **nonadditive value:** Gene combination value.

nonparent EPD: An EPD for an animal without progeny data. Nonparent EPDs typically do not come with associated accuracy measures.

nonrandom mating: Any mating system in which males are not randomly assigned to females.

normal distribution: The statistical distribution that appears graphically as a symmetric, bell-shaped curve. In an animal breeding context, the values along the horizontal axis represent levels of performance, breeding value, etc. in a population, and the height of the curve at any point represents the relative frequency of that value in the population.

nuclear fusion: Artificial fertilization—a laboratory technique for combining the nuclei of two gametes.

nuclear transplantation: Surgical removal of an egg's nucleus followed by insertion of an individual cell extracted from an embryo.

nucleus breeding scheme: A cooperative breeding program in which elite animals are concentrated in a nucleus herd or flock and superior germ plasm is then distributed among cooperating herds or flocks.

open nucleus breeding scheme: A nucleus breeding scheme in which the flow of germ plasm is bidirectional—from the nucleus to cooperating herds or flocks and from cooperating herds or flocks to the nucleus.

outbreeding or **outcrossing:** The mating of unrelated individuals.

overdominance: A form of dominance in which the expression of the heterozygote is outside the range defined by the expressions of the homozygous genotypes and most closely resembles the expression of the homozygous dominant genotype.

own performance data: Information on an individual's own phenotype.

parent EPD: An EPD for an animal with progeny data. Parent EPDs typically come with associated accuracy measures.

partial dominance: A form of dominance in which the expression of the heterozygote is intermediate to the expressions of the homozygous genotypes and more closely resembles the expression of the homozygous dominant genotype.

paternal breed: A breed that excels in paternal traits.

paternal component or **paternal effect:** In rare instances, the effect of genes in the sire of an individual that influence the performance of the individual through the environment provided by the sire. Traits of the dam or offspring that are affected by a male's fertility and physical ability to breed are also said to have a paternal component.

paternal hybrid vigor: Hybrid vigor for the paternal component of a trait.

paternal trait: A trait especially important in market offspring. Examples include rate and efficiency of gain, meat quality, and carcass yield.

path method: A method for calculating inbreeding and relationship coefficients that simulates the "paths" taken by identical genes as they flow from ancestors to descendants.

pedigree data: Information on the genotype or performance of ancestors and(or) collateral relatives of an individual.

pedigree estimate: A genetic prediction based solely on pedigree data.

pedigree relationship: Relationship between animals due to kinship. Examples include full-sib, half-sib, and parent-offspring relationships.

permanent environmental effect (E_p): An environmental effect that permanently influences an individual's performance for a repeated trait.

phenotype: An observed category or measured level of performance for a trait in an individual.

phenotypic correlation (r_{P_X,P_Y}): A measure of the strength (consistency, reliability) of the relationship between performance in one trait and performance in another trait.

phenotypic selection: Selection based solely on an individual's own phenotype(s).

phenotypic selection differential (S): The difference between the mean performance of those individuals selected to be parents and the average performance of all potential parents, expressed in units of the trait.

phenotypic selection index: A form of economic selection index used with phenotypic selection. In the classic form of phenotypic index, the traits in the index are identical to the traits in the breeding objective.

phenotypic value (P): A measure of performance for a trait in an individual—a performance record.

pleiotropy: The phenomenon of a single gene affecting more than one trait.

polled: Naturally without horns.

polygenic trait: A trait affected by many genes, no single gene having an overriding influence.

polymorphic: (with respect to genes and DNA fragments) Having at least two alternative forms or alleles in a population.

population: A group of intermating individuals. The term can refer to a breed, an entire species, a single herd or flock, or even a small group of animals within a herd.

population genetics: The study of factors affecting gene and genotypic frequencies in a population.

population mean (μ): The average phenotypic value of all individuals in a population.

population measure: Any measure applied to a population as opposed to an individual.

population parameter: A true (as opposed to estimated) population measure. Examples are true population means, variances, and standard deviations; true correlations between traits; and true heritabilities.

positive assortative mating: The mating of similar individuals.

possible change (PC) or **standard error of prediction:** A measure of accuracy indicating the potential amount of future change in a prediction.

predicted difference (PD): *See* **expected progeny difference.**

predicted value: A prediction of a true value. The most common predicted values are estimated breeding value (EBV or $\hat{BV}$), expected progeny difference (EPD or $\hat{PD}$), and most probable producing ability (MPPA or $\hat{PA}$).

prediction equation: A mathematical equation used to calculate a predicted value based (usually) on phenotypic data.

prepotency: The ability of an individual to produce progeny whose performance is especially like its own and(or) is especially uniform.

producing ability (*PA*): The performance potential of an individual for a repeated trait.

progeny data: Information on the genotype or performance of descendants of an individual.

progeny difference (*PD*) or **transmitting ability (*TA*):** Half an individual's breeding value—the expected difference between the mean performance of the individual's progeny and the mean performance of all progeny (assuming randomly chosen mates).

progeny test: A test used to help predict an individual's breeding values involving multiple matings of that individual and evaluation of its offspring.

proportion saved (*p*): The number of individuals chosen to be parents as a proportion of the number of potential parents.

Punnett square: A two-dimensional grid used to determine the possible zygotes obtainable from a mating.

purebred: Wholly of one breed or line (as opposed to crossbred).

purebreeding or **straightbreeding:** The mating of purebreds of the same breed.

pure composite system: A mating system limited to matings within a single composite breed.

qualitative or **categorical trait:** A trait in which phenotypes are expressed in categories.

quantitative genetics: The branch of genetics concerned with influences on, measurement of, relationships among, genetic prediction for, and rate of change in traits that are or can be treated as quantitative.

quantitative trait: A trait in which phenotypes show continuous (numerical) expression.

quantitative trait locus (QTL): A locus that affects a quantitative trait.

random mating: A mating system in which mates are chosen at random.

rate of genetic change ($\Delta_{BV}/_t$) or **response to selection:** The rate of change in the mean breeding value of a population caused by selection.

recessiveness: *See* **dominance.**

recombination: The formation of a new combination of genes on a chromosome as a result of crossing over.

recombination loss: A loss in gene combination value caused by the gradual breaking up of favorable epistatic blocks of linked loci in advanced generations of certain hybrids.

regression or **regression coefficient ($b_{Y \cdot X}$):** The expected or average change in one variable (Y) per unit change in another (X).

regression for amount of information: The mathematical process causing genetic predictions to be more or less "conservative" (closer to the mean) depending on the amount of information used in calculating them.

repeatability (*r*): (1) A measure of the strength of the relationship between repeated records (repeated phenotypic values) for a trait in a population. (2) A measure of the strength of the relationship between single performance records (phenotypic values) and producing abilities for a trait in a population. (3) In dairy publications, accuracy of prediction.

repeated backcrossing or **introgression:** A mating system used to incorporate an allele or alleles existing in one population into another population. An initial cross is followed by successive generations of backcrossing combined with selection for the desired allele(s).

repeated trait: A trait for which individuals commonly have more than one performance record.

replacement rate: The rate at which newly selected individuals replace existing parents in a population.

replacement selection: The process that determines which individuals will become parents for the first time.

response to selection: *See* **rate of genetic change.**

retained hybrid vigor or **retained heterosis:** Hybrid vigor remaining in later generations of hybrids—generations subsequent to the first-cross (F_1) generation. Retained hybrid vigor is commonly expressed as a proportion of F_1 (maximum) vigor.

rotational crossbreeding system: A crossbreeding system in which generations of females are "rotated" among sire breeds in such a way that they are mated to sires whose breed composition is most different from their own.

rotational/terminal system: A crossbreeding system combining a maternal rotation for producing replacement females with terminal sires for producing market offspring.

rotation in time: A rotational crossbreeding system in which sire breeds are not used simultaneously, but are introduced in sequence.

same-sex mating: A reproductive technology for mating individuals of the same sex.

sample statistic: An estimate of a population parameter.

seedstock: Breeding stock, animals whose role is to be a parent or, in other words, to contribute genes to the next generation.

segregation: The separation of paired genes during germ cell formation.

selection: The process that determines which individuals become parents, how many offspring they may produce, and how long they remain in the breeding population.

selection criterion (SC): An EBV, EPD, phenotypic value, or other piece of information forming the basis for selection decisions.

selection differential: The difference between the mean selection criterion of those individuals selected to be parents and the average selection criterion of all potential parents, expressed in units of the selection criterion.

selection index: A linear combination of phenotypic information and weighting factors that is used for genetic prediction when performance data come from generally similar contemporary groups. *See also* **economic selection index.**

selection intensity (*i*): (1) A measure of how "choosey" breeders are in deciding which individuals are selected. (2) The difference between the mean selection criterion of those individuals selected to be parents and the average selection criterion of all potential parents, expressed in standard deviation units.

selection risk: The risk that the true breeding values of replacements will be significantly poorer than expected.

selection target: A level of breeding value considered optimal in an absolute or practical sense.

selfing: The mating of an individual to itself.

simply-inherited trait: A trait affected by only a few genes.

single-sex system: A proposed production system for beef cattle that increases overall feed efficiency by combining sex control (all females) with slaughter at young ages.

single-trait selection: Selection for one trait.

sire: A male parent.

sire × dam complementarity: The classic form of complementarity produced by mating sires strong in paternal traits to dams strong in maternal traits. Offspring inherit superior market characteristics from their sires and benefit from the maternal environment provided by their dams. *See also* **complementarity.**

sire summary: A list of genetic predictions, accuracy values, and other useful information about sires in a breed.

spatial rotation: A rotational crossbreeding system in which all sire breeds are used simultaneously—they are *spatially* separated. Replacement females leave the location of their birth to be mated to sires with different breed composition.

standard deviation (σ): A mathematical measure of variation that can be thought of as an average deviation from the mean. The square root of the variance.

standard error of prediction: *See* **possible change.**

static terminal system: A terminal sire crossbreeding system in which replacement females are either purchased or produced from separate purebred populations within the system.

statistic: *See* **sample statistic.**

statistical model: A mathematical representation of animal performance that includes various genetic and environmental effects and is used for genetic prediction.

straightbreeding: *See* **purebreeding.**

superovulation: The administration of a hormone causing a female to develop and release more eggs than normal.

system: A group of interdependent component parts.

tabular method: A method for calculating inbreeding and relationship coefficients involving construction and updating of a table relating all members of a population.

tandem selection: Selection first for one trait, then another.

temporary environmental effect (E_t): An environmental effect that influences a single performance record of an individual but does not permanently affect the individual's performance potential for a repeated trait.

terminal sire: A paternal-breed sire used in a terminal sire crossbreeding system.

terminal sire crossbreeding system: A crossbreeding system in which maternal-breed females are mated to paternal-breed sires to efficiently produce progeny that are especially desirable from a market standpoint. Terminally sired females are not kept as replacements, but are sold as slaughter animals.

test cross or **test mating:** A mating designed to reveal the genotype of an individual for a small number of loci.

threshold: A point on the continuous lability scale for a threshold trait above which animals exhibit one phenotype and below which they exhibit another.

threshold trait: A polygenic trait in which phenotypes are expressed in categories.

topcrossing: *See* **grading up.**

total maternal value (BV_{tm}): A combination of breeding values for both the direct and maternal components of a trait. A female's total maternal value represents the heritable part of her ability to influence a trait measured in her offspring.

trait: Any observable or measurable characteristic of an individual.

trait of the dam: A trait in which each progeny record is attributed to the dam, not the offspring.

trait of the offspring: A trait in which each record is attributed to an offspring, not to its dam.

trait ratio: An expression of relative performance—the ratio of an individual's performance to the average performance of all animals in the individual's contemporary group.

transgenic: (1) Referring to gene transfer. (2) An individual that has received genetic material by gene transfer.

transmitting ability (*TA*) or **progeny difference (*PD*):** Half an individual's breeding value—the expected difference between the mean performance of the individual's progeny and the mean performance of all progeny (assuming random mating).

true value: An unknown, underlying attribute that affects animal performance. Examples include breeding value (*BV*), progeny difference (*PD*), gene combination value (*GCV*), producing ability (*PA*), environmental effect (*E*), etc.

truncation selection: Selection on the basis of a distinct division in the selection criterion (point of truncation) above which individuals are selected and below which they are rejected.

unbiased: A genetic prediction is considered unbiased if, as more information is used in subsequent predictions for the same animal, those predictions are as likely to change in a positive direction as they are to change in a negative direction.

value: Any measure applied to an individual as opposed to a population. Examples are phenotypic value, genotypic value, breeding value, and environmental effect.

variable: Any quantity that can take on different numerical values. All elements (except μ) of the genetic model for quantitative traits—*P*, *BV*, *E*, etc.—are considered variables.

variance (σ^2): A mathematical measure of variation.

variation: (in most animal breeding applications) Differences among individuals within a population.

Wright's coefficient of relationship (R_{XY}): A measure of pedigree relationship: (1) The probable proportion of one individual's genes that are identical by descent to genes of a second individual. (2) The correlation between the breeding values of two individuals due to pedigree relationship alone.

zygote: A cell formed from the union of male and female gametes. A zygote has a full complement of genes—half from the sperm and half from the egg.

Appendix

The Algebra of Variances and Covariances

A number of formulas in this book are presented without derivation. That is because most students do not care where the formulas came from. If you are one of those exceptional students who likes to know the "why" behind the formulas, however, you can prove many of them—particularly those presented in Chapters 9 and 11—using what some call the algebra of variances and covariances. It consists of a fairly simple set of arithmetic rules for manipulating variances and covariances. Once you know the rules, you can derive a good share of the formulas in this book and many more that are not shown here.

The rules are as follows. The first three rules are for variances; the last five are for covariances.

1. The variance of a constant is zero.

   Example

   $$\text{var}(\mu) = 0$$

2. The variance of a constant times a random variable is the constant squared times the variance of the random variable.

   Example

   $$\text{var}(bP) = b^2\sigma_P^2$$

   A consequence of this rule is that the standard deviation of a constant times a random variable is the constant times the standard deviation of the random variable.

   Example

   $$\sigma_{bP} = b\sigma_P$$

3. The variance of a sum of random variables is the sum of the variances of each random variable plus two times the covariance of each pair of random variables.

Example

$$\text{var}(BV + GCV + E_p) = \sigma^2_{BV} + \sigma^2_{GCV} + \sigma^2_{E_p} + 2\text{cov}(BV,GCV) + 2\text{cov}(BV,E_p) + 2\text{cov}(GCV,E_p)$$

4. The covariance between a constant and a random variable is zero.

Example

$$\text{cov}(\mu,BV) = 0$$

5. The covariance between two random variables with constant coefficients is the product of the constants times the covariance between the random variables.

Example

$$\text{cov}\left(\frac{1}{2}BV_i,\frac{1}{2}BV_{i'}\right) = \frac{1}{2}\left(\frac{1}{2}\right)\text{cov}(BV_i,BV_{i'}) = \frac{1}{4}\text{cov}(BV_i,BV_{i'})$$

6. The covariance between a random variable and itself is its variance.

Example

$$\text{cov}(PA,PA) = \sigma^2_{PA}$$

7. If two random variables are independent, the covariance between them is zero.

Example

$$\text{cov}(G,E) = 0$$

8. The covariance between a random variable (or sum of random variables) and a sum of random variables is the sum of the individual covariances.

Examples

$$\text{cov}(PA,PA + E_t) = \text{cov}(PA,PA) + \text{cov}(PA,E_t)$$

and

$$\text{cov}(PA + E_{t_1},PA + E_{t_2}) = \text{cov}(PA,PA) + \text{cov}(PA,E_{t_2}) + \text{cov}(E_{t_1},PA) + \text{cov}(E_{t_1},E_{t_2})$$

A Sampling of Proofs Using the Algebra of Variances and Covariances

The following proofs use the algebra of variances and covariances to derive some common formulas in animal breeding. Study them for insight into the formulas and for practice using the algebra of variances and covariances.

Heritability as a Ratio of Variances

Given:

$$h^2 = r^2_{BV,P}$$

then

$$\begin{aligned} h^2 &= r^2_{BV,P} \\ &= \left(\frac{\text{cov}(BV,P)}{\sigma_{BV}\sigma_P}\right)^2 \\ &= \left(\frac{\text{cov}(BV,\mu + BV + GCV + E)}{\sigma_{BV}\sigma_P}\right)^2 \\ &= \left(\frac{\text{cov}(BV,\mu) + \text{cov}(BV,BV) + \text{cov}(BV,GCV) + \text{cov}(BV,E)}{\sigma_{BV}\sigma_P}\right)^2 \\ &= \left(\frac{0 + \sigma^2_{BV} + 0 + 0}{\sigma_{BV}\sigma_P}\right)^2 \\ &= \left(\frac{\sigma^2_{BV}}{\sigma_{BV}\sigma_P}\right)^2 \\ &= \frac{\sigma^4_{BV}}{\sigma^2_{BV}\sigma^2_P} \\ &= \frac{\sigma^2_{BV}}{\sigma^2_P} \end{aligned}$$

Heritability as a Regression Coefficient

Given:

$$h^2 = \frac{\sigma^2_{BV}}{\sigma^2_P}$$

then

$$\begin{aligned} b_{BV \cdot P} &= \frac{\text{cov}(BV,P)}{\sigma^2_P} \\ &= \frac{\text{cov}(BV,\mu + BV + GCV + E)}{\sigma^2_P} \\ &= \frac{\text{cov}(BV,\mu) + \text{cov}(BV,BV) + \text{cov}(BV,GCV) + \text{cov}(BV,E)}{\sigma^2_P} \\ &= \frac{0 + \sigma^2_{BV} + 0 + 0}{\sigma^2_P} \\ &= \frac{\sigma^2_{BV}}{\sigma^2_P} \\ &= h^2 \end{aligned}$$

Repeatability as a Ratio of Variances

Given:

$$r = r_{P_1,P_2}$$

and

$$PA = BV + GCV + E_p$$

then

$$\begin{aligned}
r &= r_{P_1,P_2} \\
&= \frac{\text{cov}(P_1,P_2)}{\sigma_{P_1}\sigma_{P_2}} \\
&= \frac{\text{cov}(\mu + BV + GCV + E_p + E_{t_1},\mu + BV + GCV + E_p + E_{t_2})}{\sigma_P\sigma_P} \\
&= \frac{\text{cov}(\mu + PA + E_{t_1},\mu + PA + E_{t_2})}{\sigma_P^2} \\
&= \frac{\begin{array}{c}\text{cov}(\mu,\mu + PA + E_{t_2}) + \text{cov}(PA,\mu) + \text{cov}(PA,PA) \\ + \text{cov}(PA,E_{t_2}) + \text{cov}(E_{t_1},\mu) + \text{cov}(E_{t_1},PA) + \text{cov}(E_{t_1},E_{t_2})\end{array}}{\sigma_P^2} \\
&= \frac{0 + 0 + \sigma_{PA}^2 + 0 + 0 + 0 + 0}{\sigma_P^2} \\
&= \frac{\sigma_{PA}^2}{\sigma_P^2}
\end{aligned}$$

Predicting Producing Ability from an Individual's Own Records

Prediction Equation and Regression Coefficient

The prediction equation is

$$\hat{PA} + b_{PA \cdot \overline{P}}\overline{P}$$

where $\bar{P}$ is the average of n records on the individual (expressed as deviations from contemporary group means). The regression coefficient is then

$$b_{PA \cdot \overline{P}} = \frac{\text{cov}(PA,\overline{P})}{\sigma_{\overline{P}}^2}$$

First the numerator:

$$\begin{aligned}
\operatorname{cov}(PA,\bar{P}) &= \operatorname{cov}\left(PA,\frac{1}{n}\sum_{i=1}^{n}P_i\right) \\
&= \frac{1}{n}\operatorname{cov}\left(PA,\sum_{i=1}^{n}P_i\right) \\
&= \frac{1}{n}\operatorname{cov}(PA,P_1 + P_2 + \cdots + P_n) \\
&= \frac{1}{n}(\operatorname{cov}(PA,P_1) + \operatorname{cov}(PA,P_2) + \cdots + \operatorname{cov}(PA,P_n))
\end{aligned}$$

All of these covariances are the same, and each one can be represented as:

$$\begin{aligned}
\operatorname{cov}(PA,P_i) &= \operatorname{cov}(PA,\mu + PA + E_{t_i}) \\
&= \operatorname{cov}(PA,\mu) + \operatorname{cov}(PA,PA) + \operatorname{cov}(PA,E_{t_i}) \\
&= 0 + \sigma^2_{PA} + 0 \\
&= \sigma^2_{PA}
\end{aligned}$$

Therefore

$$\begin{aligned}
\operatorname{cov}(PA,\bar{P}) &= \frac{1}{n}n\sigma^2_{PA} \\
&= \sigma^2_{PA}
\end{aligned}$$

Now the denominator:

$$\begin{aligned}
\sigma^2_{\bar{P}} &= \operatorname{var}(\bar{P}) \\
&= \operatorname{var}\left(\frac{1}{n}\sum_{i=1}^{n}P_i\right) \\
&= \frac{1}{n^2}\operatorname{var}(P_1 + P_2 + \cdots + P_n) \\
&= \frac{1}{n^2}\left(\begin{array}{l}\sigma^2_{P_1} + \sigma^2_{P_2} + \cdots + \sigma^2_{P_n} \\ + 2\operatorname{cov}(P_1,P_2) + \cdots + 2\operatorname{cov}(P_1,P_n) \\ + \cdots + 2\operatorname{cov}(P_2,P_n) + \cdots + 2\operatorname{cov}(P_{n-1},P_n)\end{array}\right) \\
&= \frac{1}{n^2}(n\sigma^2_P + (n^2 - n)\operatorname{cov}(P_i,P_{i'})) \\
&= \frac{1}{n^2}(n\sigma^2_P + n(n-1)\operatorname{cov}(P_i,P_{i'})) \\
&= \frac{1}{n}(\sigma^2_P + (n-1)\operatorname{cov}(P_i,P_{i'}))
\end{aligned}$$

From the earlier proof of repeatability as a ratio of variances we know that

$$\operatorname{cov}(P_i,P_{i'}) = \sigma^2_{PA}$$

So

$$\sigma^2_{\overline{P}} = \frac{1}{n}(\sigma^2_P + (n-1)\sigma^2_{PA})$$

Combining numerator and denominator,

$$b_{PA\cdot\overline{P}} = \frac{\operatorname{cov}(PA,\overline{P})}{\sigma^2_{\overline{P}}}$$

$$= \frac{\sigma^2_{PA}}{\frac{1}{n}(\sigma^2_P + (n-1)\sigma^2_{PA})}$$

$$= \frac{n\sigma^2_{PA}}{\sigma^2_P + (n-1)\sigma^2_{PA}}$$

Given $r = \dfrac{\sigma^2_{PA}}{\sigma^2_P}$, dividing both numerator and denominator by σ^2_P, we have:

$$b_{PA\cdot\overline{P}} = \frac{(n\sigma^2_{PA})/\sigma^2_P}{(\sigma^2_P + (n-1)\sigma^2_{PA})/\sigma^2_P}$$

$$= \frac{n\dfrac{\sigma^2_{PA}}{\sigma^2_P}}{\dfrac{\sigma^2_P}{\sigma^2_P} + (n-1)\dfrac{\sigma^2_{PA}}{\sigma^2_P}}$$

$$= \frac{nr}{1 + (n-1)r}$$

The prediction equation is then:

$$\hat{PA} = b_{PA\cdot\overline{P}}\overline{P}$$

$$= \frac{nr}{1 + (n-1)r}\overline{P}$$

Accuracy

The specific formula for accuracy of prediction in this case can be similarly derived, but let's save some steps by doing a more generic derivation first. Let T represent any true value, $\hat{T}$ a prediction of that value, and X the evidence used for prediction. Then

$$\hat{T} = b_{T\cdot X}X$$

Accuracy is then

$$\begin{aligned}
\text{ACC} &= r_{T,\hat{T}} \\
&= \frac{\text{cov}(T,\hat{T})}{\sigma_T \sigma_{\hat{T}}} \\
&= \frac{\text{cov}(T, b_{T\cdot X}X)}{\sigma_T \sigma_{b_{T\cdot X}X}} \\
&= \frac{b_{T\cdot X}\text{cov}(T,X)}{b_{T\cdot X}\sigma_T \sigma_X} \\
&= \frac{\text{cov}(T,X)}{\sigma_T \sigma_X} \\
&= r_{T,X} \\
&= b_{T\cdot X}\frac{\sigma_X}{\sigma_T}
\end{aligned}$$

and

$$\text{ACC}^2 = b^2_{T\cdot X}\frac{\sigma^2_X}{\sigma^2_T}$$

For this problem,

$$\sigma^2_T = \sigma^2_{PA}$$

and

$$\sigma^2_X = \sigma^2_{\bar{P}} = \frac{1}{n}(\sigma^2_P + (n-1)\sigma^2_{PA})$$

and a compatible version of $b_{T\cdot X}$ is

$$b_{T\cdot X} = b_{PA\cdot \bar{P}} = \frac{\sigma^2_{PA}}{\frac{1}{n}(\sigma^2_P + (n-1)\sigma^2_{PA})}$$

Then

$$\begin{aligned}
\text{ACC}^2 &= b^2_{T\cdot X}\frac{\sigma^2_X}{\sigma^2_T} \\
&= b_{PA\cdot \bar{P}}\frac{\sigma^2_{PA}}{\frac{1}{n}(\sigma^2_P + (n-1)\sigma^2_{PA})}\frac{\frac{1}{n}(\sigma^2_P + (n-1)\sigma^2_{PA})}{\sigma^2_{PA}} \\
&= b_{PA\cdot \bar{P}}
\end{aligned}$$

and

$$\begin{aligned} \text{ACC} &= \sqrt{b_{PA \cdot \overline{P}}} \\ &= \sqrt{\frac{nr}{1 + (n-1)r}} \end{aligned}$$

Predicting Breeding Value from Progeny Records

Prediction Equation and Regression Coefficient

The prediction equation is:

$$\hat{BV} = b_{BV \cdot \overline{P}} \overline{P}$$

where $\bar{P}$ is the average of single records (expressed as deviations from contemporary group means) on p progeny. The regression coefficient is then

$$b_{BV \cdot \overline{P}} = \frac{\text{cov}(BV, \overline{P})}{\sigma^2_{\overline{P}}}$$

First the numerator:

$$\begin{aligned} \text{cov}(BV, \overline{P}) &= \text{cov}\left(BV, \frac{1}{p}\sum_{i=1}^{p} P_i\right) \\ &= \frac{1}{p}\text{cov}\left(BV, \sum_{i=1}^{p} P_i\right) \\ &= \frac{1}{p}\text{cov}(BV, P_1 + P_2 + \cdots + P_p) \\ &= \frac{1}{p}(\text{cov}(BV, P_1) + \text{cov}(BV, P_2) + \cdots + \text{cov}(BV, P_p)) \end{aligned}$$

All of these covariances are the same, and each one can be represented as:

$$\begin{aligned} \text{cov}(BV, P_i) &= \text{cov}(BV, \mu + BV_i + GCV_i + E_i) \\ &= \text{cov}(BV, \mu) + \text{cov}(BV, BV_i) + \text{cov}(BV, GCV_i) + \text{cov}(BV, E_{t_i}) \\ &= 0 + \text{cov}(BV, BV_i) + 0 + 0 \\ &= \text{cov}(BV, BV_i) \end{aligned}$$

This is the covariance between breeding values of parent and offspring—a function of pedigree relationship—and is $\frac{1}{2}\sigma^2_{BV}$.

Therefore

$$\begin{aligned} \text{cov}(BV,\overline{P}) &= \frac{1}{p}p\frac{1}{2}\sigma^2_{BV} \\ &= \frac{1}{2}\sigma^2_{BV} \end{aligned}$$

Now the denominator:

$$\begin{aligned} \sigma^2_{\overline{P}} &= \text{var}(\overline{P}) \\ &= \text{var}\left(\frac{1}{p}\sum_{i=1}^{p}P_i\right) \\ &= \frac{1}{p^2}\text{var}(P_1 + P_2 + \cdots + P_p) \\ &= \frac{1}{p^2}\begin{pmatrix} \sigma^2_{P_1} + \sigma^2_{P_2} + \cdots + \sigma^2_{P_p} \\ + 2\text{cov}(P_1,P_2) + \cdots + 2\text{cov}(P_1,P_p) \\ + \cdots + 2\text{cov}(P_2,P_p) + \cdots + 2\text{cov}(P_{p-1},P_p) \end{pmatrix} \\ &= \frac{1}{p^2}(p\sigma^2_P + (p^2 - p)\text{cov}(P_i,P_{i'})) \\ &= \frac{1}{p^2}(p\sigma^2_P + p(p-1)\text{cov}(P_i,P_{i'})) \\ &= \frac{1}{p}(\sigma^2_P + (p-1)\text{cov}(P_i,P_{i'})) \end{aligned}$$

The covariance in the last expression is the covariance between phenotypic records of half sibs. Keeping in mind that the relationship between half sibs is $\frac{1}{4}$.

$$\text{cov}(P_i,P_{i'}) = \text{cov}(\mu + BV_i + GCV_i + E_i,\mu + BV_{i'} + GCV_{i'} + E_{i'})$$

Nearly all of the component covariances are zero, leaving

$$\begin{aligned} \text{cov}(P_i,P_{i'}) &= \text{cov}(BV_i,BV_{i'}) \\ &= \frac{1}{4}\sigma^2_{BV} \end{aligned}$$

So

$$\sigma^2_{\overline{P}} = \frac{1}{p}\left(\sigma^2_P + (p-1)\frac{1}{4}\sigma^2_{BV}\right)$$

Combining numerator and denominator,

$$b_{BV\cdot\overline{P}} = \frac{\text{cov}(BV,\overline{P})}{\sigma^2_{\overline{P}}}$$

$$= \frac{\frac{1}{2}\sigma^2_{BV}}{\frac{1}{p}\left(\sigma^2_P + (p-1)\frac{1}{4}\sigma^2_{BV}\right)}$$

$$= \frac{\frac{1}{2}p\sigma^2_{BV}}{\sigma^2_P + (p-1)\frac{1}{4}\sigma^2_{BV}}$$

Multiplying both numerator and denominator by $4/\sigma^2_P$, we have:

$$b_{PA\cdot\overline{P}} = \frac{4\left(\frac{1}{2}p\sigma^2_{BV}\right)\Big/\sigma^2_P}{4\left(\sigma^2_P + (p-1)\frac{1}{4}\sigma^2_{BV}\right)\Big/\sigma^2_P}$$

$$= \frac{2p\dfrac{\sigma^2_{BV}}{\sigma^2_P}}{4\left(\dfrac{\sigma^2_P}{\sigma^2_P} + (p-1)\dfrac{1}{4}\dfrac{\sigma^2_{BV}}{\sigma^2_P}\right)}$$

$$= \frac{2ph^2}{4\left(1 + (p-1)\frac{1}{4}h^2\right)}$$

$$= \frac{2ph^2}{4 + (p-1)h^2}$$

The prediction equation is then:

$$\hat{BV} = b_{BV\cdot\overline{P}}\overline{P}$$

$$= \frac{2ph^2}{4 + (p-1)h^2}\overline{P}$$

Accuracy

For this problem,

$$\sigma^2_T = \sigma^2_{BV}$$

and

$$\sigma^2_X = \sigma^2_{\overline{P}} = \frac{1}{p}\left(\sigma^2_P + (p-1)\frac{1}{4}\sigma^2_{BV}\right)$$

and a compatible version of $b_{T \cdot X}$ is

$$b_{T \cdot X} = b_{BV \cdot \overline{P}} = \frac{\frac{1}{2}\sigma_{BV}^2}{\frac{1}{p}\left(\sigma_P^2 + (p-1)\frac{1}{4}\sigma_{BV}^2\right)}$$

Then

$$\begin{aligned} \text{ACC}^2 &= b_{T \cdot X}^2 \frac{\sigma_X^2}{\sigma_T^2} \\ &= b_{BV \cdot \overline{P}} \frac{\frac{1}{2}\sigma_{BV}^2}{\frac{1}{p}(\sigma_P^2 + (p-1)\frac{1}{4}\sigma_{BV}^2)} \frac{\frac{1}{p}\left(\sigma_P^2 + (p-1)\frac{1}{4}\sigma_{BV}^2\right)}{\sigma_{BV}^2} \\ &= \frac{1}{2} b_{BV \cdot \overline{P}} \end{aligned}$$

and

$$\begin{aligned} \text{ACC} &= \sqrt{\frac{1}{2} b_{BV \cdot \overline{P}}} \\ &= \sqrt{\frac{1}{2}\left(\frac{2ph^2}{4 + (p-1)h^2}\right)} \\ &= \sqrt{\frac{ph^2}{4 + (p-1)h^2}} \end{aligned}$$

Answers to Problems

ANSWERS TO CHAPTER 1 PROBLEMS

1.1 Listed in the following table are net profits ($) for equivalent beef cattle operations given three biological types: large, medium, and small mature size; and three economic scenarios: standard cost/price relationships, doubled cow herd feed costs, and doubled feedlot feed costs.

	Biological Type		
Economic Scenario	Large	Medium	Small
Standard cost/price relationships	24,510	18,825	15,990
Doubled cow herd feed costs	−4,973	−2,552	−20,157
Doubled feedlot feed costs	−15,819	−9,986	−11,336

Graph profit versus economic scenario for the different biological types.

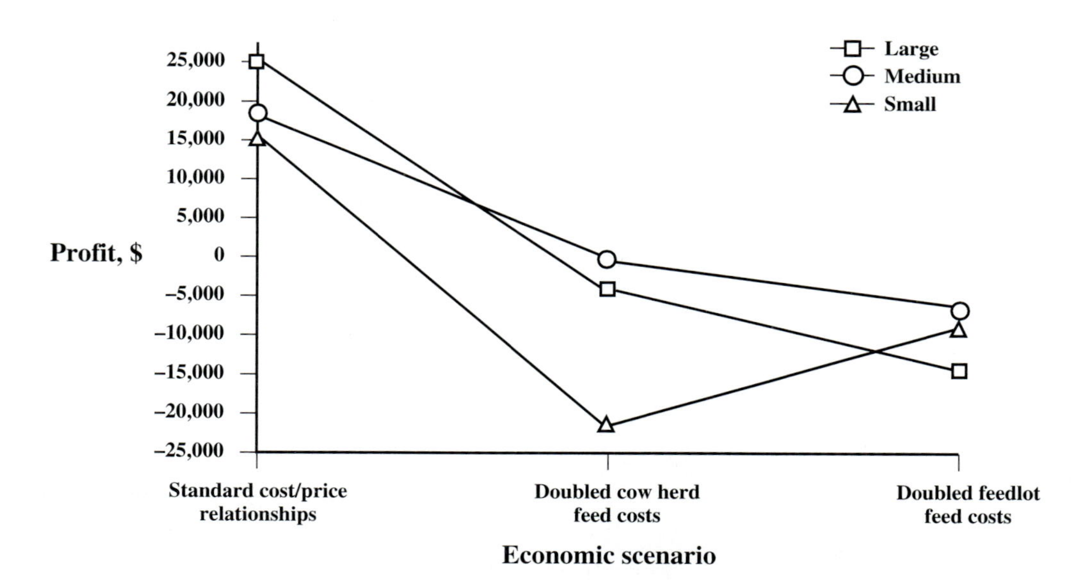

a. Do genotype by economics interactions exist? If so, describe them.

Yes, genotype by economics interactions do exist. Large cows are more profitable when costs are low in general. Medium-sized cows are more profitable (less costly) when costs are high. Small cows are very unprofitable when cow herd feed costs are high.

b. How would you explain these interactions? (Background information: No cattle are sold before slaughter, and all young nonreplacement animals are fed in the feedlot to a constant degree of fatness. Cattle are managed in such a way that the operation can run 222 large cows (but no yearling steers), 204 medium-sized cows, and 284 small cows. Large, medium, and small cows produce a herdwide total of 111, 97, and 106 tons of beef, respectively.)

(Explanations of these interactions are open to interpretation. Here are my thoughts.) Large cows are most profitable when costs are low in general because they produce the most product to offset fixed costs. They are less profitable when cow herd feed costs are high because each large cow eats so much. They are even less profitable when feedlot feed costs are high because large steers and heifers require more time and feed in the lot than smaller biological types. Small cows are very unprofitable when cow herd feed costs are high because they collectively eat a lot—there are many of them.

c. How should breeding objectives for operations like these change as economic scenarios change?

If it appears that feed costs will increase considerably, breeders of large cows should consider breeding for smaller size (especially if feedlot costs increase the most), and breeders of small cows should consider breeding for larger size (especially if cow herd costs increase the most).

1.2 Listed in the table below are typical survival percentages for newborn calves varying in birth weight.

Birth Weight, lb	Survival, %
30	15
40	42
50	63
60	93
70	98
80	94
90	90
100	82
110	71
120	58
130	35
140	26

Plot survival versus birth weight.

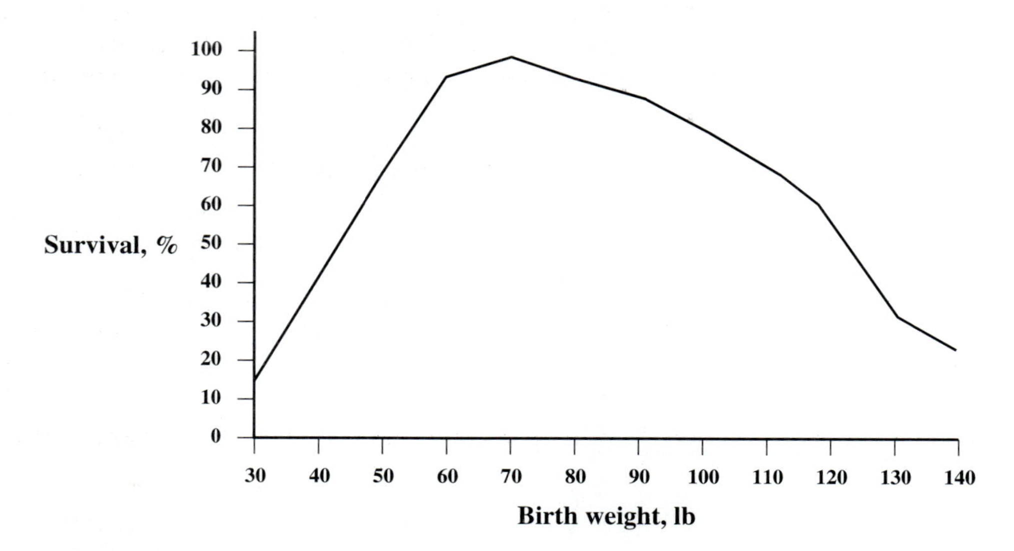

a. What concept is illustrated here?

An intermediate optimum.

b. Describe logical selection objectives for birth weight if the average birth weight in a herd is:

i. 60 lb

Select (with moderate intensity) for heavier birth weight.

ii. 75 lb

Pay little attention to birth weight.

iii. 97 lb

Select (fairly intensely) for lighter birth weight.

ANSWERS TO CHAPTER 3 PROBLEMS

3.1 Given the following alleles at the J locus: *J*, *J´*, *j*, and *j´*, list all possible:

a. homozygous combinations.

JJ, *J´J´*, *jj*, *j´j´*

b. heterozygous combinations.

JJ´, *Jj*, *Jj´*, *J´j*, *J´j´*, *jj´*

3.2 If you made 32 matings of roan Shorthorn cattle, how many calves would you *expect* to be red? Roan? White? If 32 calves result, will their coat colors match expectations? Why or why not?

Because roans are heterozygotes, we expect a 1:2:1 ratio in the offspring—i.e., 1 red to 2 roans to 1 white. With 32 calves, that would be 8 reds, 16 roans, and 8 whites.

Chances are we will not get exactly these proportions because the random process known as Mendelian sampling causes samples of offspring to differ.

3.3 A sire's five-locus genotype is *AaBBCcDdee*. A dam's genotype is *AABbCcDdEe*. Considering just these five loci:

a. How many unique gametes can the sire produce?

$$\text{Number of unique gametes} = 2^n$$

where n is the number of loci at which an individual is heterozygous.

$$= 2^3$$
$$= 8$$

b. How many unique gametes can the dam produce?

$$\text{Number of unique gametes} = 2^n$$
$$= 2^4$$
$$= 16$$

c. How many unique zygotes can be produced from this mating?

$$\text{Number of unique zygotes} = 3^n \times 2^m$$

where n is the number of loci at which both parents are heterozygous, and m is the number of loci at which only one parent is heterozygous.

$$= 3^2 \times 2^3$$
$$= 9 \times 8$$
$$= 72$$

3.4 Consider a hypothetical locus for tuberculosis resistance/susceptibility with alleles T^r (resistant) and T^s (susceptible). When exposed to the tuberculosis pathogen, T^rT^r individuals survive 90% of the time, and T^sT^s individuals survive 30% of the time. What is the value (range of values) for survival percentage of T^rT^s individuals if the locus exhibits:

a. complete dominance and T^r is the dominant allele?

Let $\%S_H$ stand for survival percentage of the heterozygote (T^rT^s). Then:

$$\%S_H = 90$$

b. complete dominance and T^s is the dominant allele?

$$\%S_H = 30$$

c. no dominance?

$$\%S_H = \frac{(90 + 30)}{2} = 60$$

d. partial dominance and T^r is the dominant allele?

$$60 < \%S_H < 90$$

e. partial dominance and T^s is the dominant allele?

$$30 < \%S_H < 60$$

f. overdominance and T^r is the dominant allele?

$$90 < \%S_H \leq 100$$

g. overdominance and T^s is the dominant allele?

$$0 \leq \%S_H < 30$$

3.5 A roan, heterozygous polled bull (*RrPp*) is mated to a roan, horned cow. What are the possible *phenotypes* and their expected proportions from this mating?

Possible Phenotypes	Expected Proportions
Red, polled	$\frac{1}{8}$
Red, horned	$\frac{1}{8}$
Roan, polled	$\frac{2}{8} = \frac{1}{4}$
Roan, horned	$\frac{2}{8} = \frac{1}{4}$
White, polled	$\frac{1}{8}$
White, horned	$\frac{1}{8}$

♂ \ ♀	*Rp*	*rp*
RP	**Red, polled** *RRPp*	**Roan, polled** *RrPp*
Rp	**Red, horned** *RRpp*	**Roan, horned** *Rrpp*
rP	**Roan, polled** *RrPp*	**White, polled** *rrPp*
rp	**Roan, horned** *Rrpp*	**White, horned** *rrpp*

3.6 Your prize chocolate Labrador is from a mixed litter of black, yellow, and chocolate pups. He is by a yellow dog and out of a black bitch. What colors in what proportions would you expect if you mated your dog to his dam?

Coat color in Labradors is determined by two loci that exhibit epistasis such that:

*B_E_*⇒ black
*bbE_*⇒ chocolate
_ _*ee*⇒ yellow

Because my dog is chocolate, we know that his genotype is *bbE_*, and because he must have received an *e* allele from his yellow sire, we know that he is *bbEe.*

His black dam has produced all three colors of puppies, so she must be *BbEe.* The mating of my dog to his dam can be represented:

♂ \ ♀	***BE***	***Be***	***bE***	***be***
bE	**Black** ***BbEE***	**Black** ***BbEe***	**Chocolate** ***bbEE***	**Chocolate** ***bbEe***
be	**Black** ***BbEe***	**Yellow** ***Bbee***	**Chocolate** ***bbEe***	**Yellow** ***bbee***

Possible Phenotypes	Expected Proportions
Black	$3/8$
Chocolate	$3/8$
Yellow	$2/8 = 1/4$

ANSWERS TO CHAPTER 4 PROBLEMS

4.1 At the C locus in horses, chestnuts (sorrels) are CC, palominos are Cc^{cr}, and cremellos are $c^{cr}c^{cr}$. In a herd of 10 horses, there are 3 chestnuts, 6 palominos, and 1 cremello. What are the gene and genotypic frequencies at the C locus in this herd?

$$\text{Gene frequency of the } C \text{ allele} = p = \frac{2(3) + 6}{2(10)} = \frac{12}{20} = .6$$

$$\text{Gene frequency of the } c^{cr} \text{ allele} = q = \frac{2(1) + 6}{2(10)} = \frac{8}{20} = .4$$

Genotypic frequencies:

$$P = \frac{3}{10} = .3$$

$$H = \frac{6}{10} = .6$$

$$Q = \frac{1}{10} = .1$$

4.2 Skin color in Bohemian bullfrogs is determined by a single locus (S) such that:
$SS \Rightarrow$ solid
$Ss \Rightarrow$ striped
$ss \Rightarrow$ spotted

Of the Bohemians entered in the Backcounty Bullfrog Bounce, a regional frog jumping contest, 70% were solid colored, 20% were striped, and 10% were spotted. What were the gene and genotypic frequencies at the S locus for the frogs at the contest?

If we call S the "dominant" allele, then:

$$P = .7$$
$$H = .2$$
$$Q = .1$$

and therefore

$$p = P + \frac{1}{2}H = .7 + \frac{1}{2}(.2) = .8$$
$$q = 1 - p = 1 - .8 = .2$$

4.3 Construct a pedigree and arrow diagram for:
a. a sire × daughter mating.
b. a full-sib mating.
Identify the common ancestor(s) in each case.

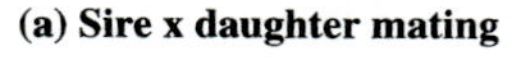

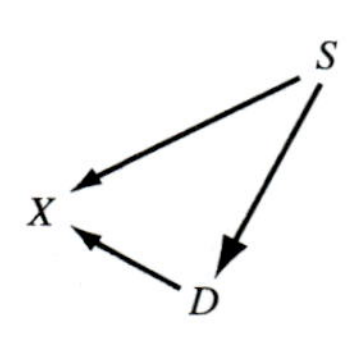

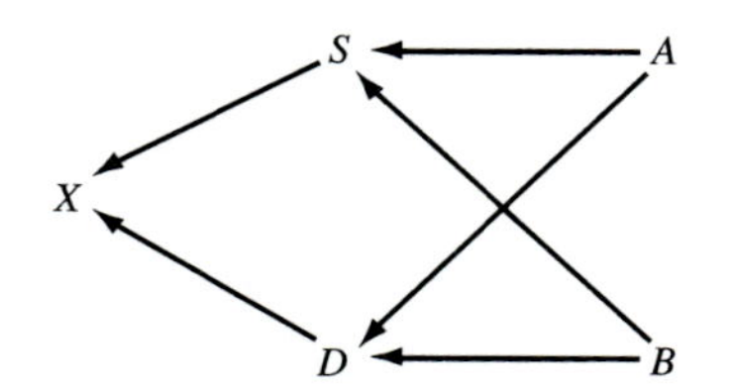

Common ancestors are:
(a) S
(b) A and B

4.4 Two large populations of horses are being systematically crossed (mares from one population bred to stallions of the other and vice versa). Coat color is *not* a factor in determining which animals are selected and which individual matings are made. Frequencies of coat color genes at the C locus are:

Population 1		Population 2	
C	c^{cr}	C	c^{cr}
.8	.2	.3	.7

a. What will be the gene and genotypic frequencies at the C locus in the offspring (F_1) population?

The first cross can be represented by the following Punnett square:

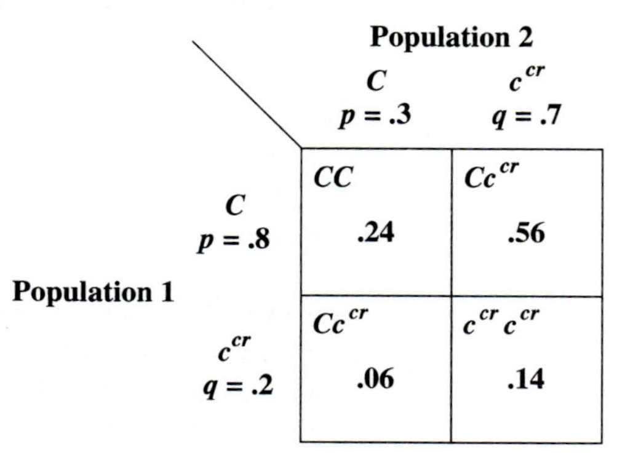

In F_1 offspring:

$$P = .24$$
$$H = .56 + .06 = .62$$
$$Q = .14$$

Therefore:

$$p = P + \frac{1}{2}H = .24 + \frac{1}{2}(.62) = .55$$
$$q = 1 - p = 1 - .55 = .45$$

b. If the crossbred offspring are mated among themselves, what will be the gene and genotypic frequencies in the F_2 generation? (Assume random mating, no selection, etc.)

$F_1 \times F_1 = F_2$

F_1 \ F_1	C $p = .55$	c^{cr} $q = .45$
C $p = .55$	CC .3025	Cc^{cr} .2475
c^{cr} $q = .45$	Cc^{cr} .2475	$c^{cr}c^{cr}$ .2025

In F_2 offspring:

$$P = .3025$$
$$H = .2475 + .2475 = .495$$
$$Q = .2025$$

Therefore:

$$p = P + \frac{1}{2}H = .3025 + \frac{1}{2}(.495) = .55$$
$$q = 1 - p = 1 - .55 = .45$$

Alternatively, because all Hardy-Weinberg requirements are met in the F_2 generation, we could simply apply the Hardy-Weinberg formulas. Thus:

$$p = .55$$
$$q = .45$$
$$P = p^2 = (.55)^2 = .3025$$
$$H = 2pq = 2(.55)(.45) = .495$$
$$Q = q^2 = (.45)^2 = .2025$$

c. In the F_3 and subsequent generations?

By the F_2 generation, the population is in Hardy-Weinberg equilibrium. Gene and genotypic frequencies in the F_3 and later generations should therefore be the same as in the F_2s.

4.5 In a population of Garden-digging Armadillos in Hardy-Weinberg equilibrium, the *C* allele for long claws is completely dominant to the *c* allele for clawlessness. Extensive sampling of this population showed 16% of the armadillos to be clawless. What are the gene and genotypic frequencies at the C locus for these varmints?

We know that the genotypic frequency (Q) of the homozygous recessive, clawless type (*cc*) is .16.

$$Q = .16$$

Therefore, applying the Hardy-Weinberg formulas:

$$q = \sqrt{Q} = \sqrt{.16} = .4$$
$$p = 1 - q = 1 - .4 = .6$$
$$P = p^2 = (.6)^2 = .36$$
$$H = 2pq = 2(.6)(.4) = .48$$

ANSWERS TO CHAPTER 6 PROBLEMS

6.1 A Labrador breeder analyzed the pedigrees of two of her dogs and determined that the black male has a 50% chance of having the genotype *BBEe* and a 50% chance of having the genotype *BbEe.* The yellow female has a 75% chance of having the genotype *BBee* and a 25% chance of having the genotype *Bbee.* For matings of animals with probable genotypes like these, what proportion of puppies is expected to be chocolate? (See Chapter 3 for an explanation of coat color in Labs.)

We can answer this question using a Punnett square, but first we'll need to know the probabilities that particular gametes are contributed.

The male has a 50% chance of being *BBEe*. From this we can deduce the following:

Gamete	Probability of Being Contributed
BE	.25
Be	.25

He also has a 50% chance of being *BbEe*. From this we can deduce the following:

Gamete	Probability of Being Contributed
BE	.125
Be	.125
bE	.125
be	.125

Combining information on the male, we have:

Gamete	Probability of Being Contributed
BE	.25 + .125 = .375
Be	.25 + .125 = .375
bE	.125
be	.125

The female has a 75% chance of being *BBee*. From this we can deduce the following:

Gamete	Probability of Being Contributed
Be	.75

She also has a 25% chance of being *Bbee*. From this we can deduce the following:

Gamete	Probability of Being Contributed
Be	.125
be	.125

Combining information on the female, we have:

Gamete	Probability of Being Contributed
Be	.75 + .125 = .875
be	.125

Mating these two:

♂ \ ♀	*Be* .875	*be* .125
BE .375	**Black** *BBEe* **.3281**	**Black** *BbEe* **.0469**
Be .375	**Yellow** *BBee* **.3281**	**Yellow** *Bbee* **.0469**
bE .125	**Black** *BbEe* **.1094**	**Chocolate** *bbEe* **.0156**
be .125	**Yellow** *Bbee* **.1094**	**Yellow** *bbee* **.0156**

Less than 2% of puppies should be chocolate.

6.2 A large artificial insemination stud has just purchased a promising bull. Management is concerned, however, that the bull might be a carrier of osteopetrosis (marble bone disease), a recessive lethal condition. Five percent of all cows are thought to be carriers of the osteopetrosis allele.

a. A.I. matings to randomly selected cows have already produced 100 normal calves (and no homozygous recessive calves). What is management's level of confidence that the bull is not a carrier of the gene for osteopetrosis?

$$P[D_n] = 1 - \left(P_{BB} + \frac{3}{4}P_{Bb} + \frac{1}{2}P_{bb}\right)^n$$

Because 5% of mates are thought to be carriers, $P_{Bb} = .05$. $P_{bb} = 0$ because the homozygous recessive condition is lethal. Therefore, $P_{BB} = 1 - .05 = .95$. Then:

$$P[D_n] = 1 - \left(.95 + \frac{3}{4}(.05) + \frac{1}{2}(0)\right)^{100}$$
$$= 1 - (.9875)^{100}$$
$$\approx .72$$

b. How many successful A.I. matings to randomly selected cows are required to be 99% sure he does not carry the gene?

$$n = \frac{\log(1 - P[D_n])}{\log\left(P_{BB} + \frac{3}{4}P_{Bb} + \frac{1}{2}P_{bb}\right)}$$
$$= \frac{\log(1 - .99)}{\log\left(.95 + \frac{3}{4}(.05) + \frac{1}{2}(0)\right)}$$
$$= \frac{\log(.01)}{\log(.9875)}$$
$$= \frac{-2}{-.00546}$$
$$\approx 366 \text{ matings}$$

c. If the bull sires a calf with marble bone disease next year, what will be management's level of confidence that the bull does not carry the osteopetrosis gene? What will be their level of confidence that he *does* carry the gene?

Management will be 0% confident that the bull does not carry the osteopetrosis gene and 100% confident that he does carry it.

6.3 A ram was bred to eight of his daughters to see if he carries any undesirable recessive genes. Four daughters produced twins, three produced singles, and one produced triplets. All lambs were normal.

a. How confident are we that the ram does not carry any undesirable recessives?

There are a couple of ways to approach this question. The easiest way is to consider all the daughters part of one uniform group averaging $\frac{4(2) + 3(1) + 1(3)}{8} = \frac{14}{8} = 1.75$ lambs per ewe. Then:

$$P[D_n^m] = 1 - \left(P_{BB} + \left(\frac{3}{4}\right)^m P_{Bb} + \left(\frac{1}{2}\right)^m P_{bb}\right)^n$$
$$= 1 - \left(\frac{1}{2} + \left(\frac{3}{4}\right)^{1.75}\left(\frac{1}{2}\right) + \left(\frac{1}{2}\right)^{1.75}(0)\right)^8$$
$$= 1 - (.8022)^8$$
$$\approx .83$$

A more precise approach would be to divide the daughters into groups according to number of lambs produced. Then:

$$P[D_n^m] = 1 - \prod_{i=1}^{k}\left(P_{BB_i} + \left(\frac{3}{4}\right)^{m_i} P_{Bb_i} + \left(\frac{1}{2}\right)^{m_i} P_{bb_i}\right)^{n_i}$$
$$= 1 - \prod_{i=1}^{k}\left(\frac{1}{2} + \left(\frac{3}{4}\right)^{m_i}\left(\frac{1}{2}\right) + \left(\frac{1}{2}\right)^{m_i}(0)\right)^{n_i}$$
$$= 1 - \left(\frac{1}{2} + \left(\frac{3}{4}\right)^2\left(\frac{1}{2}\right)\right)^4\left(\frac{1}{2} + \left(\frac{3}{4}\right)^1\left(\frac{1}{2}\right)\right)^3\left(\frac{1}{2} + \left(\frac{3}{4}\right)^3\left(\frac{1}{2}\right)\right)^1$$
$$= 1 - (.78125)^4(.875)^3(.71094)$$
$$\approx .82$$

b. The ram was also bred to three known carriers of the recessive allele for spider syndrome. Each of these ewes produced a set of normal twin lambs. How confident are we that the ram does not carry the spider allele?

$$P[D_n^m] = 1 - \prod_{i=1}^{k}\left(P_{BB_i} + \left(\frac{3}{4}\right)^{m_i} P_{Bb_i} + \left(\frac{1}{2}\right)^{m_i} P_{bb_i}\right)^{n_i}$$
$$= 1 - (.78125)^4(.875)^3(.71094)\left(0 + \left(\frac{3}{4}\right)^2(1) + \left(\frac{1}{2}\right)^2(0)\right)^3$$
$$= 1 - (.78125)^4(.875)^3(.71094)(.5625)^3$$
$$\approx .97$$

6.4 Consider a herd of 100 Hampshire sows in Hardy-Weinberg equilibrium.

a. What is the expected ratio of heterozygotes (white belted carriers of the allele for solid color) to homozygous recessive (solid colored) animals if:

i. 25 sows are solid colored.

$$Q = \frac{25}{100} = .25$$
$$q = \sqrt{Q} = \sqrt{.25} = .5$$
$$H = 2pq = 2(1 - .5)(.5) = .5$$
$$H{:}Q = .5 : .25 = 2 : 1$$

ii. Four sows are solid colored.

$$Q = \frac{4}{100} = .04$$
$$q = \sqrt{Q} = \sqrt{.04} = .2$$
$$H = 2pq = 2(1 - .2)(.2) = .32$$
$$H{:}Q = .32 : .04 = 8{:}1$$

iii. One sow is solid colored.

$$Q = \frac{1}{100} = .01$$
$$q = \sqrt{Q} = \sqrt{.01} = .1$$
$$H = 2pq = 2(1 - .1)(.1) = .18$$
$$H{:}Q = .18 : .01 = 18{:}1$$

b. What is the ratio of recessive genes found in heterozygotes to recessive genes found in homozygotes for (i), (ii), and (iii) above?

i. Recessives in heterozygotes : recessives in homozygotes = 50:25(2) = 1:1

ii. Recessives in heterozygotes : recessives in homozygotes = 32:4(2) = 4:1

iii. Recessives in heterozygotes : recessives in homozygotes = 18:1(2) = 9:1

c. What relationship is evident here?

The lower the frequency of the recessive allele in the population, the greater the ratio of heterozygotes to homozygous recessive types, and therefore the greater the number of recessive alleles found in heterozygotes relative to the number found in homozygous recessives.

6.5 There are two alleles at the J locus: J_1 and J_2. J_2 is the less desirable allele—in homozygous form it is lethal—and its frequency in the current generation is .2. What will be the frequency of J_2 in the next generation if:

a. with respect to fitness, J_1 is completely dominant to J_2?

J_1J_1 and J_1J_2 are the most fit genotypes, and because of complete dominance for fitness, they are equally fit. Therefore, $s_1 = s_2 = 0$. J_2J_2 individuals do not survive, so $s_3 = 1$. Then:

$$q_1 = \frac{(1 - s_2)q + (s_2 - s_3)q^2}{1 - s_1 + 2(s_1 - s_2)q + (2s_2 - s_1 - s_3)q^2}$$

$$= \frac{(1 - 0)(.2) + (0 - 1)(.2)^2}{1 - 0 + 2(0 - 0)(.2) + (2(0) - 0 - 1)(.2)^2}$$

$$= \frac{.2 - (.2)^2}{1 - (.2)^2}$$

$$= .167$$

b. J_2 is partially dominant such that J_1J_2 individuals produce 70% fewer offspring than J_1J_1 types?

$$s_1 = 0$$
$$s_2 = .7$$
$$s_3 = 1$$

and

$$q_1 = \frac{(1 - s_2)q + (s_2 - s_3)q^2}{1 - s_1 + 2(s_1 - s_2)q + (2s_2 - s_1 - s_3)q^2}$$

$$= \frac{(1 - .7)(.2) + (.7 - 1)(.2)^2}{1 - 0 + 2(0 - .7)(.2) + (2(.7) - 0 - 1)(.2)^2}$$

$$= \frac{.3(.2) - .3(.2)^2}{1 - 1.4(.2) + .4(.2)^2}$$

$$= .065$$

c. overdominance for fitness exists at the J locus such that J_1J_1 individuals produce 25% fewer offspring than J_1J_2 types?

In this scenario, J_1J_2 is the most fit genotype, so:

$$s_1 = .25$$
$$s_2 = 0$$
$$s_3 = 1$$

and

$$
\begin{aligned}
q_1 &= \frac{(1 - s_2)q + (s_2 - s_3)q^2}{1 - s_1 + 2(s_1 - s_2)q + (2s_2 - s_1 - s_3)q^2} \\
&= \frac{(1 - 0)(.2) + (0 - 1)(.2)^2}{1 - .25 + 2(.25 - 0)(.2) + (2(0) - .25 - 1)(.2)^2} \\
&= \frac{.2 - (.2)^2}{.75 + .5(.2) - 1.25(.2)^2} \\
&= .2
\end{aligned}
$$

d. Compare your results for (a), (b), and (c) above and explain why they differ.

Selection against J_2 is relatively ineffective when J_2 is completely recessive (a) because most J_2 genes are "hidden" in the heterozygote and are not selected against. When J_2 is partially dominant (b), selection against J_2 is quite effective because selection works against J_2 genes whether they occur in homozygous or heterozygous form. When overdominance exists at the J locus (c), gene frequencies tend toward an intermediate equilibrium value—which just happens to be .2.

ANSWERS TO CHAPTER 7 PROBLEMS

7.1 **a.** Construct to scale a diagram like Figure 7.1 showing the following sample of records for milk production in dairy cows (μ = 13,600 lb):

Cow #	*P*	*G*	*E*
1	12,100	−300	−1,200
2	14,600	+1,200	−200
3	14,600	−400	+1,400

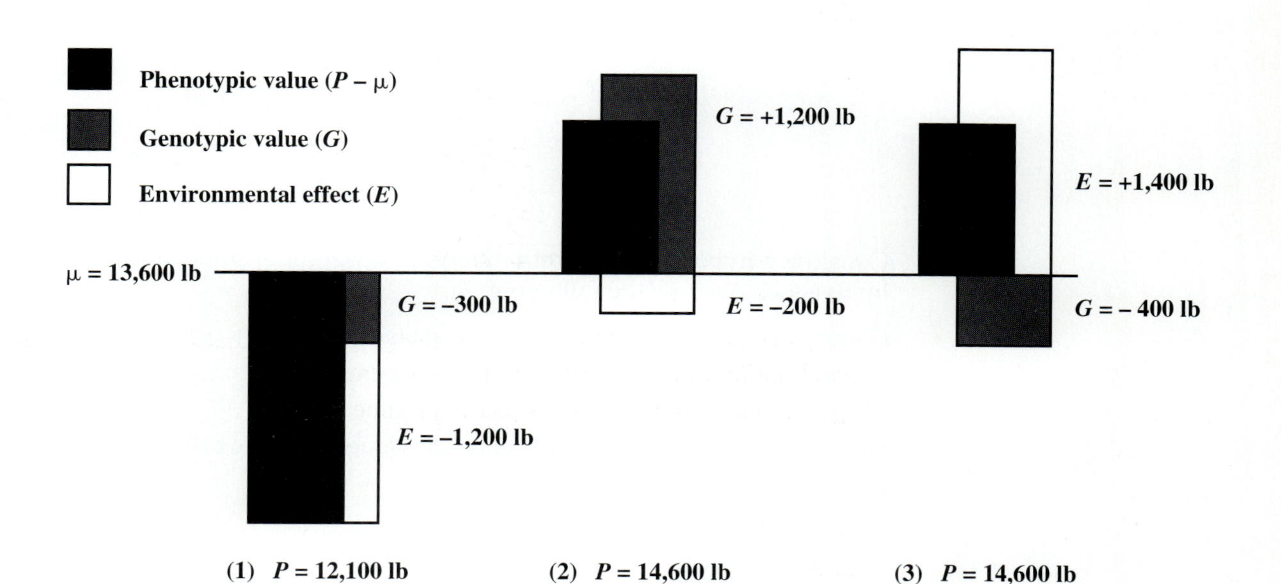

b. Cows 2 and 3 have similar records, but for very different reasons. Explain.

Cows 2 and 3 have the same phenotypes—14,600 lb of milk. Cow 2 is genetically superior, but cow 3 is not. Cow 3's record is good because she experienced a favorable environment.

7.2 A famous draft horse was mated to a large number of *randomly selected* mares. On average, offspring of these matings pull 200 lb more than the average horse in major contests.

a. What is the sire's progeny difference for pulling power?

Progeny difference is simply the difference between the average performance of an individual's progeny and the population average (assuming randomly chosen mates). So in this case, $PD = +200$ lb.

b. What is his breeding value for pulling power?

$$PD = \frac{1}{2}BV$$

so

$$\begin{aligned} BV &= 2PD \\ &= 2(200) \\ &= +400 \text{ lb} \end{aligned}$$

c. The horse was later mated to a large number of mares handpicked for pulling power. Offspring of these matings pull 300 lb more than the average horse. What is the mean breeding value of their *dams* for pulling power?

$$\overline{P}_{Offspring} = \mu + \overline{BV}_{Offspring} = \mu + \frac{1}{2}BV_{Sire} + \frac{1}{2}\overline{BV}_{Dams}$$

so

$$\begin{aligned} \overline{P}_{Offspring} - \mu &= \frac{1}{2}BV_{Sire} + \frac{1}{2}\overline{BV}_{Dams} \\ \overline{BV}_{Dams} &= 2\left(\overline{P}_{Offspring} - \mu - \frac{1}{2}BV_{Sire}\right) \\ &= 2\left(300 - \frac{1}{2}(400)\right) \\ &= +200 \text{ lb} \end{aligned}$$

7.3 Consider a hypothetical quantitative trait (a weight of some kind) affected by five loci. Assume the following:

Complete dominance at all loci. No epistasis.
The independent effect of each dominant gene is +10 lb.
The independent effect of each recessive gene is −4 lb.
For homozygous combinations, genotypic values are equal to breeding values.
$\mu = 600$ lb.

a. Fill in the following table:

Genotype	*BV*	*G*	*GCV*	*E*	*P*
1. *AaBbCcDdEe*	5(10)+5(−4)=+30	5(20)=+100	100−30=+70	+17	600+30+70+17=717
2. *AAbbCCddEE*	6(10)+4(−4)=+44	3(20)+2(−8)=+44	44−44=0	−21	600+44+0−21=623

b. Which individual is the heaviest? Explain.

Individual 1 is the heaviest—it has the largest phenotypic value.

c. Which would produce the heaviest offspring (on average)? Explain.

Individual 2 would produce the heaviest offspring—it has the largest breeding value.

7.4 Consider the Thoroughbred stallions Raise-A-Ruckus and Presidium. Raise-a-Ruckus's breeding value for racing time is −8 seconds. He was particularly well trained, having a permanent environmental effect of −6 seconds. Presidium's breeding value is −12 seconds, but his permanent environmental effect is +2 seconds. Assuming both horses have gene combination values of 0,

a. Calculate progeny difference for each horse.

$$PD = \frac{1}{2}BV$$

Raise-a-Ruckus:

$$\begin{aligned} PD &= \frac{1}{2}(-8) \\ &= -4 \text{ seconds} \end{aligned}$$

Presidium:

$$\begin{aligned} PD &= \frac{1}{2}(-12) \\ &= -6 \text{ seconds} \end{aligned}$$

b. Calculate producing ability for each horse.

$$PA = BV + GCV + E_p$$

Raise-a-Ruckus:

$$\begin{aligned} PA &= -8 + 0 + (-6) \\ &= -14 \text{ seconds} \end{aligned}$$

Presidium:

$$\begin{aligned} PA &= -12 + 0 + 2 \\ &= -10 \text{ seconds} \end{aligned}$$

c. Which horse would you bet on in a race? Why?

I would bet on Raise-a-Ruckus because he has the better producing ability (horse value).

d. Which horse would you breed mares to? Why?

I would use Presidium because he has the better breeding value.

7.5 Calving difficulty in beef cattle is a threshold trait, and breeders often record just two categories of calving difficulty scores: assisted and unassisted. Using Figure 7.6 as a guide, show the distributions of liability for calving difficulty if:

a. about 90% of the cows in a population calve unassisted.

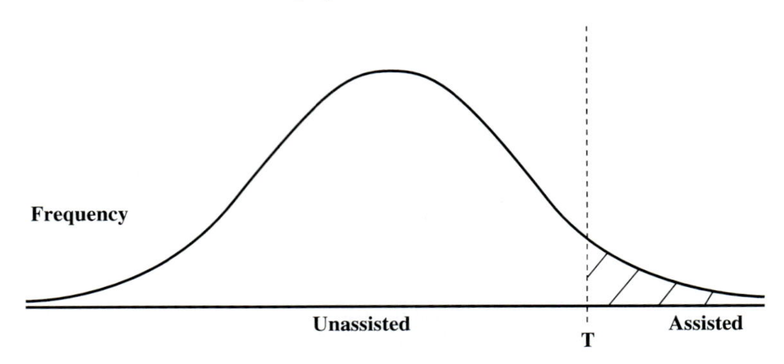

b. only 50% of the cows in a population calve unassisted.

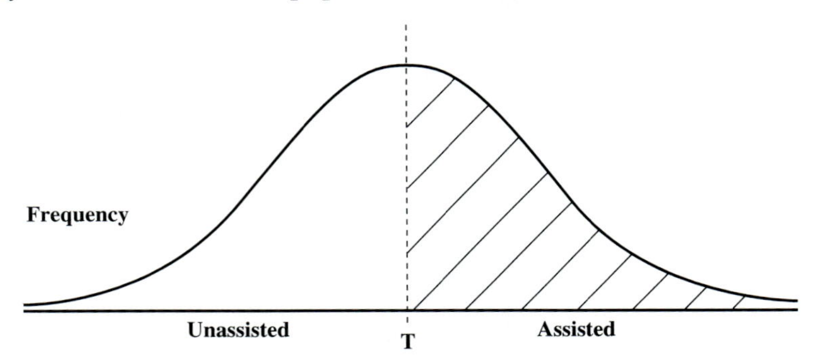

ANSWERS TO CHAPTER 8 PROBLEMS

8.1 Given the following set of data on days to 230 lb (D230) and backfat thickness (BF) in pigs:

Pig #	Days to 230 lb, days	Backfat Thickness, in
1	164	1.1
2	181	1.2
3	158	1.3
4	160	1.5
5	198	1.3
6	172	1.4
7	187	1.2
8	180	1.4
9	178	1.4
10	186	1.0

a. Calculate:

i. $\hat{\mu}_{P_{D230}}$

$$\hat{\mu}_{P_{D230}} = \frac{1}{n}\sum_{i=1}^{n} P_{D230_i} = \frac{1}{10}\sum_{i=1}^{10} P_{D230_i} = 176.4 \text{ days}$$

ii. $\hat{\mu}_{P_{BF}}$

$$\hat{\mu}_{P_{BF}} = \frac{1}{n}\sum_{i=1}^{n} P_{BF_i} = \frac{1}{10}\sum_{i=1}^{10} P_{BF_i} = 1.28 \text{ in}$$

iii. $\hat{\sigma}^2_{P_{D230}}$

$$\hat{\sigma}^2_{P_{D230}} = \frac{\sum_{i=1}^{n}(P_{D230_i} - \hat{\mu}_{P_{D230}})^2}{n-1} = \frac{\sum_{i=1}^{10}(P_{D230_i} - 176.4)^2}{10-1} = 165.38 \text{ days}^2$$

iv. $\hat{\sigma}^2_{P_{BF}}$

$$\hat{\sigma}^2_{P_{BF}} = \frac{\sum_{i=1}^{n}(P_{BF_i} - \hat{\mu}_{P_{BF}})^2}{n-1} = \frac{\sum_{i=1}^{10}(P_{BF_i} - 1.28)^2}{10-1} = .024 \text{ in}^2$$

v. $\hat{\sigma}_{P_{D230}}$

$$\hat{\sigma}_{P_{D230}} = \sqrt{\hat{\sigma}^2_{P_{D230}}} = \sqrt{165.38} = 12.86 \text{ days}$$

vi. $\hat{\sigma}_{P_{BF}}$

$$\hat{\sigma}_{P_{BF}} = \sqrt{\hat{\sigma}^2_{P_{BF}}} = \sqrt{.024} = .155 \text{ in}$$

b. Calculate $\text{cov}(P_{D230}, P_{BF})$. What is implied by the *sign* of the covariance? What will be the signs of $\hat{r}_{P_{D230}, P_{BF}}$ and $\hat{b}_{P_{BF} \cdot P_{D230}}$?

$$\text{cov}(P_{D230}, P_{BF}) = \frac{\sum_{i=1}^{n} (P_{D230_i} - \hat{\mu}_{P_{D230}})(P_{BF_i} - \hat{\mu}_{P_{BF}})}{n - 1}$$

$$= \frac{\sum_{i=1}^{10} (P_{D230_i} - 176.4)(P_{BF_i} - 1.28)}{10 - 1}$$

$$= -.569 \text{ days} \cdot \text{in}$$

The negative sign of the covariance indicates that an increase in the number of days required to reach 230 lb is associated with a decrease in backfat thickness. In other words, faster growing pigs are fatter—slower growing pigs are less fat.

The signs of $\hat{r}_{P_{D230}, P_{BF}}$ and $\hat{b}_{P_{BF} \cdot P_{D230}}$ will be negative because the covariance determines their sign.

c. Calculate $\hat{r}_{P_{D230} P_{BF}}$. Characterize this correlation.

$$\hat{r}_{P_{D230}, P_{BF}} = \frac{\hat{\text{cov}}(P_{D230}, P_{BF})}{\hat{\sigma}_{P_{D230}} \hat{\sigma}_{P_{BF}}}$$

$$= \frac{-.569}{12.86(.155)}$$

$$= -.29$$

There is a moderate, negative phenotypic correlation between days to 230 lb and backfat thickness in these data.

d. Calculate $\hat{b}_{P_{BF} \cdot P_{D230}}$. Interpret this regression in your own words.

$$\hat{b}_{P_{BF} \cdot P_{D230}} = \frac{\hat{\text{cov}}(P_{D230}, P_{BF})}{\hat{\sigma}^2_{P_{D230}}}$$

$$= \frac{-.569}{165.38}$$

$$= -.0034 \text{ in per day}$$

On average, for every one-day increase in the number of days required to reach 230 lb, backfat thickness decreases .0034 in.

e. Using the means and regression coefficient you calculated already, predict the backfat thickness of a pig that reached 230 lb in

i. 156 days.

$$\hat{P}_{BF_i} = \hat{\mu}_{P_{BF}} + \hat{b}_{P_{BF} \cdot P_{D230}} (P_{D230_i} - \hat{\mu}_{P_{D230}})$$

$$= 1.28 + (-.0034)(156 - 176.4)$$

$$= 1.35 \text{ in}$$

ii. 200 days.

$$\hat{P}_{BF_i} = \hat{\mu}_{P_{BF}} + \hat{b}_{P_{BF} \cdot P_{D230}} (P_{D230_i} - \hat{\mu}_{P_{D230}})$$

$$= 1.28 + (-.0034)(200 - 176.4)$$

$$= 1.20 \text{ in}$$

ANSWERS TO CHAPTER 9 PROBLEMS

9.1 Siberian racing muskrats are the Russian equivalent of American jumping frogs. The following genetic parameters for time to swim 50 meters (T) and lifetime winnings (W) have been estimated by animal scientists at the Smyatogorsk Polytechnic Institute:

$$\sigma^2_{BV_T} = 4 \text{ sec}^2 \quad \sigma^2_{GCV_T} = 1 \text{ sec}^2 \quad \sigma^2_{E_T} = 11 \text{ sec}^2$$

$$\sigma^2_{BV_W} = 100 \text{ rubles}^2 \quad \sigma^2_{GCV_W} = 0 \text{ rubles}^2 \quad \sigma^2_{E_W} = 2400 \text{ rubles}^2$$

a. Calculate:

i. phenotypic variance of 50-m time.

$$\begin{aligned} \sigma^2_{P_T} &= \sigma^2_{BV_T} + \sigma^2_{GCV_T} + \sigma^2_{E_T} \\ &= 4 + 1 + 11 \\ &= 16 \text{ sec}^2 \end{aligned}$$

ii. phenotypic variance of lifetime winnings.

$$\begin{aligned} \sigma^2_{P_W} &= \sigma^2_{BV_W} + \sigma^2_{GCV_W} + \sigma^2_{E_W} \\ &= 100 + 0 + 2{,}400 \\ &= 2{,}500 \text{ rubles}^2 \end{aligned}$$

iii. heritability of 50-m time.

$$\begin{aligned} h^2_T &= \frac{\sigma^2_{BV_T}}{\sigma^2_{P_T}} \\ &= \frac{4}{16} \\ &= .25 \end{aligned}$$

iv. heritability of lifetime winnings.

$$\begin{aligned} h^2_W &= \frac{\sigma^2_{BV_W}}{\sigma^2_{P_W}} \\ &= \frac{100}{2{,}500} \\ &= .04 \end{aligned}$$

b. Is a muskrat's single record for 50-m time a good indicator of its breeding value for the trait? Why or why not?

It is a reasonable indicator. A heritability of .25 is moderate and suggests a relationship between BV and P that is moderate in strength.

c. Would you expect 50-m time to respond well to selection? Why or why not?

50-m time should respond reasonably well to selection due to its moderate heritability.

d. Are a muskrat's lifetime winnings a good indicator of its breeding value for the trait? Why or why not?

No. Heritability is much too low.

e. Would you expect lifetime winnings to respond well to selection? Why or why not?

No—for the same reason.

f. A young male muskrat, Pyotr's Oski Doski, swam 50 meters 8 seconds faster than average. Predict his breeding value for 50-m time.

$$\begin{aligned}\hat{BV}_{T_i} &= h_T^2(P_{T_i} - \hat{\mu}_{P_T}) \\ &= .25(-8) \\ &= -2 \text{ sec}\end{aligned}$$

9.2 H. Cushman (Cushy) Pearson IV raises Thoroughbreds. His six two-year-olds posted the following records for lengths behind at the finish in their first two races:

	Lengths Behind	
Horse #	Race 1	Race 2
1	0.0	0.0
2	4.5	3.0
3	9.0	10.5
4	4.0	0.0
5	13.0	9.5
6	5.5	7.0

a. Calculate repeatability for lengths behind from this admittedly small sample.

$$\begin{aligned}\hat{r}_{LB} &= \hat{r}_{P_{LB_1},P_{LB_2}} \\ &= \frac{\hat{\text{cov}}(P_{LB_1},P_{LB_2})}{\hat{\sigma}_{P_{LB_1}}\hat{\sigma}_{P_{LB_2}}} \\ &= \frac{18}{4.48(4.66)} \\ &= .86\end{aligned}$$

b. Is first race performance a good indicator of second race performance? How do you know?

Yes. Repeatability of lengths behind is very high.

c. Should Cushy have sold any of his horses after their first race? Why or why not?

Yes. With such a high repeatability, first race performance predicts second race performance quite well. So Cushy should have sold the horses with poor first race performance.

d. If phenotypic variance for lengths behind is approximately 18 lengths^2, what is the variance of producing ability (horse value) for this trait? What is temporary environmental variance for the trait?

$$r_{LB} = \frac{\sigma^2_{PA_{LB}}}{\sigma^2_{P_{LB}}}$$

Therefore

$$\begin{aligned}\sigma^2_{PA_{LB}} &= r_{LB}\sigma^2_{P_{LB}} \\ &= .86(18) \\ &= 15.48 \text{ lengths}^2\end{aligned}$$

And

$$\sigma^2_{P_{LB}} = \sigma^2_{PA_{LB}} + \sigma^2_{E_{t_{LB}}}$$

so

$$\begin{aligned}\sigma^2_{E_{t_{LB}}} &= \sigma^2_{P_{LB}} - \sigma^2_{PA_{LB}} \\ &= 18 - 15.48 \\ &= 2.52 \text{ lengths}^2\end{aligned}$$

e. Are differences in performance in this trait due more to differences in horse value or differences in temporary environmental effects? How do you know?

Horse value. Differences in horse value account for 86% of differences in racing performance. Differences in temporary environmental effects account for only $\frac{2.52}{18} = 14\%$.

9.3 Vasily Yevshenko is widely recognized as a master muskrat breeder. A true perfectionist, Vasily has so standardized the management and training of his animals that the variance of environmental effects on 50-m time in his pack is just 6 sec^2. Assuming other genetic parameters are those listed for Problem 9.1, what is the heritability of 50-m time in Yevshenko's pack? What principle is illustrated here?

$$\begin{aligned}h^2_T &= \frac{\sigma^2_{BV_T}}{\sigma^2_{P_T}} \\ &= \frac{\sigma^2_{BV_T}}{\sigma^2_{BV_T} + \sigma^2_{GCV_T} + \sigma^2_{E_T}} \\ &= \frac{4}{4 + 1 + 6} \\ &= .36\end{aligned}$$

The more uniform the environment, the higher the heritability.

9.4 Birth weights in a breed of beef cattle average 76.8 lb for heifer calves and 82.2 lb for bull calves. Calculate:

a. the additive adjustment factor needed to adjust heifer birth weights to a bull basis.

The additive adjustment factor is just the difference between the averages:

$$82.2 - 76.8 = 5.4 \text{ lb}$$

b. the equivalent *multiplicative* adjustment factor.

The multiplicative adjustment factor is the number that, when multiplied by the heifer average, equals the bull average. In this case,

$$\frac{82.2}{76.8} = 1.07$$

9.5 Age-of-dam adjustment factors for weaning weight in a particular breed of beef cattle are:

Cow Age, yr	Adjustment, lb
2	+60
3	+40
4	+20
5+	0

Assume $\sigma_{BV_{WW}} = 30$ lb.

a. Calculate $\sigma_{P_{WW}}$ and h^2_{WW} using unadjusted weights from the following data set.

Calf #	Age of Dam, yr	Unadjusted Weight, lb
1	2	440
2	2	470
3	3	530
4	4	630
5	5	520
6	7	560
7	9	570
8	10	460

Before adjustment for age of dam,

$$\hat{\sigma}^2_{P_{WW}} = 4{,}107 \ \text{lb}^2$$

and

$$\begin{aligned}\hat{\sigma}_{P_{WW}} &= \sqrt{\hat{\sigma}^2_{P_{WW}}} \\ &= \sqrt{4{,}107} \\ &= 64.1 \text{ lb}\end{aligned}$$

Then

$$\begin{aligned}\hat{h}^2_{WW} &= \frac{\hat{\sigma}^2_{BV_{WW}}}{\hat{\sigma}^2_{P_{WW}}} \\ &= \frac{(30)^2}{4{,}107} \\ &= .22\end{aligned}$$

b. Do the same using weights that have been adjusted for age of dam.

After adjustment for age of dam,

$$\hat{\sigma}^2_{P_{WW}} = 3{,}229 \text{ lb}^2$$

and

$$\hat{\sigma}_{P_{WW}} = \sqrt{\hat{\sigma}^2_{P_{WW}}}$$
$$= \sqrt{3{,}229}$$
$$= 56.8 \text{ lb}$$

Then

$$\hat{h}^2_{WW} = \frac{\hat{\sigma}^2_{BV_{WW}}}{\hat{\sigma}^2_{P_{WW}}}$$
$$= \frac{(30)^2}{3{,}229}$$
$$= .28$$

c. What effects did adjustment for age of dam have?

Adjusting for age of dam decreased variance due to known environmental effects. That caused a reduction in phenotypic variance which, in turn, caused an increase in heritability.

9.6 Given the following average daily gain (ADG) data on a contemporary group of pigs, calculate an ADG ratio for each pig.

Pig #	ADG, lb/day
1	1.82
2	1.49
3	1.23
4	1.54
5	1.60
6	1.29
7	1.43
8	1.62

$$\text{ADG ratio}_i = \left(\frac{P_{ADG_i}}{\overline{P}_{cg_{ADG}}}\right) \times 100$$
$$\overline{P}_{cg_{ADG}} = 1.5025 \text{ lb/day}$$

For pig 1,

$$\text{ADG ratio} = \left(\frac{1.82}{1.5025}\right) \times 100$$
$$= 121$$

For the whole group of pigs,

Pig #	ADG, lb/day	Ratio
1	1.82	121
2	1.49	99
3	1.23	82
4	1.54	102
5	1.60	106
6	1.29	86
7	1.43	95
8	1.62	108

ANSWERS TO CHAPTER 10 PROBLEMS

10.1 Calculate the rate of genetic change in feed conversion in a swine population given the following:
Heritability of feed conversion (h^2) = .35
Phenotypic standard deviation (σ_P) = .2 lb
Accuracy of male selection ($r_{BV_m \hat{BV}_m}$) = .8
Accuracy of female selection ($r_{BV_f \hat{BV}_f}$) = .5
Intensity of male selection (i_m) = -2.4
Intensity of female selection (i_f) = -1.5
Generation interval for males (L_m) = 1.8 years
Generation interval for females (L_f) = 1.8 years

$$h^2 = \frac{\sigma^2_{BV}}{\sigma^2_P}$$

so

$$\begin{aligned}\sigma_{BV} &= \sqrt{h^2\sigma^2_P} \\ &= \sqrt{.35(.2)^2} \\ &= .118 \text{ lb}\end{aligned}$$

Then

$$\begin{aligned}\Delta_{BV}\Big/t &= \frac{(r_{BV_m,\hat{BV}_m} i_m + r_{BV_f,\hat{BV}_f} i_f)\sigma_{BV}}{L_m + L_f} \\ &= \frac{(.8(-2.4) + .5(-1.5))(.118)}{1.8 + 1.8} \\ &= -.088 \text{ lb}\Big/\text{year}\end{aligned}$$

10.2 In the beef cattle population described in Problem 9.5, adjusting weaning weights for age of dam caused the heritability of weaning weight to increase from .22 to .28. Assuming phenotypic selection for weaning weight alone:

a. What will be the percentage improvement in the rate of genetic change in weaning weight when selection is based on adjusted weights instead of unadjusted weights?

With phenotypic selection, accuracy equals the square root of heritability. All else being equal, the percentage improvement (%Δ) in the rate of genetic change in weaning weight when selection is based on adjusted weights instead of unadjusted weights is just the percentage increase in accuracy—a simple function of heritability. Mathematically,

$$\%\Delta = \frac{h_a - h_u}{h_u} \times 100$$

where a represents adjusted weights and u represents unadjusted weights. Then

$$\%\Delta = \frac{h_a - h_u}{h_u} \times 100$$
$$= \frac{\sqrt{.28} - \sqrt{.22}}{\sqrt{.22}} \times 100$$
$$= 12.8\%$$

b. What element of the key equation has been changed by adjusting weights?

Accuracy of selection.

10.3 A sheep breeder has determined that her ewes are not producing enough lambs and has decided to cull them heavily for twinning, a threshold trait. She keeps the top 35% of her ewes based on number of lambs born. What will be the effective proportion saved and selection intensity (culling intensity) for these ewes if:

a. under current conditions, 36% of the ewes produce twins, 56% produce singles, and 8% fail to breed.

The breeder is selecting a random sample of ewes from those that produce twins. She is therefore selecting from the top 36%. Effective proportion saved (p_e) is then .36 and, from Table 10.1, selection intensity (i) is 1.04 standard deviations.

b. management is improved so that 56% of the ewes produce twins, 41% produce singles, and only 3% fail to breed.

The breeder is now selecting a random sample of ewes from the top 56%. Effective proportion saved (p_e) is therefore .56, and, from Table 10.1, selection intensity (i) is .71 standard deviations.

c. Why did improving management reduce selection intensity?

Improved management caused a greater proportion of ewes to have twins. The breeder cannot now distinguish between the 36% that produce twins under current conditions and the 20% that twin only under improved conditions. She must therefore select from the top 56% rather than the top 36%—less intense selection.

10.4 The Dairy Board of Eastern Serbo-Slavonia is reexamining its dairy improvement program. The current program has the following attributes:

Path	Accuracy of Selection ($r_{BV,\hat{BV}}$)	Proportion Saved (p)	Selection Intensity (i)	Generation Interval (L)
(1) Sires to produce future sires	.85	3%	2.27	6 years
(2) Sires to produce future dams	.85	15%	1.55	7 years
(3) Dams to produce future sires	.5	1%	2.67	5 years
(4) Dams to produce future dams	.5	90%	.20	6 years

The phenotypic standard deviation of milk yield in this population is 2,160 lb, and heritability of milk yield is .25.

a. What is the rate of increase in milk yield under the current program?

$$h^2 = \frac{\sigma^2_{BV}}{\sigma^2_P}$$

so

$$\sigma_{BV} = \sqrt{h^2\sigma^2_P}$$
$$= \sqrt{.25(2{,}160)^2}$$
$$= 1{,}080 \text{ lb}$$

Then

$$\Delta_{BV}\Big/t = \frac{(r_{BV_1,\hat{BV}_1}i_1 + r_{BV_2,\hat{BV}_2}i_2 + r_{BV_3,\hat{BV}_3}i_3 + r_{BV_4,\hat{BV}_4}i_4)\sigma_{BV}}{L_1 + L_2 + L_3 + L_4}$$
$$= \frac{(.85(2.27) + .85(1.55) + .5(2.67) + .5(.20))(1{,}080)}{6 + 7 + 5 + 6}$$
$$= 211 \text{ lb/year}$$

b. Serbo-Slavonian dairy scientists are considering requiring dams of future sires to have an additional lactation record. They anticipate that this would increase accuracy of selection for these dams from .5 to .6. It would, of course, increase their generation interval by a year. What should be the rate of increase in milk yield under the revised program?

$$\Delta_{BV}\Big/t = \frac{(r_{BV_1,\hat{BV}_1}i_1 + r_{BV_2,\hat{BV}_2}i_2 + r_{BV_3,\hat{BV}_3}i_3 + r_{BV_4,\hat{BV}_4}i_4)\sigma_{BV}}{L_1 + L_2 + L_3 + L_4}$$
$$= \frac{(.85(2.27) + .85(1.55) + .6(2.67) + .5(.20))(1{,}080)}{6 + 7 + 6 + 6}$$
$$= 214 \text{ lb/year}$$

c. Which program should work better? Why?

Requiring dams of future sires to have an additional lactation record results in slightly faster genetic gain. Increased accuracy of selection outweighs the benefit of a shorter generation interval for this path. Realistically, however, the difference in results between the two programs is negligible.

10.5 A rancher runs a closed herd of breeding cattle. He normally keeps and breeds the top 3% of his bull calves based on individual performance for yearling weight (YW). His sires average three years of age when their offspring are born. He is studying two female replacement strategies:

a. saving the top 20% of his heifers based on YW (L_f= 6.2 years)

b. saving the top 60% based on YW (L_f= 3.2 years)

i. If $h^2_{YW} = .5$, and $\sigma_{P_{YW}} = 60$ lb, calculate the expected rate of genetic change in yearling weight for each strategy.

$$\sigma_{BV} = \sqrt{h^2\sigma^2_P}$$
$$= \sqrt{.5(60)^2}$$
$$= 42.4 \text{ lb}$$

ii. Saving the top 20% of heifers:

$$\Delta_{BV}\Big/t = \frac{(r_{BV_m,\hat{BV}_m} i_m + r_{BV_f,\hat{BV}_f} i_f)\sigma_{BV}}{L_m + L_f}$$
$$= \frac{(\sqrt{.5}(2.27) + \sqrt{.5}(1.40))(42.4)}{3 + 6.2}$$
$$= 12.0 \text{ lb/year}$$

iii. Saving the top 60% of heifers:

$$\Delta_{BV}\Big/t = \frac{(r_{BV_m,\hat{BV}_m} i_m + r_{BV_f,\hat{BV}_f} i_f)\sigma_{BV}}{L_m + L_f}$$
$$= \frac{(\sqrt{.5}(2.27) + \sqrt{.5}(.64))(42.4)}{3 + 3.2}$$
$$= 14.1 \text{ lb/year}$$

a. What elements of the key equation is the rancher experimenting with?

Female selection intensity and generation interval.

b. What element appears to be more important?

Generation interval appears to be more important. In this example, it is better to sacrifice female selection intensity in order to decrease generation interval.

10.6 Of the 9.5 lb per year improvement in yearling weight expected using all of Sarah's breeding recommendations, over 80% can be attributed to sire selection. Prove it. (See text for details.)

The 9.5 lb per year improvement in yearling weight resulted from the following calculation:

$$\Delta_{BV}\Big/t = \frac{(r_{BV_m,\hat{BV}_m} i_m + r_{BV_f,\hat{BV}_f} i_f)\sigma_{BV}}{L_m + L_f}$$
$$= \frac{[.8(2.27) + .8(.35)](35)}{4 + 3.7}$$
$$\approx 9.5 \text{ lb per year}$$

The male contribution to the numerator is

$$r_{BV_m,\hat{BV}_m} i_m \sigma_{BV}$$
$$= .8(2.27)(35)$$

The male contribution to the denominator (generation interval) is difficult to determine exactly, but at worst—i.e., if we use male generation interval—the denominator is

$$L_m + L_m$$
$$= 4 + 4$$
$$= 8$$

Then, the overall contribution of sire selection to genetic change is at least

$$\frac{.8(2.27)(35)}{8}$$
$$= 7.9 \text{ lb/year}$$

Expressed as a proportion of total genetic change per year, that is

$$\frac{7.9}{9.5}$$
$$=.83 \text{ or } 83\%$$

ANSWERS TO CHAPTER 11 PROBLEMS

11.1 **a.** Gay Blade, a promising Thoroughbred stallion, was recently retired to stud. His first six foals have just completed their maiden races, averaging two seconds faster than their contemporary group means.

i. Use a single-source selection index to predict Gay Blade's breeding value for racing time. (Assume $h^2 = .35$.)

$$\hat{BV} = \frac{2ph^2}{4 + (p - 1)h^2}\overline{P}$$
$$= \frac{2(6)(.35)}{4 + (6 - 1)(.35)}(-2)$$
$$= -1.46 \text{ sec}$$

ii. Calculate accuracy of prediction.

$$r_{BV,\hat{BV}} = \sqrt{\frac{ph^2}{4 + (p - 1)h^2}}$$
$$= \sqrt{\frac{6(.35)}{4 + (6 - 1)(.35)}}$$
$$= .60$$

b. Megabuck, the old champion, has sired 120 foals. They have run, on average, 4.2 races apiece, averaging one second faster than contemporary group means.

i. Predict Megabuck's breeding value for racing time. (Assume $h^2 = .35$ and $r = .57$.)

$$\hat{BV} = \frac{\frac{1}{2}ph^2}{\frac{1 + (n - 1)r}{n} + (p - 1)\frac{h^2}{4}}\overline{P}$$
$$= \frac{\frac{1}{2}(120)(.35)}{\frac{1 + (4.2 - 1)(.57)}{4.2} + (120 - 1)\frac{.35}{4}}(-1)$$
$$= -1.89 \text{ sec}$$

ii. Calculate accuracy of prediction.

$$r_{BV,\hat{BV}} = \sqrt{\frac{\frac{1}{4}ph^2}{\frac{1 + (n - 1)r}{n} + (p - 1)\frac{h^2}{4}}}$$

$$= \sqrt{\frac{\frac{1}{4}(120)(.35)}{\frac{1 + (4.2 - 1)(.57)}{4.2} + (120 - 1)\frac{.35}{4}}}$$

$$= .97$$

c. Why is Megabuck's EBV better than Gay Blade's even though Gay Blade's offspring have run faster than Megabuck's?

Megabuck's prediction has been regressed less because he has more information.

d. All else being equal, which sire would you use? Why?

The EBVs of the two horses can be directly compared because they have both been adjusted for amount of information. And because Megabuck has the best prediction, I'll use him. Furthermore, I don't like taking risks. So I'll go with the known quantity—Megabuck.

11.2 A boar has sired 20 litters averaging 7.8 weaned pigs each. Postweaning average daily gains of his progeny average .06 lb per day above contemporary group means. Heritability of postweaning gain is .28 and c^2_{FS}, a measure of covariation among postweaning gains of littermates that is caused by common environment (in this case, covariation due to a common dam), is estimated to be about .07.

a. Use a single-source selection index to estimate the boar's breeding value for postweaning average daily gain.

$$\hat{BV} = \frac{2lkh^2}{4 + (k - 1)(2h^2 + 4c^2_{FS}) + (l - 1)kh^2}\overline{P}$$

$$= \frac{2(20)(7.8)(.28)}{4 + (7.8 - 1)(2(.28) + 4(.07)) + (20 - 1)(7.8)(.28)}(.06)$$

$$= +.102 \text{ lb/day}$$

b. Estimate the boar's breeding value assuming there is no environmental covariation among postweaning gains of littermates.

$$\hat{BV} = \frac{2lkh^2}{4 + (k - 1)(2h^2 + 4c^2_{FS}) + (l - 1)kh^2}\overline{P}$$

$$= \frac{2(20)(7.8)(.28)}{4 + (7.8 - 1)(2(.28) + 4(0)) + (20 - 1)(7.8)(.28)}(.06)$$

$$= +.106 \text{ lb/day}$$

c. Why did his EBV increase when c^2_{FS} was assumed to be zero?

When full-sib records are not environmentally correlated, each record provides more independent information. The more information, the less regression for amount of information, and the further the prediction from zero.

11.3 **a.** Calculate accuracy of breeding value prediction given the following information:

i. a single performance record on the individual; $h^2 = .25$

$$r_{BV,\hat{BV}} = h$$
$$= \sqrt{h^2}$$
$$= \sqrt{.25}$$
$$= .5$$

ii. a single performance record on the individual; $h^2 = .5$

$$\begin{aligned} r_{BV,\hat{BV}} &= h \\ &= \sqrt{h^2} \\ &= \sqrt{.5} \\ &= .71 \end{aligned}$$

iii. five repeated records on the individual; $h^2 = .25$; $r = .3$

$$\begin{aligned} r_{BV,\hat{BV}} &= \sqrt{\frac{nh^2}{1 + (n - 1)r}} \\ &= \sqrt{\frac{5(.25)}{1 + (5 - 1)(.3)}} \\ &= .75 \end{aligned}$$

iv. five repeated records on the individual; $h^2 = .25$; $r = .6$

$$\begin{aligned} r_{BV,\hat{BV}} &= \sqrt{\frac{nh^2}{1 + (n - 1)r}} \\ &= \sqrt{\frac{5(.25)}{1 + (5 - 1)(.6)}} \\ &= .61 \end{aligned}$$

v. single records on five half sibs; $h^2 = .25$

$$\begin{aligned} r_{BV,\hat{BV}} &= \sqrt{\frac{\frac{1}{4}mh^2}{4 + (m - 1)h^2}} \\ &= \sqrt{\frac{\frac{1}{4}(5)(.25)}{4 + (5 - 1)(.25)}} \\ &= .25 \end{aligned}$$

vi. single records on 500 half sibs; $h^2 = .25$

$$\begin{aligned} r_{BV,\hat{BV}} &= \sqrt{\frac{\frac{1}{4}mh^2}{4 + (m - 1)h^2}} \\ &= \sqrt{\frac{\frac{1}{4}(500)(.25)}{4 + (500 - 1)(.25)}} \\ &= .49 \end{aligned}$$

vii. single records on five progeny; $h^2 = .25$

$$\begin{aligned} r_{BV,\hat{BV}} &= \sqrt{\frac{ph^2}{4 + (p - 1)h^2}} \\ &= \sqrt{\frac{5(.25)}{4 + (5 - 1)(.25)}} \\ &= .5 \end{aligned}$$

viii. single records on 500 progeny; $h^2 = .25$

$$r_{BV,\hat{BV}} = \sqrt{\frac{ph^2}{4 + (p - 1)h^2}}$$
$$= \sqrt{\frac{500(.25)}{4 + (500 - 1)(.25)}}$$
$$= .99$$

b. Why did accuracy change the way it did in the above scenarios?

i. a single performance record on the individual; $h^2 = .25$

($r_{BV,\hat{BV}} = .5$) This is the starting point—the basis for comparison. No change in accuracy yet.

ii. a single performance record on the individual; $h^2 = .5$

($r_{BV,\hat{BV}} = .71$) Accuracy increased due to increased heritability.

iii. five repeated records on the individual; $h^2 = .25$; $r = .3$

($r_{BV,\hat{BV}} = .75$) Accuracy increased due to the increased number of records.

iv. five repeated records on the individual; $h^2 = .25$; $r = .6$

($r_{BV,\hat{BV}} = .61$) Accuracy less than in (iii) because higher repeatabil ity means less independent information per record.

v. single records on five half sibs; $h^2 = .25$

($r_{BV,\hat{BV}} = .25$) Lower accuracy than in (iii) or (iv) because records on half sibs convey less information than records on the individual itself.

vi. single records on 500 half sibs; $h^2 = .25$

($r_{BV,\hat{BV}} = .49$) Higher accuracy than in (v), but still limited because half-sib information does not account for breeding values of mates or Mendelian sampling.

vii. single records on five progeny; $h^2 = .25$

($r_{BV,\hat{BV}} = .5$) Higher accuracy than in (v) because progeny are more closely related to the individual than are half sibs.

viii. single records on 500 progeny; $h^2 = .25$

($r_{BV,\hat{BV}} = .99$) Very high accuracy with large progeny numbers. The ultimate test of an individual's breeding value.

11.4 **a.** Use your answers to Problem 11.1 to calculate 68% confidence ranges for Gay Blade's and Megabuck's EBVs for racing time. (Assume $\sigma_P = 1.3$ sec.)

$$\text{Confidence range} = \hat{BV} \pm \sqrt{(1 - r^2_{BV,\hat{BV}})\sigma^2_{BV}}$$

and

$$h^2 = \frac{\sigma^2_{BV}}{\sigma^2_P}$$

so

$$\sigma^2_{BV} = h^2\sigma^2_p$$
$$= .35(1.3)^2$$
$$= .59 \text{ sec}^2$$

Then, for Gay Blade:

$$\text{Confidence range} = -1.46 \pm \sqrt{(1 - (.60)^2)(.59)}$$
$$= -1.46 \pm .61$$
$$= -2.07 \text{ to } -.85 \text{ sec}$$

and for Megabuck:

$$\text{Confidence range} = -1.89 \pm \sqrt{(1 - (.97)^2)(.59)}$$
$$= -1.89 \pm .19$$
$$= -2.08 \text{ to } -1.70 \text{ sec}$$

b. Which sire represents the greater selection risk? Why?

Gay Blade represents the greater selection risk. The upper limit of the confidence range for his EBV (−.85 sec) is considerably higher (slower) than the upper limit for Megabuck (−1.70 sec).

11.5 A young beef bull is being genetically evaluated for weaning weight. The information used in the analysis includes the bull's own weaning weight, lots of paternal half-sib data, a couple of maternal half-sib records, and limited progeny data. All this information is combined in the following selection index:

$$I = b_1x_1 + b_2x_2 + b_3x_3 + b_4x_4$$

or

$$\text{EBV} = b_1P_{\text{IND}} + b_2\overline{P}_{\text{PHS}} + b_3\overline{P}_{\text{MHS}} + b_4\overline{P}_{\text{PROG}}$$

Weighting factors calculated from simultaneous equations are:

$$b_1 = .169$$
$$b_2 = .500$$
$$b_3 = .075$$
$$b_4 = .624$$

Given the following performance data (expressed as deviations from contemporary group means):

$$P_{\text{IND}} = +128 \text{ lb}$$
$$\overline{P}_{\text{PHS}} = +22 \text{ lb}$$
$$\overline{P}_{\text{MHS}} = +35 \text{ lb}$$
$$\overline{P}_{\text{PROG}} = +26 \text{ lb}$$

a. Calculate the bull's EBV for weaning weight.

$$\widehat{BV} = b_1P_{\text{IND}} + b_2\overline{P}_{\text{PHS}} + b_3\overline{P}_{\text{MHS}} + b_4\overline{P}_{\text{PROG}}$$
$$= (.169)(128) + (.500)(22) + (.075)(35) + (.624)(26)$$
$$= +51.5 \text{ lb}$$

b. Calculate accuracy of prediction.

$$r_{BV,\hat{BV}} = \sqrt{b_1 + \frac{1}{4}b_2 + \frac{1}{4}b_3 + \frac{1}{2}b_4}$$
$$= \sqrt{.169 + \frac{1}{4}(.500) + \frac{1}{4}(.075) + \frac{1}{2}(.624)}$$
$$= .79$$

11.6 A ewe's direct and maternal EPDs for weaning weight (a trait of the lamb) are:

$$\text{EPD}_d = -1 \text{ lb}$$
$$\text{EPD}_m = +2.5 \text{ lb}$$

a. Given just these EPDs, what do we expect her future lambs to weigh (expressed as a deviation from the population mean)?

We want to know her total maternal EBV ($\hat{BV}_{tm}$)

$$\hat{BV}_{tm} = 2(\hat{PD}_{tm})$$
$$= 2\left(\hat{PD}_m + \frac{1}{2}\hat{PD}_d\right)$$
$$= 2\left(2.5 + \frac{1}{2}(-1)\right)$$
$$= +4 \text{ lb}$$

b. The ewe's permanent environmental effect (E_p) is predicted to be +3 lb. What is her MPPA?

$$\hat{PA} = \hat{BV}_{tm} + \hat{E}_p$$
$$= 4 + 3$$
$$= +7 \text{ lb}$$

c. Estimates of contemporary group effects ($\mu + \hat{E}_{cg}$) for her flock average 39 lb. What do we expect her future lambs to weigh?

$$\hat{P} = \hat{PA} + (\mu + \hat{E}_{cg})$$
$$= 7 + 39$$
$$= 46 \text{ lb}$$

Answers to Chapter 13 Problems

13.1 Recall from Problem 9.1 the following genetic parameters for time to swim 50 meters (T) and lifetime winnings (W) in Siberian racing muskrats:

$$\sigma^2_{BV_T} = 4 \text{ sec}^2 \quad \sigma^2_{GCV_T} = 1 \text{ sec}^2 \quad \sigma^2_{E_T} = 11 \text{ sec}^2$$
$$\sigma^2_{BV_W} = 100 \text{ rubles}^2 \quad \sigma^2_{GCV_W} = 0 \text{ rubles}^2 \quad \sigma^2_{E_W} = 2{,}400 \text{ rubles}^2$$

Animal scientists at the Smyatogorsk Polytechnic Institute have also estimated the following covariances between the two traits:

$$\text{cov}(BV_T, BV_W) = -2.0 \text{ sec·rubles}$$
$$\text{cov}(E_T, E_W) = -118.8 \text{ sec·rubles}$$
$$\text{cov}(P_T, P_W) = -120.8 \text{ sec·rubles}$$

Calculate:

a. the genetic correlation between 50-meter time and lifetime winnings.

$$r_{BV_T,BV_W} = \frac{\text{cov}(BV_T,BV_W)}{\sigma_{BV_T}\sigma_{BV_W}} = \frac{-2}{\sqrt{4}\sqrt{100}} = -.10$$

b. the environmental* correlation between the traits. (*For this problem, do *not* include gene combination effects in the environmental category.)

$$r_{E_T,E_W} = \frac{\text{cov}(E_T,E_W)}{\sigma_{E_T}\sigma_{E_W}} = \frac{-118.8}{\sqrt{11}\sqrt{2{,}400}} = -.73$$

c. the phenotypic correlation between the traits.

$$r_{P_T,P_W} = \frac{\text{cov}(P_T,P_W)}{\sigma_{P_T}\sigma_{P_W}} = \frac{-120.8}{\sqrt{16}\sqrt{2{,}500}} = -.60$$

13.2 Ace Maverick wants to shorten gestation length (GL), and reduce birth weight (BW) in his herd of registered beef cattle. EPDs for these traits are available, and Ace will use them for both male and female selection. Because many fewer gestation lengths are reported than birth weights, average accuracy of selected animals for gestation length is only .40 compared to .80 for birth weight. Given the following:

$$\sigma_{BV_{GL}} = 2.8 \text{ days} \quad \sigma_{BV_{BW}} = 6.3 \text{ lb} \quad r_{BV_{GL},BV_{BW}} = .25$$

$$i_{GL} = i_{BW} = -1.0 \quad L = 5 \text{ years}'$$

a. Calculate:

i. $\Delta BV_{GL}/t$

$$\Delta BV_{GL}/t = \frac{r_{BV_{GL},\hat{BV}_{GL}}\, i_{GL}\, \sigma_{BV_{GL}}}{L} = \frac{.4(-1)(2.8)}{5} = -.22 \text{ days/year}$$

ii. $\Delta_{BV_{BW}} \Big/ t$

$$\Delta_{BV_{BW}} \Big/ t = \frac{r_{BV_{BW},\hat{BV}_{BW}} i_{BW} \sigma_{BV_{BW}}}{L}$$
$$= \frac{.8(-1)(6.3)}{5}$$
$$= -1.0 \text{ lb/year}$$

iii. $\Delta_{BV_{GL|BW}} \Big/ t$

$$\Delta_{BV_{GL|BW}} \Big/ t = \frac{r_{BV_{GL},BV_{BW}} r_{BV_{BW},\hat{BV}_{BW}} i_{BW} \sigma_{BV_{GL}}}{L}$$
$$= \frac{.25(.8)(-1)(2.8)}{5}$$
$$= -.11 \text{ days/year}$$

iv. $\Delta_{BV_{BW|GL}} \Big/ t$

$$\Delta_{BV_{BW|GL}} \Big/ t = \frac{r_{BV_{GL},BV_{BW}} r_{BV_{GL},\hat{BV}_{GL}} i_{GL} \sigma_{BV_{BW}}}{L}$$
$$= \frac{.25(.4)(-1)(6.3)}{5}$$
$$= -.13 \text{ lb/year}$$

v. $\dfrac{\Delta_{BV_{GL|BW}}}{\Delta_{BV_{GL}}}$

$$\frac{\Delta_{BV_{GL|BW}}}{\Delta_{BV_{GL}}} = \frac{-.11}{-.22}$$
$$= .5$$

vi. $\dfrac{\Delta_{BV_{BW|GL}}}{\Delta_{BV_{BW}}}$

$$\frac{\Delta_{BV_{BW|GL}}}{\Delta_{BV_{BW}}} = \frac{-.13}{-1.0}$$
$$= .13$$

b. Interpret your results for (v) and (vi).

Selecting for birth weight is half as effective in changing gestation length as direct selection for gestation length.

Selecting for gestation length is only 13% as effective in changing birth weight as direct selection for birth weight.

c. If you were Ace, which trait would you select for? Why?

Given the choice of one trait or the other, I'd select for lower birth weight. That way I would get maximum response in birth weight and respectable change in gestation length too. The alternative, selecting for gestation length, is less appealing. Lower accuracy of selection for gestation length limits the progress that can be made in either trait.

13.3 Slim Maverick, Ace's brother, runs commercial cattle just south of Ace's place. Slim wants to improve the probability of conception (PC) in his yearling heifers. But conception is an all-or-none threshold trait for which Slim can achieve little selection intensity. The bulls that Slim buys have EPDs for scrotal circumference (SC), a trait known to be related to heifer conception rate. Slim is contemplating two strategies: (1) practicing phenotypic selection for probability of conception by retaining only those heifers that conceive in their first season, and (2) selecting bulls for scrotal EPD. The following genetic parameters have been estimated for these traits:

$$h^2_{PC} = .10 \quad h^2_{SC} = .5 \quad \sigma_{P_{PC}} = .46 \quad \sigma_{P_{SC}} = 2.0 \text{ cm} \quad r_{BV_{SC},BV_{PC}} = .25$$

Assume:

Slim breeds all his heifers and keeps 40% for replacements.

Typical heifer conception rate is 70%.

The bulls that Slim buys represent the equivalent of the top 5% of his herd for scrotal circumference.

$L_m = L_f = 5$ years

a. Calculate annual selection response in probability of heifer conception using strategy 1.

This is simple phenotypic selection. Using the two-path method for males and females,

$$\Delta_{BV_{PC}}/t = \frac{h^2{}_{PC}(i_{m_{PC}} + i_{f_{PC}})\sigma_{P_{PC}}}{L_m + L_f}$$

Because there is no male selection, $i_{m_{PC}} = 0$, and because Slim is selecting from the top 70% of his heifers, $p_{e_{f_{PC}}} = .7$. From Table 10.1, $i_{f_{PC}} = .5$ standard deviations. Then

$$\begin{aligned}\Delta_{BV_{PC}}/t &= \frac{h^2{}_{PC}(i_{m_{PC}} + i_{f_{PC}})\sigma_{P_{PC}}}{L_m + L_f} \\ &= \frac{.10(0 + .5)(.46)}{5 + 5} \\ &= .0023 \text{ or } .23\%/\text{year}\end{aligned}$$

b. Calculate annual selection response in probability of heifer conception using strategy 2.

For this strategy we can use the following equation for correlated response to selection:

$$\Delta_{BV_{PC|SC}}\Big/ t = r_{BV_{SC},BV_{PC}}\left(\frac{\sigma_{BV_{PC}}}{\sigma_{BV_{SC}}}\right) \Delta_{BV_{SC}}\Big/ t$$

Now,

$$\Delta_{BV_{SC}}\Big/ t = \frac{(r_{BV_{m_{sc}},\hat{BV}_{m_{sc}}} i_{m_{sc}} + {}_{,BV_{f_{sc}},\hat{BV}_{f_{sc}}} i_{f_{sc}})\sigma_{BV_{sc}}}{L_m + L_f}$$

For females, accuracy and intensity of selection are both zero, so

$$r_{BV_{f_{sc}},\hat{BV}_{f_{sc}}} = 0$$

and

$$i_{f_{SC}} = 0$$

Also

$$\begin{aligned}\sigma_{BV_{SC}} &= \sqrt{h^2_{SC}\sigma^2_{P_{sc}}}\\ &= \sqrt{.5(2)^2}\\ &= 1.41 \text{ cm}\end{aligned}$$

$p_{m_{SC}} = .05$, so from Table 10.1, $i_{m_{SC}} = 2.06$ standard deviations. Then

$$\begin{aligned}\Delta_{BV_{SC}}\Big/ t &= \frac{(r_{BV_{m_{sc}},\hat{BV}_{m_{sc}}} i_{m_{sc}} + r_{BV_{f_{sc}},\hat{BV}_{f_{sc}}} i_{f_{sc}})\sigma_{BV_{sc}}}{L_m + L_f}\\ &= \frac{(.9(2.06) + 0(0))(1.41)}{5 + 5}\\ &= .26 \text{ cm/year}\end{aligned}$$

We need one more parameter:

$$\begin{aligned}\sigma_{BV_{PC}} &= \sqrt{h^2_{PC}\sigma^2_{P_{PC}}}\\ &= \sqrt{.10(.46)^2}\\ &= .145\end{aligned}$$

Then

$$\begin{aligned}\Delta_{BV_{PC|SC}}\Big/ t &= r_{BV_{SC},BV_{PC}}\left(\frac{\sigma_{BV_{PC}}}{\sigma_{BV_{SC}}}\right) \Delta_{BV_{SC}}\Big/ t\\ &= .25\left(\frac{.145}{1.41}\right).26\\ &= .0067 \text{ or } .67\%/\text{year}\end{aligned}$$

c. Which strategy works better? Why?

Selecting sires for scrotal circumference works better for two reasons. Accuracy is higher for scrotal circumference in sires than for probability of conception in heifers because of the higher heritability of scrotal circumference and the use of EPDs. Selection intensity is greater for scrotal circumference in sires than for probability of conception in heifers because relatively few sires are needed, sires come from outside the herd (i.e., from a larger population), and scrotal circumference is a continuous trait (unlike conception, which is a threshold trait).

13.4 Pyotr—remember him, Oski Doski's breeder—has selected for decreased 50-meter time in his muskrats for many generations. By selection alone he has improved the average time of his muskrats by 10 seconds. Use information from Problem 13.1 to answer the following:

a. How much more money should Pyotr's muskrats be winning now than before as a result of genetic improvement?

To answer this question, we can use a prediction equation regressing the change in breeding value for winnings on the change in breeding value for 50-meter time.

$$\Delta_{BV_{W|T}} = b_{BV_W \cdot BV_T} \Delta_{BV_T}$$

or

$$\begin{aligned}\Delta_{BV_{W|T}} &= r_{BV_W,BV_T}\left(\frac{\sigma_{BV_W}}{\sigma_{BV_T}}\right)\Delta_{BV_T} \\ &= -.10\left(\frac{\sqrt{100}}{\sqrt{4}}\right)(-10) \\ &= +5 \text{ rubles}\end{aligned}$$

b. Pyotr's training regimen has resulted in a 10-second improvement in 50-meter time also. How much more money should Pyotr's muskrats be winning now than before as a result of better training?

To answer this question, we can use a prediction equation regressing the change in environmental effect for winnings on the change in environmental effect for 50-meter time.

$$\Delta_{E_{W|T}} = b_{E_W \cdot E_T} \Delta_{E_T}$$

or

$$\begin{aligned}\Delta_{E_{W|T}} &= r_{E_W,E_T}\left(\frac{\sigma_{E_w}}{\sigma_{E_T}}\right)\Delta_{E_T} \\ &= -.73\left(\frac{\sqrt{2{,}400}}{\sqrt{11}}\right)(-10) \\ &= +108 \text{ rubles}\end{aligned}$$

c. Would Pyotr be better advised to concentrate on his breeding program or his training program? Why?

If past success is any indicator, he really should emphasize his training program. Because the environmental correlation between 50-meter time and lifetime winnings is so much higher than the genetic correlation, training will do more for Pyotr's bottom line than selection will.

ANSWERS TO CHAPTER 14 PROBLEMS

14.1 A swine breeder is selecting for increased number of pigs weaned (NW) and reduced backfat (BF) in her pigs. She plans to choose three out of the following eight boars based on EBVs for these traits.

Boar #	EBV for Number of Pigs Weaned	EBV for Backfat, in
1	+1.1	−.11
2	+.8	−.25
3	+2.4	−.05
4	+.3	−.36
5	−.5	−.10
6	+3.0	+.20
7	+1.0	+.05
8	−.6	−.40

a. Which boars would she initially select using tandem selection:

i. when NW is the first trait under selection?

Boars 6, 3, and 1

ii. when BF is the first trait under selection?

Boars 8, 4, and 2

b. Which boars would she select using independent culling levels if the levels were set at 0 pigs for NW and −.1 in for BF?

Boars 1, 2, and 4 are the only boars that qualify.

c. Which boars would she select using an economic selection index if an independent one-pig increase in NW is worth $100 and an independent 1-in decrease in BF is worth $1,000?

Here we can use the EBVs as the *x*s and the economic weights as the *b*s. Then

$$I = 100\,\hat{BV}_{NW} - 1{,}000\,\hat{BV}_{BF}$$

or

$$I = \hat{BV}_{NW} - 10\,\hat{BV}_{BF}$$

Applying this index to all eight boars, we get:

Boar #	Index Value
1	2.2
2	3.3
3	2.9
4	3.9
5	.5
6	1.0
7	.5
8	3.4

Boars 4, 8, and 2 are the best boars based on the index.

d. Why was boar 8 selected with the index but not with independent culling levels?

Boar 8's EBV for NW was too low for the independent culling levels but, with the index, his excellent EBV for BF offset his weak EBV for NW.

14.2 The following genetic parameters were used in the yearling weight (YW)/birth weight(BW) example in the boxed section entitled *Calculating the Classic Form of Economic Selection Index:*

$$\sigma^2_{P_{YW}} = 3{,}600 \text{ lb}^2 \qquad \sigma^2_{P_{BW}} = 100 \text{ lb}^2 \qquad cov(P_{YW}, P_{BW}) = 210 \text{ lb}^2$$
$$\sigma^2_{BV_{YW}} = 1{,}440 \text{ lb}^2 \qquad \sigma^2_{BV_{BW}} = 40 \text{ lb}^2 \qquad cov(BV_{YW}, BV_{BW}) = 168 \text{ lb}^2$$

Conditions have changed so that a 1-lb increase in yearling weight is now worth \$1.22 and a 1-lb increase in birth weight is worth \$−4.35. Recalculate the economic selection index accordingly.

$$H = 1.22BV_{YW} - 4.35\, BV_{BW}$$

Then

$$\sigma^2_{P_{YW}} b_1 + cov(P_{YW}, P_{BW}) b_2 = \sigma^2_{BV_{YW}} v_1 + cov(BV_{YW}, BV_{BW}) v_2$$
$$cov(P_{BW}, P_{YW}) b_1 + \sigma^2_{P_{BW}} b_2 = cov(BV_{BW}, BV_{YW}) v_1 + \sigma^2_{BV_{BW}} v_2$$

or

$$3{,}600b_1 + 210b_2 = 1{,}440(1.22) + 168(-4.35)$$
$$210b_1 + 100b_2 = 168(1.22) + 40(-4.35)$$

or

$$3{,}600b_1 + 210b_2 = 1{,}026$$
$$210b_1 + 100b_2 = 30.96$$

Solve the first equation for b_1 in terms of b_2.

$$b_1 = \frac{1{,}026 - 210b_2}{3{,}600}$$
$$= .285 - .0583b_2$$

Now substitute this value for b_1 into the second equation and solve for b_2.

$$210(.285 - .0583b_2) + 100b_2 = 30.96$$
$$59.85 - 12.25b_2 + 100b_2 = 30.96$$
$$b_2 = \frac{30.96 - 59.85}{100 - 12.25}$$
$$= -.33$$

Now substitute the solution for b_2 into the first equation and solve for b_1.

$$3{,}600b_1 + 210(-.33) = 1{,}026$$
$$b_1 = \frac{1{,}026 + 69.14}{3{,}600}$$
$$= .30$$

The selection index is now

$$I = .30YW - .33BW$$

Dividing both economic weights by .30 (actually .3042), we have

$$I = YW - 1.08BW$$

14.3 A horse breeder is selecting for a number of equally important, uncorrelated traits. He needs to replace 10% of his mares each year (i.e., keep 20% of his fillies). Calculate effective proportion of females saved (p_{e_f}) and female selection intensity (i_f) for each trait if the number of traits is:

a. 2

$$p_{e_f} = \sqrt{.20}$$
$$= .45$$

From Table 10.1,

$$i_f = .88$$

b. 3

$$p_{e_f} = \sqrt[3]{.20}$$
$$= .58$$
$$i_f = .67$$

c. 4

$$p_{e_f} = \sqrt[4]{.20}$$
$$= .67$$
$$i_f = .54$$

d. 10

$$p_{e_f} = \sqrt[10]{.20}$$
$$= .85$$
$$i_f = .27$$

Answers to Chapter 15 Problems

15.1 In the Labrador example in this chapter, we decided to mate the chocolate bitch, Rachel, to the chocolate dog, Murray, in order to minimize the chances of producing yellow puppies. We have just received new information about Murray's dam, Georgie Girl; at one time she produced a litter containing yellow pups. Should we change mating plans? Support your answer mathematically.

Yes. Because Georgie Girl is a proven yellow carrier, the information in her pedigree (see following page) is of no importance. Murray's chances of being a carrier are now 50%, not 12.5% as we thought. A mating of Rachel to Murray has a .25(.5) = .125 or 12.5% probability of being the kind of mating that could produce yellow pups. Better to mate her to Phantom. The corresponding probability for that mating is only 6.25%.

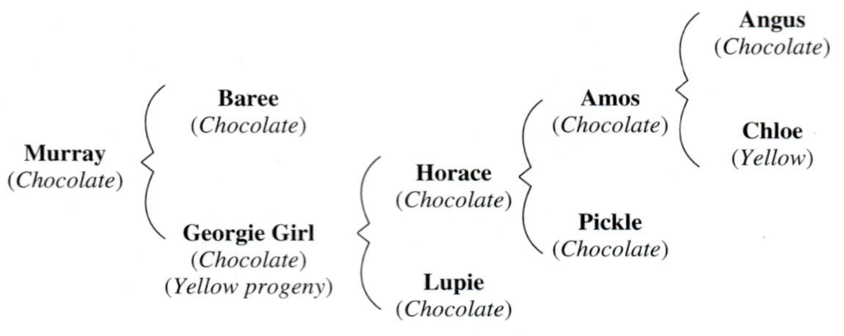

15.2 Cushy Pearson has a thing for bay colored horses. He purchased a single service of a stylish (and expensive) bay stallion in hopes of producing a bay foal. He has four mares available: one brown, one mouse colored, one black, and one chestnut. Assume the following:

The inheritance of coat color in horses is no more complicated than it appears in Table 15.1.

Cushy has no information on the genotypes of his horses other than their phenotypes.

No linkage exists among the four loci shown in Table 15.1.

Frequencies of coat color alleles in the Thoroughbred population are estimated to be:

Allele	Frequency
A	.6
a	.3
a^t	.1
C	.7
c^{cr}	.3
D	.2
d	.8
E	.3
e	.7

a. To which mare should Cushy mate his bay stallion in order to maximize the likelihood of producing a bay foal?

The brown mare.

b. To which mare should Cushy be sure *not* to mate his stallion?

The mouse colored mare.

c. Prove your answers mathematically.[1]

(This problem involves some logic and a lot of arithmetic. Students may want to work in teams.)

Because these loci are independent (not linked), we can save time and work by determining, locus by locus, the probabilities that a particular parent combination will produce the desired one-locus genotype for a bay, then multiply probabilities across loci to determine the overall probability that a given mating will produce a bay.

A locus:

(You can solve this problem graphically using Punnett squares. I prefer to think it out in words.) The desired genotype is $A\_$. Another way to

[1]You may want to review the subsection of Chapter 6 entitled *Probabilities of Outcomes of Matings.*

think of this genotype is "anything but two non-*A* genes." The probability of "anything but two non-*A* genes" is 1 minus the probability of two non-*A* genes. And that is simply the product of each parent's probability of contributing a non-*A* allele.

Let P_c [non-*A*] be the probability that a parent will contribute a non-*A* allele to its offspring. For *A*_ types, that probability is half the probability that, if the *A* gene is contributed, a non-*A* gamete results—that probability is, of course, zero—plus half the probability that, if the "_" allele is contributed, a non-*A* gamete results. This probability is just the sum of the frequencies of the *a* and a^t alleles in the general population. If we let *P*[*A*_] be the probability of getting the desired (*A*_) genotype in the offspring, then:

For *A*_ types, i.e., the bay stallion:

$$P_c[\text{non-}A] = \frac{1}{2}(0) + \frac{1}{2}(.3 + .1)$$
$$= .2$$

For a^t_ types:

(We can use similar logic here.)

$$P_c[\text{non-}A] = \frac{1}{2}(1) + \frac{1}{2}(.3 + .1)$$
$$= .7$$

and

$$P[A\_] = 1 - .2(.7)$$
$$= .86$$

For *aa* types:

$$P_c[\text{non-}A] = \frac{1}{2}(1) + \frac{1}{2}(1)$$
$$= 1$$

and

$$P[A\_] = 1 - .2(1)$$
$$= .8$$

For _ _ types:

$$P_c[\text{non-}A] = \frac{1}{2}(.3 + .1) + \frac{1}{2}(.3 + .1),$$
$$= .4$$

and

$$P[A\_] = 1 - .2(.4)$$
$$= .92$$

E locus:

(This is similar to the A locus because the desired genotype is *E*_.)

For *E*_ types:

$$P_c[\text{non-}E] = \frac{1}{2}(0) + \frac{1}{2}(.7)$$
$$= .35$$

and

$$P[E\_] = 1 - .35(.35)$$
$$= .8775$$

For *ee* types:

$$P_c[\text{non-}E] = \frac{1}{2}(1) + \frac{1}{2}(1)$$
$$= 1$$

and

$$P[E\_] = 1 - .35(1)$$
$$= .65$$

C locus:
For this locus, $P[CC]$ is just the product of the probabilities that each parent contributes a C allele.

For *CC* types:

$$P_c[C] = 1$$

and

$$P[CC] = 1(1)$$
$$= 1$$

For *C_* types:

$$P_c[C] = \frac{1}{2}(1) + \frac{1}{2}(.7),$$
$$= .85$$

and

$$P[CC] = 1(.85)$$
$$= .85$$

D locus:
For this locus, $P[dd]$ is just the product of the probabilities that each parent contributes a d allele.

For *dd* types:

$$P_c[d] = 1$$

and

$$P[dd] = 1(1)$$
$$= 1$$

For *D_* types:

$$P_c[d] = \frac{1}{2}(0) + \frac{1}{2}(.8)$$
$$= .4$$

and

$$P[dd] = 1(.4) = .4$$

All of the above is summarized in the following table.

Color	Genotype	Probability That a Mating Will Produce the Desired One-Locus Genotype: $A\_$	CC	dd	$E\_$	Probability of a Bay Foal
Brown	$a^t\_CCddE\_$	.86	1	1	.8775	.75
Mouse	$aaC\_D\_E\_$	.8	.85	.4	.8775	.24
Black	$aaC\_ddE\_$	.8	.85	1	.8775	.60
Chestnut	$\_\_CCddee$	.92	1	1	.65	.60

15.3 J.F. Turner owns an exceptional herd of Black Angus (BA) cows that she wants to develop into a herd of red cows, yet retain as much of her original breeding as possible. She will use purebred Red Angus bulls for one generation to supply the red allele, then backcross repeatedly via artificial insemination to black bulls from her foundation herd. Assume the following:

Foundation cows and bulls are homozygous (*BB*) at the black/red color locus.

J.F. breeds 50% of her replacement heifers.

100% conception, no death loss.

Molecular geneticists have located a reliable genetic marker near the black/red locus enabling J.F. to test black animals to see if they are homozygous or heterozygous.[2]

Show the effects of repeated backcrossing on J.F.'s herd by filling in the following chart. (If there are multiple genotypes or colors within a generation, include the expected proportions.)

Offspring Generation	Sires: Genotype	Sires: Color	Sires: % BA	Dams: Genotype(s)	Dams: Color(s)	Dams: % BA	Offspring: Genotype(s)	Offspring: Color(s)	Offspring: % BA
F_1	*bb*	Red	0	*BB*	Black	100	*Bb*	Black	50
BC_1	*BB*	Black	100	*Bb*	Black	50	½ *BB* ½ *Bb*	Black	75
BC_2	*BB*	Black	100	*Bb*	Black	75	½ *BB* ½ *Bb*	Black	88
BC_3	*BB*	Black	100	*Bb*	Black	88	½ *BB* ½ *Bb*	Black	94
Inter se	*Bb*	Black	94	*Bb*	Black	94	¼ *BB* ½ *Bb* ¼ *bb*	¾ black ¼ red	94
Inter se	*bb*	Red	94	½ *Bb* ½ *bb*	½ black ½ red	94	¼ *Bb* ¾ *bb*	¼ black ¾ red	94
Inter se	*bb*	Red	94	*bb*	Red	94	*bb*	Red	94

[2]See Chapter 20 for more information on genetic markers.

ANSWERS TO CHAPTER 17 PROBLEMS

17.1 **a.** Which individual is likely to be more prepotent—one with the genotype:

$$AaBbCCDdEeFf$$

or one with the genotype:

$$AABbccDDeeFF?$$

The individual with the second genotype (*AABbccDDeeFF*). This individual has more homozygous loci and can produce fewer unique gametes.

b. Prove your answer mathematically. (See the section of Chapter 3 entitled *The Randomness of Inheritance* for hints.)

The number of unique gametes that an individual can produce is 2^n, where n is the number of heterozygous loci possessed by the individual. If we let N_G represent number of unique gametes, then
For the first individual,

$$\begin{aligned} N_G &= 2^n \\ &= 2^5 \\ &= 32 \end{aligned}$$

For the second individual,

$$\begin{aligned} N_G &= 2^n \\ &= 2^1 \\ &= 2 \end{aligned}$$

Because the second individual can produce fewer unique gametes than the first individual, the second individual should be more prepotent.

17.2 A dog is a carrier of the recessive allele (h) for diaphragmatic hernia, having inherited it from his grandsire. The frequency of the h allele is .05 in this dog's breed. Use Punnett squares to:

a. estimate the frequency of affected (hh), noninbred pups from matings of the dog to unrelated females.

Matings of a carrier male to randomly chosen females

♂ \ ♀	*H* $p = .95$	*h* $q = .05$
H $p = .5$	Normal (*HH*) .475	Normal (*Hh*) .025
h $q = .5$	Normal (*Hh*) .475	Affected (*hh*) .025

Only 2.5% of pups should be affected.

b. estimate the frequency of affected, inbred pups from matings of the dog to his first cousins (*HH* or *Hh*—granddaughters of the carrier grandsire). Assume these females, if they carry the h allele at all, received it from no other source but their grandsire.

One in two sons and daughters of the carrier grandsire is expected to carry the h allele. Therefore, one in four granddaughters is expected to carry the allele.

The probability that a granddaughter will contribute the h allele is $\frac{1}{4}\left(\frac{1}{2}\right) = \frac{1}{8}$ or .125. Then

Matings of a carrier male to his first cousins

♂ \ ♀	*H* $p = .875$	*h* $q = .125$
H $p = .5$	Normal (*HH*) .4375	Normal (*Hh*) .0625
h $q = .5$	Normal (*Hh*) .4375	Affected (*hh*) .0625

6.25% of pups should be affected.

17.3 Two maximally inbred (completely homozygous) mice differ at five loci. Assume the following:

Each dominant allele contributes +3 mg to six-week weight.

Each recessive allele contributes −3 mg to six-week weight.

Partial dominance exists such that each heterozygous locus gains 2 mg in gene combination value (i.e., the genotypic value of each heterozygous locus is 2 mg greater than the breeding value of that locus).

Genetic values for homozygous combinations are the same as breeding values for those combinations.

Environmental effects are as shown.

No epistasis.

a. Fill in the missing values in the following table.

Genotype	*BV*	*GCV*	*G*	*E*	$P-\mu$
1. *AAbbCCddEE*	$6(3) + 4(-3) = +6$	$6 - 6 = 0$	$+6$	0 mg	$6 + 0 = +6$
2. *aaBBccDDee*	$4(3) + 6(-3) = -6$	$-6 - (-6) = 0$	-6	−3 mg	$-6 + (-3) = -9$
3. An offspring of (1) × (2)	$5(3) + 5(-3) = 0$	$10 - 0 = 10$	$0 + 5(2) = 10$	2 mg	$10 + 2 = +12$

b. Which mouse is the heaviest at six weeks?

Mouse 3. ($P - \mu = +12$ mg)

c. Which mouse is the lightest?

Mouse 2. ($P - \mu = -9$ mg)

d. Which mouse enjoys the most hybrid vigor?

Mouse 3. (*GCV* = +10 mg)

e. Which mouse is the best bet to produce offspring with heavy six-week weights?

Mouse 1. (*BV* = +6 mg)

f. Assume complete dominance at all loci and fill in the table again.

Genotype	*BV*	*GCV*	*G*	*E*	*P*−μ
1. *AAbbCCddEE*	6(3) + 4(−3) = +6	6 − 6 = 0	+6	0 mg	6 + 0 = +6
2. *aaBBccDDee*	4(3) + 6(−3) = −6	−6 − (−6) = 0	−6	−3 mg	−6 + (−3) = −9
3. An offspring of (1) × (2)	5(3) + 5(23) = 0	30 − 0 = 30	5(6) = 30	2 mg	30 + 2 = 132

g. Do any of your answers for (*b*) through (*e*) change?

No.

h. What does change?

Because of the greater degree of dominance, *GCV* for Mouse 3 increases from 10 to 30 mg (i.e., more hybrid vigor).

17.4 Given the following pedigree:

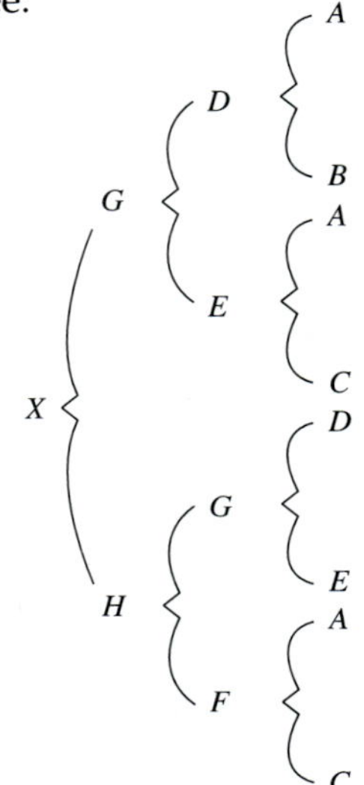

a. Use the path method to calculate F_X.

Step 1. Convert the pedigree to an arrow diagram.

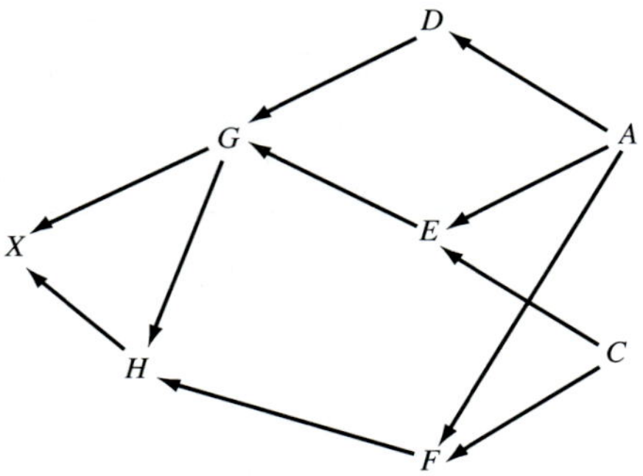

B does not appear in the arrow diagram because, with only one offspring and unknown parents, she does not contribute to inbreeding or relationship of other animals in the pedigree.

Step 2. Locate common ancestors of the sire and dam of *X*.

Common ancestors of *G* and *H* are *A*, *C*, and *G*. *D* and *E* are also common ancestors, but only through *G*. They don't count.

Step 3. Locate inbred common ancestors and calculate the inbreeding coefficient for each.

G is the only inbred common ancestor. Because he is the product of a simple half-sib mating, $F_G = .125$.

Step 4. Fill in the table.

Common Ancestor	Paths	$(\frac{1}{2})^{n_1+n_2+1}$	$1+F_{CA}$	Product of Last Two Columns
A	$G \leftarrow D \leftarrow A \rightarrow F \rightarrow H$	$(\frac{1}{2})^5$	$1+0$	$(\frac{1}{2})^5$
	$G \leftarrow E \leftarrow A \rightarrow F \rightarrow H$	$(\frac{1}{2})^5$	$1+0$	$(\frac{1}{2})^5$
C	$G \leftarrow E \leftarrow C \rightarrow F \rightarrow H$	$(\frac{1}{2})^5$	$1+0$	$(\frac{1}{2})^5$
G	$G \leftarrow G \rightarrow H$	$(\frac{1}{2})^{0+1+1}$	$1+.125$	$1.125(\frac{1}{2})^2$

Step 5. Sum the last column.

$$F_X = (\tfrac{1}{2})^5 + (\tfrac{1}{2})^5 + (\tfrac{1}{2})^5 + 1.125(\tfrac{1}{2})^2$$
$$= .375$$

b. Use the path method to calculate R_{XA}.

Step 1. Convert the pedigree to an arrow diagram.
Done already.

Step 2. Locate common ancestors of X and A.
The only common ancestor of X and A is A himself.

Step 3. Locate inbred common ancestors, calculate the inbreeding coefficient for each, and compute the inbreeding coefficients for X and A as well. A is not inbred. We already calculated the inbreeding coefficient for X. $F_X = .375$.

Step 4. Fill in the table.

Common Ancestor	Paths	$(\frac{1}{2})^{n_1+n_2}$	$1+F_{CA}$	Product of Last Two Columns
A	$X \leftarrow G \leftarrow D \leftarrow A \rightarrow A$	$(\frac{1}{2})^{3+0}$	$1+0$	$(\frac{1}{2})^3$
	$X \leftarrow G \leftarrow E \leftarrow A \rightarrow A$	$(\frac{1}{2})^{3+0}$	$1+0$	$(\frac{1}{2})^3$
	$X \leftarrow H \leftarrow F \leftarrow A \rightarrow A$	$(\frac{1}{2})^{3+0}$	$1+0$	$(\frac{1}{2})^3$
	$X \leftarrow H \leftarrow G \leftarrow D \leftarrow A \rightarrow A$	$(\frac{1}{2})^{4+0}$	$1+0$	$(\frac{1}{2})^4$
	$X \leftarrow H \leftarrow G \leftarrow E \leftarrow A \rightarrow A$	$(\frac{1}{2})^{4+0}$	$1+0$	$(\frac{1}{2})^4$

Step 5. Sum the last column to compute the numerator of R_{XA}.

$$\text{Numerator of } R_{XA} = (\tfrac{1}{2})^3 + (\tfrac{1}{2})^3 + (\tfrac{1}{2})^3 + (\tfrac{1}{2})^4 + (\tfrac{1}{2})^4$$
$$= .5$$

Step 6. Divide the sum by $\sqrt{1 + F_X}\sqrt{1 + F_A}$

$$R_{XA} = \frac{.5}{\sqrt{1 + F_X}\sqrt{1 + F_A}}$$
$$= \frac{.5}{\sqrt{1 + .375}\sqrt{1 + 0}}$$
$$= .4264$$

c. Use the tabular method to calculate F_X.

Step 1. Order all animals by birth date, earliest to latest.
The appropriate order is *A, B, C, D, E, F, G, H, X*.

Step 2. Create a table listing all animals across the top and down the left side.

	A	*B*	*C*	*D*	*E*	*F*	*G*	*H*	*X*
A									
B									
C									
D									
E									
F									
G									
H									
X									

Step 3. Write in the parents of each animal above it at the top of the table.

	– –	– –	– –	*AB*	*AC*	*AC*	*DE*	*GF*	*GH*
	A	*B*	*C*	*D*	*E*	*F*	*G*	*H*	*X*
A									
B									
C									
D									
E									
F									
G									
H									
X									

Step 4. Proceeding from left to right, fill in the first row of numerator relationships.

To get the first diagonal element, use Rule 2. *A's* diagonal element is then

$$\begin{aligned} r_{AA} &= 1 + F_A \\ &= 1 + 0 \\ &= 1 \end{aligned}$$

We can use Rule 1 to compute the remaining elements in the first row. Thus

$$\begin{aligned} r_{AB} &= \frac{1}{2}r_{A\_} + \frac{1}{2}r_{A\_} \\ &= \frac{1}{2}(0) + \frac{1}{2}(0) \\ &= 0 \\ r_{AD} &= \frac{1}{2}r_{AA} + \frac{1}{2}r_{AB} \\ &= \frac{1}{2}(1) + \frac{1}{2}(0) \\ &= .5 \end{aligned}$$

and so on.

	– – A	– – B	– – C	AB D	AC E	AC F	DE G	GF H	GH X
A	1	0	0	.5	.5	.5	.5	.5	.5
B									
C									
D									
E									
F									
G									
H									
X									

Step 5. Copy the values in the first row of the table to the first column of the table.

	– – *A*	– – *B*	– – *C*	*AB* *D*	*AC* *E*	*AC* *F*	*DE* *G*	*GF* *H*	*GH* *X*
A	1	0	0	.5	.5	.5	.5	.5	.5
B	0								
C	0								
D	.5								
E	.5								
F	.5								
G	.5								
H	.5								
X	.5								

Step 6. Fill in the next row of numerator relationships.

	– – *A*	– – *B*	– – *C*	*AB* *D*	*AC* *E*	*AC* *F*	*DE* *G*	*GF* *H*	*GH* *X*
A	1	0	0	.5	.5	.5	.5	.5	.5
B	0	1	0	.5	0	0	.25	.125	.1875
C	0								
D	.5								
E	.5								
F	.5								
G	.5								
H	.5								
X	.5								

Step 7. Copy the values from this row into the corresponding column.

	— —	— —	— —	AB	AC	AC	DE	GF	GH
	A	B	C	D	E	F	G	H	X
A	1	0	0	.5	.5	.5	.5	.5	.5
B	0	1	0	.5	0	0	.25	.125	.1875
C	0	0							
D	.5	.5							
E	.5	0							
F	.5	0							
G	.5	.25							
H	.5	.125							
X	.5	.1875							

Step 8. Repeat Steps 6 and 7 until the table is complete. Off-diagonal elements are computed using Rule 1. Diagonal elements require Rules 2 and 3. For example,

$$
\begin{aligned}
r_{GG} &= 1 + F_G \\
&= 1 + \frac{1}{2} r_{DE} \\
&= 1 + \frac{1}{2}(.25) \\
&= 1.125
\end{aligned}
$$

Altogether,

	— —	— —	— —	AB	AC	AC	DE	GF	GH
	A	B	C	D	E	F	G	H	X
A	1	0	0	.5	.5	.5	.5	.5	.5
B	0	1	0	.5	0	.0	.25	.125	.1875
C	0	0	1	0	.5	.5	.25	.375	.3125
D	.5	.5	0	1	.25	.25	.625	.4375	.5313
E	.5	0	.5	.25	1	.5	.625	.5625	.5938
F	.5	0	.5	.25	.5	1	.375	.6875	.5313
G	.5	.25	.25	.625	.625	.375	1.125	.75	.9375
H	.5	.125	.375	.4375	.5625	.6875	.75	1.1875	.9688
X	.5555	.1875	.3125	.5313	.5938	.5313	.9375	.9688	1.375

Step 9. To determine an individual's inbreeding coefficient, find the individual's diagonal element and subtract 1.

$$\begin{aligned} F_X &= r_{XX} - 1 \\ &= 1.375 - 1 \\ &= .375 \end{aligned}$$

d. Use the tabular method to calculate R_{XA}.

Step 10. To determine Wright's coefficient of relationship between X and A, find the appropriate off-diagonal element and divide by $\sqrt{1 + F_X}\sqrt{1 + F_A}$.

$$\begin{aligned} R_{XA} &= \frac{r_{XA}}{\sqrt{1 + F_X}\sqrt{1 + F_A}} \\ &= \frac{.5}{\sqrt{1 + .375}\sqrt{1 + 0}} \\ &= .4264 \end{aligned}$$

17.5 The B locus is a representative locus for a polygenic trait. Gene frequencies at the B locus in three pure breeds are:

Breed	Frequency of B (p)	Frequency of b (q)
X	.4	.6
Y	.9	.1
Z	.8	.2

Assuming that hybrid vigor in crosses of these breeds is proportional to heterozygosity at the B locus, use Punnett squares to determine which of the three crosses of purebreds produces the most hybrid vigor.

$X \times Y$:

Breed X \ Breed Y	**B** $p = .9$	**b** $q = .1$
B $p = .4$	**BB** .36	**Bb** .04
b $q = .6$	**Bb** .54	**bb** .06

$X \times Z$:

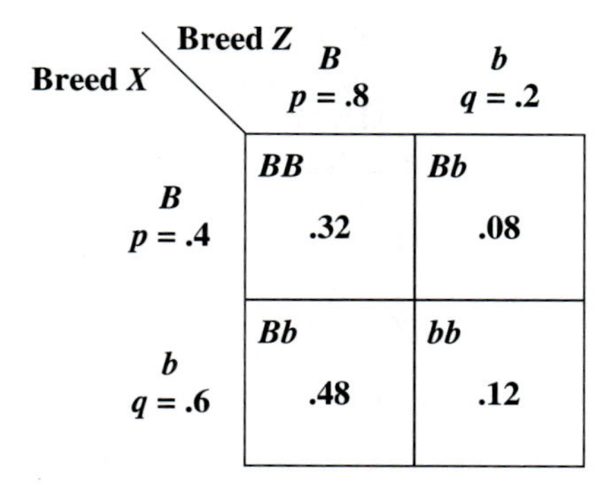

$Y \times Z$:

Breed Y \ Breed Z	B $p = .8$	b $q = .2$
B $p = .9$	BB .72	Bb .18
b $q = .1$	Bb .08	bb .02

$H = .18 + .08 = .26$

Heterozygosity at the B locus is greatest for crosses of breeds X and Y. If the same is true at most other loci, we can expect more hybrid vigor from this cross than from the other two crosses.

ANSWERS TO CHAPTER 18 PROBLEMS

18.1 Peregrine falcons and gyrfalcons are two species that are prized by breeders but suffer from low reproductive rates. Given that peregrines average 2.3, gyrfalcons 3.2, and F_1 crosses of the two species 3.8 viable offspring per clutch, calculate hybrid vigor and percent hybrid vigor for number of viable chicks per clutch.

$$\begin{aligned} \text{HV} &= \overline{P}_{F_1} - \overline{\overline{P}}_P \\ &= 3.8 - \frac{2.3 + 3.2}{2} \\ &= 1.05 \text{ chicks} \end{aligned}$$

$$\begin{aligned} \%\text{HV} &= \frac{\overline{P}_{F_1} - \overline{\overline{P}}_P}{\overline{\overline{P}}_P} \times 100 \\ &= \frac{3.8 - \dfrac{2.3 + 3.2}{2}}{\dfrac{2.3 + 3.2}{2}} \times 100 \\ &= 38\% \end{aligned}$$

18.2 Swine breeds A, B, and C average 1.36, 1.40, and 1.48 lb per day for post-weaning average daily gain (ADG), respectively. Percent hybrid vigor for ADG is typically about 5%.

a. Rank the pure breeds and two-breed crosses for ADG.

$$\%\text{HV} = \frac{\overline{P}_{F_1} - \overline{\overline{P}}_P}{\overline{\overline{P}}_P} \times 100$$

Therefore

$$\overline{P}_{F_1} = \frac{\overline{\overline{P}}_P(\%\text{HV})}{100} + \overline{\overline{P}}_P$$

For $A \times B$ crosses:

$$\overline{P}_{F_1} = \frac{\overline{\overline{P}}_P(\%\text{HV})}{100} + \overline{\overline{P}}_P$$
$$= \frac{\frac{1.36 + 1.40}{2}(5)}{100} + \frac{1.36 + 1.40}{2}$$
$$= 1.45 \text{ lb/day}$$

For $A \times C$ crosses:

$$\overline{P}_{F_1} = \frac{\overline{\overline{P}}_P(\%\text{HV})}{100} + \overline{\overline{P}}_P$$
$$= \frac{\frac{1.36 + 1.48}{2}(5)}{100} + \frac{1.36 + 1.48}{2}$$
$$= 1.49 \text{ lb/day}$$

For $B \times C$ crosses:

$$\overline{P}_{F_1} = \frac{\overline{\overline{P}}_P(\%\text{HV})}{100} + \overline{\overline{P}}_P$$
$$= \frac{\frac{1.40 + 1.48}{2}(5)}{100} + \frac{1.40 + 1.48}{2}$$
$$= 1.51 \text{ lb/day}$$

Ranked from highest to lowest ADG:

1	$B \times C$	1.51
2	$A \times C$	1.49
3	C	1.48
4	$A \times B$	1.45
5	B	1.40
6	A	1.36

b. Did crossbreds always have the best ADG? What principle is illustrated here?

No. Having maximum hybrid vigor (gene combination value) does not necessarily mean maximum performance. Breeding value is important too. Purebreds of Breed *C*, the breed with the highest breeding value, actually performed better than $A \times B$ crosses, and the only groups that performed better than purebred *C* animals were crosses involving Breed *C*.

18.3 A sheep breeder has decided to mate his Columbia ewes to Targhee rams. Columbias are ½ Rambouillet and ½ Lincoln. Targhees are ¾ Rambouillet and ¼ Lincoln (approximately).

a. What proportion of F_1 hybrid vigor is retained in this cross?

$$\begin{aligned}\%\hat{RHV} &= \left(1 - \sum_{i=1}^{n} p_{s_i} p_{d_i}\right) \times 100 \\ &= \left(1 - \left(\frac{3}{4}\left(\frac{1}{2}\right) + \frac{1}{4}\left(\frac{1}{2}\right)\right)\right) \times 100 \\ &= 50\%\end{aligned}$$

b. Based upon your answer to (*a*), would you advise a different choice of sire breed? Why or why not?

Yes. There is so much backcrossing in this mating that relatively little hybrid vigor is retained. I'd look for a less related sire breed.

18.4 A rancher has a herd of ½ Red Angus, ¼ Angus, ¼ Hereford cows (daughters of Red Angus sires and Angus × Hereford dams). She plans to mate them to ½ Charolais, ½ Angus bulls. Typical F_1 individual hybrid vigor for weaning weight is 27 lb. Typical F_1 maternal hybrid vigor for weaning weight is 44 lb.

a. Assuming all four breeds are equally unrelated, calculate the amount of individual, maternal, and total hybrid vigor retained in this cross.

Individual hybrid vigor:

$$\begin{aligned}\hat{RHV}_I &= \left(1 - \sum_{i=1}^{n} p_{s_i} p_{d_i}\right) F_1\hat{HV}_I \\ &= \left(1 - \left((0)\left(\frac{1}{2}\right) + \frac{1}{2}\left(\frac{1}{4}\right) + (0)\left(\frac{1}{4}\right) + \frac{1}{2}(0)\right)\right)(27) \\ &= 23.6 \text{ lb}\end{aligned}$$

Maternal hybrid vigor:

$$\begin{aligned}\hat{RHV}_M &= \left(1 - \sum_{i=1}^{n} p_{s_i} p_{d_i}\right) F_1\hat{HV}_M \\ &= \left(1 - \left(1(0) + (0)\left(\frac{1}{2}\right) + (0)\left(\frac{1}{2}\right)\right)\right)(44) \\ &= 44 \text{ lb}\end{aligned}$$

Total hybrid vigor:

$$\begin{aligned}\hat{RHV} &= \hat{RHV}_I + \hat{RHV}_M \\ &= 23.6 + 44 \\ &= 67.6 \text{ lb}\end{aligned}$$

b. Calculate the amount of individual, maternal, and total hybrid vigor retained again—this time accounting for the fact that Angus and Red Angus are so closely related that F_1 individual hybrid vigor for weaning weight in Angus × Red Angus crosses is only 5 lb, and F_1 maternal hybrid vigor for weaning weight in these crosses is only 8 lb.

For this we need to use the more sophisticated formula presented in the boxed section at the end of Chapter 18.

Individual hybrid vigor:

$$
\begin{aligned}
R\hat{H}V_I &= \sum_{i=1}^{n}\sum_{j=1}^{n} p_{s_i} p_{d_j} F_1 \hat{H}V_{I_{ij}} \\
&= ((0)(.5)(0) + (0)(.25)(5) + (0)(.25)(27) + (0)(0)(27)) \\
&+ ((.5)(.5)(5) + (.5)(.25)(0) + (.5)(.25)(27) + (.5)(0)(27)) \\
&+ ((0)(.5)(27) + (0)(.25)(27) + (0)(.25)(0) + (0)(0)(27)) \\
&+ ((.5)(.5)(27) + (.5)(.25)(27) + (.5)(.25)(27) + (.5)(0)(0)) \\
&= 18.1 \text{ lb}
\end{aligned}
$$

Maternal hybrid vigor:

$$
\begin{aligned}
R\hat{H}V_M &= \sum_{i=1}^{n}\sum_{j=1}^{n} p_{s_i} p_{d_j} F_1 \hat{H}V_{I_{ij}} \\
&= (1(0)(0) + 1(.5)(8) + 1(.5)(44)) \\
&+ ((0)(0)(8) + (0)(.5)(0) + (0)(.5)(44)) \\
&+ ((0)(0)(44) + (0)(.5)(44) + (0)(.5)(0)) \\
&= 26.0 \text{ lb}
\end{aligned}
$$

Total hybrid vigor:

$$
\begin{aligned}
R\hat{H}V &= R\hat{H}V_I + R\hat{H}V_M \\
&= 18.1 + 26 \\
&= 44.1 \text{ lb}
\end{aligned}
$$

ANSWERS TO CHAPTER 19 PROBLEMS

19.1 Rank the following rotational crossbreeding systems for proportion of F_1 hybrid vigor retained at equilibrium. (For systems using hybrid sires, assume that the breeds in each sire type are present in equal proportions.)

a. Four-way rotation using purebred *A*, *B*, *C*, and *D* sires.

$$
\begin{aligned}
\%R\hat{H}V &= \left(\frac{2^n - 2}{2^n - 1}\right) \times 100 \\
&= \left(\frac{2^4 - 2}{2^4 - 1}\right) \times 100 \\
&= 93.3\%
\end{aligned}
$$

b. Three-way rotation using hybrid $A \times B \times C \times D$, $E \times F \times G \times H$, and $I \times J \times K \times L$ sires.

If we let *n* be the number of sire types in the system and m be the number of breeds present in each sire type, then

$$
\begin{aligned}
\%R\hat{H}V &= \left(\frac{m(2^n - 1) - 1}{m(2^n - 1)}\right) \times 100 \\
&= \left(\frac{4(2^3 - 1) - 1}{4(2^3 - 1)}\right) \times 100 \\
&= 96.4\%
\end{aligned}
$$

c. Two-way rotation using hybrid $A \times B \times C \times D \times E \times F \times G \times H$ and $I \times J \times K \times L \times M \times N \times O \times P$ sires.

$$\%\hat{RHV} = \left(\frac{m(2^n - 1) - 1}{m(2^n - 1)}\right) \times 100$$
$$= \left(\frac{8(2^2 - 1) - 1}{8(2^2 - 1)}\right) \times 100$$
$$= 95.8\%$$

All these systems produce similar amounts of retained hybrid vigor. Theoretically, (*b*) produces the most, followed by (*c*), then (*a*).

19.2 Assuming negligible inbreeding and recombination loss, calculate the proportion of F_1 hybrid vigor retained in advanced generations of each of the following composite beef breeds:

a. Brangus (5/8 Angus, 3/8 Brahman)

$$\%\hat{RHV} = \left(1 - \sum_{i=1}^{n} p_i^2\right) \times 100$$
$$= \left(1 - \left(\left(\frac{5}{8}\right)^2 + \left(\frac{3}{8}\right)^2\right)\right) \times 100$$
$$= 46.9\%$$

b. Charbray (13/16 Charolais, 3/16 Brahman)

$$\%\hat{RHV} = \left(1 - \sum_{i=1}^{n} p_i^2\right) \times 100$$
$$= \left(1 - \left(\left(\frac{13}{16}\right)^2 + \left(\frac{3}{16}\right)^2\right)\right) \times 100$$
$$= 30.5\%$$

c. RX_3 (1/2 Red Angus, 1/4 Hereford, 1/4 Red Holstein)

$$\%\hat{RHV} = \left(1 - \sum_{i=1}^{n} p_i^2\right) \times 100$$
$$= \left(1 - \left(\left(\frac{1}{2}\right)^2 + \left(\frac{1}{4}\right)^2 + \left(\frac{1}{4}\right)^2\right)\right) \times 100$$
$$= 62.5\%$$

d. Barzona (approximately 1/4 Hereford, 5/64 Angus, 1/4 Afrikaner, 15/64 Shorthorn, 3/16 Brahman)

$$\%\hat{RHV} = \left(1 - \sum_{i=1}^{n} p_i^2\right) \times 100$$
$$= \left(1 - \left(\left(\frac{1}{4}\right)^2 + \left(\frac{5}{64}\right)^2 + \left(\frac{1}{4}\right)^2 + \left(\frac{15}{64}\right)^2 + \left(\frac{3}{16}\right)^2\right)\right) \times 100$$
$$= 77.9\%$$

19.3 Use the genetic and environmental sheep data and assumptions listed in the last boxed section of this chapter to rank the following pure and composite breeds for 60-day weaning weight performance:

a. *A*

Here are the givens: $(\mu \hat{+} E_{cg}) = 42$ lb

$$\overline{BV^{\star}_{d_A}} = -1 \text{ lb}$$
$$\overline{BV^{\star}_{d_B}} = +4 \text{ lb}$$
$$\overline{BV^{\star}_{d_C}} = +6 \text{ lb}$$
$$\overline{BV^{\star}_{m_A}} = +4 \text{ lb}$$
$$\overline{BV^{\star}_{m_B}} = -2 \text{ lb}$$
$$\overline{BV^{\star}_{m_C}} = +1 \text{ lb}$$

Also,we know F_1 hybrid vigor (both individual and maternal) for $A \times B$ crosses, 2.7 lb; for $A \times C$ crosses, 5.5 lb; and for $B \times C$ crosses, 4.0 lb. Then

$$\hat{\bar{P}} = (\mu \hat{+} E_{cg}) + \hat{\overline{BV}} + \hat{RHV}$$

For a pure breed, there is no retained hybrid vigor, so

$$\hat{\bar{P}} = (\mu \hat{+} E_{cg}) + \hat{\overline{BV}}$$

For Breed *A*,

$$\begin{aligned}\hat{\overline{BV}} &= \hat{\overline{BV}}_d + \hat{\overline{BV}}_m \\ &= -1 + 4 \\ &= +3 \text{ lb}\end{aligned}$$

Then

$$\begin{aligned}\hat{\bar{P}} &= (\mu \hat{+} E_{cg}) + \hat{\overline{BV}} \\ &= 42 + 3 \\ &= 45 \text{ lb}\end{aligned}$$

b. *B*

$$\begin{aligned}\hat{\overline{BV}} &= \hat{\overline{BV}}_d + \hat{\overline{BV}}_m \\ &= 4 + (-2) \\ &= +2 \text{ lb}\end{aligned}$$

So

$$\begin{aligned}\hat{\bar{P}} &= (\mu \hat{+} E_{cg}) + \hat{\overline{BV}} \\ &= 42 + 2 \\ &= 44 \text{ lb}\end{aligned}$$

c. *C*

$$\begin{aligned}\hat{\overline{BV}} &= \hat{\overline{BV}}_d + \hat{\overline{BV}}_m \\ &= 6 + 1 \\ &= +7 \text{ lb}\end{aligned}$$

So

$$\begin{aligned}\hat{\bar{P}} &= (\mu \hat{+} E_{cg}) + \hat{\overline{BV}} \\ &= 42 + 7 \\ &= 49 \text{ lb}\end{aligned}$$

d. ½ *A*, ½ *C*

$$\begin{aligned}\overline{\hat{BV}_d} &= \sum_{i=1}^{n} p_i \overline{BV^*_{d_i}} \\ &= \frac{1}{2}(-1) + \frac{1}{2}(6) \\ &= +2.5 \text{ lb}\end{aligned}$$

and

$$\begin{aligned}\overline{\hat{BV}_m} &= \sum_{i=1}^{n} p_i \overline{BV^*_{m_i}} \\ &= \frac{1}{2}(4) + \frac{1}{2}(1) \\ &= +2.5 \text{ lb}\end{aligned}$$

Then

$$\begin{aligned}\overline{\hat{BV}} &= \overline{\hat{BV}_d} + \overline{\hat{BV}_m} \\ &= 2.5 + 2.5 \\ &= +5 \text{ lb}\end{aligned}$$

Now,

$$\begin{aligned}\text{R}\hat{\text{H}}\text{V}_\text{I} &= \sum_{i=1}^{n}\sum_{j=1}^{n} p_i p_j \text{F}_1 \hat{\text{H}}\text{V}_{\text{I}_{ij}} \\ &= \left(\frac{1}{2}\left(\frac{1}{2}\right)(0) + \frac{1}{2}\left(\frac{1}{2}\right)(5.5)\right) \\ &+ \left(\frac{1}{2}\left(\frac{1}{2}\right)(5.5) + \frac{1}{2}\left(\frac{1}{2}\right)(0)\right) \\ &= +2.75 \text{ lb}\end{aligned}$$

$\text{R}\hat{\text{H}}\text{V}_\text{M}$ = 2.75 lb as well, so

$$\begin{aligned}\text{R}\hat{\text{H}}\text{V} &= \text{R}\hat{\text{H}}\text{V}_\text{I} + \text{R}\hat{\text{H}}\text{V}_\text{M} \\ &= 2.75 + 2.75 \\ &= 5.5 \text{ lb}\end{aligned}$$

Altogether,

$$\begin{aligned}\hat{P} &= (\mu \hat{+} E_{cg}) + \overline{\hat{BV}} + \text{R}\hat{\text{H}}\text{V} \\ &= 42 + 5 + 5.5 \\ &= 52.5 \text{ lb}\end{aligned}$$

e. ¼ *A*, ¼ *B*, ½ *C*

$$\begin{aligned}\overline{\hat{BV}_d} &= \sum_{i=1}^{n} p_i \overline{BV^*_{d_i}} \\ &= \frac{1}{4}(-1) + \frac{1}{4}(4) + \frac{1}{2}(6) \\ &= +3.75 \text{ lb}\end{aligned}$$

and

$$\begin{aligned}\overline{\hat{BV}_m} &= \sum_{i=1}^{n} p_i \overline{BV^*_{m_i}} \\ &= \frac{1}{4}(4) + \frac{1}{4}(-2) + \frac{1}{2}(1) \\ &= +1 \text{ lb}\end{aligned}$$

Then

$$\begin{aligned}\overline{\hat{BV}} &= \overline{\hat{BV}_d} + \overline{\hat{BV}_m} \\ &= 3.75 + 1 \\ &= +4.75 \text{ lb}\end{aligned}$$

Now,

$$\begin{aligned}\text{R}\hat{\text{H}}\text{V}_\text{I} &= \sum_{i=1}^{n}\sum_{j=1}^{n} p_i p_j \text{F}_1 \hat{\text{H}}\text{V}_{\text{I}_{ij}} \\ &= \left(\frac{1}{4}\left(\frac{1}{4}\right)(0) + \frac{1}{4}\left(\frac{1}{4}\right)(2.7) + \frac{1}{4}\left(\frac{1}{2}\right)(5.5)\right) \\ &+ \left(\frac{1}{4}\left(\frac{1}{4}\right)(2.7) + \frac{1}{4}\left(\frac{1}{4}\right)(0) + \frac{1}{4}\left(\frac{1}{2}\right)(4.0)\right) \\ &+ \left(\frac{1}{2}\left(\frac{1}{4}\right)(5.5) + \frac{1}{2}\left(\frac{1}{4}\right)(4.0) + \frac{1}{2}\left(\frac{1}{2}\right)(0)\right) \\ &= 2.71 \text{ lb}\end{aligned}$$

$\text{R}\hat{\text{H}}\text{V}_\text{M} = 2.71$ lb as well, so

$$\begin{aligned}\text{R}\hat{\text{H}}\text{V} &= \text{R}\hat{\text{H}}\text{V}_\text{I} + \text{R}\hat{\text{H}}\text{V}_\text{M} \\ &= 2.71 + 2.71 \\ &= 5.42 \text{ lb}\end{aligned}$$

Altogether,

$$\begin{aligned}\hat{P} &= (\mu \hat{+} E_{cg}) + \overline{\hat{BV}} + \text{R}\hat{\text{H}}\text{V} \\ &= 42 + 4.75 + 5.42 \\ &= 52.2 \text{ lb}\end{aligned}$$

Rankings for breed performance are then:

Rank	Breed	60-day weight, lb
1	½ *A*, ½ *C*	52.5
2	¼ *A*, ¼ *B*, ½ *C*	52.2
3	*C*	49
4	*A*	45
5	*B*	44

QUICK KEY

3.1 **(a)** *JJ, J′J′, jj, j′j′* **(b)** *JJ′, Jj, Jj′, J′j, J′j′, jj′*

3.2 8, 16, 8

3.3 **(a)** 8 **(b)** 16 **(c)** 72

3.4 **(a)** 90 **(b)** 30 **(c)** 60 **(d)** $30 < \%S_H < 90$ **(e)** $30 < \%S_H < 60$ **(f)** $90 < \%S_H \leq 100$ **(g)** $0 \leq \%S_H < 30$

3.5 Red, polled: 1/8, Red, horned: 1/8, Roan, polled: 2/8 = 1/4, Roan, horned: 2/8 = 1/4, White, polled: 1/8, White horned: 1/8

3.6 Black: 3/8, Chocolate: 3/8, Yellow: 2/8 = 1/4

4.1 $p = .6, q = .4, P = .3, H = .6, Q = .1$

4.2 $P = .7, H = .2, Q = .1, p = .8, q = .2$

4.3 **(a)** *S* **(b)** *A, B*

4.4 **(a)** $P = .24, H = .62, Q = .14, p = .55, q = .45,$ **(b)** $P = .3025, H = .495, Q = .2025, p = .55, q = .45$ **(c)** $P = .3025, H = .495, Q = .2025, p = .55, q = .45$

4.5 $p = .6, q = .4, P = .36, H = .48, Q = .16$

6.1 1.56%

6.2 **(a)** .72 **(b)** 366 **(c)** 0%, 100%

6.3 **(a)** .83 or .82 **(b)** .97

6.4 **(a)(i)** 2:1 **(ii)** 8:1 **(iii)** 18.1 **(b)(i)** 1:1 **(ii)** 4:1 **(iii)** 9:1

6.5 **(a)** .167 **(b)** .065 **(c)** .2

7.2 **(a)** +200 lb **(b)** +400 lb **(c)** +200 lb

7.3 **(b)** #1 **(c)** #2

7.4 **(a)** Raise-a-Ruckus: −4 sec., Presidium: −6 sec. **(b)** Raise-a-Ruckus: −14 sec., Presidium: −10 sec.

8.1 **(a)(i)** 176.4 days **(ii)** 1.28 in **(iii)** 165.38 days^2 **(iv)** .024 in^2 **(v)** 12.86 days **(vi)** .155 in **(b)** −.569 days · in **(c)** −.29 **(d)** −.0034 in/day **(e)(i)** 1.35 in **(ii)** 1.20 in

9.1 **(a)(i)** 16 sec^2 **(ii)** 2,500 rubles^2 **(iii)** .25 **(iv)** .04 **(f)** −2 sec

9.2 **(a)** .86 **(d)** 15.48 lengths^2, 2.52 lengths^2

9.3 .36

9.4 **(a)** 5.4 lb **(b)** 1.07

9.5 **(a)** 64.1 lb, .22 **(b)** 56.8 lb, .28

10.1 −.088 lb/lb per year

10.2 **(a)** 12.8%

10.3 **(a)** .36, 1.04 **(b)** .56, .71

10.4 **(a)** 211 lb/year **(b)** 214 lb/year

10.5 **(a)** 12.0 lb/year, 14.1 lb/year

11.1 **(a)(i)** −1.46 sec **(ii)** .60 **(b)(i)** −1.89 sec **(ii)** .97

11.2 **(a)** +.102 lb/day **(b)** +.106 lb/day

11.3 **(a)(i)** .5 **(ii)** .71 **(iii)** .75 **(iv)** .61 **(v)** .25 **(vi)** .49 **(vii)** .5 **(viii)** .99

11.4 **(a)** Gay Blade: −2.07 to −.85 sec, Megabuck: −2.08 to −1.70 sec

11.5 **(a)** +51.5 lb **(b)** .79

11.6 **(a)** +4 lb **(b)** +7 lb **(c)** 46 lb

13.1 **(a)** −.10 **(b)** −.73 **(c)** −.60

13.2 **(a)(i)** −.22 days/year **(ii)** −1.0 lb/year **(iii)** −.11 days/year **(iv)** −.13 lb/year **(v)** .5 **(vi)** .13

13.3 **(a)** .23% per year **(b)** .67% per year

13.4 **(a)** +5 rubles **(b)** +108 rubles

14.1 **(a)(i)** #6, #3, #1 **(ii)** #8, #4, #2 **(b)** #1, #2, #4 **(c)** #4, #8, #2

14.2 $I = YW - 1.08BW$

14.3 **(a)** .45, .88 **(b)** .58, .67 **(c)** .67, .54 **(d)** .85, .27

15.2 **(a)** the brown mare **(b)** the mouse colored mare

17.1 **(a)** *AABbccDDeeFF*

17.2 **(a)** 2.5% **(b)** 6.25%

17.3 **(b)** #3 **(c)** #2 **(d)** #3 **(e)** #1

17.4 **(a)** .375 **(b)** .4264 **(c)** .375 **(d)** .4264

17.5 $X \times Y$

18.1 1.05 chicks, 38%

18.2 1: $B \times C$, 2: $A \times C$, 3: C, 4: $A \times B$, 5: B, 6: A

18.3 **(a)** 50%

18.4 **(a)** 23.6 lb, 44 lb, 67.6 lb **(b)** 18.1 lb, 26 lb, 44.1 lb

19.1 1: (*b*), 2: (*c*), 3:(*a*)

19.2 **(a)** 46.9% **(b)** 30.5% **(c)** 62.5% **(d)** 77.9%

19.3 1: ½ A, ½ C, 2: ¼ A, ¼ B, ½ C, 3: C, 4: A, 5: B

Index

Page numbers of defined concepts are noted in boldface type.